AF595318

Fungal Nanotechnology 2.0

Fungal Nanotechnology 2.0

Editor

Kamel A. Abd-Elsalam

MDPI • Basel • Beijing • Wuhan • Barcelona • Belgrade • Manchester • Tokyo • Cluj • Tianjin

Editor
Kamel A. Abd-Elsalam
Plant Pathology Research
Institute
Agricultural Research Center
Giza
Egypt

Editorial Office
MDPI
St. Alban-Anlage 66
4052 Basel, Switzerland

This is a reprint of articles from the Special Issue published online in the open access journal *Journal of Fungi* (ISSN 2309-608X) (available at: www.mdpi.com/journal/jof/special_issues/fungal_nanotechnology_2022).

For citation purposes, cite each article independently as indicated on the article page online and as indicated below:

LastName, A.A.; LastName, B.B.; LastName, C.C. Article Title. *Journal Name* **Year**, *Volume Number*, Page Range.

ISBN 978-3-0365-7721-0 (Hbk)
ISBN 978-3-0365-7720-3 (PDF)

Contents

About the Editor

Kamel A. Abd-Elsalam

Kamel A. Abd-Elsalam, Ph.D., is currently a Research Professor at the PlantPathology Research Institute, Agricultural Research Center, Giza, Egypt. He has published 23 books on nano-biotechnology applications in agriculture and plant protection in major publishing houses (Springer, Tylor Frances, and Elsevier). Since 2019, he has served as the Editor-in-Chief of the Elsevier book series, titled Nanobiotechnology for Plant Protection; he also serves as the Series Editor of the Elsevier book series, titled Applications of Genome-Modified Plants and Microbes in Food and Agriculture. He has also participated as an active member of the Elsevier Advisory Panel, giving feedback and suggestions for improving Elsevier's products and services since 2020. He published more than 214 scientific research articles in international and regional specialized scientific journals with a high impact factor, and with an h-index of 39 and a i-10 index of 108, across 5911+citations. He also served as a Guest Editor for the *Journal of Fungi*, *Plants*, and *Microorganisms*, and as a Reviewer and Editor for *Frontiers* in Genomic Assay Technology, and has been refereed for several reputed journals. He was ranked in the top 2% of most influential scientists in the world in nanobiotechnology for 2020 and 2021 by Stanford University, In 2014, he was awarded the Federation of Arab Scientific Study Councils Prize for excellent scientific research in biotechnology (fungal genomics) (first ranking). Dr. Kamel earned his Ph.D. in Molecular Plant Pathology from Christian Alberchts University of Kiel (Germany) and SuezCanal University (Egypt), and, in 2008, he was awarded a postdoctoral fellowship from the same institution. Dr. Kamel was a visiting associate professor at Mac Fah Luang University in Thailand, the Institute of Microbiology at TUM in Germany, and the Laboratory of Phytopathology at Wageningen University in the Netherlands.

Preface to "Fungal Nanotechnology 2.0"

Fungal Nanotechnology offers new methods for molecular and cell biology, medicine, biotechnology, agriculture, veterinary physiology, and reproduction, offering more exciting applications in both pathogen identification and treatment, plant engineering, impressive results in animal and food systems, and much more. Myconanotechnology can be a good choice for green nanoparticle synthesis, as manufacturing via fungal resources is easy, inexpensive, and more sustainable. Mycosynthesis nanoparticles may be used in a range of fields, including agriculture, manufacturing, and medicine, and in a number of applications such as pathogen detection and diagnosis, control, wound healing, drug delivery, cosmetics, food preservation, textile fabrics, and many others. There is a growing need for a deeper understanding of molecular biology and genetic factors behind fungal nanobiosynthetic pathways. The main purpose of this Special Issue was to highlight new identification, treatment, and antifungal nanotherapy methods that are currently being implemented or under development for invasive fungal infections, including human, animal, plant, and entomopathogenic fungi. This topic also focuses on the use of nanobiofungicides as effective alternatives for the environmentally sustainable control of pathogenic and toxigenic fungi in the coming decades.

Kamel A. Abd-Elsalam

Editor

Journal of
Fungi

MDPI

Editorial

Special Issue: Fungal Nanotechnology 2

Kamel A. Abd-Elsalam 

Plant Pathology Research Institute, Agricultural Research Centre, Giza 12619, Egypt; kamelabdelsalam@gmail.com

Abstract: Fungal nanotechnology provides techniques useful for molecular and cell biology, medicine, biotechnology, agriculture, veterinary physiology, and reproduction. This technology also has exciting potential applications in pathogen identification and treatment, as well as impressive outcomes in the animal and food systems. Myconanotechnology is a viable option for the synthesis of green nanoparticles because it is simple, affordable, and more environmentally friendly to use fungal resources. Mycosynthesis nanoparticles can be used for various purposes, such as pathogen detection and diagnosis, control, wound healing, drug delivery, cosmetics, food preservation, and textile fabrics, among other applications. They can be applied to a variety of industries, such as agriculture, manufacturing, and medicine. Gaining deeper comprehension of the molecular biology and genetic components underlying the fungal nanobiosynthetic processes is becoming increasingly important. This Special Issue aims to showcase recent advancements in invasive fungal diseases caused by human, animal, plant, and entomopathogenic fungi that are being identified, treated, and treated using antifungal nanotherapy. Utilizing fungus in nanotechnology has several benefits, such as their capacity to create nanoparticles with distinctive characteristics. As an illustration, some fungi can create nanoparticles that are highly stable, biocompatible, and have antibacterial capabilities. Fungal nanoparticles may be used in a variety of industries, including biomedicine, environmental cleanup, and food preservation. Fungal nanotechnology is also a sustainable and environmentally beneficial method. Fungi are an appealing alternative to conventional chemical methods of creating nanoparticles because they are simple to cultivate using affordable substrates and may be cultivated under diverse conditions.

Keywords: fungus-derived nanoparticles; nanoparticle-based fungicides; fungal bioreactors

Citation: Abd-Elsalam, K.A. Special Issue: Fungal Nanotechnology 2. *J. Fungi* **2023**, *9*, 553. https://doi.org/10.3390/jof9050553

Received: 21 April 2023
Accepted: 10 May 2023
Published: 11 May 2023

1. Introduction

Fungal nanotechnology, also known as myconanotechnology, was coined by Rai M. from India in 2009. This term refers to the production and utilization of nanoparticles within the 1–100 nm size range using fungi. FN is particularly useful in fields such as biomedicine, agriculture, and environmental preservation [1,2]. Metal nanoparticles such as silver, gold, copper, and zinc, as well as other substances such as selenium, titanium dioxide, metal sulfides, and cellulose, have been successfully synthesized using mushrooms, Fusarium, Trichoderma, endophytic fungi, and yeast. FN studies explore various synthesis methods, environmental preservation techniques, and prospects. Investigating the process of nanoparticle production and the effect of different factors on metal ion reduction can lead to the development of cost-effective synthesis and nanoparticle extraction methods. Additionally, FN addresses risk assessment, protection, and control of mycogenic nanoparticles. Fungi can produce extracellular enzymes that hydrolyze complex macromolecules and generate hydrolytes. This metabolic ability makes fungi a strong candidate for producing various types of metallic nanoparticles through bioprocessing [3,4].

In this Special Issue, a critical and detailed analysis of the current progress on the application of metal-based nanoparticles for controlling phytopathogenic fungi in agriculture is presented. The following conclusions and future directions are proposed. The progress achieved in the use of metal nanoparticles for the control of phytopathogenic fungi is outstanding, since the studies developed thus far clearly illustrate that these nanoparticles are an excellent alternative to chemical fungicides when it comes to controlling phytopathogenic fungi in agriculture. Among the metallic nanoparticles, Ag nanoparticles have been the most studied as antifungal agents, followed by Cu nanoparticles. These nanoparticles have shown promising activity against different species of plant pathogenic fungi [5]. Additionally, various metal nanoparticles are mycosynthesised for use in agri-food applications from Trichoderma species. Mycogenic nanoparticles, which are generated from fungal cells or cell extracts, can be used as nanofertilizers, nanofungicides, plant growth regulators, and nanocoatings, among other applications. Additionally, Trichoderma-mediated NPs have been used in environmental remediation procedures such as pollutant detection and removal, including that of toxins containing heavy metals [6]. For instance, unique fungicidal activity in an in vitro experiment completely inhibited the growth of the studied plant pathogenic fungi, and under greenhouse circumstances, the symptoms of cotton seedling illness were significantly diminished. Creating a trichogenic ZnONPs form considerably improved its antifungal action. The use of biocontrol agents, such as T. harzianum, may also be a secure method for producing ZnONPs on a medium scale and applying them to the management of fungal diseases in cotton [7]. *P. indica* produced AgNPs quickly, sustainably, and in an environmentally benign manner for the treatment of mucormycosis and antioxidant activity. Results showed that *Pseudomounas indica* metabolites were used to biosynthesize AgNPs, and several cutting-edge techniques were used to describe the biosynthesized AgNPs. Additionally, *R. microsporus, M. racemosus,* and *S. racemosum* showed no resistance to the remarkable antifungal efficacy of the biosynthesized AgNPs [8]. Chi/Ag-NPs with many biological functions have been created. Gram-positive and Gram-negative bacteria were both extremely susceptible to Chi/Ag-NPs' antibacterial activity. Along with antioxidant activity, they also reported antibiofilm activity. Furthermore, both unicellular and multicellular fungi were susceptible to Chi/Ag-NPs' promising antifungal action. Due to their biocompatibility and efficient absorption of wound exudates, Chi/Ag-NPs were seen to dramatically accelerate wound healing at non-toxic doses [9]. By optimizing AgNPs concentration on the interaction with *Piriformospora indica,* for both the broth and the agar, it was possible to study chemically produced AgNPs and their impact on fungal symbiont and black rice. This led to an increase in biomass and a larger fungal colony diameter. An organism known as a "Nano-Embedded Fungus" reacted with the optimal AgNPs concentration (300 ppm). Black rice's growth and productivity were enhanced by *P. indica* inoculation and AgNPs treatment [10].

In conclusion, Fungal Nanotechnology 2 provides an updated and thorough understanding of the green and sustainable production of metal and organic-based nanostructures by various fungal species, as well as an investigation of intracellular and extracellular mechanisms, with a particular focus on the applications of fungal nanotechnology in the biomedical, environmental, and agri-food sectors. Since FN is still in its infancy, major research should be conducted in this field; plants, animals, and people will all benefit significantly from this, and effective and environmentally acceptable methods should be developed.

Conflicts of Interest: The author declares no conflict of interest.

References

1. Rai, M.; Alka, Y.; Bridge, P.; Aniket, G. Myconanotechnology: A new and emerging science. *Appl. Mycol.* **2009**, 258–267.
2. Jagtap, P.; Nath, H.; Kumari, P.B.; Dave, S.; Mohanty, P.; Das, J.; Dave, S. Mycogenic fabrication of nanoparticles and their applications in modern agricultural practices & food industries. *Fungi Bio-Prospect. Sustain. Agric. Environ. Nanotechnol.* **2021**, *3*, 475–488.
3. Alghuthaymi, M.A.; Almoammar, H.; Rai, M.; Said-Galiev, E.; Abd-Elsalam, K.A. Myconanoparticles: Synthesis and their role in phytopathogens management. *Biotechnol. Biotechnol. Equip.* **2015**, *29*, 221–236. [CrossRef]
4. Prasad, R. *Fungal Nanotechnology: Applications in Agriculture, Industry, and Medicine*; Springer International Publishing: Cham, Switzerland, 2017; ISBN 978-3-319-68423-9.

5. Cruz-Luna, A.R.; Cruz-Martínez, H.; Vásquez-López, A.; Medina, D.I. Metal nanoparticles as novel antifungal agents for sustainable agriculture: Current advances and future directions. *J. Fungi* **2021**, *7*, 1033. [CrossRef]
6. Alghuthaymi, M.A.; Abd-Elsalam, K.A.; AboDalam, H.M.; Ahmed, F.K.; Ravichandran, M.; Kalia, A.; Rai, M. *Trichoderma*: An Eco-Friendly Source of Nanomaterials for Sustainable Agroecosystems. *J. Fungi* **2022**, *8*, 367. [CrossRef] [PubMed]
7. Zaki, S.A.; Ouf, S.A.; Albarakaty, F.M.; Habeb, M.M.; Aly, A.A.; Abd-Elsalam, K.A. *Trichoderma harzianum*-mediated ZnO nanoparticles: A green tool for controlling soil-borne pathogens in cotton. *J. Fungi* **2021**, *7*, 952. [CrossRef] [PubMed]
8. Salem, S.S.; Ali, O.M.; Reyad, A.M.; Abd-Elsalam, K.A.; Hashem, A.H. *Pseudomonas indica*-mediated silver nanoparticles: Antifungal and antioxidant biogenic tool for suppressing mucormycosis Fungi. *J. Fungi* **2022**, *8*, 126. [CrossRef] [PubMed]
9. Shehabeldine, A.M.; Salem, S.S.; Ali, O.M.; Abd-Elsalam, K.A.; Elkady, F.M.; Hashem, A.H. Multifunctional silver nanoparticles based on chitosan: Antibacterial, antibiofilm, antifungal, antioxidant, and wound-healing activities. *J. Fungi* **2022**, *8*, 612. [CrossRef] [PubMed]
10. Solanki, S.; Lakshmi, G.B.V.S.; Dhiman, T.; Gupta, S.; Solanki, P.R.; Kapoor, R.; Varma, A. Co-Application of silver nanoparticles and symbiotic fungus *Piriformospora indica* improves secondary metabolite production in black rice. *J. Fungi* **2023**, *9*, 260. [CrossRef] [PubMed]

Journal of
Fungi

Article

Facile Synthesis and Characterization of Cupric Oxide Loaded 2D Structure Graphitic Carbon Nitride (g-C_3N_4) Nanocomposite: In Vitro Anti-Bacterial and Fungal Interaction Studies

Rajendran Lakshmi Priya [1], Bheeranna Kariyanna [2], Sengodan Karthi [3], Raja Sudhakaran [4], Sundaram Ganesh Babu [1,*] and Radhakrishnan Vidya [2,*]

1 Department of Chemistry, School of Advanced Sciences, Vellore Institute of Technology, Vellore 632014, Tamil Nadu, India
2 VIT School of Agricultural Innovations and Advanced Learning, Vellore Institute of Technology, Vellore 632014, Tamil Nadu, India
3 Department of Entomology, College of Agriculture, Food and Environment, University of Kentucky, Lexington, KY 40503, USA
4 School of Biosciences and Technology, Vellore Institute of Technology, Vellore 632014, Tamilnadu, India
* Correspondence: ganeshbabu.s@vit.ac.in (S.G.B.); rvidya@vit.ac.in (R.V.)

Abstract: The active and inexpensive catalyst cupric oxide (CuO) loaded foliar fertilizer of graphitic carbon nitride (g-C_3N_4) is investigated for biological applications due to its low cost and easy synthesis. The synthesized CuO NPs, bulk g-C_3N_4, exfoliated g-C_3N_4, and different weight percentages of 30 wt%, 40 wt%, 50 wt%, 60 wt%, and 70 wt% CuO-loaded g-C_3N_4 are characterized using different analytical techniques, including powder X-ray diffraction, scanning electron microscopy, energy dispersive X-ray analysis, and ultraviolet-visible spectroscopy. The nanocomposite of CuO NPs loaded g-C_3N_4 exhibits antibacterial activity against Gram-positive bacteria (*Staphylococcus aureus* and *Streptococcus pyogenes*) and Gram-negative bacteria (*Escherichia coli* and *Pseudomonas aeruginosa*). The 20 μg/mL of 70 wt% CuO/g-C_3N_4 nanocomposite showed an efficiency of 98% for Gram-positive bacteria, 80% for *E. Coli*, and 85% for *P. aeruginosa*. In the same way, since the 70 wt% CuO/g-C_3N_4 nanocomposite showed the best results for antibacterial activity, the same compound was evaluated for anti-fungal activity. For this purpose, the fungi *Fusarium oxysporum* and *Trichoderma viride* were used. The anti-fungal activity experiments were not conducted in the presence of sunlight, and no appreciable fungal inhibition was observed. As per the literature, the presence of the catalyst g-C_3N_4, without an external light source, reduces the fungal inhibition performance. Hence, in the future, some modifications in the experimental conditions should be considered to improve the anti-fungal activity.

Keywords: antibacterial; antifungal activity; CuO/g-C_3N_4 nanocomposite; fungal interactions; plant pathogens

Citation: Priya, R.L.; Kariyanna, B.; Karthi, S.; Sudhakaran, R.; Babu, S.G.; Vidya, R. Facile Synthesis and Characterization of Cupric Oxide Loaded 2D Structure Graphitic Carbon Nitride (g-C_3N_4) Nanocomposite: In Vitro Anti-Bacterial and Fungal Interaction Studies. *J. Fungi* **2023**, *9*, 310. https://doi.org/10.3390/jof9030310

Academic Editor: Kamel A. Abd-Elsalam

Received: 6 February 2023
Revised: 17 February 2023
Accepted: 18 February 2023
Published: 28 February 2023

1. Introduction

Different types of microorganisms cause disease in human beings. Recent years have seen an increase in the number of alternative antimicrobial agents or drugs (such as metal nanoparticles, polymers, and peptides) for the treatment of antibiotic-resistant infections [1]. In this study, we focused on the synthesis, characterization, antimicrobial activity, and interactions of fungi. Microorganisms produce infections in humans, animals, and plants. These infections can cause unexpected death in humans. Major causes of death, such as those related to the respiratory system, the gastrointestinal system, the central nervous system, etc., are frequently caused by bacterial infections. One of the bacteria of *P. aeruginosa* can cause serious lung problems, and it can be difficult to treat the disease due to resulting bloodstream infections. A few studies reported a rate of *P. aeruginosa* infection of up to 61%. While most studies consider plant diseases to be of great importance, bacteria-induced diseases in humans, such as cholera, diphtheria, dysentery, plague, pneumonia, tuberculosis (TB), and typhoid, among many others [2], can be deadly.

On the other hand, plant diseases have intensive effects on human civilizations because plant diseases affect food, ornamentals, and natural environments. These plant diseases are caused by pathogens or microorganisms. These pathogens cause plant diseases to spread through fungi, bacteria, and mycoplasmas, and they can easily disperse disease from an infected plant to a healthy plant. Pathogenic fungi cause plant infections such as anthracnose, leaf spot, rust, wilt, blight, coils, scab, gall, canker, damping-off, root rot, mildew, and dieback [3]. Some of the fungi (such as Clubroot, *Pythium* species, *Fusarium* species, *Rhizoctonia* species, and *Sclerotium* species) are responsible for foliar diseases and soilborne diseases [2]. Fusarium pathogens mostly enter through root wounds caused by cultivation. These diseases are effectively controlled by some medicines, but these fungicides are costly and not eco-friendly, motivating the search to find non-conventional alternatives to control plant pathogens.

Different types of metal oxide and metallic nanoparticles were fabricated and used for their antimicrobial activity. Metal oxide in the nano range contains crucial properties that vary by size, chemical composition, and surface chemistry. Transition metal oxide nanoparticles, such as CuO NPs, have been broadly used for various applications, including as sensors, catalysts, semiconductors, antimicrobial agents, supercapacitors, etc. [4–7]. Additionally, CuO nanoparticles have recently been used in biomedical and cancer applications owing to their attractive chemical properties. Currently, one of the most attractive and active metals is copper, which has a long history in biological chemistry [8,9]. CuO nanoparticles showed excellent antimicrobial activity against different bacterial strains. Previous studies described that CuO nanoparticles exhibited high potential activity for Gram-positive and Gram-negative bacteria, with high efficiency surpassing the accepted therapeutic indices. Researchers have used nanotechnology in agriculture for applications involving agrochemicals, nutrients, pesticides, etc. The uses of nanomaterials are easily adaptable to soil climates and crop conditions. Copper is an essential plant nutrient. It aids in plant growth and disease resistance. In addition, copper is necessary for the creation of key plant defense proteins, peroxidase, and copper multiple oxidases in response to pathogen diseases. The CuO NPs can affect plant nutrition, as demonstrated for the first time by Elmer et al. The Cu_2O NPs can be used as a nano-fertilizer in plant diseases, effectively suppressing the disease.

Graphitic carbon nitride (*g*-C_3N_4) is a stacked morphology of two-dimensional (2D) layered polymeric material composed of sp_2-hybridized nitrogen-substituted graphene of tris-triazine-based patterns with a small amount of hydrogen. *g*-C_3N_4 has a carbon-nitrogen bond, a covalent bond between carbon and nitrogen, and is one of the most plentiful bonds in organic chemistry and biochemistry [10,11]. These materials were prepared by the calcination method. Prepared using the exfoliated *g*-C_3N_4 process, the bulk carbon nitride forms by self-condensation of the exfoliation process [12,13]. It exhibits remarkable features including bio-friendliness, tunable electronic structures, good thermal and chemical stability, low cost, simple preparation, and large surface area [14,15]. It can be used for biocompatible medical coatings, chemically inert coatings, drug delivery, catalysis, degradation, sensing, insulators, and supercapacitors [16]. As an antibacterial agent, g-C_3N_4 nanosheets possess the ability to produce reactive oxygen species (ROS), followed by metal nanoparticles reacting to the reactive oxygen species ($\cdot O_2^-$ and $\cdot$OH), and under low-intensity light irradiation, bacteria were killed [17,18].

In the present study, the synthesis of CuO-loaded g-C_3N_4 nanocomposites is achieved using dry synthesis methods. Moreover, the antibacterial activity of the compound is investigated, and the fungal interactions with the compound are also discussed.

2. Materials and Methods

2.1. Materials

Copper acetate $Cu(CH_3COO)_2$, H_2SO_4, and melamine were acquired from Sisco Research Laboratories Pvt. Ltd., Mumbai, India. All the chemicals used in the experimental process were of pure analytical grade and used as such, without any further processing being performed.

2.2. Synthesis of g-C_3N_4

Initially, the bulk-graphitic carbon nitride (Bulk g-C_3N_4) was prepared under atmospheric air conditions by the thermal polymerization of melamine. The temperature was set to 550 °C in a muffle furnace and 6 g of melamine was placed in an alumina crucible for 4 h. After that, the sample was cooled at room temperature. Next, the obtained solid material was ground using a mortar, forming a yellow-colored powder, and the final yield was 3.94 g. During the thermal treatment, due to the decomposition of ammonia, the color changed from intense white to pale yellow. The resultant sample was named bulk-C_3N_4. Secondly, the exfoliated graphitic carbon nitride sheets (g-C_3N_4) were prepared via the thermal exfoliation method using prepared bulk g-C_3N_4. In detail, to the 2 g of prepared bulk g-C_3N_4, 20 mL of H_2SO_4 was added and stirred for 12 h. Then 200 mL of deionized water was added to the mixture. After that, the resultant mixture was washed with deionized water until it reached a neutral pH. The faint yellow-colored sample was dried at 60 °C for 12 h in a hot air oven. Finally, the faint yellow-colored powder was poured into an alumina boat, and this boat was placed into a tubular furnace at 550 °C (2 °C/min heating rate, for 4 h, under a nitrogen gas flow), and the porous sheet of exfoliated g-C_3N_4 was obtained [14] (Figure 1).

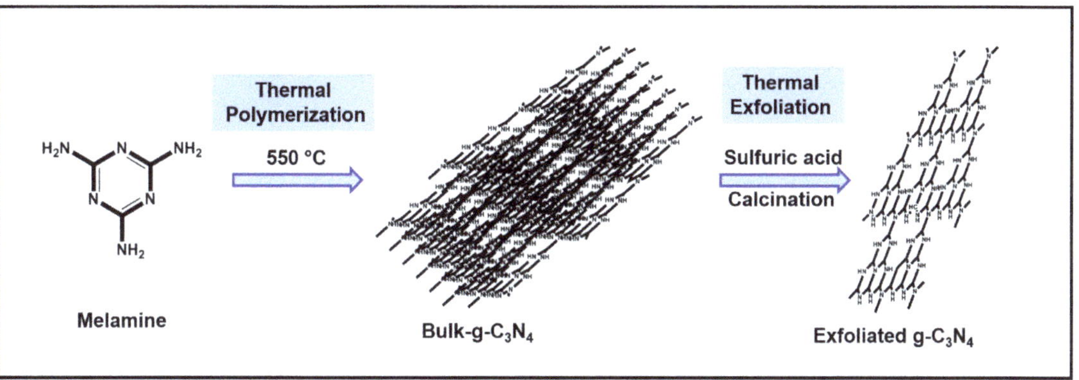

Figure 1. Simplified representation of the production of exfoliated g-C_3N_4.

2.3. Synthesis of Cupric Oxide Loaded g-C_3N_4 Nanocomposites

Synthesis of cupric oxide (CuO) loaded graphitic carbon nitride (g-C_3N_4) nanocomposites by the direct heating process of the dry synthesis method. In the detailed protocol, various amounts of g-C_3N_4 and copper acetate compounds were grained for 30 min. No solvent or capping agents were used in this method. The grained mixture was navy-blue in color. Then, the solid mixture was taken in a silica crucible and placed in a furnace for 3 h at 550 °C. The mixture was then cooled to room temperature, and we obtained a brown-colored mixture of CuO/g-C_3N_4 nanocomposites as shown in Figure 2: CuO, 30 wt% CuO/g-C_3N_4, 40 wt% CuO/g-C_3N_4, 50 wt% CuO/g-C_3N_4, 60 wt% CuO/ g-C_3N_4, and 70 wt% CuO/g-C_3N_4, respectively. The various compounds were compared to the results of the antimicrobial analysis.

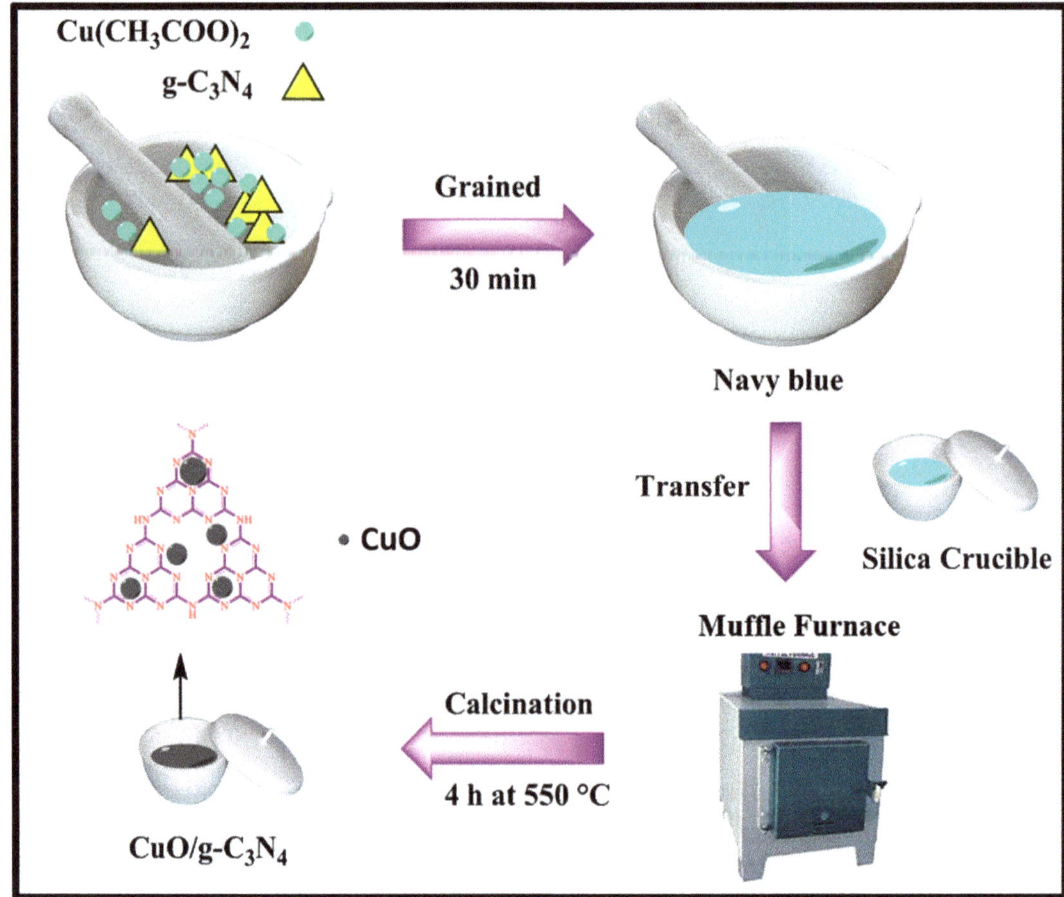

Figure 2. CuO/g-C_3N_4 nanocomposites prepared by dry synthesis method.

2.4. Characterization Techniques

The phase and crystallographic structure of as-prepared nanocrystals were observed by a Powder X-ray diffractometer (Bruker, D8 Advance). The diffractograms were recorded for 2θ in the range of 5–80°, with a time of 6 s. The SEM images were captured by scanning electron microscope SEM-ZEISS (EVO18)), Carl Zeiss Microscopy GmbH, 07745 Jena, Germany, before applying spectral coating using gold platinum metal (Quorum). The elemental analysis of the catalyst was analyzed by energy dispersive X-ray (EDX; VEGA3 XUM/TESCAN). UV-Visible absorption spectra were analyzed on JASCO (V-670 PC) equipment, with a wavelength range of 200–1000 nm at room temperature. The optical band gap of these materials was estimated using the Tauc equation.

2.5. Anti-Bacterial Activity

2.5.1. Preparation of Inoculum

The antimicrobial properties of the given samples were tested against Gram-positive bacteria (*Staphylococcus aureus* ATCC 25923, *Streptococcus pyogenes* ATCC 19615) and Gram-negative bacteria (*Escherichia coli* ATCC 25922, *Pseudomonas aeruginosa* ATCC 27853). All bacteria were pre-cultured in Mueller Hinton Broth (MHB) in a rotary shaker at 37 °C for 18 h. Next, each strain was modified at a concentration of 10^8 cells/mL using the 0.5 McFarland standard [19].

2.5.2. Agar Well Diffusion Method

The fresh bacterial culture was pipetted in a sterile petri dish. Molten-cooled Mueller Hinton Agar (MAH) was then poured into the petri dish and blended well. Upon solidification, wells were made using a sterile cork borer (6 mm in diameter) into the agar plates containing the inoculums. Then, 50 µL of the sample (20 µg/µL concentration) was added to the respective wells. These plates were then incubated at 37 °C for 18 h. After the incubation period, the antibacterial activity was obtained by measuring the zone of inhibition (including the diameter of the wall). Saline was applied as a negative control.

2.6. Fungal Activity

The antifungal activity of the nanocomposites prepared from a facile dry synthesis method was tested against 2 fungi, namely *Fusarium oxysporum* and *Trichoderma viride.*

Agar Well Diffusion Method

The fungal cultures *Fusarium oxysporum* and *Trichoderma viride* were cultured in potato dextrose broth. PDA agar plates were prepared. Four wells were bored in each plate. Overnight fungal cultures were swabbed into the PDA plates. The prepared compounds were pipetted into the wells in volumes of 20, 40, 60, and 80 µL. The plates were incubated at room temperature for 2–3 days to check the formation of the zone around the well. The plates maintained were kept in triplicate, and zones around the wells were closely monitored.

3. Results

3.1. X-ray Diffraction Method (XRD)

To confirm the structural properties of the synthesized cupric oxide nanoparticles (CuO NPs), bulk and exfoliated g-C_3N_4, and 30 wt%, 40 wt%, 50 wt%, 60 wt%, 70 wt% of CuO loaded g-C_3N_4 nanocomposites were evaluated by powder X-ray diffraction (P-XRD) analysis. The XRD pattern of g-C_3N_4, with hexagonal symmetry, is also presented in Figure 3A. The diffraction weak peak at 13.1° (110) and the strong peak at 27.4° (200) represent the g-C_3N_4 surfaces. Figure 3B represents the XRD spectra of the CuO nanoparticles. The diffraction spectra of pure CuO NPs at 32.35°, 35.26°, 39.35°, 48.97°, 53.15°, 58.90°, 61.78°, 66.35°, 68.88°, and 72.13° corresponded to the (-110), (111), (111), (202), (020), (202), (113), (004), (220), and (311) crystal facets and the lattice parameters a = 4.68 Å, b = 3.43 Å, c = 5.13 Å, β = 99.47°. Both diffracted peaks of the *g*-C_3N_4 and CuO NPs could be seen clearly for all different weight percentages of CuO-loaded *g*-C_3N_4 nanocomposites (Figure 3C).

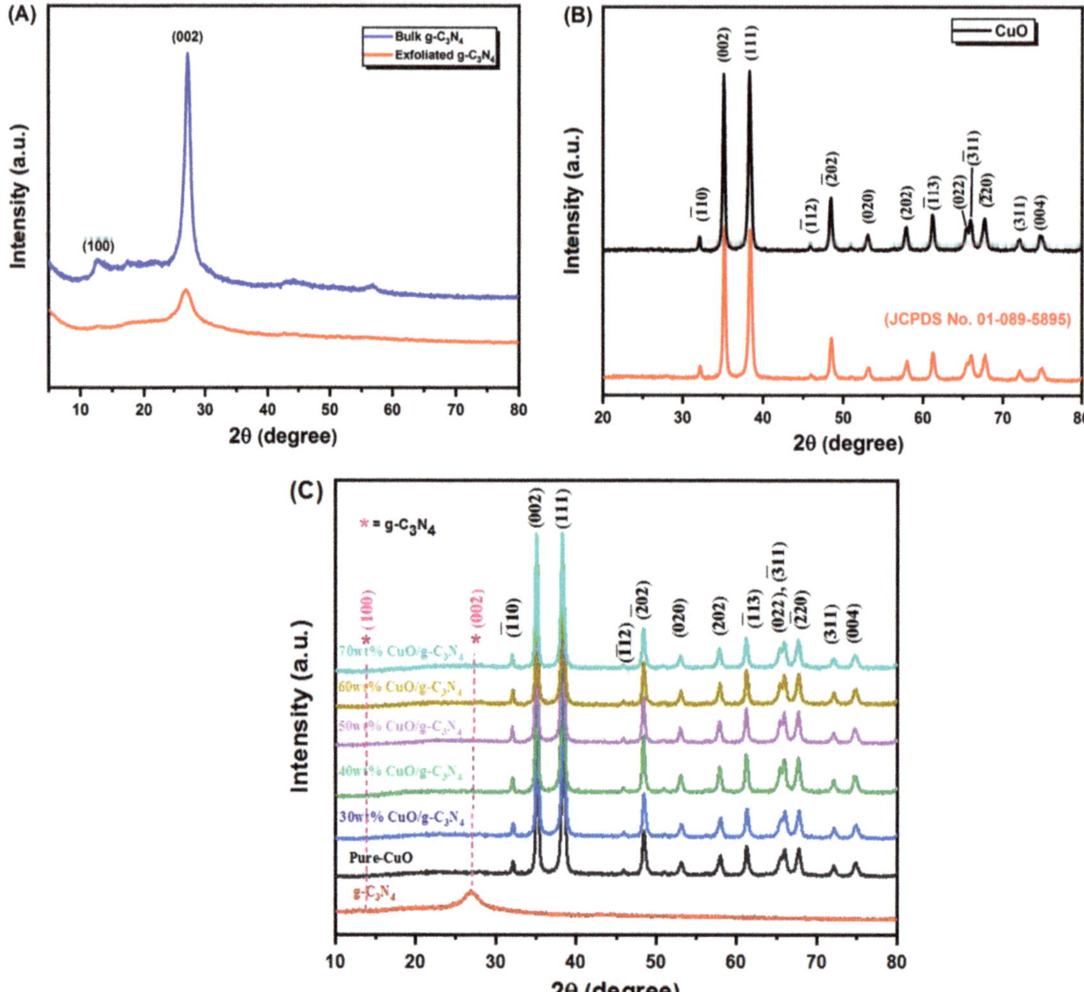

Figure 3. XRD diffractogram of (**A**) bulk g-C_3N_4 and exfoliated g-C_3N_4, (**B**) bare CuO and standard peak, (**C**) bare CuO and exfoliated g-C_3N_4, and various weight percentages of CuO loaded g-C_3N_4.

3.2. Scanning Electron Microscopy (SEM)

The morphology and elemental composition of the products were analyzed through SEM and EDS. Figure 4 displays the SEM images of the obtained CuO sample from the dry synthesis method of the copper acetate precursor, the bulk g-C_3N_4, the exfoliated g-C_3N_4, and the CuO-loaded g-C_3N_4 nanocomposite. From Figure 4A,B, the sheet-like structure of bulk g-C_3N_4 (Figure 4A) and exfoliated g-C_3N_4 (Figure 4B) is observed. Moreover, thermal treatment resulted in a sheet-like g-C_3N_4 structure. Figure 4C–F shows the SEM images of the CuO nanoparticles and CuO loaded onto the g-C_3N_4 nanocomposites. This analysis confirms that the shape of as-prepared CuO NPs shows solid agglomeration, as seen in Figure 4C,D. Furthermore, the successful loading of CuO Nps on exfoliated g-C_3N_4 is confirmed in Figure 4E,F. The SEM-EDS measurements of the presence of C and N atoms in the nanosheets are shown in Figure 4G.

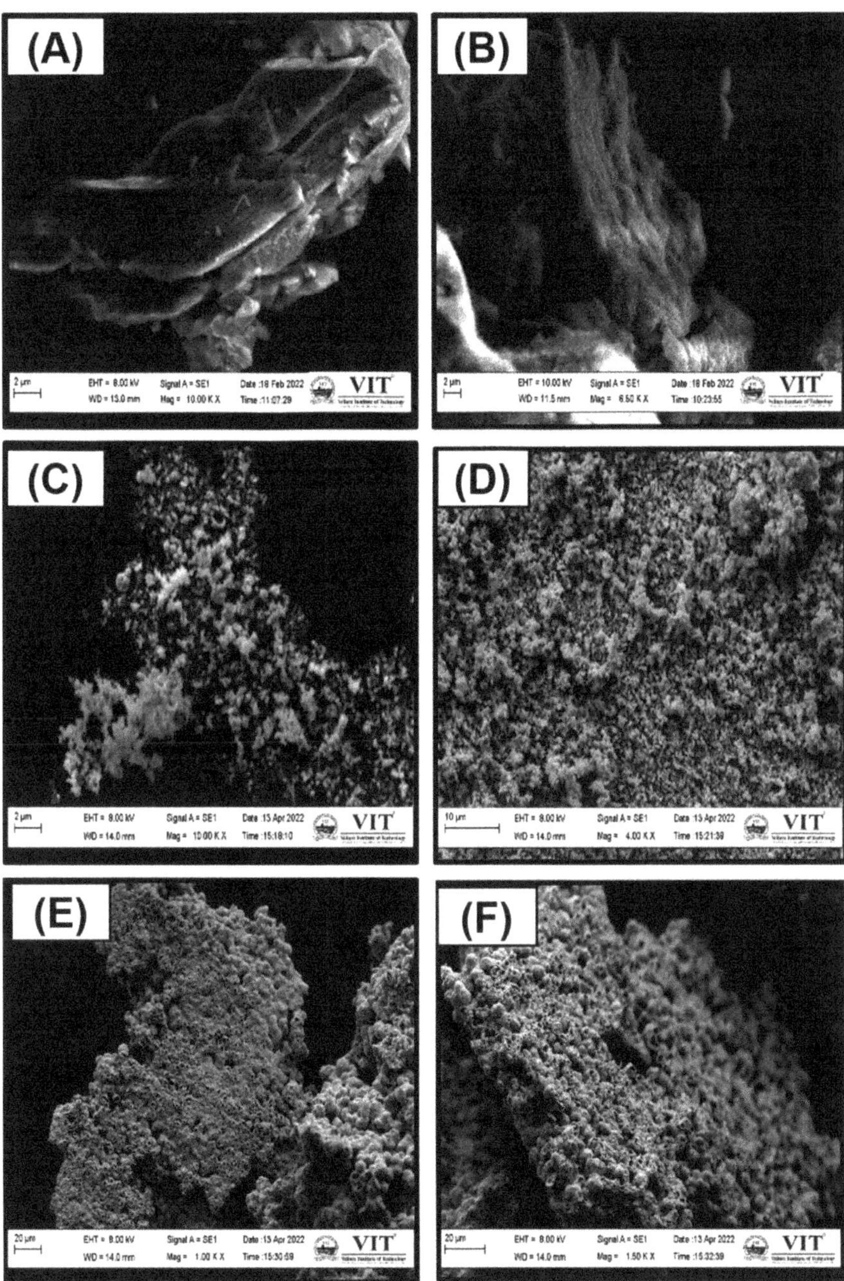

Figure 4. *Cont.*

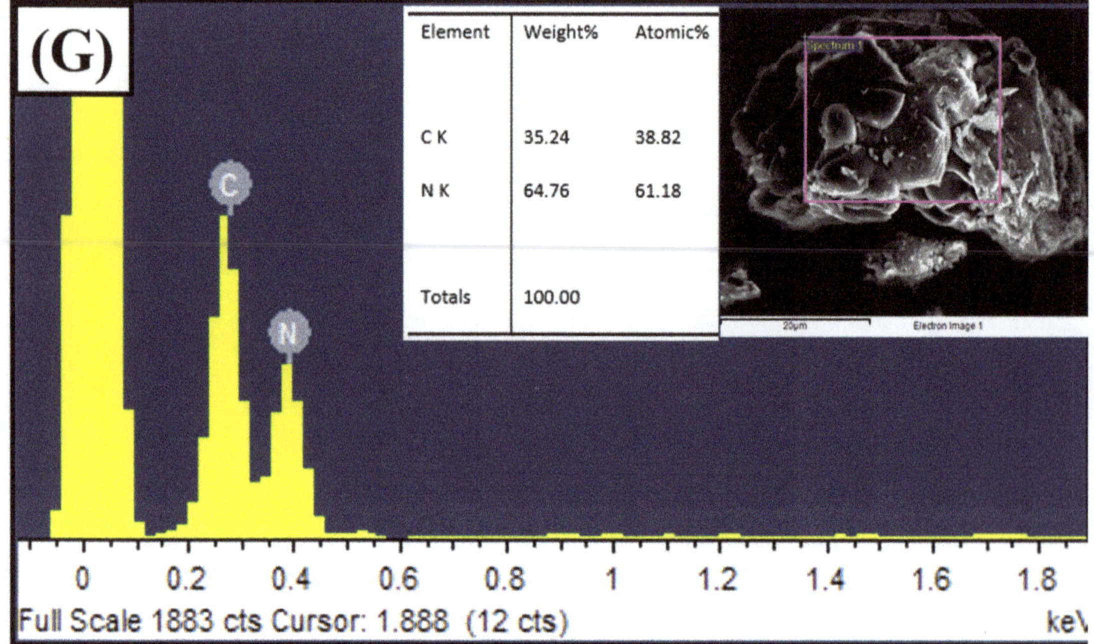

Figure 4. SEM figure of (**A**) bulk g-C_3N_4, (**B**) exfoliated g-C_3N_4, (**C**,**D**) CuO, (**E**,**F**) CuO/g-C_3N_4 nanocomposites, and (**G**) EDS spectra of bulk g-C_3N_4.

3.3. UV-Visible Diffuse Reflectance Spectroscopy (UV-Vis DRS)

The optical absorption properties of CuO, g-C_3N_4, and CuO-loaded g-C_3N_4 are presented in Figure 5. The estimated band gap is obtained by extrapolating the straight portion of $(\alpha hv)^2$ against the hv plot to the point $\alpha = 0$, i.e., 1.10, 2.89, 1.17, 1.06, 1.03, 1.30, and 1.16 eV for the bare CuO, g-C_3N_4, 30 wt% CuO loaded g-C_3N_4, 40 wt% CuO loaded g-C_3N_4, 50 wt% CuO loaded g-C_3N_4, 60 wt% CuO loaded g-C_3N_4, and 70 wt% CuO loaded g-C_3N_4, respectively, as shown in Figure 5A–G. The UV-Vis spectrum of the sphere-like structured CuO nanoparticles (inset of Figure 5B) showed two absorptions at 290 and 355 nm. The g-C_3N_4 and 30 wt% CuO/g-C_3N_4, 40 wt% CuO/g-C_3N_4, 50 wt% CuO/g-C_3N_4, 60 wt% CuO/g-C_3N_4, 70 wt% CuO/g-C_3N_4 manifest the absorption maxima in the region of 310 nm and 230–380 nm, respectively (inset of Figure 5A,C–G).

3.4. Antimicrobial Activity

The antibacterial activity of bare CuO, bulk g-C_3N_4, exfoliated g-C_3N_4, and CuO-loaded g-C_3N_4 was analyzed against various bacterial strains of Gram-positive bacteria (*Staphylococcus aureus* ATCC 25923, *Streptococcus pyogenes* ATCC 19615) and Gram-negative bacteria (*Escherichia coli* ATCC 25922, *Pseudomonas aeruginosa* ATCC 27853). The CuO NPs strains manifest a larger zone of inhibition, and a smaller zone of inhibition is manifested by the resistant strains. In regards to the zone of inhibition, the Gram-positive bacteria exhibited the highest activity toward CuO, while the Gram-negative bacteria manifested the lowest activity [1]. The inactivation rates of *Staphylococcus aureus* and *Streptococcus pyogenes* are not significantly different from each other; however, compared to that of *E. coli*, they are significantly higher. The antibacterial activity exhibited by CuO NPs and different weight percentages of 40 wt%, 50 wt%, 60 wt%, and 70 wt% cupric oxide-loaded g-C_3N_4 were effective against both Gram-positive and Gram-negative bacteria (Figure 6).

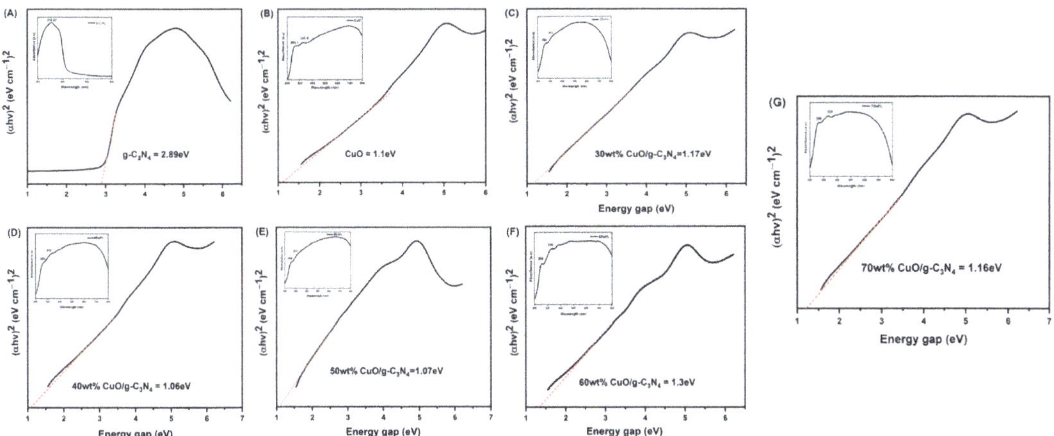

Figure 5. UV-DRS spectra exhibit band gap energy of exfoliated g-C_3N_4 (**A**), CuO (**B**), and CuO loaded g-C_3N_4 (**C–G**); (inset: absorption peak). (Bold black line represented by UV-DRS and dotted line noted by bandgap measurement line).

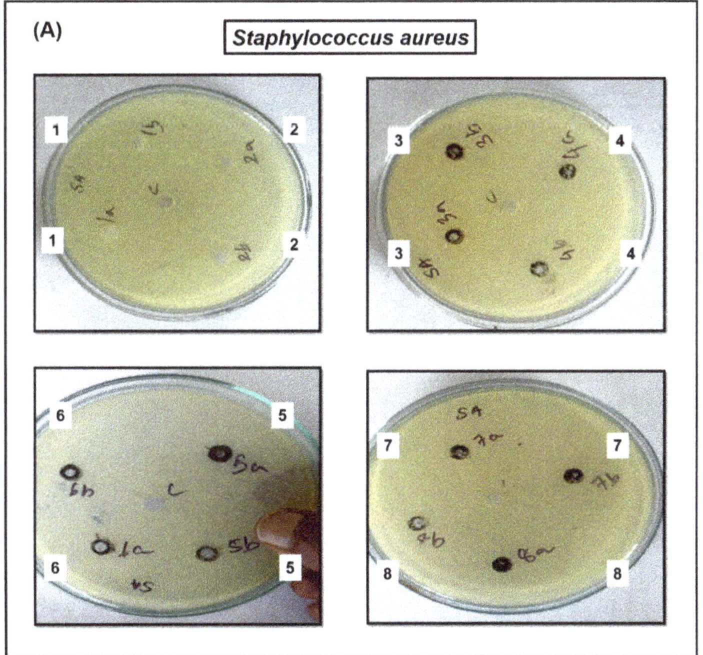

Figure 6. *Cont.*

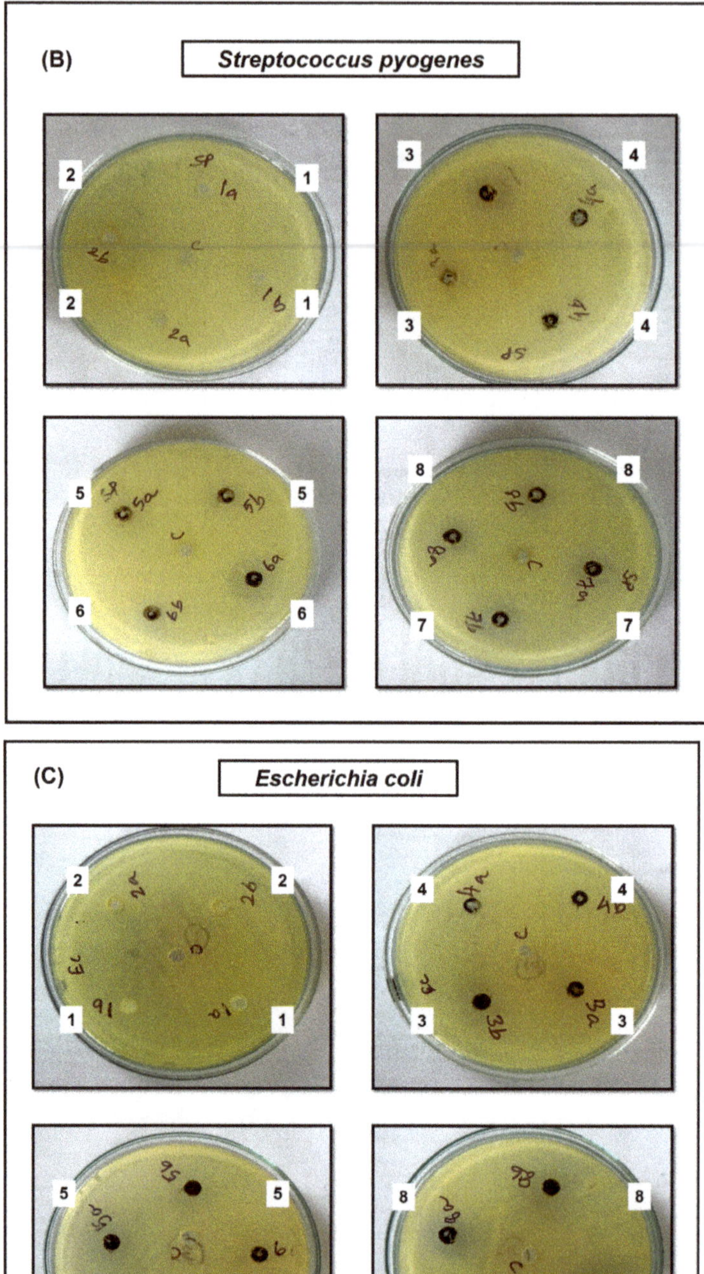

Figure 6. *Cont.*

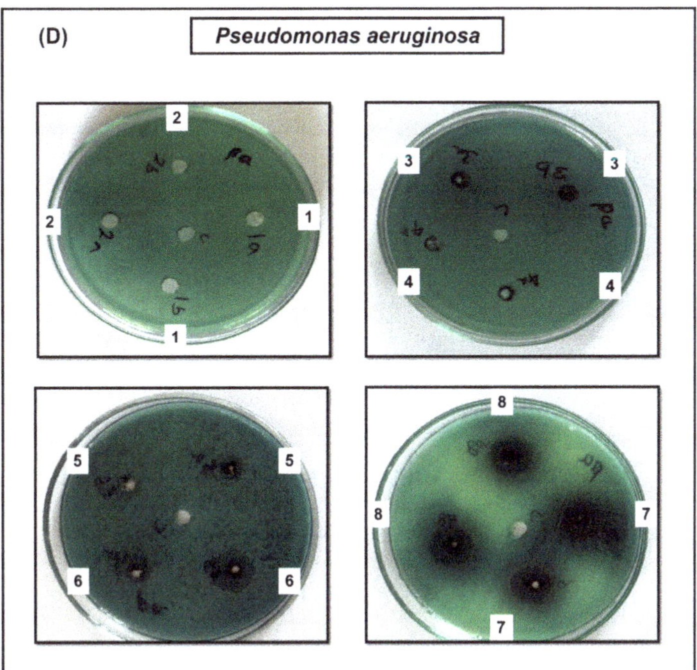

Figure 6. Antimicrobial activity of bulk *g*-C_3N_4 (1), exfoliated *g*-C_3N_4 (2), CuO (3), and different weight percentages of CuO loaded *g*-C_3N_4 nanocomposite, 30 wt% (4), 40 wt% (5), 50 wt% (6), 60 wt% (7), and 70 wt% (8) against *Staphylococcus aureus* (**A**), *Streptococcus pyogenes* (**B**), *Escherichia coli* (**C**), and *Pseudomonas aeruginosa* (**D**).

The effect of 70 wt% CuO-loaded *g*-C_3N_4 nanocomposites on the development of *Fusarium oxysporum* and *Trichoderma viride* as determined by the diameter of the fungal colonies on the samples, non-nanocomposite samples, and other samples of 1 mg/10 mL 70 wt% of CuO loaded *g*-C_3N_4 solution with the various concentrations of 20, 40, 60, 80 µL, respectively, after 4 days of incubation. In addition, Figure 7 shows the change in the diameter of fungal colonies for various concentrations of CuO/*g*-C_3N_4 solutions. These results show that 70 wt% of CuO-loaded g-C_3N_4 nanocomposite does not satisfactorily inhibit the fungal species.

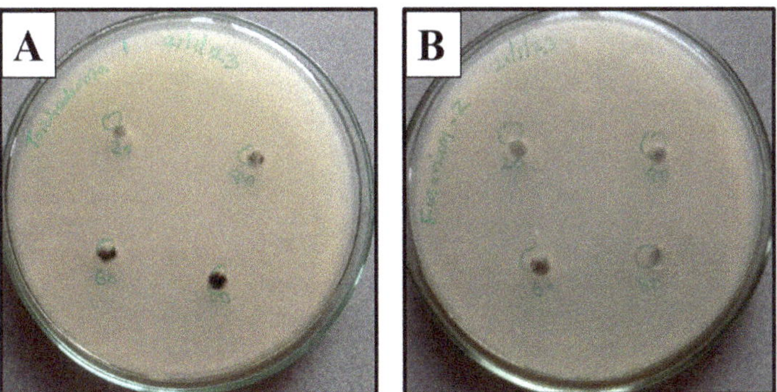

Figure 7. Antifungal activity of 70 wt% of CuO loaded g-C_3N_4 nanocomposite solution with the various concentrations of 20, 40, 60, and 80 µL (**A**), *Trichoderma viride*, and (**B**) *Fusarium oxysporum*.

4. Discussion

In this study, for the first time, cupric (II) oxide (CuO) nanoparticles loaded with exfoliated graphitic carbon nitride (g-C_3N_4) nanocomposite also showed antimicrobial activity. The current trending field of nanotechnology is critical and requires environmentally safe methods for the synthesis of nanoparticles. Here, a facile, rapid route and a low-cost approach for the preparation of stable cupric oxide (CuO NPs) nanoparticles and graphitic carbon nitride using a dry synthesis method is reported. The characteristics of the formed cupric (II) oxide loaded on graphitic carbon nitride nanocomposites were confirmed using XRD, UV-Vis (DRS), EDX, and SEM analyses.

The XRD patterns of CuO-loaded g-C_3N_4 nanosheets demonstrated their sheet-like structure [20,21]. The XRD patterns of CuO-loaded g-C_3N_4 nanocomposite revealed their crystalline structure. The XRD pattern of g-C_3N_4 was also presented in Figure 3A. The diffraction weak peak at 13.1° (110) and the strong peak at 27.4° (200) represent the g-C_3N_4 surface, arising from the in-plane structure of triazine, with the typical interplanar staking peaks of the inner layer structural packing. Figure 3B represents the XRD spectra of the CuO nanospheres particles. The diffraction spectra of pure CuO NPs at 32.35°, 35.26°, 39.35°, 48.97°, 53.15°, 58.90°, 61.78°, 66.35°, 68.88°, and 72.13° corresponded to the (-110), (111), (111), (202), (020), (202), (113), (004), (220), and (311) crystal facets of CuO [22]. In the diffractogram, all the strong and sharp peaks were in accord with the standard JCPDS card no 01-089-5895. The lattice parameters a = 4.68 Å, b = 3.43 Å, c = 5.13 Å, and β = 99.47° of the materials are indicated by a purely monoclinic phase and are additionally confirmed, without any phase impurity present in the synthesis materials. Both diffracted peaks of the g-C_3N_4 and CuO NPs could be seen clearly for all different weight percentages of the CuO-loaded g-C_3N_4 nanocomposites. The peak pertaining to g-C_3N_4 can be found as the mass ratio of g-C_3N_4 in CuO-loaded g-C_3N_4 nanocomposites of 30 wt%. While low intensities of g-C_3N_4 diffraction peaks are observed, this can be attributed to the low content of g-C_3N_4, which confirms the presence of g-C_3N_4 and CuO in the CuO-loaded g-C_3N_4 nanocomposites (Figure 3C).

The morphology and structure of bulk, exfoliated g-C_3N_4, CuO, and CuO-loaded g-C_3N_4 nanocomposites were analyzed by SEM. From Figure 3A,B, the sheet-like structure of bulk g-C_3N_4 (Figure 3A) and exfoliated g-C_3N_4 (Figure 3B) is clearly observed. Besides, thermal treatment resulted in a sheet-like g-C_3N_4 structure, and the layers became thinner, with a more detached and several-layer structure characteristic of two-dimensional exfoliated materials in the thermally exfoliated processed nanosheets. Furthermore, scanning electron microscopy analysis confirmed that the as-prepared CuO NPs exhibit nanoparticles of uniform size, as shown in Figure 3C,D. In CuO NPs loaded g-C_3N_4, Figure 3E,F shows that the nanoparticles were loaded onto the crumpled and rippled surface of the g-C_3N_4 sheets.

One of the most important electronic parameters for metal oxide nanomaterials is band gap energy. The band gap energy impacts the activity of the molecular adsorption sites and affects adsorption activity. The band gap energy for these samples was calculated from the optical absorption experiments using the Tauc equation, in which

$$(\alpha h v) = A \times (h v - E_g) \quad (1)$$

where the optical absorption coefficient is indicated by α. A is the constant, hv is an incident photon energy, and Eg is the energy gap, respectively. α can be replaced with the absorbance of the sample. The estimated band gap was obtained by extrapolating the straight portion of $(\alpha hv)^2$ against the hv plot to the point α = 0, i.e., 1.10, 2.89, 1.17, 1.06, 1.03, 1.30, and 1.16 eV for the samples of bare CuO NSs, g-C_3N_4, 30 wt% CuO loaded g-C_3N_4, 40 wt% CuO loaded g-C_3N_4, 50 wt% CuO loaded g-C_3N_4, 60 wt% CuO loaded g-C_3N_4, and 70 wt% CuO loaded g-C_3N_4, respectively, as shown in Figure 5A–G. The reason for the absorption was that the 3p state was located at the conduction band in the g-C_3N_4, resulting in a decrease in the band gap. Its band gap energy is 2.89 eV, as shown in Figure 5A. This indicates that g-C_3N_4 enhances the visible light utilization ability. The UV-Vis spectrum of the sphere-like structured CuO nanoparticles (inset of Figure 5B) showed two absorptions at 290 and 355 nm. The g-C_3N_4 and 30 wt% CuO/g-C_3N_4, 40 wt% CuO/g-C_3N_4, 50 wt% CuO/g-C_3N_4, 60 wt% CuO/g-C_3N_4, and 70 wt% CuO/g-C_3N_4 manifest the absorption maxima in the region of 310 nm and 230–380 nm, respectively (inset of Figure 5A,C–G).

The antibacterial activity of various nanoparticles for various bacteria using a well diffusion assay is shown in Figure 6A–D. The results illustrated an increase in CuO weight percentage directly proportionally to the antibacterial activity. The CuO NPs strains manifest a larger zone of inhibition, and a smaller zone of inhibition is manifested by the resistant strains. In regards to the zone of inhibition, the Gram-positive bacteria exhibited the highest activity toward CuO, while the Gram-negative bacteria manifest the lowest activity. The inactivation rates of *Staphylococcus aureus* and *Streptococcus pyogenes* are not significantly different from each other; however, compared to that of *E. coli*, they are significantly higher. These results indicated that Gram-positive bacteria are more resistant to the antibacterial activity of CuO-loaded g-C_3N_4 than Gram-negative bacteria, such as *Escherichia coli* and *Pseudomonas aeruginosa* [21]. The antibacterial activity results for CuO-loaded g-C_3N_4 indicated a superior resistance to Gram-positive bacteria compared to Gram-negative bacteria such as *Escherichia coli* and *Pseudomonas aeruginosa* [1,23–25]. These results demonstrated that susceptibility to inactivation is dependent upon the microorganism, as shown in Table 1. Future work is necessary to draw firm conclusions about the improved g-C_3N_4 inactivation performance for Gram-negative over Gram-positive microorganisms.

Table 1. Antimicrobial activity of cupric oxide/g-C_3N_4nanocomposite.

Samples	*S. aureus* (mm)	*S. pyogenes* (mm)	*E. coli* (mm)	*P. aeruginosa* (mm)
Bulk-g-C_3N_4	R	R	R	R
Exfoliated g-C_3N_4	R	R	R	R
CuO	8	12	6	8
30 wt% CuO/g-C_3N_4	R	R	R	R
40 wt% CuO/g-C_3N_4	16	16	10	12
50 wt% CuO/g-C_3N_4	16	16	10	14
60 wt% CuO/g-C_3N_4	18	18	12	14
70 wt% CuO/g-C_3N_4	18	18	12	14

(R)—resistant.

The photocatalytic activities are depended on the amount of active radical species (such as h^+, e^-, $\cdot OH$, $\cdot O_2^-$) produced in the reactions [21]. Previous studies confirmed that radicals play important roles in photoreactions. Figure 8 shows an increase in active metal oxide, indicating that g-C_3N_4 under light irradiation can easily produce $\cdot OH$, due to the oxidation between H^+ and H_2O. Subsequently, the $\cdot OH$ reacts with the bacterial cell membrane and cleaves the linkages. It is commonly known that g-C_3N_4 is a photocatalytic agent; therefore, the application of light and well-produced radicals, such as $\cdot O_2^-$ and $\cdot OH$, play important roles in the photocatalytic process [14–16,20,26]. This compound is highly active under light irradiation, but in this study, dark conditions were maintained. However, it is still moderately active against Gram-positive and Gram-negative bacteria, compared to metal oxide nanoparticles [27,28], but CuO/g-C_3N_4 nanocomposites exhibit good antibacterial activity. Finally, the bacteria cell was distorted. If its cell membrane is damaged, the intracellular contents (such as DNA, protein, and mitochondria) can leak from the extracellular suspensions of the cell, resulting in the destruction of the cell structure.

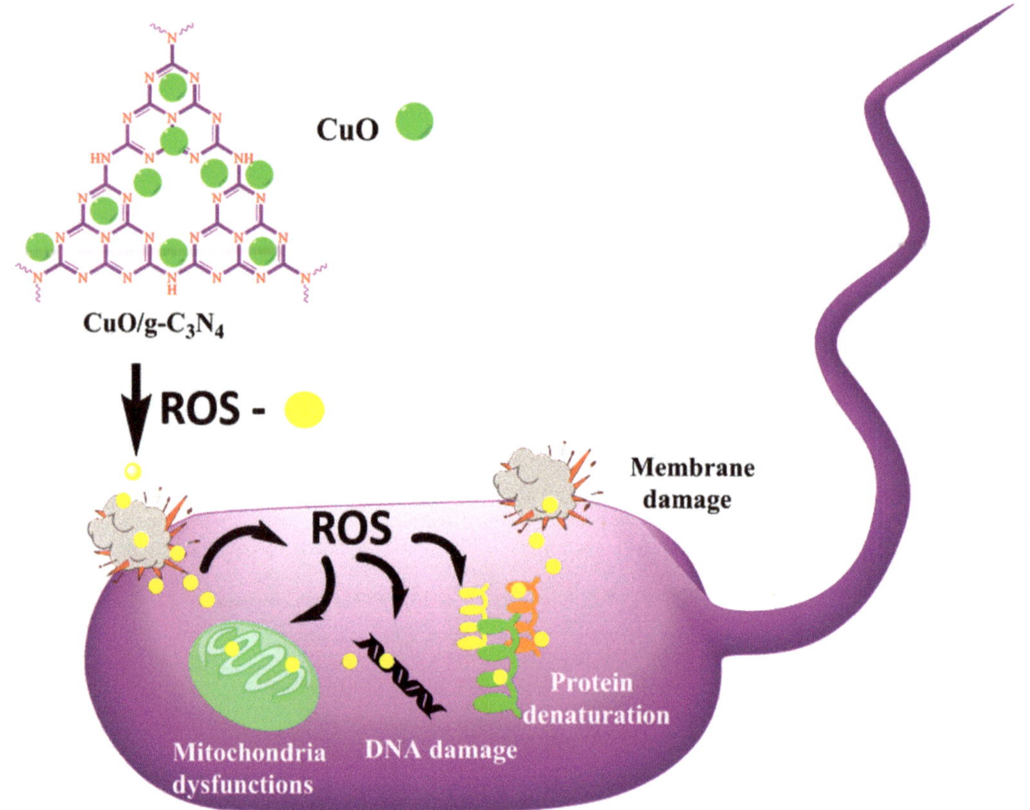

Figure 8. Plausible mechanisms of antimicrobial activity.

In previous reports, g-C_3N_4 exhibited excellent activity in visible light irradiation due to the production of $\cdot O_2^-$ and $\cdot OH$ active radical species in photocatalytic disinfection [27]. The g-C_3N_4 exhibits good photoelectric properties and band structure. In our case, however, since this study was performed under room conditions in vitro, the results showed that fungal growth was not inhibited [21,29,30]. Thus, future studies should be conducted in such a way that a fungal species should be subjected to photocatalytic irradiation [1]. Here, this study was not performed in the presence of sunlight. Under lab conditions, the pathogenic fungi (*Fusarium oxysporum* and *Trichoderma viride*) were allowed to interact with 1 mg/mL in different concentrations of 20, 40, 60, and 80 µL (triplicates), and the compounds at these particular concentrations did not inhibit the fungi in the presence of this CuO/g-C_3N_4 nanocomposite [31,32]. This is because g-C_3N_4 masks the effect of the CuO molecules and does not inhibit fungi. According to Lin et al., g-C_3N_4 responds to nitrogen fixation and acts as a foliar fertilizer [26]. Since effective nitrogen fixation results in prominent nitrogen fixations for a good yield, as well as predominant protein synthesis, the result is the growth of fungi, rather than fungal inhibition. Therefore, further studies should be conducted in such a way that the CuO/g-C_3N_4 compounds are subjected to photocatalytic exposure in the presence of sunlight. In this way, the inhibition of fungi may be possible for these molecules.

5. Conclusions

The various biological processes, including apoptosis, anti-angiogenesis, oxidative stress, chemotherapy, and inflammation, are modulated by inorganic nanoparticles. In this study, cupric (II) oxide (CuO) nanoparticles loaded with exfoliated graphitic carbon nitride (g-C_3N_4) nanocomposite assisted in the suppression or inhibition of bacterial growth and the interaction of fungi. The current trending field of nanotechnology is critical and requires environmentally safe methods for the synthesis of nanoparticles. Here, a facile, rapid route and a low-cost approach for the preparation of stable cupric oxide (CuO) nanospheres and graphitic carbon nitride using a dry synthesis method was reported. The characteristics of the formed copper (II) oxide loaded on graphitic carbon nitride nanocomposites were confirmed using XRD, UV-Vis (DRS), EDX, and SEM analyses. The XRD patterns of CuO-loaded g-C_3N_4 nanosheets demonstrated the sheet-like structure. The XRD patterns of CuO-loaded g-C_3N_4 nanocomposite revealed the crystalline structure. The SEM images of the synthesis CuO-loaded g-C_3N_4 nanocomposite morphology shows a two-dimensional sheet-like structure. The antibacterial activity was exhibited by CuO NPs, with different weight percentages of 40 wt%, 50 wt%, 60 wt%, and 70 wt% cupric oxide-loaded g-C_3N_4 against both Gram-positive and Gram-negative bacteria. The synthesized bulk g-C_3N_4, exfoliated g-C_3N_4, and lower weight percentages of the nanocomposite (30 wt%) showed low antibacterial activity. The optimum dose of 70wt % CuO-loaded g-C_3N_4 showed high antibacterial activity but did not exhibit satisfactory activity in antifungal studies conducted in dark fields. This is because g-C_3N_4 masks the effect of CuO molecules and does not inhibit fungi. The scope of using CuO-loaded g-C_3N_4 nanocomposites as antimicrobials needs to be further explored in the presence of sunlight.

Author Contributions: R.L.P.—methodology, analysis, writing—original draft preparation; S.K.—methodology, writing—review and editing; B.K.—methodology, writing—review and editing; R.S.—methodology, writing—review and editing; R.V.—methodology, antibacterial activity, anti-angiogenesis analysis, and writing—review, and editing; S.G.B.—design, conceptualization, writing—review, and editing. All authors have read and agreed to the published version of the manuscript.

Funding: This work was supported by the VIT School of Agricultural Innovations and Advanced Learning (VAIAL), Vellore Institute of Technology, Vellore, India.

Institutional Review Board Statement: The article was reviewed by the institutions.

Informed Consent Statement: Not applicable.

Data Availability Statement: Not applicable.

Acknowledgments: The authors wish to thank VIT management for characterization support and infrastructure. S.G.B. thanks VIT Seed Grand (SG20210159) for financial support. V.R. acknowledges VIT for financial support to perform biology studies. R.L.P. acknowledges VIT for providing the fellowship grant.

Conflicts of Interest: The authors declare no competing financial support.

References

1. Shinde, R.S.; More, R.A.; Adole, V.A.; Koli, P.B.; Pawar, T.B.; Jagdale, B.S.; Desale, B.S.; Sarnikar, Y.P. Design, fabrication, antitubercular, antibacterial, antifungal and antioxidant study of silver doped ZnO and CuO nano candidates: A comparative pharmacological study. *Curr. Opin. Green Sustain. Chem.* **2021**, *4*, 100138. [CrossRef]
2. Peng, Y.; Li, S.J.; Yan, J.; Tang, Y.; Cheng, J.P.; Gao, A.J.; Yao, X.; Ruan, J.J.; Xu, B.L. Research progress on phytopathogenic fungi and their role as biocontrol agents. *Front. Microbiol.* **2021**, *12*, 670135. [CrossRef]
3. Gomes, E.V.; Costa, M.d.N.; de Paula, R.G.; Ricci de Azevedo, R.; da Silva, F.L.; Noronha, E.F.; José Ulhoa, C.; Neves Monteiro, V.; Elena Cardoza, R.; Gutiérrez, S. The Cerato-Platanin protein Epl-1 from Trichoderma harzianum is involved in mycoparasitism, plant resistance induction and self cell wall protection. *Sci. Rep.* **2015**, *5*, 17998. [CrossRef]
4. Verma, A.; Anand, P.; Kumar, S.; Fu, Y.-P. Cu-cuprous/cupric oxide nanoparticles towards dual application for nitrophenol conversion and electrochemical hydrogen evolution. *Appl. Surf. Sci.* **2022**, *578*, 151795. [CrossRef]

5. Giridasappa, A.; Rangappa, D.; Shanubhoganahalli Maheswarappa, G.; Marilingaiah, N.R.; Kagepura Thammaiah, C.; Shareef, I.M.; Kanchugarakoppal Subbegowda, R.; Doddakunche Shivaramu, P. Phytofabrication of cupric oxide nanoparticles using Simarouba glauca and Celastrus paniculatus extracts and their enhanced apoptotic inducing and anticancer effects. *Appl. Nanosci.* **2021**, *11*, 1393–1409. [CrossRef]
6. Jithul, K.; Samra, K.S. Cupric Oxide based Supercapacitors: A Review. In Proceedings of the Journal of Physics: Conference Series, Wuhan, China, 18 September 2022; p. 012120.
7. Sumathi, S.; Nehru, M.; Vidya, R. Synthesis, characterization and effect of precipitating agent on the antibacterial properties of cobalt ferrite nanoparticles. *Trans. Indian Ceram. Soc.* **2015**, *74*, 79–82. [CrossRef]
8. Cuong, H.N.; Pansambal, S.; Ghotekar, S.; Oza, R.; Hai, N.T.T.; Viet, N.M.; Nguyen, V.-H. New frontiers in the plant extract mediated biosynthesis of copper oxide (CuO) nanoparticles and their potential applications: A review. *Environ. Res.* **2022**, *203*, 111858. [CrossRef]
9. Rangarajan, S.; Verekar, S.; Deshmukh, S.K.; Bellare, J.R.; Balakrishnan, A.; Sharma, S.; Vidya, R.; Chimote, G. Evaluation of anti-bacterial activity of silver nanoparticles synthesised by coprophilous fungus PM0651419. *IET Nanobiotechnol.* **2018**, *12*, 106–115. [CrossRef]
10. Hafeez, H.Y.; Lakhera, S.K.; Bellamkonda, S.; Rao, G.R.; Shankar, M.; Bahnemann, D.; Neppolian, B. Construction of ternary hybrid layered reduced graphene oxide supported g-C_3N_4-TiO_2 nanocomposite and its photocatalytic hydrogen production activity. *Int. J. Hydrog. Energy* **2018**, *43*, 3892–3904. [CrossRef]
11. Moradi, M.; Hasanvandian, F.; Isari, A.A.; Hayati, F.; Kakavandi, B.; Setayesh, S.R. CuO and ZnO co-anchored on g-C_3N_4 nanosheets as an affordable double Z-scheme nanocomposite for photocatalytic decontamination of amoxicillin. *Appl. Catal.* **2021**, *285*, 119838. [CrossRef]
12. Thiyagarajan, K.; Samuel, S.; Kumar, P.S.; Babu, S.G.J.B.o.M.S. C_3N_4 supported on chitosan for simple and easy recovery of visible light active efficient photocatalysts. *Bull. Mater. Sci.* **2020**, *43*, 1–7. [CrossRef]
13. Liu, Q.; Wang, X.; Yang, Q.; Zhang, Z.; Fang, X. Mesoporous g-C_3N_4 nanosheets prepared by calcining a novel supramolecular precursor for high-efficiency photocatalytic hydrogen evolution. *Appl. Surf. Sci.* **2018**, *450*, 46–56. [CrossRef]
14. Bhuvaneswari, C.; Babu, S.G. Nanoarchitecture and surface engineering strategy for the construction of 3D hierarchical CuS-rGO/g-C_3N_4 nanostructure: An ultrasensitive and highly selective electrochemical sensor for the detection of furazolidone drug. *J. Electroanal. Chem.* **2022**, *907*, 116080. [CrossRef]
15. Liu, Y.; Shen, S.; Li, Z.; Ma, D.; Xu, G.; Fang, B. Mesoporous g-C_3N_4 nanosheets with improved photocatalytic performance for hydrogen evolution. *Mater. Charact.* **2021**, *174*, 111031. [CrossRef]
16. Wen, J.; Xie, J.; Chen, X.; Li, X. A review on g-C3N4-based photocatalysts. *Appl. Surf. Sci.* **2017**, *391*, 72–123. [CrossRef]
17. Li, R.; Ren, Y.; Zhao, P.; Wang, J.; Liu, J.; Zhang, Y. Graphitic carbon nitride (g-C_3N_4) nanosheets functionalized composite membrane with self-cleaning and antibacterial performance. *J. Hazard. Mater.* **2019**, *365*, 606–614. [CrossRef] [PubMed]
18. Sher, M.; Shahid, S.; Javed, M. Synthesis of a novel ternary (g-C_3N_4 nanosheets loaded with Mo doped ZnO nanoparticles) nanocomposite for superior photocatalytic and antibacterial applications. *J. Photochem. Photobiol. B Biol.* **2021**, *219*, 112202. [CrossRef]
19. Gonelimali, F.D.; Lin, J.; Miao, W.; Xuan, J.; Charles, F.; Chen, M.; Hatab, S.R. Antimicrobial properties and mechanism of action of some plant extracts against food pathogens and spoilage microorganisms. *Front. Microbiol.* **2018**, *9*, 1639. [CrossRef] [PubMed]
20. Induja, M.; Sivaprakash, K.; Karthikeyan, S.J.M.R.B. Facile green synthesis and antimicrobial performance of Cu_2O nanospheres decorated g-C_3N_4 nanocomposite. *Mater. Res. Bull.* **2019**, *112*, 331–335. [CrossRef]
21. Kublik, N.; Gomes, L.E.; Plaça, L.F.; Lima, T.H.; Abelha, T.F.; Ferencz, J.A.; Caires, A.R.; Wender, H.J.P. Metal-free g-C_3N_4/nanodiamond heterostructures for enhanced photocatalytic pollutant removal and bacteria photoinactivation. *Photochem.* **2021**, *1*, 302–318. [CrossRef]
22. Bai, B.; Saranya, S.; Dheepaasri, V.; Muniasamy, S.; Alharbi, N.S.; Selvaraj, B.; Undal, V.S.; Gnanamangai, B.M. Biosynthesized copper oxide nanoparticles (CuO NPs) enhances the anti-biofilm efficacy against K. pneumoniae and S. aureus. *J. King Saud Univ.-Sci.* **2022**, *34*, 102120. [CrossRef]
23. Selvaraj, S.P. Enhanced surface morphology of copper oxide (CuO) nanoparticles and its antibacterial activities. *Mater. Today Proc.* **2022**, *50*, 2865–2868. [CrossRef]
24. Du, B.D.; Phu, D.V.; Quoc, L.A.; Hien, N.Q. Synthesis and investigation of antimicrobial activity of Cu_2O nanoparticles/zeolite. *J. Nanoparticles Res.* **2017**, *2017*, 705684. [CrossRef]
25. Salem, S.S.; Ali, O.M.; Reyad, A.M.; Abd-Elsalam, K.A.; Hashem, A.H. Pseudomonas indica-mediated silver nanoparticles: Antifungal and antioxidant biogenic tool for suppressing mucormycosis fungi. *J. Fungus.* **2022**, *8*, 126. [CrossRef]
26. Liu, Y.; Huang, X.; Yu, Z.; Yao, L.; Guo, S.; Zhao, W.J.C. Environmentally Friendly Non-Metal Solar Photocatalyst C_3N_4 for Efficient Nitrogen Fixation as Foliar Fertilizer. *ChemistrySelect* **2020**, *5*, 7720–7727. [CrossRef]
27. Kalia, A.; Abd-Elsalam, K.A.; Kuca, K. Zinc-based nanomaterials for diagnosis and management of plant diseases: Ecological safety and future prospects. *J. Fungus* **2020**, *6*, 222. [CrossRef] [PubMed]
28. Chen, J.; Mao, S.; Xu, Z.; Ding, W. Various antibacterial mechanisms of biosynthesized copper oxide nanoparticles against soilborne Ralstonia solanacearum. *RSC Adv.* **2019**, *9*, 3788–3799. [CrossRef]
29. Soni, A.; Kaushal, D.; Kumar, M.; Sharma, A.; Maurya, I.K.; Kumar, S. Synthesis, characterizations and antifungal activities of copper oxide and differentially doped copper oxide nanostructures. *Mater. Today Proc.* **2022**, in press. [CrossRef]

30. Kumar, P.; Inwati, G.K.; Mathpal, M.C.; Ghosh, S.; Roos, W.; Swart, H.C. Defects induced enhancement of antifungal activities of Zn doped CuO nanostructures. *Appl. Surf. Sci.* **2021**, *560*, 150026. [CrossRef]
31. Zaki, S.A.; Ouf, S.A.; Albarakaty, F.M.; Habeb, M.M.; Aly, A.A.; Abd-Elsalam, K.A. Trichoderma harzianum-mediated ZnO nanoparticles: A green tool for controlling soil-borne pathogens in cotton. *J. Fungus* **2021**, *7*, 952. [CrossRef] [PubMed]
32. Alghuthaymi, M.A.; Abd-Elsalam, K.A.; AboDalam, H.M.; Ahmed, F.K.; Ravichandran, M.; Kalia, A.; Rai, M.J. Trichoderma: An eco-friendly source of nanomaterials for sustainable agroecosystems. *J. Fungus* **2022**, *8*, 367. [CrossRef] [PubMed]

Journal of
Fungi

Article

Co-Application of Silver Nanoparticles and Symbiotic Fungus *Piriformospora indica* Improves Secondary Metabolite Production in Black Rice

Shikha Solanki [1], G. B. V. S. Lakshmi [2], Tarun Dhiman [2], Samta Gupta [3], Pratima R. Solanki [2], Rupam Kapoor [3] and Ajit Varma [1,*]

[1] Amity Institute of Microbial Technology, Amity University, Sector-125, Noida 201303, India
[2] Special Centre for Nanoscience, Jawaharlal Nehru University, New Delhi 110067, India
[3] Department of Botany, University of Delhi, New Delhi 110007, India
* Correspondence: ajitvarma@amity.edu

Abstract: In the current research, unique Nano-Embedded Fungus (NEF), made by the synergic association of silver nanoparticles (AgNPs) and endophytic fungus (*Piriformospora indica*), is studied, and the impact of NEF on black rice secondary metabolites is reported. AgNPs were synthesized by chemical reduction process using the temperature-dependent method and characterized for morphological and structural features through UV visible absorption spectroscopy, zeta potential, XRD, SEM-EDX, and FTIR spectroscopy. The NEF, prepared by optimizing the AgNPs concentration (300 ppm) in agar and broth media, showed better fungal biomass, colony diameter, spore count, and spore size than the control *P. indica*. Treatment with AgNPs, *P. indica*, and NEF resulted in growth enhancement in black rice. NEF and AgNPs stimulated the production of secondary metabolites in its leaves. The concentrations of chlorophyll, carotenoids, flavonoids, and terpenoids were increased in plants inoculated with *P. indica* and AgNPs. The findings of the study highlight the synergistic effect of AgNPs and the fungal symbionts in augmenting the secondary metabolites in leaves of black rice.

Keywords: *Serendipita indica*; endosymbiont; nano-embedded fungus; confocal microscopy; scanning electron microscopy; *Oryza sativa* L. *indica*; nano-bioformulation

Citation: Solanki, S.; Lakshmi, G.B.V.S.; Dhiman, T.; Gupta, S.; Solanki, P.R.; Kapoor, R.; Varma, A. Co-Application of Silver Nanoparticles and Symbiotic Fungus *Piriformospora indica* Improves Secondary Metabolite Production in Black Rice. *J. Fungi* **2023**, *9*, 260. https://doi.org/10.3390/jof9020260

Academic Editor: Kamel A. Abd-Elsalam

Received: 28 December 2022
Revised: 21 January 2023
Accepted: 25 January 2023
Published: 15 February 2023

1. Introduction

The evolution of nanotechnology has strengthened the effective use of nanoparticles and has emerged as a radical tool for enhancing agricultural practices and moving to a more sustainable and resilient farming sector [1–3]. Nanoscale Particles (NSPs) are atomic or molecular-sized particles of dimension between 1 nm and 100 nm that can modify their physiochemical properties compared to bulk material [4,5]. Agricultural nanotechnology has presented a wide range of possibilities to utilize nanoparticles to enrich crops and increase their productivity. It has the potential to revolutionize the agro sector and its allied fields through its varied applications [6,7]. There is a wide range of nanoparticles conferring plant growth-promoting abilities, providing diverse implications in the field of agriculture. In recent studies, FeO, ZnO, and Zn-Cu-Fe-oxide nanoparticles have shown a positive growth response to the seedlings of mung (*Vigna radiata*) [8]. The use of TiO_2 nanoparticles has improved the metabolism of nitrogen as well as photosynthetic activity in spinach plants while enhancing drought tolerance capacity in wheat, along with the increase in the content of starch and gluten present in its seed [9–11]. Nanoparticles of silver, iron, and carbon were evaluated and demonstrated growth-promoting capabilities such as stress tolerance, increased yield, early germination, and increased flowering rate in plants [12,13].

Silver is a naturally occurring element. Silver in any form, to a certain optimum concentration, is non-toxic to the human immune system as well as the cardiovascular, nervous, and metabolic systems. It is a good dietary source and helps in crop fortification [14]. Silver deficiency is one of the major limiting factors responsible for the low productivity and yield of crops. Nanotechnology has enabled the intensified use of AgNPs in enhancing crop productivity. AgNPs can be synthesized utilizing the chemical reduction method and do not require an expensive servo control system [15–17]. The development of AgNPs have restored interest in their positive microbial effects since the widespread use of modern mycology studies. The interaction of AgNPs with fungal symbionts is called Nano-Embedded Fungus (NEF). NEF formulation has been a focal point of study in agro nanotechnology in recent times. Few nanoparticles other than AgNPs along with endophytic fungus have been used to improve agricultural productivity [18].

Piriformospora indica, lately also referred to as *Serendipita indica*, acts as a biofertilizer, phyto-remediator, regulator of metabolic activities, and biological herbicide, as well as a bio-pesticide that promotes growth in plants [19,20]. It is a facultative root endophytic fungus belonging to the Basidiomycota order *Sebacinales*, involved in association with a wide range of plant species [21]. It can enhance the growth and yield of both monocotyledons and dicotyledonous plants by colonizing their roots and enhancing plants' resistance to biotic and abiotic stresses [21]. Due to these positive effect on plants and their ability to grow in axenic culture, *Pir. indica* has become a unique model fungus to study the molecular and physiological basis of various symbiotic plant–microbe interactions. It also helps in plant fortification due to its diverse properties as a bio-fertilizer and bio-control agent.

This research aims to evolve a nanotechnology-augmented fungal endosymbiont in order to enhance the production of secondary metabolites in black rice (*Oryza sativa* L.) leaves. Black rice is the second most consumed grain and one of the main crops that provide food and energy for over half of the world's population [22]. Due to its effectiveness in maintaining good health, it is consumed as a functional food. It has a wide range of bioactive and nutritive elements including metabolites such as phenols, flavonoids, terpenes, steroids, alkaloids, and carotenoids, which act either as defensive agents or as plant growth regulators [23]. Black rice, which has been cultivated for more than 4000 years, first appeared in China and, at present, is consumed as functional food across the world, especially in Asian countries [24]. Black rice is a superfood of the 20th century, according to research. The fact that it extends life has earned it the nickname "long-life rice" because of its high nutritional value. In addition, it can potentially be utilized to make nutritious foods and drinks, such as gluten-free cereals, giving consumers additional health benefits [25]. *P. indica* enhances biomass production and is tolerant to numerous biotic and abiotic stresses. This research assesses the effect of AgNPs on black rice in the presence of endophytic fungi—*P. indica.* It seeks to develop a nanotechnology-embedded endosymbiont that enhances secondary metabolite production and can provide innovative solutions to the challenges that conventional farming practices cannot address.

2. Methods

2.1. Materials and Reagents

All the analytical reagents, such as silver nitrate, sodium borohydrate PVP Polyvinyl pyrrolidine, and nutrient agar, were purchased from SRL labs. Organic jaggery was obtained from Dhampur. Root endophyte of *Piriformospora indica* was discovered, screened, and identified as an endophyte from orchid plants of the Thar deserts in Rajasthan, India, by Prof. Ajit Varma and the research group (Verma et al. (1998) [26] and maintained at Amity Institute of Microbial Technology (AIMT) as accession number DSM 11827. The pure inoculum was obtained from the (AIMT) laboratory and maintained in 4% jaggery media following the methodology of). Attri et al. (2018) [27] Grains of the black rice variety Chakhao Poireiton were obtained from Manipur, Northeast India. Milli-Q water was used in all the Sconducted experiments. Before use in experiments, the glass wares were washed, autoclaved, and dried rinsed with Milli-Q water.

2.2. Synthesis and Characterization of AgNPs

The AgNPs were synthesized using the temperature-dependent chemical reduction method [28]. Briefly, 20 mL of 1.5 mM silver nitrate was taken in the burette at room temperature. Then, 300 mL of 2.0 mM sodium borohydrate was taken in the beaker and stirred in an ice bath. After the solution was completely cooled, silver nitrate was added to it drop by drop. Once, the color change was observed, 16 drops of 0.3% PVP were added under continuous stirring for 5 min. In this process of synthesis, the sharp SPR peak was recorded by the UV visible spectroscopy (UV-Vis) to confirm the formation of AgNPs. The reduction of Ag ions occurred when the silver nitrate solution was treated with sodium borohydrate. Characterization of chemically synthesized AgNPs was performed by X-ray Diffraction (XRD), UV visible (UV-Vis) absorption spectroscopy, Zeta potential, Scanning Electron Microscopy–Energy Dispersive X-ray analysis (SEM–EDX), and Fourier transform infrared resonance spectroscopy (FTIR). The mycelial morphology, cellular ultrastructure, and internal machinery were examined using SEM and confocal microscopy techniques.

X-ray Diffraction analysis was performed, and Rigaku MiniFlex600 X-ray Diffractometer Riga, Lativa was used with Cu Kα radiation at wavelength $\lambda = 1.54$ A° operating at 40 kV and 15 mA. The XRD pattern of the sample silver nanoparticles was obtained in the 2θ range of 10–90° at a scan rate of 4° per minute with a 0.02-degree step size. The UV visible absorption spectrum was obtained with a 1 cm path length quartz cell using a T90+ UV/VIS Spectrometer by PG Instruments Limited Leicestershire, United Kingdom. Perkin-Elmer Spectrum T machine Shelton, USA was used to carry out the FTIR spectrum. Zeta Potential Analyser ZEE COM Microtech Co., Limited, Takidai, Japan was used to study the zeta potential of silver nanoparticles. The scanning electron microscope (EVO-18, Zeiss, Jena, Germany), at the Amity Institute of Advanced Research and Studies-Materials and Devices (AIARS-M&D), Amity University, Noida, was used for Scanning Electron Microscopy. The Nikon A1 confocal microscope under 40× magnification at AIMT, Amity University, and Uttar Pradesh was used for the confocal study. The flavonoid, terpenoid (LCMSMS-6470 MODEL), chlorophyll, carotenoid (HPLC 1100 SERIES), and saffrole (GCMS-7890(GC), 5975(MS) MODEL) were analyzed.

2.3. Fungus Piriformospora indica Culture Conditions

Piriformospora indica was grown on 4% jaggery media (Dhampur jaggery) at pH 6.7 with 8–14 days of incubation at 28–30 °C. Similarly, the broth medium was prepared without adding the solidifying agent. Characterization conditions of P. indica spores by SEM-EDX and confocal microscopy SEM (EVO-18, ZEISS, Germany) for SEM analysis and (Oxford instruments, 51-ADD0048) using SMARTSEM software for EDX elemental analysis was carried out to validate the presence of silver in the treated sample.

A culture of *P. indica* spores (control and silver embedded *P. indica* at 300 ppm), grown on 4% jaggery medium plates, was incubated at 28–30 °C for 1 day. The fixation of the specimen was performed in 4% glutaraldehyde for 3 h to preserve its structure and treated for 1 h with 1 mM phosphate-buffered saline (PBS) (pH 7.4) and followed the process by centrifugation for further isolation. After washing with distilled water, the specimen was dehydrated with ascending gradient ethanol series from 20 to 100% EtOH for 10 min each and dried. The specimen was observed under SEM (EVO-18, ZEISS, Germany) for structural analysis, which was carried out at a 10 kV accelerating voltage.

The methodology of Hilbert et al. (2013) with some modifications was followed for confocal microscopy using the Nikon A1 [29]. The control, as well as the AgNPs-treated *P. indica* fungal spores, were gently scrapped from the agar plate using a spreader and cleaned with 0.002% (Tween 20) solution to remove all impurities for confocal microscopy. Then, the samples were collected in a 1 mL centrifuge tube followed by centrifugation at 5000 rpm for 10 min Finally, after washing the pellet with 1 mL autoclaved distilled water, it was observed under 40× magnification with a confocal microscope (Model: Nikon A1, Tokyo, Japan) at AIMT, Amity University, India.

2.4. Black Rice Growth Conditions

The NEF formed by *P. indica* treated with chemically synthesized AgNPs showed a significant increase in the spore size, spore count, germination percentage, and biomass in both broth as well as agar media when compared with control *P. indica*. In addition, black rice-treated AgNPs showed positive growth at the 80 ppm-optimized concentration (Figure 1). Further inoculations were performed with the AgNPs, *P. indica*, NEF (both), targeting secondary metabolites production in black rice leaves in pots.

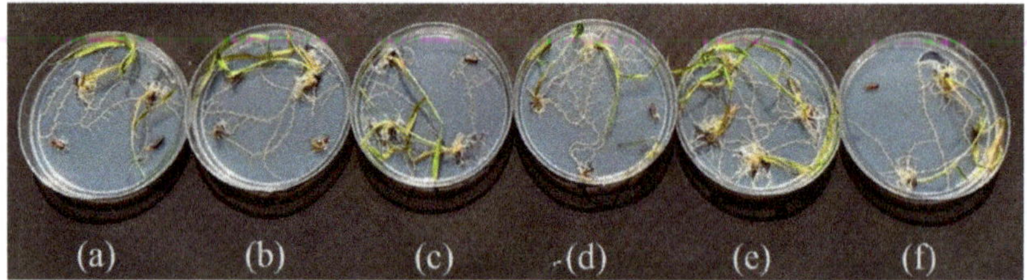

Figure 1. In vitro germination of AgNPs treated black rice seeds showing (**a**) control, (**b**) 20 ppm, (**c**) 40 ppm, (**d**) 60 ppm, (**e**) 80 ppm, and (**f**) 100 ppm.

A pot-based experiment was performed in the greenhouse of the Amity Institute of Microbial Technology. Modified Morishige and Skoog medium was used for in vitro black rice seed germination. After hardening for 15 days, seedlings were transferred to bigger earthen pots (25 cm diameter) containing sand, vermiculite, and sterile soil (1:1:1). The experiment consisted of four treatments, namely (a) untreated black rice (control), (b) only AgNPs-treated (80 ppm) black rice, (c) only *P. indica* (5×10^5 spores mL^{-1})-treated black rice, and (d) Nano-Embedded Fungus (5×10^5 spores mL^{-1})-treated black rice. Whereas the selection of the concentration of AgNPs was optimized for the experiment, the selection of concentration of spores of the *P. indica* was based on a previous study (Dabral et al., 2019). The efficiency of *P. indica* and AgNPs on the growth and secondary metabolite production in black rice was evaluated. Plants were grown for six months under controlled greenhouse conditions (RH 85%; temperature 28 °C, Figure 2).

Figure 2. Pot culture experiment of black rice (*Oryza sativa* L.) showing (**a**) control, (**b**) only AgNPs-treated, (**c**) only *P. indica*-treated, and (**d**) NEF (Nano-Embedded Fungus).

2.5. Secondary Metabolite Analysis

2.5.1. Carotenoids

Carotenoids were extracted using the methodology of Saini (2015) [30]. In total, 1 g of the study specimen was homogenized, of which 3 mL of acetone was taken. Then, 0.1% Butylated hydroxy toluene (BHT) solution in acetone was added to the study specimen as an antioxidant. Buchner's funnel of 5-micron porosity was used to filter the extract. Washing of the residue with acetone was performed twice until it turned colorless. The filtrate was mixed with 2 gm of anhydrous sodium sulfate. Filtration was performed to remove anhydrous sodium sulfate, and the extract volume was reduced by the Nitrogen evaporator. The extract was moved quantitatively to a 10 mL volumetric flask, and acetone with water was used to ensure the volume reached the mark so that the final extract contains 80% of acetone. The condition of the experiment included an immobile column of Agilent ZORBAX Eclipse Plus C18 with a dimension of 3.0X × 100 mm, Mobile Gradient of 1.8 mm, volume injection of 5 mL, a Detector A red-sensitive FLD, and an excitation of 430 nm/emission of 670 nm. The mobile phase preparation was performed in a mixture of Water–Methanol–Acetonitrile–Dichloromethane (10 + 20 + 70).

2.5.2. Chlorophyll

Chlorophyll concentrations were detected using the methodology of Hornero-Mendez (2005) [31]. In total, 300 mg of black rice leaves were dissolved in petroleum ether (1 mL) and then vortexed. The sample solution was then placed on the cartridge (Phenomenex, Torrance, CA, USA) to rinse it twice with the 1 mL petroleum ether.As the solvent drained to the top of the column packing, the nonpolar substances were quickly eluted with 5 mL petroleum ether/diethyl ether (90:10, v/v) twice and then discarded. After that, the chlorophyll was eluted with 5 mL of acetone then collected in a glass tube of 10 mL volume and covered with foil. The solvent was evaporated and dried on a rotary evaporator at 20 °C. Then dried residue was then dissolved in another 1 mL acetone for further HPLC analysis. The HPLC conditions were similar to the carotenoid analysis mentioned above.

2.5.3. Saffrole

The homogenized sample (2.0 gm) was taken in a 50 mL polypropylene tube, dissolved in 10 mL MilliQ water, and mixed well. Chloroform (10 mL) was added to the sample and vortexed for 1 min followed by shaking for 30 min on a mechanical shaker. Sodium chloride (3 gm) was added and mixed with vigorous shaking. The sample was further centrifuged for 15 min at 4000 rpm. After centrifugation, the lower layer of chloroform was taken and filtered with a 0.2 μm microfilter and injected into GC-MS [32]. The GC-MS was performed using Agilent GC-7890 with an auto sampler. The column used was HP-5MS UI 15 m, 0.25 μm, 25 mm ID capillary column at 350 °C, with a run time of 12 min. The Inlet program was in split less mode, with 2.7324 mL/min of constant flow with helium gas type. MS conditions included the Agilent 6470 Triple Quad mass spectrometer with 2 filaments at 300 °C temperature and positive-mode electron ionization.

2.5.4. Flavonoids and Terpenoids

Agilent 1290LCMSMS-6470 MODEL was used for the flavonoid and terpenoid analysis following the methodology of Giese (2015) [33]. In total, 50 mg of black rice leaves were extracted in 10 mL of N-Hexane for 24 h in the Soxhlet apparatus. The sample was air dried, extracted with methanol (10 mL), and allowed to dehydrate at 40 °C on a rotatory evaporator. The dehydrated sample was dissolved in methanol (1 mL) and injected into LC-MS/MS. The LC-MS was performed using ESI electron spray ionization mode positive and negative mode at 300 °C ion source temperature with auto sampler. The column ZORBAX RRHD SB C18, 2.1 × 150 mm was used at 30 °C with following conditions: nebulizer pressure 45 psi, drying gas flow rate 9 L/min, fragmentor voltage 250 V, and capillary voltage 3500 V.

2.6. Statistical Analysis

The Statistical Package for the Social Sciences Statistics software version 21.0 (SPSS Inc., IBM Corporation, Armonk, NY, USA was used to statistically analyze the data. A one-way analysis of variance (ANOVA) was performed to compare the differences between individual means using Tukey's honest significant difference (HSD) post-hoc test at $p \leq 0.05$. All the values are represented as mean of three biological replicates ± standard deviation (SD).

3. Results

3.1. Characterizations of Silver Nanoparticles

XRD analysis was performed to verify the crystallinity of AgNPs. The obtained XRD spectra showed the XRD peaks 2θ = 38°, 44.14°, 64.24°, 77.24°, 81.36°, 97.66°, respectively, with an intense peak at 2θ = 38°. Figure 3a shows the face-centered cubic (FCC) crystal reflection planes of the four faces of crystalline AgNPs [1,2,16]. XRD peaks confirmed the formation of FCC crystallographic planes which aligned with the standard silver peaks (Figure 3a). UV-Vis absorption spectrophotometry showed a peak around the 400 nm wavelength, characteristic of well-dispersed AgNPs (Figure 3b). The SPR peak of chemically synthesized AgNPs has been obtained at the 400 nm wavelength [1]. Figure 3c shows the Zeta potential of the silver nanoparticles, which was around −30 Mv. This aligns with the standardized Zeta potential readings of chemically synthesized silver nanoparticles.

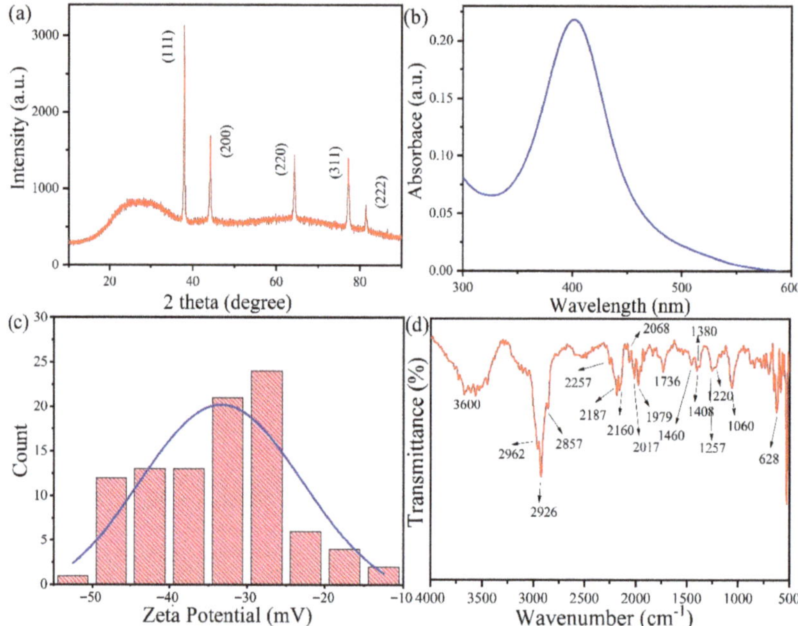

Figure 3. (**a**) X-ray Diffraction Spectroscopy of chemically synthesized silver nanoparticle, (**b**) UV-Vis's absorption spectrum of silver nanoparticles, (**c**) Zeta potential of chemically synthesized silver nanoparticle, and (**d**) FTIR spectrum scan of chemically synthesized silver nanoparticle.

The FTIR spectrum of silver nanoparticles is shown in Figure 3d. The vibration modes around 3600 cm^{-1} showed the -OH bond vibration. The peak at 2962 and 2926 cm^{-1} were assigned to the symmetric and asymmetric stretching vibrations of -CH bonds in PVP, respectively. The C-N stretch vibration was found at 2187 cm^{-1}, and the peaks at 2017, 2068, and 1979, 1739 cm^{-1} were assigned to the -C-C aliphatic stretch and -C-O stretch vibrations, respectively [34]. The symmetric deformation of -CH_3 in aliphatic compounds was shown by a peak at 1380 cm^{-1}. The peaks at 1260 and 1220 cm^{-1} indicate the C-O-C antisymmetric stretch in vinyl compounds. The peak at 1060 cm^{-1} was of the C-O stretch of CH_2-OH bonds. The peak at 628 cm^{-1} represents the pyridines. All these functional groups of PVP played a vital role in stabilizing the AgNPs [35].

The prepared AgNPs were analyzed using SEM-EDX. The average size of silver nanoparticles varied from 80 to 120 nm. The silver nanoparticles were analyzed using computer image analysis tools, which confirmed hexagonal-shaped silver nanoparticles, as shown in Figure 4. The EDX spectroscopy shows the presence of the Ag element in the sample.

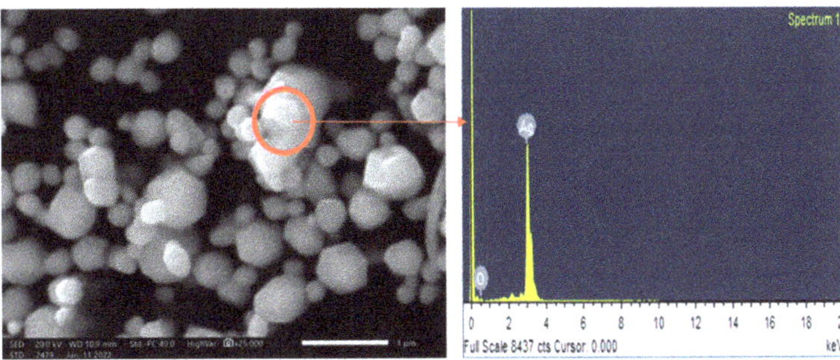

Figure 4. Scanning Electron Micrograph (SEM) of chemically synthesized silver nanoparticles showing hexagonal-shaped nanoparticles, and its EDX spectra showing the Ag element.

3.2. *Effects of Silver Nanoparticles on the Growth of P. indica*

In the broth medium, the growth of *P. indica* was visibly enhanced by the silver nanoparticles in a concentration-dependent manner (Figure 5). The size and biomass of the colonies increased with the increasing concentration of silver nanoparticles from 100 ppm to 300 ppm and then eventually retarded at 400 ppm as confirmed by fungal colony diameter in agar and dry cell weight in broth. Fungal colony diameters for *P. indica* grown in agar were taken at regular intervals of three days. On the 10th day, the mycelium colony diameter was reported to reach its maximum at 300 ppm AgNPs in comparison to the control (Figure 6).

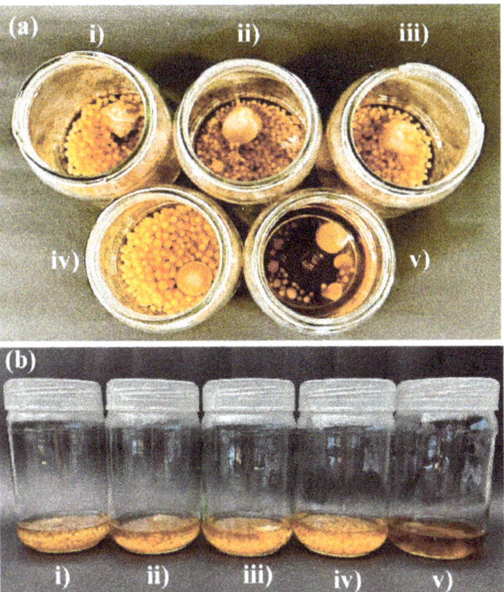

Figure 5. The growth enhancement of *P. indica* when treated with silver nanoparticles in 4% jaggery broth media prepared at pH 6.7: (**a**) Top view, (**b**) front view. Note: (i) Control, (ii) 100 ppm, (iii) 200 ppm, (iv) 300 ppm, and (v) 400 ppm. Growth retardation was observed in higher concentrations (400 ppm) of AgNPs.

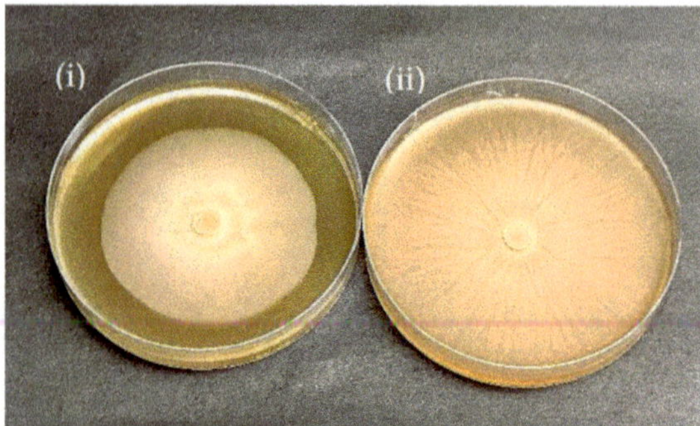

Figure 6. Growth enhancement of silver nanoparticle treated *P. indica* on agar media showing comparative growth enhancement with AgNPs untreated control (i) and treated *P. indica* (ii).

The increase in (a) fungal colony diameter with an increase in AgNPs concentration in agar and (b) dry cell weight with an increase in AgNPs concentration in broth is seen Figure 7. In comparison to the control (3.6 cm fungal colony diameter) Figure 6(i), maximum growth was observed at 300 ppm (8.4 cm treated colony diameter) Figure 6(ii) in agar after AgNPs treatment and inoculation (Figure 7a). Similarly, the dry cell weight of broth treated with AgNPs at 300 ppm was reported as 6.4 gm (Figure 7b), whereas in the control, it was just 3.8 g. The growth promotion of AgNPs-treated *P. indica* occurred in an increased concentration-dependent method where maximum growth was observed at 300 ppm concentration of AgNPs, as confirmed by the dry cell weight measured after 10 days.

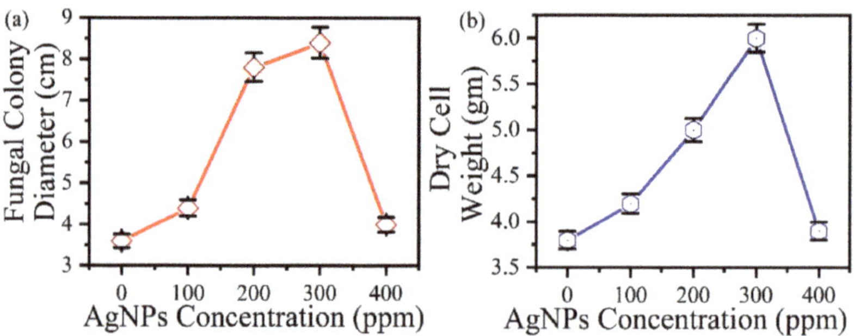

Figure 7. (**a**) The increase in fungal colony diameter with an increase in AgNPs concentration in agar media, (**b**) the increase in dry cell weight with an increase in AgNPs concentration in broth.

3.3. Microscopy Studies

The surface spore and hyphae morphology of *P. indica* using SEM-EDX and confocal microscopy was performed (Figure 8), which showed that in the control, *P. indica spores* were less in number, non-germinating, and small-sized (Figure 8a), whereas those treated with 300 ppm AgNPs were more in number, in the germinating phase with a peculiar germ tube, and larger in size (Figure 8b). The EDX result of the control *P. indica* showed the presence of Carbon (C) and Oxygen (O) (Figure 8c), while the EDX of the AgNPs-treated *P. indica* confirmed the presence of silver along with C and O (Figure 8d). Similar results can be seen in confocal microscopy, as shown in Figure 8e,f, where growth enhancement in *P. indica* spores can be seen. These results confirm the positive growth of *P. indica* spores caused due to the presence of Ag in the treated sample, thereby leading to the formation of NEF.

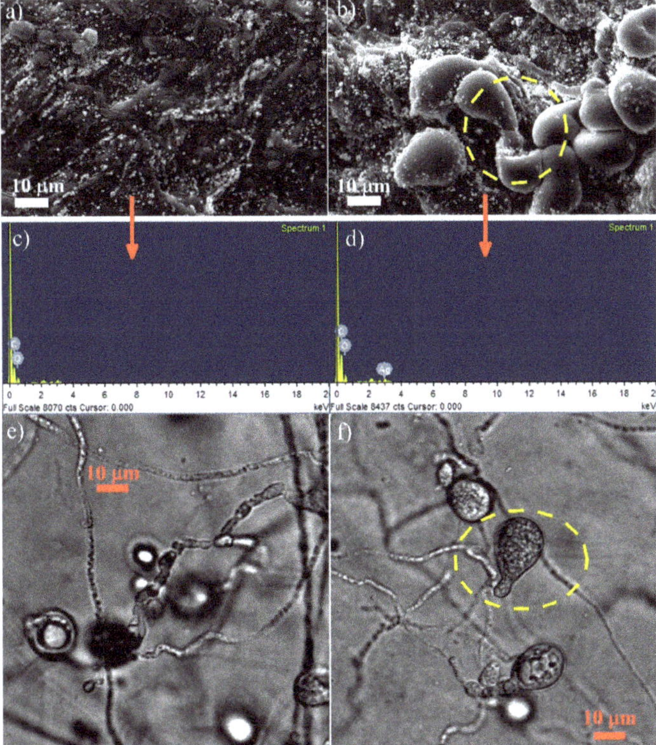

Figure 8. (**a**,**b**) The Scanning Electron Microscopy of the controlled and treated sample, respectively; (**c**,**d**) the EDX graphs of the controlled and treated sample, respectively; (**e**,**f**) the confocal microscopy image of the controlled and treated sample, respectively; evaluating the surface, spores, and hyphae morphology of *P. indica*.

3.4. Secondary Metabolite Analysis in Black Rice Leaves

The optimization of AgNPs in black rice in vitro seed germination was performed as shown in Figure 1, where six different concentrations of AgNPs were taken: control (untreated), 20 ppm, 40 ppm, 60 ppm, 80 ppm, and 100 ppm respectively. Maximum growth was observed at 80 ppm, as seen by the increased root and shoot length after 12 days at 28 °C.

The qualitative and quantitative analysis of secondary metabolites, such as chlorophyll (a and b), carotenoids, flavonoids, terpenoids, and safrole, was carried out in the black rice leaves. Four experimental sets of untreated, only AgNPs-treated, only *P. indica*-treated, and NEF (interaction of both)-optimized *P. indica* with AgNPs were studied in pot culture. The effect NEF on the secondary metabolite composition in black rice (*Oryza sativa L.*) was reported and statistically analyzed (Table 1). Concentrations of majority of secondary compounds were significantly ($p \leq 0.05$) increased in AgNPs-, *P. indica*-, and NEF-treated plants when compared with the leaves of control plants. Chlorophyll showed an increase of 24.22 %, 58.19 %, and 100% in AgNPs-, *P. indica*-, and NEF-treated plants, respectively. Among the flavonoids, while the concentration of quercetin, apigenin, myricetin, catechin, kaempferol, isorhamnetin, luteolin, and tricine were increased significantly ($p \leq 0.05$) in all the treatments when compared with the control, the effect of the treatments was non-significant on concentrations of luteolin-7-0-glucosides and isorhamnetin-3-0-glucosides. Terpenoids such as beta cymene, gamma terpinene, terpinene-4-ol, alpha elemene, linalool, caryophyllene, beta ocimene, trans linalool, and myrcene were significantly ($p \leq 0.05$) increased in AgNPs-, *P. indica-*, and NEF-treated plants. Interestingly, saffrole, a toxic benzodioxole element, was not detected in any of the treatments.

Table 1. Effect of silver nanoparticle and *Piriformospora indica* (*Serendipita indica*) and their co-application on secondary metabolite composition of black rice *Oryza sativa* L.

Compound	Control (µg/g)	AgNPs (µg/g)	*Piriformospora indica* (µg/g)	NEF (µg/g)
Chlorophyll (a + b)	2928.33 ± 33.93 d	3637.73 ± 42.15 c	4632.38 ± 53.68 b	5852.61 ± 67.82 a
Carotenoid				
Alpha Carotenoid	54.07 ± 0.62 d	71.25 ± 0.82 c	90.31 ± 1.04 b	118.35 ± 1.36 a
Beta Carotenoid	203.71 ± 2.36 d	267.29 ± 3.09 c	343.96 ± 3.98 b	455.69 ± 5.28 a
Safrole	nd	nd	Nd	nd
Flavonoids				
Quercetin	177.85 ± 2.05 d	234.77 ± 2.72 c	297 ± 3.44 b	421.75 ± 4.88 a
Apigenin	237.58 ± 2.75 d	313.60 ± 3.63 c	397.77 ± 4.61 b	564.84 ± 6.54 a
Myricetin	199.51 ± 2.31 d	262.35 ± 3.04 c	334.18 ± 3.87 b	474.55 ± 5.5 a
Catechin	208.53 ± 2.4 d	275.26 ± 3.16 c	352.92 ± 4.09 b	494.53 ± 5.69 a
Kaempferol	138.17 ± 1.58 d	187.34 ± 3.71 c	235.86 ± 2.73 b	339.46 ± 6.72 a
Isorhamnetin	183.61 ± 3.63 d	242.43 ± 4.8 c	302.54 ± 3.5 b	435.41 ± 8.62 a
Luteolin	259.99 ± 5.14 d	343.20 ± 6.79 c	427.40 ± 4.95 b	606.91 ± 7.03 a
Luteolin-7-0-glucosides	71.56 ± 1.42 a	83.52 ± 11.14 a	71.48 ± 7.55 a	76.48 ± 3.19 a
Apigenin-7-0-glucoside	61.66 ± 0.57 a	56.68 ± 0.60 b	56.34 ± 1.06 b	61.96 ± 2.70 a
Quercetin-3-0-rutinosides	51.83 ± 1.03 a	51.05 ± 0.94 a	45.78 ± 0.22 b	48.35 ± 0.75 b
Isorhamnetin-3-0-glucosides	67.28 ± 0.75 a	66.14 ± 1.08 a	63.54 ± 1.60 a	63.20 ± 2.46 a
Quercetin-3-0-glucosides	73.61 ± 1.44 a	63.49 ± 0.87 c	67.77 ± 1.19 b	66.72 ± 1.40 bc
Tricin	1330.87 ± 26.35 d	1757.78 ± 34.81 c	2222.56 ± 44.01 b	3156.05 ± 62.49 a
Tricin-4-0 erythro-B guaiacylglyceryl	59.27 ± 0.87 b	59.25 ± 1.11 b	60.52 ± 1.63 b	65.48 ± 1.42 a
Terpenoids				
Beta Cymene	370.52 ± 7.33 d	479.25 ± 1.03 c	610.13 ± 1.00 b	821.66 ± 16.27 a
Gamma-Terpinene	99.21 ± 2 d	130.99 ± 2.64 c	173.48 ± 3.43 b	229 ± 4.53 a
Terpinen-4-ol	135.69 ± 2.68 d	179.18 ± 3.55 c	227.97 ± 4.51 b	300.92 ± 5.96 a
Beta-Pinene	47.97 ± 0.95 c	64.25 ± 1.27 ab	80.6 ± 1.59 ab	139.68 ± 59.49 a
Alpha-Elemene	121.56 ± 2.46 d	160.39 ± 3.24 c	209.75 ± 2.40 b	281.63 ± 5.57 a
Linalool	53.06 ± 1.05 d	70.14 ± 1.39 c	89.15 ± 1.76 b	117.88 ± 2.33 a

Table 1. *Cont.*

Compound	Control (μg/g)	AgNPs (μg/g)	*Piriformospora indica* (μg/g)	NEF (μg/g)
10-acetylmethyl-3-carene	97.96 ± 1.94 [d]	129.41 ± 2.56 [c]	164.58 ± 3.26 [b]	217.27 ± 4.30 [a]
Carveol	13.74 ± 0.28 [c]	15.73 ± 1.21 [bc]	17.32 ± 1.14 [b]	31.73 ± 0.63 [a]
Limonene	308.74 ± 6.11 [d]	407.79 ± 8.07 [c]	518.8 ± 10.27 [b]	684.66 ± 13.55 [a]
Caryophyllene	70.41 ± 1.39 [d]	93.37 ± 1.85 [c]	118.3 ± 2.34 [b]	156.17 ± 3.09 [a]
Sabinene	70.23 ± 1.41 [b]	68.03 ± 1.78 [b]	70.68 ± 1.14 [b]	157.32 ± 1.55 [a]
Beta-Ocimene	202 ± 4 [d]	266.91 ± 5.28 [c]	339.46 ± 6.72 [b]	447.97 ± 8.87 [a]
Trans-Linalool	135.89 ± 2.69 [d]	179.40 ± 3.55 [c]	228.29 ± 4.52 [b]	302.62 ± 5.99 [a]
Cis linalool	59.81 ± 1.21 [b]	69.91 ± 3.40 [b]	66.24 ± 3.80 [b]	134.99 ± 3.92 [a]
Myrcene	198.95 ± 3.93 [d]	232.06 ± 4.59 [c]	334.24 ± 6.61 [b]	442.60 ± 8.76 [a]

Values represent means of three biological replicates ± SD. Different letters within each column represent significant difference among the treatments at $p \leq 0.05$, derived from Tukey's HSD. AgNPs—silver nanoparticles; NEF—Nano-Embedded Fungi; nd—not detected.

4. Discussion

Previous research on the bacterial and viral microbes as an effect of AgNPs on root endophytes is scarce. Here, we formulated nano-embedded *P. indica* that can be utilized to enhance the production of secondary metabolites in black rice (*Oryza sativa* L.) leaves. Morphological changes in the spore and hyphae of *P. indica* after AgNPs treatment were confirmed by SEM and confocal microscopy. The study clearly demonstrated the enhanced fungal growth in response to AgNPs treatment. The interaction of optimized concentrations of AgNPs with NEF showed the production of secondary metabolites in black rice. Similarly, a recent publication with Zn nanoparticles showed enhanced fungal growth as well as enhanced seed germination and root development [36]. Our microscopy analysis also showed that AgNPs enhance the spores as well as hyphae growth. The inoculation of this Nano-Embedded Fungus (NEF) into black rice demonstrated an increased production of secondary metabolites in black rice leaves. For the present research, AgNPs were synthesized, and the impact of fungal endosymbiont was studied by optimizing AgNPs concentration on the interaction with symbiotic mycorrhizal fungus, *P. indica*, which resulted in enhanced productivity of secondary metabolites. As we have observed, the secondary metabolites in black rice leaves showed enormous growth upon being treated with NEF, which shows that NEF has strong growth-promoting effects on the plant system.

Black rice is enriched in anthocyanin pigments, vitamins, proteins, phytochemicals, antioxidants, and secondary metabolites [37]. Some genera in the core microbiome of the black rice plant are well-established plant growth-promoting rhizobacteria PGPRs, which are well known to enhance synthesis of secondary metabolites in plants and improve the antioxidant activity [38,39]. The effective integration of black rice with NEF stimulates the production of secondary metabolites in its leaves. Microscopy analysis also showed that AgNPs enhance the spores as well as hyphae growth of *P. indica*. It was also found that chemically synthesized AgNPs enhanced the growth of fungal spores at 300 ppm concentration in both agar and broth. A positive effect of ZnO nanoparticles on the growth of *Brassica oleracea* var. *botrytis* Broccoli with *P. indica* has also been reported by Singhal et. al (2017) [34]. In addition, ZnO nano materials showed positive growth in medicinal black rice, which may be due to the symplast and apoplast transport channel [18]. The inoculation of the Nano-Embedded Fungus (NEF) in black rice demonstrated positive growth patterns (Figure 2). According to Ferrer et al. (2008), there are about 10,000 flavonoids which have been identified in plants, and their synthesis appears to be pervasive to date. A steady increase of secondary metabolites, namely chlorophyll, carotenoids, flavonoids, and terpenoids in our findings, implies the compositional and essential functional role of endophytic fungal community and its positive correlation or association with AgNPs in black rice, which further contributes to its agricultural implication. The synthesis of active secondary metabolites may be induced due to the endophytic interactions and their hosts [40].

Secondary metabolites are compounds formed by plants through metabolic biosynthetic pathways that make them competitive in their own environment. They induce flowering and fruiting, as well as maintain growth, regulation and signaling in plants. Modern agriculture, as well as the pharmaceutical and nutraceutical sector, depend on plant secondary metabolites for their functioning. According to Vogt et al. (2010), the phenylpropanoid pathway is the pivotal point that initiates various secondary compounds synthesis in the plants [41]. The first enzyme in the phenylpropanoid pathway, phenylalanine ammonium lyase, which turns phenylalanine into p-coumaroyl CoA, which is a key branch point causes the synthesis of plant's secondary metabolites such as flavonoids, terpenoids, cyanins etc. It can be assumed that *P. indica* along with AgNPs uses a common phenylpropanoid pathway for secondary metabolite synthesis. Khare et al. (2018) reported that an array of common secondary metabolites could be produced from similar precursors in both plants and their endophytes [42]. Our data suggest that the increase of these metabolites in black rice colonized with NEF show strong enhancing and positive effects.

Flavonoids are pivotal drivers of pharmacological action commonly known for their antioxidant activity. Moreover, flavonoids are abundant in plant kingdoms. Quercetin, an important dietary flavonoid, is a potent antioxidant with anti-allergic and anti-inflammatory properties [43]. In the current study, an increase was observed in quercetin levels in plants treated with NEF. Tricin, a potent flavonoid compound from Njavara rice, induces cytotoxic activity and apoptosis against cancer cells. [44] It was also significantly ($p \leq 0.05$) increased in our study along with apigenin, which possesses insect-resistant properties and can be used as a rice antifeedant. Terpenoids possess antiviral activities and at least 22 terpenoids have been shown to inhibit coronavirus-created havoc by [45]. Linalool, a major monoterpenoid component, is an important scent produced by orchids and was found to be increased by [46]. Therefore, NEF helps in the augmentation of secondary metabolites in black rice. The plant–microbe interaction may be modified to enhance the phytochemical and secondary metabolites production in black rice, but this remains to be explored. More extensive OMICS studies, including genomics, transcriptomics, proteomics, etc., are further needed to study the interactions and functions that occur in black rice. *P. indica* can be grown axenically; therefore, it can be used as a bio-control agent in the agricultural field to overcome the use of chemical fertilizers.

5. Conclusions

In the present research work, chemically synthesized AgNPs and their effect on fungal symbiont and black rice was studied by optimizing AgNPs concentration on the interaction with *P. indica*, both the broth as well as the agar, which resulted in an enhanced biomass and increased fungal colony diameter. The optimized AgNPs concentration (300 ppm) interacted with a fungal symbiont called a "Nano-Embedded Fungus". The inoculation of *P. indica* along with AgNPs treatment in black rice improved growth and productivity. The association of fungal endophyte and AgNPs in correlation with enhancement of secondary metabolites in black rice leaves was described. Our research aims to form a nanotechnology-assisted fungal endosymbiont that could be used as a "bio-formulation" in crop fortification for sustainable agriculture. Hence, it examined the physiological effect of AgNPs on *P. indica* spores and its implication in black rice. Confirmation of its enhanced fungal effect could be useful by integrating it into the agricultural sector.

Author Contributions: S.S. and A.V. perceived and designed the research experiments; S.S. conducted all the experimental work alongwith manuscript writing and data analysis; S.S., T.D. and G.B.V.S.L. helped in synthesis and characterization of AgNPs; S.G. performed statistical analysis P.R.S., R.K. and S.G. proofread the final draft. All authors have read and agreed to the published version of the manuscript.

Funding: This research received no external funding.

Institutional Review Board Statement: Not applicable.

Informed Consent Statement: Not applicable.

Data Availability Statement: Not applicable.

Acknowledgments: The authors are thankful to DST-FIST for providing a confocal microscope, AIMT, AUUP, and Noida for microscopic and Special Centre for Nanosciences (SCNS), JNU, New Delhi for nano studies.

Conflicts of Interest: The authors declare no conflict of interest.

References

1. Dhiman, T.K.; Lakshmi, G.; Dave, K.; Roychoudhury, A.; Dalal, N.; Jha, S.K.; Kumar, A.; Han, K.-H.; Solanki, P.R. Rapid and label-free electrochemical detection of fumonisin-b1 using microfluidic biosensing platform based on Ag-CeO_2 nanocomposite. *J. Electrochem. Soc.* **2021**, *168*, 077510. [CrossRef]
2. Dhiman, T.K.; Lakshmi, G.; Roychoudhury, A.; Jha, S.K.; Solanki, P.R. Ceria-nanoparticles-based microfluidic nanobiochip electrochemical sensor for the detection of ochratoxin-A. *ChemistrySelect* **2019**, *4*, 4867–4873. [CrossRef]
3. Garimella, L.B.; Dhiman, T.K.; Kumar, R.; Singh, A.K.; Solanki, P.R. One-step synthesized ZnO np-based optical sensors for detection of aldicarb via a photoinduced electron transfer route. *ACS Omega* **2020**, *5*, 2552–2560. [CrossRef] [PubMed]
4. Salama, H.M. Effects of silver nanoparticles in some crop plants, common bean (*Phaseolus vulgaris* L.) and corn (*Zea mays* L.). *Int. Res. J. Biotechnol.* **2012**, *3*, 190–197.
5. Dalal, N. Microfluidic based electrochemical detection of fumonisin-b1 using silver-ceria nanocomposite. *SPAST Abstr.* **2021**, *1*. Available online: https://spast.org/techrep/article/view/1387 (accessed on 30 November 2022).
6. Lakshmi, G.; Poddar, M.; Dhiman, T.K.; Singh, A.K.; Solanki, P.R. Gold-ceria nanocomposite based highly sensitive and selective aptasensing platform for the detection of the chlorpyrifos in *Solanum tuberosum*. *Colloids Surf. A Physicochem. Eng. Asp.* **2022**, *653*, 129819. [CrossRef]
7. Singh, A.K.; Ahlawat, A.; Dhiman, T.K.; Lakshmi, G.; Solanki, P.R. Degradation of methyl parathion using manganese oxide (MnO_2) nanoparticles through photocatalysis. *SPAST Abstr.* **2021**, *1*. Available online: https://spast.org/techrep/article/view/1027 (accessed on 30 November 2022).
8. Dhoke, S.K.; Mahajan, P.; Kamble, R.; Khanna, A. Effect of nanoparticles suspension on the growth of mung (*Vigna radiata*) seedlings by foliar spray method. *Nanotechnol. Dev.* **2013**, *3*, e1. [CrossRef]
9. Rane, M.; Bawskar, M.; Rathod, D.; Nagaonkar, D.; Rai, M. Influence of calcium phosphate nanoparticles, *Piriformospora indica* and *Glomus mosseae* on growth of *Zea mays*. *Adv. Nat. Sci. Nanosci. Nanotechnol.* **2015**, *6*, 045014. [CrossRef]
10. Yang, R.Y.; Chang, L.C.; Hsu, J.C.; Weng, B.B.; Palada, M.C.; Chadha, M.L.; Levasseur, V. Nutritional and functional properties of Moringa leaves—From germplasm, to plant, to food, to health. In Proceedings of the Moringa and Other Highly Nutritious Plant Resources: Strategies, Standards and Markets for a Better Impact on Nutrition in Africa, Accra, Ghana, 16–18 November 2006; pp. 1–9.
11. Zheng, L.; Hong, F.; Lu, S.; Liu, C. Effect of nano-TiO_2 on strength of naturally aged seeds and growth of spinach. *Biol. Trace Elem. Res.* **2005**, *104*, 83–91. [CrossRef]
12. Sharma, R.; Patel, S.; Pargaien, K.C. Synthesis, characterization and properties of Mn-doped ZnO nanocrystals. *Adv. Nat. Sci. Nanosci. Nanotechnol.* **2012**, *3*, 035005. [CrossRef]
13. Sheykhbaglou, R.; Sedghi, M.; Shishevan, M.T.; Sharifi, R.S. Effects of nano-iron oxide particles on agronomic traits of soybean. *Not. Sci. Biol.* **2010**, *2*, 112–113. [CrossRef]
14. Dawadi, S.; Katuwal, S.; Gupta, A.; Lamichhane, U.; Thapa, R.; Jaisi, S.; Lamichhane, G.; Bhattarai, D.P.; Parajuli, N. Current research on silver nanoparticles: Synthesis, characterization, and applications. *J. Nanomater.* **2021**, *2021*, 6687290. [CrossRef]
15. Verma, A.K.; Mishra, A.; Dhiman, T.K.; Sardar, M.; Solanki, P. Experimental and in silico interaction studies of alpha amylase-silver nanoparticle: A nano-bio-conjugate. *bioRxiv* **2022**. [CrossRef]
16. Verma, D.; Dhiman, T.K.; Mukherjee, M.D.; Solanki, P.R. Electrophoretically deposited green synthesized silver nanoparticles anchored in reduced graphene oxide composite based electrochemical sensor for detection of bisphenol A. *J. Electrochem. Soc.* **2021**, *168*, 097504. [CrossRef]
17. Verma, D.; Dhiman, T.K.; Mukherjee, M.D.; Solanki, P.R. The development of green synthesized AgNps anchored RGO nanocomposite based amperometric sensor for BPA detection. *SPAST Abstr.* **2021**, *1*. Available online: https://spast.org/techrep/article/view/180 (accessed on 30 November 2022).

18. Mahajan, S.; Barthwal, S.; Attri, M.K.; Bajpai, S.; Dabral, S.; Khanuja, M.; Varma, A. Impact of ZnO nano materials on medicinal black rice seed germination. *J. Miner. Mater. Charact. Eng.* **2019**, *7*, 180. [CrossRef]
19. Gill, S.S.; Gill, R.; Trivedi, D.K.; Anjum, N.A.; Sharma, K.K.; Ansari, M.W.; Ansari, A.A.; Johri, A.K.; Prasad, R.; Pereira, E.; et al. *Piriformospora indica*: Potential and significance in plant stress tolerance. *Front. Microbiol.* **2016**, *7*, 332. [CrossRef]
20. Prasad, R.; Swamy, V.S. Antibacterial activity of silver nanoparticles synthesized by bark extract of *Syzygium cumini*. *J. Nanopart.* **2013**, *2013*, 431218. [CrossRef]
21. Dabral, S.; Yashaswee; Varma, A.; Choudhary, D.K.; Bahuguna, R.N.; Nath, M. Biopriming with *Piriformospora indica* ameliorates cadmium stress in rice by lowering oxidative stress and cell death in root cells. *Ecotoxicol. Environ. Saf.* **2019**, *186*, 109741. [CrossRef]
22. Rowndel, K.; Chongtham, S.; Bhupenchandra, I.; Singh, T.B.; Choudhary, S. Nutritional advantage of black rice—A heirloom rice. *Indian Farming* **2020**, *69*. Available online: https://epubs.icar.org.in/index.php/IndFarm/article/view/96997 (accessed on 30 November 2022).
23. Wang, W.; Li, Y.; Dang, P.; Zhao, S.; Lai, D.; Zhou, L. Rice secondary metabolites: Structures, roles, biosynthesis, and metabolic regulation. *Molecules* **2018**, *23*, 3098. [CrossRef] [PubMed]
24. Pereira-Caro, G.; Cros, G.; Yokota, T.; Crozier, A. Phytochemical profiles of black, red, brown, and white rice from the camargue region of france. *J. Agric. Food Chem.* **2013**, *61*, 7976–7986. [CrossRef] [PubMed]
25. Pratiwi, R.; Purwestri, Y.A. Black rice as a functional food in Indonesia. *Funct. Foods Health Dis.* **2017**, *7*, 182–194. [CrossRef]
26. Verma, S.; Varma, A.; Rexer, K.H.; Hassel, A.; Kost, G.; Sarbhoy, A.; Bisen, P.; Bütehorn, B.; Franken, P. *Piriformospora indica*, gen. et sp. nov., a new root-colonizing fungus. *Mycologia* **1998**, *90*, 896–903. [CrossRef]
27. Attri, M.K.; Varma, A. Comparative study of growth of Piriformospora indica by using different sources of jaggery. *J. Pure Appl. Microbiol.* **2018**, *12*, 933–942. [CrossRef]
28. Mirzaei, H.; Darroudi, M. Zinc oxide nanoparticles: Biological synthesis and biomedical applications. *Ceram. Int.* **2017**, *43*, 907–914. [CrossRef]
29. Hilbert, M.; Novero, M.; Rovenich, H.; Mari, S.; Grimm, C.; Bonfante, P.; Zuccaro, A. Mlo differentially regulates barley root colonization by beneficial endophytic and mycorrhizal fungi. *Front. Plant Sci.* **2020**, *10*, 1678. [CrossRef]
30. Saini, R.K.; Nile, S.H.; Park, S.W. Carotenoids from fruits and vegetables: Chemistry, analysis, occurrence, bioavailability and biological activities. *Food Res. Int.* **2015**, *76*, 735–750. [CrossRef]
31. Hornero-Méndez, D.; Gandul-Rojas, B.; Mínguez-Mosquera, M.I. Routine and sensitive SPE-HPLC method for quantitative determination of pheophytin A and pyropheophytin A in olive oils. *Food Res. Int.* **2005**, *38*, 1067–1072. [CrossRef]
32. Jiménez-Durán, A.; Barrera-Cortés, J.; Lina-García, L.P.; Santillan, R.; Soto-Hernández, R.M.; Ramos-Valdivia, A.C.; Ponce-Noyola, T.; Ríos-Leal, E. Biological activity of phytochemicals from agricultural wastes and weeds on *Spodoptera frugiperda* (je Smith) (Lepidoptera: Noctuidae). *Sustainability* **2021**, *13*, 13896. [CrossRef]
33. Giese, M.W.; Lewis, M.A.; Giese, L.; Smith, K.M. Method for the analysis of cannabinoids and terpenes in cannabis. *J. AOAC Int.* **2015**, *98*, 1503–1522. [CrossRef] [PubMed]
34. Kowalska, M.; Woźniak, M.; Kijek, M.; Mitrosz, P.; Szakiel, J.; Turek, P. Management of validation of HPLC method for determination of acetylsalicylic acid impurities in a new pharmaceutical product. *Sci. Rep.* **2022**, *12*, 1. [CrossRef] [PubMed]
35. Kumar, M.; Devi, P.; Kumar, A. Structural analysis of PVP capped silver nanoparticles synthesized at room temperature for optical, electrical and gas sensing properties. *J. Mater. Sci. Mater. Electron.* **2017**, *28*, 5014–5020. [CrossRef]
36. Singhal, U.; Khanuja, M.; Prasad, R.; Varma, A. Impact of synergistic association of ZnO-nanorods and symbiotic fungus *Piriformospora indica* DSM 11827 on *Brassica oleracea* var. Botrytis (broccoli). *Front. Microbiol.* **2017**, *8*, 1909. [CrossRef]
37. Ichikawa, H.; Ichiyanagi, T.; Xu, B.; Yoshii, Y.; Nakajima, M.; Konishi, T. Antioxidant activity of anthocyanin extract from purple black rice. *J. Med. Food* **2001**, *4*, 211–218. [CrossRef]
38. Mishra, P.K.; Mishra, H.; Ekielski, A.; Talegaonkar, S.; Vaidya, B. Zinc oxide nanoparticles: A promising nanomaterial for biomedical applications. *Drug Discov. Today* **2017**, *22*, 1825–1834. [CrossRef]
39. Taghinasab, M.; Jabaji, S. Cannabis microbiome and the role of endophytes in modulating the production of secondary metabolites: An overview. *Microorganisms* **2020**, *8*, 355. [CrossRef]
40. Wang, Y.; Dai, C.-C. Endophytes: A potential resource for biosynthesis, biotransformation, and biodegradation. *Ann. Microbiol.* **2011**, *61*, 207–215. [CrossRef]
41. Vogt, T. Phenylpropanoid biosynthesis. *Mol. Plant* **2010**, *3*, 2–20. [CrossRef]
42. Khare, E.; Mishra, J.; Arora, N.K. Multifaceted interactions between endophytes and plant: Developments and prospects. *Front. Microbiol.* **2018**, *9*, 2732. [CrossRef] [PubMed]
43. Li, X.; Liu, Y.; Yu, Y.; Chen, W.; Liu, Y.; Yu, H. Nanoformulations of quercetin and cellulose nanofibers as healthcare supplements with sustained antioxidant activity. *Carbohydr. Polym.* **2019**, *207*, 160–168. [CrossRef] [PubMed]
44. Mohanlal, S.; Parvathy, R.; Shalini, V.; Mohanan, R.; Helen, A.; Jayalekshmy, A. Chemical indices, antioxidant activity and anti-inflammatory effect of extracts of the medicinal rice "njavara" and staple varieties: A comparative study. *J. Food Biochem.* **2013**, *37*, 369–380. [CrossRef]

45. Khan, T.; Khan, M.A.; Karam, K.; Ullah, N.; Mashwani, Z.-u.-R.; Nadhman, A. Plant in vitro culture technologies; a promise into factories of secondary metabolites against COVID-19. *Front. Plant Sci.* **2021**, *12*, 610194. [CrossRef]
46. Mohd-Hairul, A.; Namasivayam, P.; Cheng Lian, G.E.; Abdullah, J.O. Terpenoid, benzenoid, and phenylpropanoid compounds in the floral scent of *Vanda* mimi palmer. *J. Plant Biol.* **2010**, *53*, 358–366. [CrossRef]

Journal of *Fungi*

MDPI

Article

Trichoderma-Mediated ZnO Nanoparticles and Their Antibiofilm and Antibacterial Activities

Balagangadharaswamy Shobha [1], Bagepalli Shivaram Ashwini [2], Mohammed Ghazwani [3], Umme Hani [3], Banan Atwah [4], Maryam S. Alhumaidi [5], Sumanth Basavaraju [1], Srinivas Chowdappa [1], Tekupalli Ravikiran [1], Shadma Wahab [6], Wasim Ahmad [7], Thimappa Ramachandrappa Lakshmeesha [1,*] and Mohammad Azam Ansari [8,*]

1 Department of Microbiology and Biotechnology, Bangalore University, Jnana Bharathi Campus, Bengaluru 560056, India
2 Department of Microbiology, Sri Siddhartha Medical College, Tumkur 572107, India
3 Department of Pharmaceutics, College of Pharmacy, King Khalid University, Abha 62529, Saudi Arabia
4 Laboratory Medicine Department, Faculty of Applied Medical Sciences, Umm Al-Qura University, Makkah 24382, Saudi Arabia
5 Department of Biology, College of Science, University of Hafr Al Batin, Hafr Al Batin 31991, Saudi Arabia
6 Department of Pharmacognosy, College of Pharmacy, King Khalid University, Abha 61421, Saudi Arabia
7 Department of Pharmacy, Mohammed Al-Mana College for Medical Sciences, Dammam 34222, Saudi Arabia
8 Department of Epidemic Disease Research, Institute for Research and Medical Consultations (IRMC), Imam Abdulrahman Bin Faisal University, Dammam 31441, Saudi Arabia
* Correspondence: lakshmeesha@bub.ernet.in (T.R.L.); azammicro@gmail.com (M.A.A.)

Abstract: Antimicrobial resistance is a major global health concern and one of the gravest challenges to humanity today. Antibiotic resistance has been acquired by certain bacterial strains. As a result, new antibacterial drugs are urgently required to combat resistant microorganisms. Species of *Trichoderma* are known to produce a wide range of enzymes and secondary metabolites that can be exploited for the synthesis of nanoparticles. In the present study, *Trichoderma asperellum* was isolated from rhizosphere soil and used for the biosynthesis of ZnO NPs. To examine the antibacterial activity of ZnO NPs against human pathogens, *Escherichia coli* and *Staphylococcus aureus* were used. The obtained antibacterial results show that the biosynthesized ZnO NPs were efficient antibacterial agents against the pathogens *E. coli* and *S. aureus*, with an inhibition zone of 3–9 mm. The ZnO NPs were also effective in the prevention of *S. aureus* biofilm formation and adherence. The current work shows that the MIC dosages of ZnO NPs (25, 50, and 75 μg/mL) have effective antibacterial activity and antibiofilm action against *S. aureus*. As a result, ZnO NPs can be used as a part of combination therapy for drug-resistant *S. aureus* infections, where biofilm development is critical for disease progression.

Keywords: myconanotechnology; ZnO nanoparticles; nanofabrication; antimicrobial resistance; biofilm; Trichoderma; green synthesis

Citation: Shobha, B.; Ashwini, B.S.; Ghazwani, M.; Hani, U.; Atwah, B.; Alhumaidi, M.S.; Basavaraju, S.; Chowdappa, S.; Ravikiran, T.; Wahab, S.; et al. Trichoderma-Mediated ZnO Nanoparticles and Their Antibiofilm and Antibacterial Activities. *J. Fungi* **2023**, *9*, 133. https://doi.org/10.3390/jof9020133

Academic Editor: Kamel A. Abd-Elsalam

Received: 19 December 2022
Revised: 15 January 2023
Accepted: 16 January 2023
Published: 18 January 2023

1. Introduction

Antimicrobial resistance is a major worldwide health concern and one of humanity's most severe threats today [1]. Antibiotic resistance has been acquired by certain bacterial strains. As a result, new antibacterial drugs are urgently required to combat resistant microorganisms [2]. The human body acts as a host to a wide range of microbial communities [3], and the complex interactions that exist between the host and its microbial community play a crucial role in human health and diseases [4].

The facultative non-pathogenic human microflora is dominated by the Gram-negative *E. coli* bacterium. However, some *E. coli* strains have acquired the property to cause urinary tract infections, gastrointestinal illnesses, and central nervous system illnesses in the most resistant human hosts [5]. The bacterium *S. aureus* is a Gram-positive bacterium, and it can grow in both aerobic and anaerobic environments as a facultative bacterium [6]. *S. aureus* is the most common bacterial pathogen and continues to endanger public health [7,8], causing a variety of symptoms in infected people and being found in normal human flora and the environment. *S. aureus* can be isolated from the skin and mucous membranes (most often the nasal parts) of healthy individuals.

Myconanotechnology is a term that combines mycology and nanotechnology [9], and it offers a lot of potential in nanoparticle synthesis [10]. As it is simple to grow fungi in bulk and the extracellular release of enzymes provides an advantage in downstream processing, the production of nanoparticles employing fungi has drawn a lot of attention. Fungi synthesize more proteins than bacteria, leading to increased nanoparticle production. Fungi have been extensively explored due to their characteristics and their rapid and eco-friendly production of metal nanoparticles [11]. It has been observed that the nanoparticles produced by fungi have homogeneous diameters and monodispersity. Nanoparticles are synthesized via a biological approach using intracellular and extracellular mechanisms [12]. Because the intracellular process requires an additional step to obtain pure nanoparticles, the extracellular method is preferable over the intracellular approach [13]. Microorganisms act as reducing and capping agents [14].

Nanotechnology is the study and application of particles with a size range of 1 to 1000 nm [15]. Nanoparticles have unique features due to their small size compared to their bulk counterpart, making them excellent for applications in domains such as electronics, energy, the environment, and health [10]. Physical, chemical, and biological processes can be used to synthesize nanoparticles with desired characteristics, such as size and form [16]. However, because of the high cost and toxicity of the chemicals employed in synthesis, physical and chemical procedures are rarely used. As a result, several studies on the biological or green synthesis of metallic nanoparticles, such as silver, gold, titanium dioxide, iron oxide, magnesium oxide, zinc oxide, copper, and aluminum oxide nanoparticles, have been conducted [17]. Scientists all around the globe have been attracted to zinc oxide nanoparticles (ZnO NPs) because of their therapeutic properties. Pathogenic bacteria can be killed with the use of zinc oxide nanoparticles as an antimicrobial treatment. Several plants and microbes have been reported to synthesize ZnO NPs through the biosynthesis process [18,19]. The antibacterial activity of ZnO NPs could be increased via doping with ions [20]. As the nanoparticles fill the gaps between larger particles and atomic or molecular structures, they are of tremendous scientific interest [21]. Controlling the size and form of nanoparticles to adjust their optical, electronic, and electrical properties is a challenge in nanotechnology. The ideal metal nanoparticles are perfect monodispersed metal nanoparticles [22]. Green synthesis, also known as biosynthesis, is a biological technique of synthesizing ZnO NPs that involves the employment of microorganisms, such as algae, fungus, yeast, bacteria, and plant extracts, as the reducing agents [23]. Despite the benefits of using microbes as reducing and stabilizing agents during the biosynthesis of ZnO NPs, extra caution is necessary due to the toxicity of certain bacteria, as well as incubation concerns.

Plant-growth-promoting microbes (PGPMs) are rhizosphere microorganisms that can colonize the root environment. Bacteria and fungi that can colonize the roots and rhizosphere soil are among the microorganisms that live in this zone. The group of plant-growth-promoting fungi (PGPFs) includes certain *Trichoderma* species that have been identified with plant roots, where they create a symbiotic association or act as endophytes [24]. Plants' root-driven beneficial activities are mostly determined by their interactions with a wide range of microbial populations in the environment. Weindling was the first to describe the antibiotic synthesis of *Trichoderma* spp. [25]. There are several reports available regarding the *Trichoderma* species compounds proved to have antibacterial properties, volatile compounds (e.g., hydrogen cyanide, ethylene, monoterpenes, and alcohol), and non-volatile compounds (e.g., diketopiperazine, such as gliotoxin and gliovirin, and peptaibols) [26]. *Trichoderma* species are well-known for their capability to secrete a large number of secondary metabolites, such as plant growth promoters and a variety of enzymes [27]. The fungus genus *Trichoderma* is one of the most investigated groups of fungi utilized as biological control agents [28]. *Trichoderma* species are known to produce antibiotics or low-molecular-weight compounds [29]. Fungi, particularly *Trichoderma* species, are known to produce metabolites with antibacterial, anticancer, antioxidant, and antifungal properties among microbes. *Trichoderma* has been studied extensively as a biocontrol agent, biofungicide, biofertilizer, and plant growth enhancer. However, research into the medicinal potential of *Trichoderma* metabolites has received limited attention [30]. Therefore, in this study, ZnO NPs were synthesized from *Trichoderma* spp.

Traditional antimicrobial or antibiotic therapies used for bacterial diseases rely on the use of antimicrobial compounds or antibiotics that can inhibit or destroy microbial cell development. Pathogenic microbes, however, can build biofilms to defend themselves against inhibitory chemicals [31]. A colony of bacteria living in a self-produced matrix of biopolymers adhered to surfaces is referred to as a "biofilm". Microbes prefer to form biofilms on surfaces, avoiding the detrimental effects of antibiotics and detergents, and they persist in hospitals, generating a high number of hospital-acquired diseases [32]. Given that *S. aureus* is found in the skin's natural flora, it is likely that it is one of the most prevalent causal agents in hospital-acquired infections involving medical implants [33]. Furthermore, *S. aureus* has been shown to be resistant to larger dosages of antibiotics, perhaps contributing to the development of antibiotic-resistant insusceptible strains [34]. Based on our previous study [35], *Trichoderma* spp. isolated from rhizosphere soil was chosen for the biosynthesis of ZnO NPs and to determine their antibacterial activity against human pathogens. The present study is focused on the biosynthesis of ZnO NPs from *Trichoderma* spp. and their antibacterial action against human pathogens *S. aureus* and *E. coli*, as well as the inhibition of biofilm formation (*S. aureus*) by using different dosages of ZnO NPs.

2. Materials and Methods

2.1. Collection of Rhizosphere Soil

For this study, 10 g of rhizosphere soil was collected by uprooting a plant. The soil samples were collected and stored in polythene bags at 4 °C until further use [36].

2.2. Isolation and Identification of Fungi from Rhizosphere Soil

The collected rhizosphere soil sample was serially diluted into different concentrations and vortexed well. The supernatants were then transferred to sterilized potato dextrose agar (PDA) medium obtained from Himedia, India, and they were incubated for 7 days at 26 ± 2 °C on plates with the standard antibiotic chloramphenicol. Following incubation, the fungal colonies obtained on the PDA plates were isolated and re-inoculated into freshly prepared sterile plates. The obtained single-spore colonies were subjected to morphological and molecular characterization procedures after a 7-day incubation period [37]. For the isolation of genomic DNA from the fungi, a DNA isolation kit was procured from Chromous Biotech, India, and it was used by following the method described in the manufacturer's manual. 18S rRNA gene amplification was carried out using PCR, and the obtained gene sequence of the test strain was compared against the nucleotide collection (nr/nt) database using the BLAST program [38,39]. The gene sequences were deposited in GenBank, and the following accession number was obtained: OL826855. A similar study was carried out, where medicinal plants were used against *Streptococcus pyogenes* [40].

2.3. Green Synthesis of Zinc Oxide Nanoparticles (ZnO NPs)

In the present study, a green synthesis protocol was used for the synthesis of ZnO NPs from the rhizosphere soil fungus *T. asperellum*. The substrate zinc nitrate hexahydrate ($Zn(NO_3)2.6H_2O$) was purchased from Sigma-Aldrich (analytical grade), and it was used for the synthesis of ZnO NPs without any further purification. The substrate solutions were prepared by dissolving 1 g of $Zn(NO_3)2.6H_2O$ in 10 mL of double-distilled water, then adding 2 mL of a fungal extract of *T. asperellum*, and stirring the reaction mixture for ~5–10 min using a magnetic stirrer. A preheated muffle furnace, maintained at 400 ± 10 °C, was used where the obtained mixture was stored, and at that temperature, the reaction mixture boils, bubbles, and foam dehydrates in less than 3 min. The resulting product was calcinated for 2 h at 700 °C, and the final obtained product was used for further studies [41,42].

2.4. Characterization of Zinc Oxide Nanoparticles (ZnO NPs)

The formation of ZnO NPs was confirmed by recording UV–visible absorption spectra with a UV–visible spectrophotometer (SL 159 ELICO). The chemical composition and surface functional groups of the sample were evaluated by using Fourier transform infrared spectroscopy (FTIR) (Varian 3100). A powder X-ray diffractometer (PXRD) (Shimadzu) with Cu Kα (1.5418 Å) radiation and a nickel filter was used to examine the phase purity and crystalline nature of the ZnO NPs. The shape and surface morphology of the synthesized nanoparticles were analyzed by using scanning electron microscopy (SEM) (Hitachi Tabletop TM-3000) and an energy-dispersive analysis of X-rays (EDAX). To determine the size of the synthesized nanoparticles, high-resolution transmission electron microscopy (HRTEM) (JEOL JEM 2100), along with selected area electron diffraction (SAED), was used [43].

2.5. In Vitro Screening of ZnO NPs for Their Antibacterial Property against Human Pathogens

The pathogenic cultures of *S. aureus* (NCIM No. 2079) and *E. coli* (NCIM No. 2556) used in this study were National Collection of Industrial Microorganisms (NCIM) cultures. The antibacterial property of biogenic ZnO NPs against the selected pathogens was assessed by using the disc diffusion method. The standard 0.5 McFarland concentration of the selected bacterial culture was used to prepare a culture on Mueller–Hinton agar medium plates (procured from Himedia, India) using sterile swabs. Different concentrations of biogenic ZnO NPs (25, 50, and 75 μg/mL) were loaded onto the sterile discs. The test plate was placed with a positive control disc, a negative control disc, and biologically synthesized ZnO NPs. The standard antibiotic tetracycline (100 μg/mL) was used as a positive control, and sterile distilled water was used as a negative control before the plates were incubated at 28 ± 2 °C for 24 h. After incubation, the plates were examined for the inhibition zone, and the results were recorded in mm [44].

To determine the minimum inhibitory concentration (MIC) of the biosynthesized ZnO NPs against the selected pathogens, the broth microdilution protocol was followed with minor modifications. Different ZnO NP concentrations were prepared by diluting the stock solution of the ZnO NPs (1 mg/mL) with the dilution in the Mueller–Hinton Broth (MHB) medium and then by loading them onto sterile 96-well microtiter plates. To each well of the microtiter plate, 10 μL of bacterial suspension was added and incubated for 24 h at 28 °C. MHB was kept as a negative control, and the standard tetracycline at a concentration of 100 μg/mL was used as a positive control; all experimental tests were performed in triplicate. After 24 h of incubation, 20 μL of iodonitrotetrazolium chloride dye (INT) (0.5 μg/mL) was added to each well, followed by the incubation of the microtiter plates at 28 °C for 60 min. The MIC value indicates the sample concentration that prevents the color shift from colorless to red, where the colorless tetrazolium salt works as an electron acceptor and is reduced by active organisms to a red-colored formazan product [45].

2.6. Fluorescence Microscopy and Scanning Electron Microscopic Analysis

The biosynthesized ZnO NPs at a concentration of 75 μg/mL showed promising antibacterial activity. The dead and live cells of the treated *E. coli* and *S. aureus* were monitored by using fluorescent microscopy. The pathogenic bacteria *E. coli* and *S. aureus* were treated with the ZnO NPs (75 μg/mL) and incubated for 24 h at 37 °C. After incubation, the bacteria were stained with 1 μL of ethidium bromide and acridine orange, and then they were incubated in the dark for 10–15 min. The nuclei of the bacterial cells were stained green with acridine orange procured from Thermofisher, US, whereas the nuclei of the bacterial cells were stained orange with ethidium bromide procured from Thermofisher, US. Then, 10 μL of each of the test strains was put on a slide and viewed under a fluorescent microscope (Carl Zeiss, barrier filter O 515, excitation filter BP 490) at 40X magnification. SEM microscopy was used to investigate the morphological properties of the ZnO-NP-treated bacterial cells [43].

2.7. Antiadherence Assay

Inoculate 50 µL of the ZnO NPs at various concentrations (25, 50, and 75 µg/mL) into selected wells. From a 24 h bacterial culture, make a bacterial suspension (1×10^8 CFU/mL) in 15 mL. To obtain a 10^6 CFU/mL bacterial suspension, carry out a 1:100 dilution in a separate centrifuge tube. Using a suitable pipette, add 50 µL of the diluted bacterial concentration to the relevant wells. In this study, the organism was only permitted to grow without the ZnO NPs in the control set. To avoid the evaporation of the water from the test wells, fill the neighboring wells of the microplate with sterile distilled water. Water evaporation in the test wells might tamper with the results. Place the plate in an incubator at 37 °C for 18 to 24 h, covered with the lid. Remove the 96-well plate from the incubator and slowly decant or pipette the nutrient broth (NB). Allow the plate to air dry beneath the biosafety cabinet after rinsing it three times with sterile, double-distilled water. Turn the plates upside down to speed up the drying process. Before moving onto the next stage, make sure that they are completely dry. Fill the test wells with 100 mL of aqueous crystal violet (1 percent *w/v*), and let it stain the bacterial cell walls for 10 to 15 min. Into a sink, decant the crystal violet. Allow the test wells to dry beneath the biosafety cabinet after three rinses with sterile, double-distilled water. To solubilize the crystal violet, dissolve 30 % (*v/v*) glacial acid in water, and let it stand for 15 min. In each of the test wells, make sure that there is a clear blue/violet solution with no apparent residue. Examine the UV absorbance at 570 nm of each well. Using the formula below, calculate the antiadherence activity of the ZnO NPs and tetracycline [45,46].

$$\text{Antiadherence activity }\% = \frac{\text{Absorbance of control} - \text{Absorbance of ZnO NPs}}{\text{Absorbance of control}} \times 100$$

2.8. Antibiofilm Assay

To make a 10^6 CFU/mL bacterial suspension, follow procedures similar to those of the antiadherence assay. In a fresh 96-well microplate, inoculate 100 µL of the diluted bacterial suspension in NB into each well and incubate the plate for 24 h at 37 °C. Remove all of the NB broth from the microplate, and wash the wells three times with sterile phosphate-buffered saline (PBS) to avoid damaging the biomass growing on the bottom and the walls of the wells. In this study, we use 100 µL of sterile NB as a control, the ZnO NPs suspended in NB with the test concentrations of 25, 50, and 75 µg/mL, and NB containing a concentration of tetracycline of 100 µg/mL. Remove the 96-well plate from the incubator, and slowly decant or pipette the NB. Allow the plate to air dry beneath a biosafety cabinet after rinsing it three times with sterile, double-distilled water. Turn the plates upside down to speed up the drying process. Before moving on to the next stage, make sure that they are completely dry. Fill the test wells with 100 mL of aqueous crystal violet (1% *w/v*), let it stain the bacterial cell walls for 10 to 15 min, and then decant the crystal violet. Allow the test wells to dry beneath the biosafety cabinet after three rinses with sterile, double-distilled water. To solubilize the crystal violet, dissolve 30% (*v/v*) glacial acid in the water, and allow it to stand for 15 min. In each of the test wells, make sure that there is a clear blue/violet solution with no apparent residue. The UV-spec absorbance at 570 nm of each well was examined. Using the formula below, calculate the antibiofilm activity of the test material and tetracycline [47–50].

$$Antibiofilm\ \% = \frac{Absorbance\ of\ control - Absorbance\ of\ ZnONPs}{Absorbance\ of\ control} \times 100$$

2.9. Microscopic Studies

2.9.1. Bright-Field Microscopic Studies

After biofilm formation, the Gram-stained sample of S. aureus was examined under a bright-field microscope with a 100X objective in an oil immersion medium. Similarly, the samples with the ZnO NPs and tetracycline treatment were examined under a microscope [51].

2.9.2. Scanning Electron Microscopy (SEM)

The biofilm morphology was studied using a modified SEM approach. The test samples were fixed in 3% glutaraldehyde at 4–6 °C for 24 h. At each interval, the cells were washed three times with 0.1 M PBS for 10 min each time. The cultures were dehydrated for 10 min in the different gradient alcohol concentrations (50–100%). To keep the specimen from drying out, it was placed in 100% alcohol and attached to an aluminum stub with carbon tape before being sputter-coated with gold [52].

2.10. Statistical Analysis

All the obtained results of the antibacterial experiments were analyzed statistically using SPSS software (version 20.0) and Microsoft Excel.

3. Results and Discussions

3.1. Collection, Isolation, and Identification of Fungi from Rhizosphere Soil

The rhizosphere soil sample was collected from Eleusine coracana (Ragi) in Ramanagara district, southern Karnataka. Only *Trichoderma* spp. were isolated from the collected rhizosphere soil. The fungal colonies were grown on potato dextrose agar (PDA) medium plates supplemented with antibiotics, and the plates were incubated at 28 °C for 6–7 days. The isolated fungi were identified based on cultural and colony characteristics and a microscopic observation by using standard manuals Barnett and Hunter [53]; based on this, the fungi were identified as *Trichoderma* spp.

3.2. Molecular Characterization of Fungi Isolated from Rhizosphere Soil

The fungi isolated from Eleusine coracana (Ragi) were subjected to molecular characterization, which was validated by database searches using BLAST tools at the National Centre for Biotechnology Information (Bethesda, MD, USA); this revealed the fungi to be *Trichoderma asperellum* with a 99% similarity. Similar studies were conducted by Tomah et al. [29]; in this report, *Trichoderma* spp. were isolated from the soil sample and identified by using molecular methods.

3.3. Characterization of Biosynthesized Zinc Oxide Nanoparticles (ZnO NPs)

UV–Vis spectroscopy is a commonly used approach for the characterization and confirmation of synthesized ZnO NPs based on surface plasmon resonance (SPR) peaks. A characteristic absorption peak was observed at 373 nm by using UV–Vis spectroscopy for the synthesized ZnO nanoparticle suspension. The reported absorption peaks of the ZnO NPs are in good accordance with those in previous research, where absorption peaks have been found between 355 and 380 nm. Our results agree with those of Shaikhaldein et al. [54], who obtained an absorption peak at 380 nm. In a study conducted by Mahamuni et al. [55], the absorption peak recorded in each spectrum was found to be in the range of 360–380 nm, which is a characteristic feature of pure ZnO. The study conducted by Wang et al. [56] obtained a maximum peak at 330 nm, confirming that the synthesized nanoparticles in this study were ZnO NPs (Figure 1).

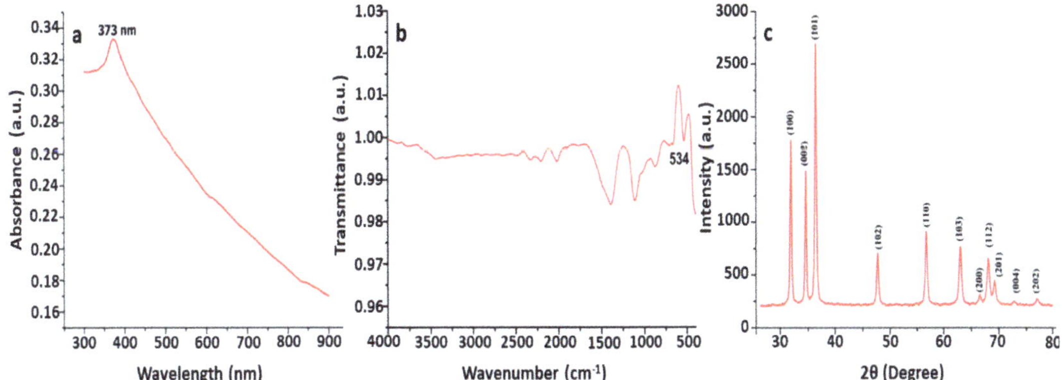

Figure 1. Characterization of biosynthesized ZnO NPs using *T. asperellum*: (**a**) UV—vis spectra; (**b**) FT-IR spectrogram; (**c**) PXRD patterns.

The application of the FTIR approach in the analyses of biosynthesized ZnO NPs has been useful in identifying the biomolecules involved in the formation of ZnO NPs [57]. The absorbance was recorded at 400 cm^{-1} to 600 cm^{-1} in the FTIR spectrum, confirming that ZnO nanoparticles were synthesized using *T. asperellum* (Figure 1). Our results agree with the results of Pillai et al. [58], who found a noticeable peak at 482 cm^{-1}, confirming the presence of ZnO NPs in this study. In a study conducted by Selim et al. [59], the FTIR spectra showed a peak at 442 cm^{-1} for biosynthesized ZnO NPs. Jayappa et al. [60] reported that absorption peaks found at 475 cm^{-1}, 486 cm^{-1}, and 473 cm^{-1} depict the inter-atomic vibrations that cause stretching vibrations in metallic ZnO (Figure 1).

ZnO NPs are subjected to intense rays from XRD machines during XRD examinations, and these rays pierce through the ZnO NPs and offer crucial details about their structure [55]. The PXRD graph's unique peaks confirmed and validated the biosynthesized ZnO nanostructure with the *Trichoderma* spp. fungus extract. The obtained PXRD patterns of the biosynthesized ZnO NPs revealed the presence of distinct peaks and are well-matched with JCPDS No. 89-7102. The crystalline structure of the synthesized ZnO NPs was observed to have stiff and narrow diffraction peaks, with no substantial variations in the diffraction peaks, indicating that the crystalline product was impurity-free. Peaks were absent from other phases or contaminants, indicating that the product was pure-phase ZnO NPs. Scherrer's formula was applied to the first intense PXRD peaks, and the sizes of the green synthesized ZnO NPs were found to be between 44 nm and 78 nm. The XRD peaks found in this work are comparable to the XRD patterns previously reported for ZnO nanoparticles produced using *Xylaria acuta*. The results are equivalent to those of nanoparticles produced with a crystallographic hexagonal wurtzite structure [61] (Figure 1).

$$D = 0.9 \times \lambda/(\beta \cos\theta)$$

where λ is the wavelength of the X-ray (1.542 Å), β refers to the full width at half maximum (FWHM in radian) caused by the crystallites, and θ refers to the Bragg angle.

Signals were created and recorded by the detector when the ZnO NPs were subjected to electron beams. Information on the shape, orientation, and crystalline structure of ZnO NPs may be derived from the recorded signal [56]. The surface morphologies of the ZnO NPs synthesized by *T. asperellum* were recorded at different magnifications and depicted in SEM images, which revealed the form and size of the zinc oxide nanoparticles. The SEM micrographs revealed a variety of nanoparticle combinations, as well as unique ZnO NPs. The produced forms of the ZnO NPs with various surface morphologies were also demonstrated using SEM images. An EDAX analysis was used to determine the qualitative and quantitative differences between the materials involved in the making of the nanoparticles. The synthesized nanoparticles were found to have the greatest proportions of zinc and oxygen in the analysis. The additional elements, as well as zinc, were detected in the synthesized ZnO NPs using an EDAX spectrum. The presence of metallic zinc oxide in the biosynthesized ZnO NPs was confirmed by the EDAX analysis. Our results are in good agreement with the results obtained by Shobha et al. [35], indicating that the biologically synthesized ZnO NPs had a similar elemental proportion of zinc and oxygen (Figure 2).

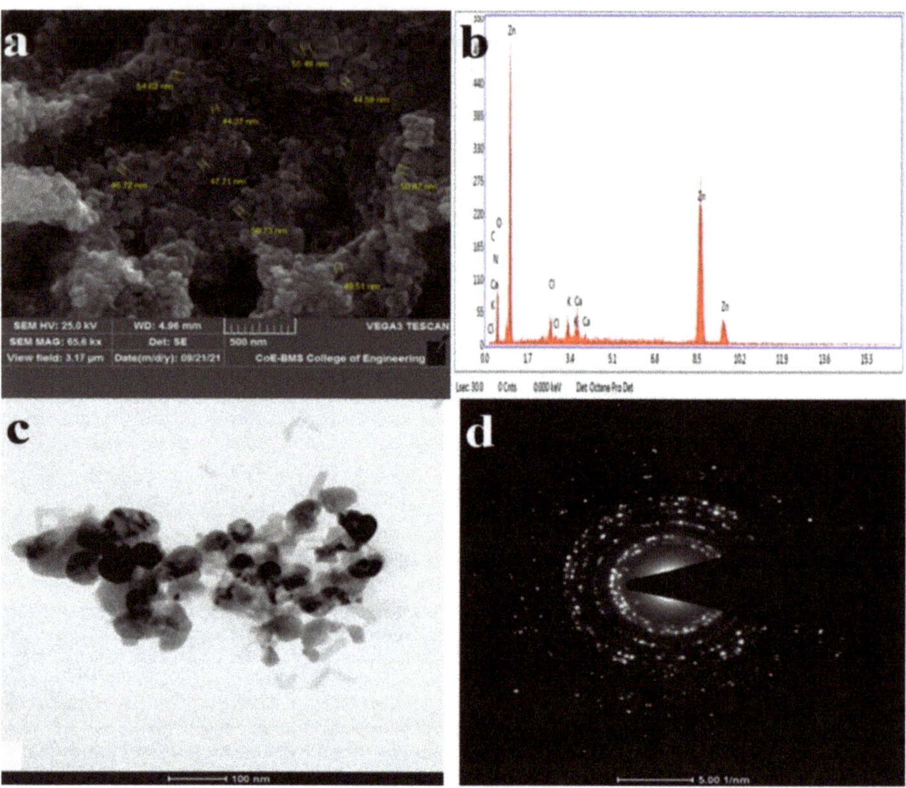

Figure 2. (**a**) High-magnification SEM image of ZnO NPs synthesized from *T. asperellum*: (**b**) energy-dispersive X-ray spectroscopy (EDAX) analysis; (**c**) TEM images of ZnO NPs synthesized from *T. asperellum*; and (**d**) SAED pattern of ZnO NPs synthesized from *T. asperellum*.

The TEM images revealed agglomerated ZnO NPs that were composed of well-dispersed minute particles. The size of the synthesized ZnO NPs ranged from 44 nm to 78 nm, according to the TEM investigation. The HRTEM and SAED patterns were found to correspond to ZnO compounds. The TEM picture depicts agglomerated, tiny ZnO NPs. Well-defined crystal planes can be seen in the high-resolution TEM picture. The particle size of the biosynthesized ZnO NPs from *T. asperellum* ranged from 44 nm to 78 nm. The (hkl) values corresponding to the significant peaks in the PXRD profiles fit the SAED patterns very well [60] (Figure 2).

3.4. In Vitro Screening of Zinc Oxide Nanoparticles (ZnO NPs) for Their Antibacterial Activity against Human Pathogens

A wide variety of metabolites, such as alkaloids, terpenoids, and flavonoids, are known to be produced by fungi and exhibit various properties, such as antibacterial, antiviral, anti-inflammatory, antitumor, and antifungal properties [62]. In a similar study, gold nanoparticles were synthesized from Bauhinia tomentosa Linn leaf extracts and tested for antibacterial activity against *E. coli* and *S. aureus* [63]. In the present study, *S. aureus* and *E. coli* were used to examine the antibacterial activity of ZnO NPs. By measuring the inhibition zone surrounding the disc, the antibacterial activity of the ZnO NPs was studied. The disc diffusion approach was used to investigate the antibacterial activity of the biosynthesized ZnO NPs, where the discs were placed on Mueller–Hinton agar (MHA) medium plates that were pre-swabbed with the bacteria. The inhibition zone was measured and tabulated. The MIC values of the biosynthesized ZnO NPs were determined using the 96-well microplate technique. Because of their smaller size and high surface-to-volume ratio, ZnO NPs have considerable antibacterial activity [61]. The current work clearly shows that ZnO NPs may be used as antibacterial agents against the human pathogens *S. aureus* and *E. coli* (Table 1).

Table 1. ZnO NP antibacterial activity against *S. aureus* and *E. coli*.

Concentrations of ZnO NPs (μg/mL)	Disc Diffusion Values (mm)		MIC Values (μg/mL)	
	S. aureus	*E. coli*	*S. aureus*	*E. coli*
25	3.18 ± 0.12	2.52 ± 0.49	25	50
50	6.23 ± 0.42	5.69 ± 0.38	12.5	25
75	9.82 ± 0.73	7.37 ± 0.27	6.25	12.5
Positive 100 μg/mL	8.37 ± 0.12	6.14 ± 0.19	50	50
Negative	0	0	0	0

The antibacterial activity of the biosynthesized ZnO NPs was further confirmed using fluorescence microscopy. The control bacterial cells were green under fluorescence microscopic inspection, which is a marker of healthy morphology in microorganisms, but the ZnO-NP-treated bacterial cells were red, with noticeable modifications in the cell wall, such as collapse, shrinkage, and a non-homologous surface. The absence of ZnO NPs (control microbial cells) resulted in the normal shape of microbial cells, as shown in the micrographs. However, the bacterial cells treated with the ZnO NPs showed morphology changes, such as ruptured cell membranes, oozed-out contents, and the aggregation of cells. Scanning electron microscopy (SEM) studies were undertaken to develop a better understanding of how the ZnO NPs cleaved the bacterial cell membrane and to predict the mechanism of the cell membrane rupture caused by the ZnO NPs. Furthermore, the effect of the ZnO NPs on the bacterial cells in comparison to the untreated bacterial cells was assessed by using SEM. The absence of ZnO NPs (control bacterial cells) resulted in the normal shape of bacterial cells, as seen in the micrographs. The bacterial cells treated with the ZnO NPs, however, displayed morphological changes, with the cell membrane rupturing, contents oozing out, and cells clumping together (Figure 3).

3.5. Determination of Biofilm Formation

The violet stains showed biofilm development according to the methodology. This indicates that the NB medium was enough for biofilm formation, whereas a 0.2% (*w/v*) crystal violet concentration was suitable for a naked-eye inspection and spectrophotometer measurement (Figure 4). The biosynthesized ZnO NPs were proved to have efficient activity against planktonic bacteria using the minimum inhibitory concentration (MIC) assay. The affinity of the ZnO NPs for planktonic microorganisms was further investigated. The particles' antibiofilm effectiveness was tested against Gram-positive microorganisms (*S. aureus*).

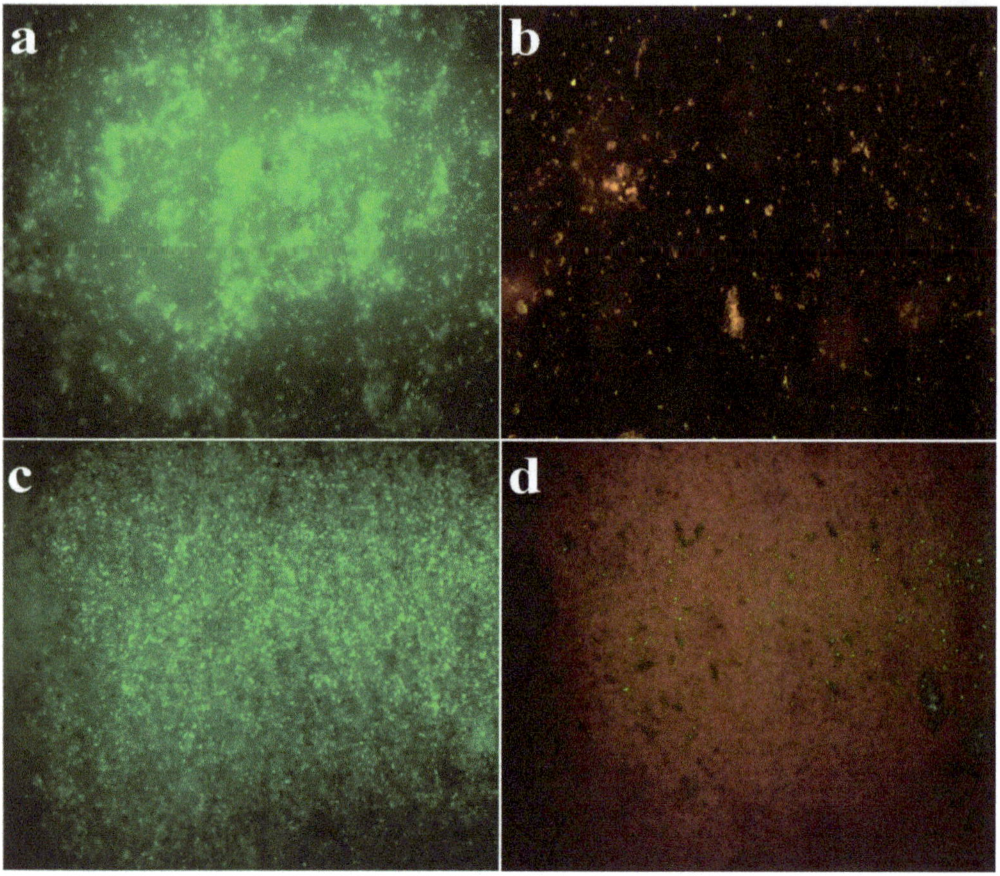

Figure 3. Fluorescence microscopic images of (**a**) *E. coli* control, (**b**) *E. coli* treated with ZnO NPs at a concentration of 75 µg/mL, (**c**) *S. aureus* control, (**d**) *S. aureus* treated with ZnO NPs at a concentration of 75 µg/mL.

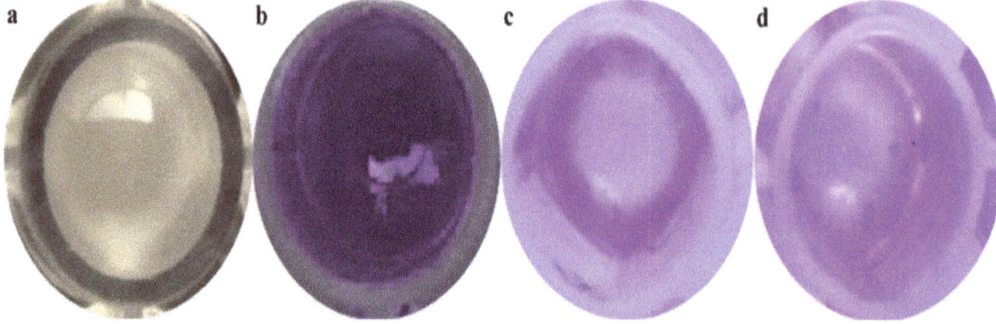

Figure 4. Crystal violet assay carried out to determine the antibiofilm activity of samples against *S. aureus* biofilms. (**a**) *S. aureus* forming a biofilm after 24 h of incubation in a microtiter well, (**b**) a well after crystal violet staining, (**c**) a control (tetracycline) well containing the crystal violet biofilm, and (**d**) a well with ZnO NPs (75 µg/mL) after decanting the crystal violet reduction of biofilms.

3.6. Assessment of Antiadherence Assay

The antiadherence properties of ZnO NPs and tetracycline has been determined by using an antiadherence assay, which uses a 96-well microplate. The biomass of the bacterial cell was quantified using a microplate reader at 570 nm, revealing a declining trend in biomass attachment as the quantity of the examined ZnO NPs increased. According to the findings, the ZnO NPs at 75 μg/mL had a strong antiadherence activity of 53.24 ± 1.37%. When compared to the control, the tetracycline at 100 μg/mL had no significant antiadherence action (Figure 5). The crystal violet staining of the biofilms was used to test the adherence of the biofilm bacteria on the microplates in order to determine whether the ZnO NPs inhibited biofilm development. At 25 μg/mL of ZnO NPs, biofilm development was dramatically reduced. The ZnO-NP-induced suppression of *S. mutans* on the oral surface has been recently established in research [64]. A similar study was carried out, where eugenol was investigated for its inhibitory property against the adherence and biofilm formation of *Streptococcus mutans* [65]. In another previous study, aquatic extracts of *Viscus album* and *Apium graveolens* showed antibiofilm and antiadherence against clinical bacterial isolates [66].

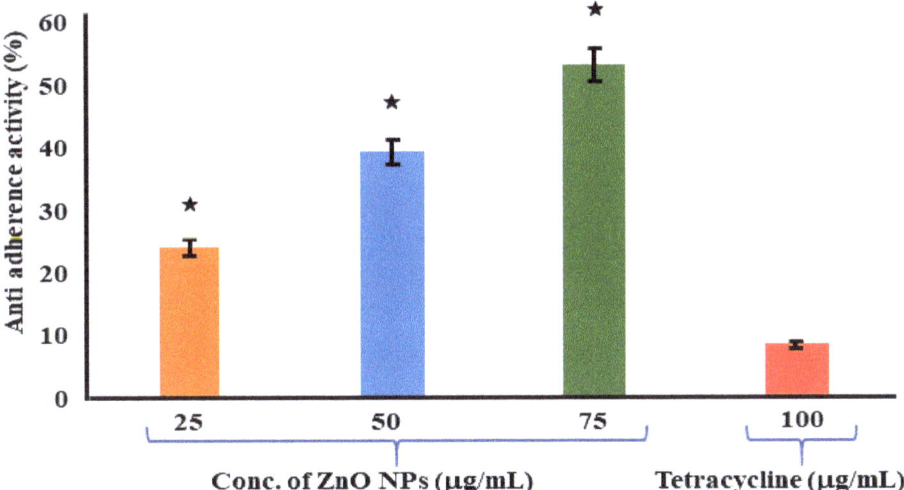

Figure 5. Antiadherence assay using tetracycline as positive control and *S. aureus* as growth control. The experiment was evaluated based on triplicate results with standard deviation ($n = 3$, $p < 0.05$). * Indicates a significant difference when compared to the negative control (NB only).

3.7. Assessment of Antibiofilm Assay

The antibiofilm properties of the ZnO NPs and tetracycline were determined using a 96-well microplate. When compared to the control (NB alone), the biomass of the bacterial cells was quantitatively examined on a microplate reader at an absorbance of 570 nm. The results revealed a considerable reduction in biofilm formation with an increase in the quantity of the biologically synthesized ZnO NPs. According to the results, the ZnO NPs at 75 μg/mL had a significant antibiofilm activity of 68.46 ± 1.72%. Meanwhile, the antibiofilm activity of the positive control tetracycline at a greater concentration of 100 μg/mL was measured to be 39.23 ± 4.61% (Figure 6). Reactive oxygen species (ROS) formation takes place when foreign particles come into contact with a bacterial solution, and this plays a crucial role in bacterial inhibition. Biofilm formation inhibition was investigated in planktonic *S. aureus* exposed to reagents for 24 h at the start. The ZnO NPs were more efficient than the tetracycline in eradicating the premade biofilm produced by *S. aureus*, according to a conventional crystal violet test for biofilm biomass. A similar study was conducted on *S. mutans*, where a solvent fraction of *Trachyspermum ammi* was used to evaluate the expressions of the genes involved in biofilm formation [67].

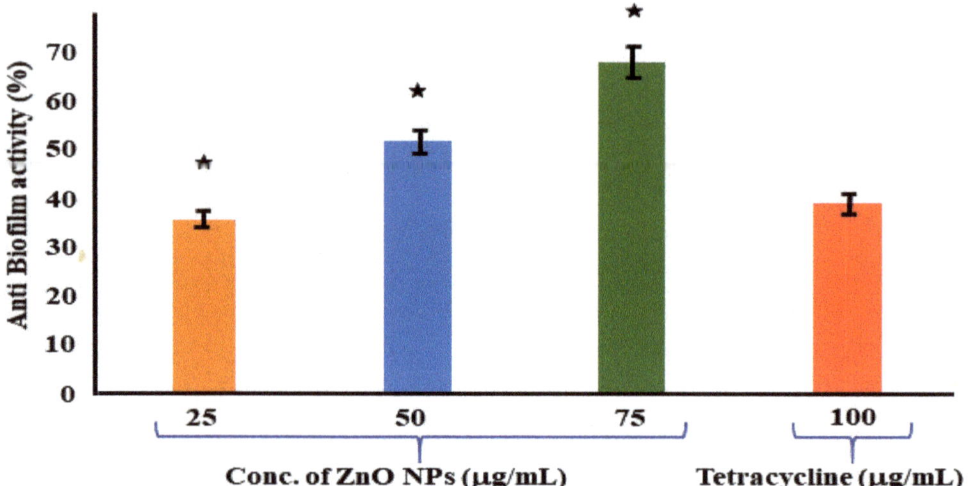

Figure 6. Antibiofilm assay using tetracycline as positive control and *S. aureus* as growth control. The experimental results were evaluated based on triplicate results with standard deviation ($n = 3$, $p < 0.05$). * Indicates a significant difference when compared to the negative control (NB only).

3.8. Microscopic Studies

The bright-field compound microscopic photographs revealed a decrease in the cells when they were Gram-stained, based on microscopic examinations. When compared to the negative control (NB alone), the positive control (tetracycline-treated) and the ZnO-NP-treated wells showed a lower number of cells (Figure 7).

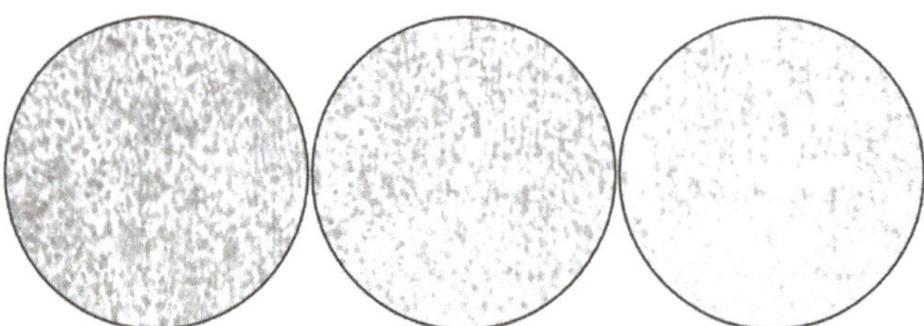

Figure 7. Micrographs of *S. aureus* biofilms. Adherence of *S. aureus* onto the coverslips; control (NB alone), tetracycline, and ZnO nanoparticles (75 µg/mL), as examined by CV staining.

After 24 h of incubation, SEM micrographs of the tested bacterium, *S. aureus,* treated with the ZnO NPs were taken. The SEM micrographs assisted in elucidating the mechanism/relationship between the bacteria and the ZnO NPs, as well as their antibacterial activity (Figure 7). In the case of *S. aureus*, it is obvious from the micrographs that the ZnO NPs initially adhered to the cell's outer membrane and then penetrated the cell entirely, perhaps resulting in cell death. The use of SEM to visualize the bacterial biofilms revealed a wide range of morphological changes in the biofilm topologies (Figures 7 and 8). There were remarkably fewer dispersed cell aggregates and fewer viable cells in the aggregates in the biofilms after 24 h of exposure to the ZnO NPs. The ZnO NPs were shown to be superior in inhibiting *S. aureus* biofilm formation when the biofilm biomass was measured [68,69].

These findings show that the ZnO NPs were more efficient than the tetracycline in inhibiting biofilm development and in disrupting the preformed biofilms of *S. aureus*.

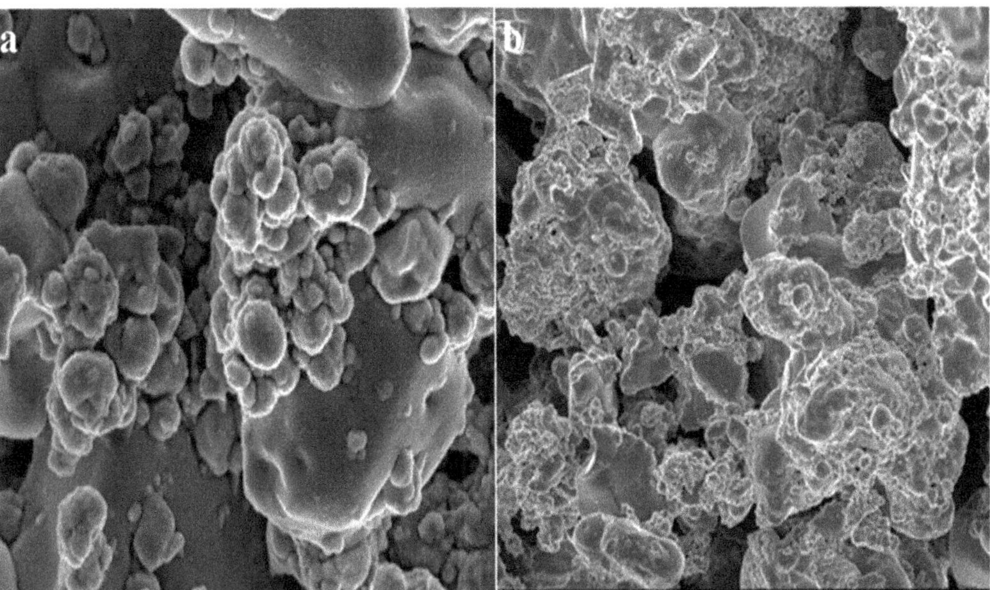

Figure 8. SEM micrographs of biofilm mass: (**a**) zinc oxide nanoparticles attached to the *S. aureus* biofilm, (**b**) biofilm of *S. aureus* disturbed after 24 h of treatment with ZnO NPs.

4. Conclusions

The present study highlights the biosynthesis of ZnO NPs from *T. asperellum*. The synthesized nanoparticles were characterized by UV–visible spectroscopy, PXRD, FTIR, SEM with EDAX, and TEM with SEAD patterns. This research focused on the biological synthesis of ZnO NPs, which is both cost-effective and environmentally friendly. Nanosized particles are more efficient in inhibiting the growth of microbes. Our studies emphasize that biologically synthesized ZnO NPs can be used as efficient antimicrobial agents in various fields, as ZnO NPs are non-toxic; have antimicrobial, barrier, and mechanical properties; and belong to the Generally Recognized as Safe (GRAS) category. Additionally, the non-central symmetry and biocompatible nature of ZnO make it the most important nanomaterial in research and applications. Based on our findings, we can conclude that the use of ZnO NPs is effective against the vast majority of pathogenic bacteria and biofilm-producing bacteria. ZnO NPs can be used for surgical instruments, which usually become colonized by bacterial biofilms. The future prospects of this study are to check the stability of these nanoparticles and their mode of action at the molecular level.

Author Contributions: Conceptualization, T.R.L. and M.A.A.; methodology, B.S., S.B. and B.S.A.; software, S.C., T.R., T.R.L. and M.A.A.; validation, T.R.L. and M.A.A.; formal analysis, M.G., U.H., B.A. and M.S.A.; investigation, B.S., B.S.A. and S.B.; resources, M.S.A.; S.W.; data curation, W.A. and M.S.A.; writing—original draft preparation, S.B., B.S., T.R.L. and M.A.A.; writing—review and editing, B.S., B.S.A., S.B., M.G., U.H., B.A., M.S.A., S.W., W.A., T.R.L. and M.A.A.; visualization, T.R.L. and M.A.A.; supervision, T.R.L. and M.A.A.; project administration, T.R.L. and M.A.A.; funding acquisition, M.G. and U.H. All authors have read and agreed to the published version of the manuscript.

Funding: The authors extend their appreciation to the Deanship of Scientific Research at King Khalid University for funding this work through the Small Groups Project under grant number (RGP. 1/343/43).

Institutional Review Board Statement: Not applicable.

Informed Consent Statement: Not applicable.

Data Availability Statement: Data is available in this manuscript.

Acknowledgments: The authors extend their appreciation to the Deanship of Scientific Research at King Khalid University for funding this work through the Small Groups Project under grant number (RGP. 1/343/43). The author would like to acknowledge the University Grants Commission (UGC), New Delhi, India, for providing financial support under the UGC startup grant No. F.30-580/2021(BSR) dated 12.11. The authors acknowledge Bangalore University for financial support, U.O. No. Dev. D2a. DU-RP.2020-21.

Conflicts of Interest: The authors declare no conflict of interest.

References

1. Barreto-Santamaría, A.; Arévalo-Pinzón, G.; Patarroyo, M.A.; Patarroyo, M.E. How to Combat Gram-Negative Bacteria Using Antimicrobial Peptides: A Challenge or an Unattainable Goal? *Antibiotics* **2021**, *10*, 1499. [CrossRef]
2. Breijyeh, Z.; Jubeh, B.; Karaman, R. Resistance of Gram-negative bacteria to current antibacterial agents and approaches to resolve it. *Molecules* **2020**, *25*, 1340. [CrossRef]
3. Martinez, J.E.; Vargas, A.; Perez-Sanchez, T.; Encio, I.J.; Cabello-Olmo, M.; Barajas, M. Human Microbiota Network: Unveiling Potential Crosstalk between the Different Microbiota Ecosystems and Their Role in Health and Disease. *Nutrients* **2021**, *13*, 2905. [CrossRef]
4. Charbonneau, M.R.; Isabella, V.M.; Li, N.; Kurtz, C.B. Developing a new class of engineered live bacterial therapeutics to treat human diseases. *Nat. Commun.* **2020**, *11*, 1738. [CrossRef]
5. Nataro, J.P.; Kaper, J.B. Diarrheagenic *Escherichia coli*. *Clin. Microbiol. Rev.* **1998**, *11*, 142–201. [PubMed]
6. Taylor, T.A.; Unakal, C.G. *Staphylococcus aureus*; StatPearls Publishing: Treasure Island, FL, USA, 2017.
7. Yang, R.; Hou, E.; Cheng, W.; Yan, X.; Zhang, T.; Li, S.; Guo, Y. Membrane-Targeting Neolignan-Antimicrobial Peptide Mimic Conjugates to Combat Methicillin-Resistant *Staphylococcus aureus* (MRSA) Infections. *J. Med. Chem.* **2022**, *65*, 16879–16892. [CrossRef] [PubMed]
8. Druvari, D.; Antonopoulou, A.; Lainioti, G.C.; Vlamis-Gardikas, A.; Bokias, G.; Kallitsis, J.K. Preparation of Antimicrobial Coatings from Cross-Linked Copolymers Containing Quaternary Dodecyl-Ammonium Compounds. *Int. J. Mol. Sci.* **2021**, *22*, 13236. [CrossRef] [PubMed]
9. Seetharaman, P.K.; Chandrasekaran, R.; Periakaruppan, R.; Gnanasekar, S.; Sivaperumal, S.; Abd-Elsalam, K.A.; Kuca, K. Functional Attributes of Myco-Synthesized Silver Nanoparticles from Endophytic Fungi: A New Implication in Biomedical Applications. *Biology* **2021**, *10*, 473. [CrossRef] [PubMed]
10. Cruz-Luna, A.R.; Cruz-Martínez, H.; Vasquez-López, A.; Medina, D.I. Metal Nanoparticles as Novel Antifungal Agents for Sustainable Agriculture: Current Advances and Future Directions. *J. Fungi* **2021**, *7*, 1033. [CrossRef] [PubMed]
11. Mughal, B.; Zaidi, S.Z.J.; Zhang, X.; Hassan, S.U. Biogenic Nanoparticles: Synthesis, Characterisation and Applications. *Appl. Sci.* **2021**, *11*, 2598. [CrossRef]
12. Santos, T.S.; Passos, E.M.D.; Seabra, M.G.; Souto, E.B.; Severino, P.; Mendonça, M.D.C. Entomopathogenic fungi biomass production and extracellular biosynthesis of silver nanoparticles for bioinsecticide action. *Appl. Sci.* **2021**, *11*, 2465. [CrossRef]
13. Chitra, K.; Annadurai, G. Bioengineered silver nanobowls using *Trichoderma viride* and its antibacterial activity against gram-positive and gram-negative bacteria. *J. Nanostructure Chem.* **2013**, *3*, 9. [CrossRef]
14. Jadoun, S.; Chauhan, N.P.S.; Zarrintaj, P.; Barani, M.; Varma, R.S.; Chinnam, S.; Rahdar, A. Synthesis of nanoparticles using microorganisms and their applications: A review. *Environ. Chem. Lett.* **2022**, *20*, 3153–3197.
15. Soltys, L.; Olkhovyy, O.; Tatarchuk, T.; Naushad, M. Green Synthesis of Metal and Metal Oxide Nanoparticles: Principles of Green Chemistry and Raw Materials. *Magnetochemistry* **2021**, *7*, 145. [CrossRef]
16. Koul, B.; Poonia, A.K.; Yadav, D.; Jin, J.O. Microbe-Mediated Biosynthesis of Nanoparticles: Applications and Future Prospects. *Biomolecules* **2021**, *11*, 886. [CrossRef]
17. Abomuti, M.A.; Danish, E.Y.; Firoz, A.; Hasan, N.; Malik, M.A. Green Synthesis of Zinc Oxide Nanoparticles Using *Salvia officinalis* Leaf Extract and Their Photocatalytic and Antifungal Activities. *Biology* **2021**, *10*, 1075. [CrossRef]
18. Chaudhary, A.; Kumar, N.; Kumar, R.; Salar, R.K. Antimicrobial activity of zinc oxide nanoparticles synthesized from *Aloe vera* peel extract. *SN Appl. Sci.* **2019**, *1*, 136. [CrossRef]
19. Abbas, A.M.; Hammad, S.A.; Sallam, H.; Mahfouz, L.; Ahmed, M.K.; Abboudy, S.M.; Mostafa, M. Biosynthesis of Zinc Oxide Nanoparticles Using Leaf Extract of *Prosopis juliflora* as Potential Photocatalyst for the Treatment of Paper Mill Effluent. *Appl. Sci.* **2021**, *11*, 11394. [CrossRef]
20. Barani, M.; Fathizadeh, H.; Arkaban, H.; Kalantar-Neyestanaki, D.; Akbarizadeh, M.R.; Turki Jalil, A.; Akhavan-Sigari, R. Recent Advances in Nanotechnology for the Management of *Klebsiella pneumoniae*—Related Infections. *Biosensors* **2022**, *12*, 1155. [CrossRef]

21. Thakkar, K.N.; Mhatre, S.S.; Parikh, R.Y. Biological synthesis of metallic nanoparticles. Nanomed.: Nanotechnol. *Biol. Med.* **2010**, *6*, 257–262.
22. Panigrahi, S.; Kundu, S.; Ghosh, S.; Nath, S.; Pal, T. General method of synthesis for metal nanoparticles. *J. Nanopart. Res.* **2004**, *6*, 411–414. [CrossRef]
23. Abdo, A.M.; Fouda, A.; Eid, A.M.; Fahmy, N.M.; Elsayed, A.M.; Khalil, A.M.A.; Soliman, A.M. Green Synthesis of Zinc Oxide Nanoparticles (ZnO-NPs) by *Pseudomonas aeruginosa* and Their Activity against Pathogenic Microbes and Common House Mosquito, Culex pipiens. *Materials* **2021**, *14*, 6983. [CrossRef]
24. Patkowska, E.; Mielniczuk, E.; Jamiołkowska, A.; Skwaryło-Bednarz, B.; Błażewicz-Wozniak, M. The Influence of *Trichoderma harzianum* Rifai T-22 and other biostimulants on rhizosphere beneficial microorganisms of carrot. *Agronomy* **2020**, *10*, 1637. [CrossRef]
25. Pandya, J.R.; Sabalpara, A.N.; Chawda, S.K. *Trichoderma*: A particular weapon for biological control of phytopathogens. *Int. J. Agric. Technol.* **2011**, *7*, 1187–1191.
26. Phupiewkham, W.; Sirithorn, P.; Saksirirat, W.; Thammasirirak, S. Antibacterial agents from *Trichoderma harzianum* strain T9 against pathogenic bacteria. *Chiang Mai J. Sci.* **2015**, *42*, 304–316.
27. Nuankeaw, K.; Chaiyosang, B.; Suebrasri, T.; Kanokmedhakul, S.; Lumyong, S.; Boonlue, S. First report of secondary metabolites, Violaceol I and Violaceol II produced by endophytic fungus, *Trichoderma polyalthiae* and their antimicrobial activity. *Mycoscience* **2020**, *61*, 16–21. [CrossRef]
28. Fan, H.; Yao, M.; Wang, H.; Zhao, D.; Zhu, X.; Wang, Y.; Chen, L. Isolation and effect of *Trichoderma citrinoviride* Snef1910 for the biological control of root-knot nematode, *Meloidogyne incognita*. *BMC Microbiol.* **2020**, *20*, 299. [CrossRef]
29. Tomah, A.A.; Abd Alamer, I.S.; Li, B.; Zhang, J.Z. A new species of Trichoderma and gliotoxin role: A new observation in enhancing biocontrol potential of *T. virens* against *Phytophthora capsici* on chili pepper. *Biol. Control* **2020**, *145*, 104261. [CrossRef]
30. Saravanakumar, K.; Chelliah, R.; Ramakrishnan, S.R.; Kathiresan, K.; Oh, D.H.; Wang, M.H. Antibacterial, and antioxidant potentials of non-cytotoxic extract of *Trichoderma atroviride*. *Microb. Pathog.* **2018**, *115*, 338–342. [CrossRef]
31. Gill, E.E.; Franco, O.L.; Hancock, R.E. Antibiotic adjuvants: Diverse strategies for controlling drug-resistant pathogens. *Chem. Biol. Drug Des.* **2015**, *85*, 56–78. [CrossRef]
32. Singh, S.; Datta, S.; Narayanan, K.B.; Rajnish, K.N. Bacterial exo-polysaccharides in biofilms: Role in antimicrobial resistance and treatments. *J. Genet. Eng. Biotechnol.* **2021**, *19*, 140. [CrossRef]
33. Gajdacs, M. The continuing threat of methicillin-resistant *Staphylococcus aureus*. *Antibiotics* **2019**, *8*, 52. [CrossRef]
34. Miklasinska-Majdanik, M.; Kępa, M.; Wojtyczka, R.D.; Idzik, D.; Wąsik, T.J. Phenolic compounds diminish antibiotic resistance of *Staphylococcus aureus* clinical strains. *Int. J. Environ. Res. Public Health.* **2018**, *15*, 2321. [CrossRef] [PubMed]
35. Shobha, B.; Lakshmeesha, T.R.; Ansari, M.A.; Almatroudi, A.; Alzohairy, M.A.; Basavaraju, S.; Chowdappa, S. Mycosynthesis of ZnO nanoparticles using *Trichoderma* spp. isolated from rhizosphere soils and its synergistic antibacterial effect against Xanthomonas oryzae pv. oryzae. *J. Fungi* **2020**, *6*, 181. [CrossRef] [PubMed]
36. Goswami, D.; Dhandhukia, P.; Patel, P.; Thakker, J.N. Screening of PGPR from saline desert of Kutch: Growth promotion in Arachis hypogea by *Bacillus licheniformis* A2. *Microbiol. Res.* **2014**, *169*, 66–75. [CrossRef] [PubMed]
37. Ru, Z.; Di, W. *Trichoderma* spp. from rhizosphere soil and their antagonism against *Fusarium sambucinum*. *Afr. J. Biotechnol.* **2012**, *11*, 4180–4186.
38. Shahid, M.; Singh, A.; Srivastava, M.; Rastogi, S.; Pathak, N. Sequencing of 28SrRNA gene for identification of *Trichoderma longibrachiatum* 28CP/7444 species in soil sample. *Int. J. Biotechnol. Wellness Ind.* **2013**, *2*, 84–90. [CrossRef]
39. Dou, K.; Lu, Z.; Wu, Q.; Ni, M.; Yu, C.; Wang, M.; Zhang, C. MIST: A multilocus identification system for *Trichoderma*. *Appl. Environ. Microbiol.* **2020**, *86*, e01532-20. [CrossRef]
40. Adil, M.; Khan, R.; Rupasinghe, H.V. Application of medicinal plants as a source for therapeutic agents against *Streptococcus pyogenes* infections. *Curr. Drug Metab.* **2018**, *19*, 695–703. [CrossRef]
41. Lakshmeesha, T.R.; Sateesh, M.K.; Prasad, B.D.; Sharma, S.C.; Kavyashree, D.; Chandrasekhar, M.; Nagabhushana, H. Reactivity of crystalline ZnO superstructures against fungi and bacterial pathogens: Synthesized using *Nerium oleander* leaf extract. *Cryst. Growth Des.* **2014**, *14*, 4068–4079. [CrossRef]
42. Shelar, G.B.; Chavan, A.M. Myco-synthesis of silver nanoparticles from *Trichoderma harzianum* and its impact on germination status of oil seed. *Biolife* **2015**, *3*, 109–113.
43. Nayak, S.; Bhat, M.P.; Udayashankar, A.C.; Lakshmeesha, T.R.; Geetha, N.; Jogaiah, S. Biosynthesis and characterization of *Dillenia indica*-mediated silver nanoparticles and their biological activity. *Appl. Organomet. Chem.* **2020**, *34*, e5567. [CrossRef]
44. Mohammed, H.A.; Al Fadhil, A.O. Antibacterial activity of *Azadirachta indica* (Neem) leaf extract against bacterial pathogens in Sudan. *Afr. J. Med. Sci.* **2017**, *3*, 246–2512.
45. Debenedetti, S.L. *TLC and PC. Isolation, Identification, and Characterization of Allelochemical/Natural Products*; CRC Press: Boca Raton, FL, USA, 2009; pp. 103–134.
46. Sangeetha, J.; Thangadurai, D. *Staining Techniques and Biochemical Methods for the Identification of Fungi In Laboratory Protocols in Fungal Biology*; Springer: New York, NY, USA, 2013; pp. 237–257.
47. Moodley, T. An In-Vitro Study of a Modified Bioactive Orthodontic Cement. 2017. Available online: http://etd.uwc.ac.za/xmlui/handle/11394/5833 (accessed on 10 August 2022).

48. Andriani, Y.; Mohamad, H.; Bhubalan, K.; Abdullah, M.I.; Amir, H. Phytochemical Analyses, Anti-Bacterial and Anti-Biofilm Activities of Mangrove-Associated *Hibiscus tiliaceus* Extracts and Fractions against *Pseudomonas aeruginosa*. *J. Sustain. Sci. Manag.* **2017**, *12*, 45–51.
49. Campbell, M.; Zhao, W.; Fathi, R.; Mihreteab, M.; Gilbert, E.S. *Rhamnus prinoides* (gesho): A source of diverse anti-biofilm activity. *J. Ethnopharmacol.* **2019**, *241*, 111955. [CrossRef]
50. Adil, M.; Baig, M.H.; Rupasinghe, H.V. Impact of citral and phloretin, alone and in combination, on major virulence traits of *Streptococcus pyogenes*. *Molecules* **2019**, *24*, 4237. [CrossRef]
51. Leoney, A.; Karthigeyan, S.; Asharaf, A.S.; Felix, A.J.W. Detection and categorization of biofilm-forming *Staphylococcus aureus*, *Viridans streptococcus*, *Klebsiella pneumoniae*, and *Escherichia coli* isolated from complete denture patients and visualization using scanning electron microscopy. *J. Int. Soc. Prev. Community Dent.* **2020**, *10*, 627. [CrossRef]
52. Li, T.; Wang, P.; Guo, W.; Huang, X.; Tian, X.; Wu, G.; Lei, H. Natural berberine-based Chinese herb medicine assembled nanostructures with modified antibacterial application. *ACS Nano* **2019**, *13*, 6770–6781. [CrossRef]
53. Barnett, H.L.; Hunter, B.B. *Illustrated Genera of Imperfect Fungi*; APS Press: St. Paul, MN, USA, 1998; p. 218.
54. Shaikhaldein, H.O.; Al-Qurainy, F.; Khan, S.; Nadeem, M.; Tarroum, M.; Salih, A.M.; Alfarraj, N.S. Biosynthesis and Characterization of ZnO Nanoparticles Using *Ochradenus arabicus* and Their Effect on Growth and Antioxidant Systems of *Maerua oblongifolia*. *Plants* **2021**, *10*, 1808. [CrossRef]
55. Mahamuni, P.P.; Patil, P.M.; Dhanavade, M.J.; Badiger, M.V.; Shadija, P.G.; Lokhande, A.C.; Bohara, R.A. Synthesis and characterization of zinc oxide nanoparticles by using polyol chemistry for their antimicrobial and antibiofilm activity. *Biochem. Biophys. Rep.* **2019**, *17*, 71–80. [CrossRef]
56. Wang, D.; Cui, L.; Chang, X.; Guan, D. Biosynthesis and characterization of zinc oxide nanoparticles from *Artemisia annua* and investigate their effect on proliferation, osteogenic differentiation and mineralization in human osteoblast-like MG-63 Cells. *J. Photochem. Photobiol. B* **2020**, *202*, 111652. [CrossRef]
57. Akintelu, S.A.; Folorunso, A.S. A review on green synthesis of zinc oxide nanoparticles using plant extracts and its biomedical applications. *Bionanoscience* **2020**, *10*, 848–863. [CrossRef]
58. Pillai, A.M.; Sivasankarapillai, V.S.; Rahdar, A.; Joseph, J.; Sadeghfar, F.; Rajesh, K.; Kyzas, G.Z. Green synthesis and characterization of zinc oxide nanoparticles with antibacterial and antifungal activity. *J. Mol. Struct.* **2020**, *1211*, 128107. [CrossRef]
59. Selim, Y.A.; Azb, M.A.; Ragab, I.; Abd El-Azim, M.H. Green synthesis of zinc oxide nanoparticles using aqueous extract of *Deverra tortuosa* and their cytotoxic activities. *Sci. Rep.* **2020**, *10*, 3445. [CrossRef]
60. Jayappa, M.D.; Ramaiah, C.K.; Kumar, M.A.P.; Suresh, D.; Prabhu, A.; Devasya, R.P.; Sheikh, S. Green synthesis of zinc oxide nanoparticles from the leaf, stem and in vitro grown callus of *Mussaenda frondosa* L.: Characterization and their applications. *Appl. Nanosci.* **2020**, *10*, 3057–3074. [CrossRef] [PubMed]
61. Sumanth, B.; Lakshmeesha, T.R.; Ansari, M.A.; Alzohairy, M.A.; Udayashankar, A.C.; Shobha, B.; Almatroudi, A. Mycogenic synthesis of extracellular zinc oxide nanoparticles from *Xylaria acuta* and its nanoantibiotic potential. *Int. J. Nanomed.* **2020**, *15*, 8519. [CrossRef]
62. Anwar, J.; Iqbal, Z. Effect of growth conditions on antibacterial activity of *Trichoderma harzianum* against selected pathogenic bacteria. *Sarhad J. Agric.* **2017**, *33*, 501–510. [CrossRef]
63. Gnanamoorthy, G.; Ramar, K.; Ali, D.; Yadav, V.K.; Kumar, G. Synthesis and effective performance of Photocatalytic and Antimicrobial activities of *Bauhinia tomentosa* Linn plants using of gold nanoparticles. *Opt. Mater.* **2022**, *123*, 111945. [CrossRef]
64. Wang, S.; Huang, Q.; Liu, X.; Li, Z.; Yang, H.; Lu, Z. Rapid antibiofilm effect of Ag/ZnO nanocomposites assisted by dental led curing light against facultative anaerobic oral pathogen *Streptococcus mutans*. *ACS Biomater. Sci. Eng.* **2019**, *5*, 2030–2040. [CrossRef]
65. Adil, M.; Singh, K.; Verma, P.K.; Khan, A.U. Eugenol-induced suppression of biofilm-forming genes in *Streptococcus mutans*: An approach to inhibit biofilms. *J. Glob. Antimicrob. Resist.* **2014**, *2*, 286–292. [CrossRef]
66. Oubaid, E.N.; Chabuck, Z.A.G.; Al-Saigh, R.J.; Hindi, N.K.K.; Kadhum, S.A. Pathogenic and antimicrobial properties of aquatic extracts of *Viscus album*. *Asian J. Plant Sci.* **2022**, *21*, 360–367. [CrossRef]
67. Khan, R.; Adil, M.; Danishuddin, M.; Verma, P.K.; Khan, A.U. In vitro and in vivo inhibition of *Streptococcus mutans* biofilm by *Trachyspermum ammi* seeds: An approach of alternative medicine. *Phytomedicine* **2012**, *19*, 747–755. [CrossRef] [PubMed]
68. Banerjee, S.; Vishakha, K.; Das, S.; Dutta, M.; Mukherjee, D.; Mondal, J.; Ganguli, A. Antibacterial, anti-biofilm activity and mechanism of action of pancreatin doped zinc oxide nanoparticles against methicillin resistant *Staphylococcus aureus*. *Colloids Surf. B* **2020**, *190*, 110921. [CrossRef] [PubMed]
69. Khan, M.I.; Behera, S.K.; Paul, P.; Das, B.; Suar, M.; Jayabalan, R.; Mishra, A. Biogenic Au@ ZnO core–shell nanocomposites kill Staphylococcus aureus without provoking nuclear damage and cytotoxicity in mouse fibroblasts cells under hyperglycemic condition with enhanced wound healing proficiency. *Med. Microbiol. Immunol.* **2019**, *208*, 609–629. [CrossRef] [PubMed]

Journal of *Fungi*

Article

Antifungal Activity of Copper Oxide Nanoparticles against Root Rot Disease in Cucumber

Said M. Kamel [1], Samah F. Elgobashy [1], Reda I. Omara [1], Aly S. Derbalah [2], Mahmoud Abdelfatah [3], Abdelhamed El-Shaer [3], Abdulaziz A. Al-Askar [4], Ahmed Abdelkhalek [5], Kamel A. Abd-Elsalam [1], Tarek Essa [1], Muhammad Kamran [6] and Mohsen Mohamed Elsharkawy [7,*]

1 Plant Pathology Research Institute, Agricultural Research Center, Giza 12619, Egypt
2 Pesticides Chemistry and Toxicology Department, Faculty of Agriculture, Kafrelsheikh University, Kafr el-Sheikh 33516, Egypt
3 Physics Department, Faculty of Science, Kafrelsheikh University, Kafr el-Sheikh 33516, Egypt
4 Department of Botany and Microbiology, College of Science, King Saud University, P.O. Box 2455, Riyadh 11451, Saudi Arabia
5 Plant Protection and Biomolecular Diagnosis Department, ALCRI, City of Scientific Research and Technological Applications, New Borg ElArab City 21934, Egypt
6 School of Agriculture, Food and Wine, The University of Adelaide, Adelaide, SA 5005, Australia
7 Agricultural Botany Department, Faculty of Agriculture, Kafrelsheikh University, Kafr el-Sheikh 33516, Egypt
* Correspondence: mohsen.abdelrahman@agr.kfs.edu.eg; Tel.: +20-1065772170

Abstract: Metal oxide nanoparticles have recently garnered interest as potentially valuable substances for the management of plant diseases. Copper oxide nanoparticles (Cu_2ONPs) were chemically fabricated to control root rot disease in cucumbers. A scanning electron microscope (SEM), X-ray diffraction (XRD) and photoluminescence (PL) were employed to characterize the produced nanoparticles. Moreover, the direct antifungal activity of Cu_2ONPs against *Fusarium solani* under laboratory, greenhouse, and field conditions were also evaluated. In addition, the induction of host-plant resistance by Cu_2ONPs was confirmed by the results of enzyme activities (catalase, peroxidase, and polyphenoloxidase) and gene expression (*PR-1* and *LOX-1*). Finally, the effect of Cu_2ONPs on the growth and productivity characteristics of the treated cucumber plants was investigated. The average particle size from all the peaks was found to be around 25.54 and 25.83 nm for 0.30 and 0.35 Cu_2O, respectively. Under laboratory conditions, the study found that Cu_2ONPs had a greater inhibitory effect on the growth of *Fusarium solani* than the untreated control. Cu_2ONP treatment considerably reduced the disease incidence of the root rot pathogen in cucumber plants in both greenhouse and field environments. Defense enzyme activity and defense genes (*PR1* and *LOX1*) transcription levels were higher in cucumber plants treated with Cu_2ONPs and fungicide than in the untreated control. SEM analysis revealed irregularities, changes, twisting, and plasmolysis in the mycelia, as well as spore shrinking and collapsing in *F. solani* treated with Cu_2ONPs, compared to the untreated control. The anatomical analysis revealed that cucumber plants treated with Cu_2ONPs had thicker cell walls, root cortex, and mesophyll tissue (MT) than untreated plants. Cucumber growth and yield characteristics were greatly improved after treatment with Cu_2ONPs and fungicide. To the best of our knowledge, employing Cu_2ONPs to treat cucumber rot root disease is a novel strategy that has not yet been reported.

Keywords: metal oxide nanoparticles; *Fusarium solani*; defense genes; control; resistance; enzymes; scanning electron microscope; anatomical structure

Citation: Kamel, S.M.; Elgobashy, S.F.; Omara, R.I.; Derbalah, A.S.; Abdelfatah, M.; El-Shaer, A.; Al-Askar, A.A.; Abdelkhalek, A.; Abd-Elsalam, K.A.; Essa, T.; et al. Antifungal Activity of Copper Oxide Nanoparticles against Root Rot Disease in Cucumber. *J. Fungi* **2022**, *8*, 911. https://doi.org/10.3390/jof8090911

Academic Editor: David S. Perlin

Received: 28 July 2022
Accepted: 25 August 2022
Published: 28 August 2022

1. Introduction

In Egypt, cucumbers (*Cucumis sativus* L.) are among the most important cucurbitaceous crops and the leading export vegetable. Cucumbers are grown under protected cultivation conditions in plastic greenhouses in two main growing seasons, i.e., autumn and winter.

Cucumbers are cultivated intensively under greenhouse conditions across a large area. Egypt ranks in 13th place with regard to cucumber productivity across the world [1]. Cucumber is rich in vitamins A, B, and C and contains 96% water, 3% carbohydrates and 1% protein, minerals, such as manganese, copper, iron, calcium and potassium, and is low in calories [2]. Unfortunately, cucumber is susceptible to infection by several soil-borne fungi, causing damping-off and root rot diseases which, of course, affect the quality and productivity of the crop. *Fusarium solani* (Mart.) App. & Wr.; *Pythium ultimum* Trow; *Rhizoctonia solani* Khun and *Sclerotium rolfsii* Sacc. are considered the most important pathogens involved [3–6].

Some chemical fungicides are effective in controlling these diseases, but these chemicals are expensive and not eco-friendly. In addition, biological control is used as an easy-to-apply strategy and does not disrupt the ecological balance [7–9]. Nonetheless, there are some issues and challenges associated with the use of biological control, such as the high specificity between biological control agents and plant pathogens [10]. Furthermore, biological control materials must adapt to soil climate and crop conditions, not to mention the possibility of pathogen resistance, which reduces the effectiveness of biological control [11]. Based on all of the above, it has become very important to find non-conventional alternatives to control plant pathogens.

The use of nanomaterials is one of the unconventional strategies for combating plant pathogens, as their unique and unusual physical, chemical, and biological properties have recently drawn the attention of members of the scientific community for their potential for a variety of purposes, including the control of plant pathogens. A straightforward, unambiguous solution to the issues with disease management is provided by nanotechnology [12–14]. The scientific and industrial fields are being revolutionized by these innovations. In the case of using nanotechnology in agriculture, the process of creating the final formulations that ensure the optimal distribution of agrochemicals, nutrients, pesticides, and even growth regulators, to increase the efficiency of use is particularly promising [15]. Various types of metallic and metal oxide nanoparticles with antimicrobial properties have been fabricated [16–18]. Metal nanoparticles, containing magnesium oxide [19], copper [20], silver [21,22], iron [23], zinc oxide [24], and nickel oxide [18,25], have shown antimicrobial properties.

Cu NPs have recently been predicted to be a key component in the next generation of nanomaterials due to their low cost [26]. Furthermore, due to their excellent efficacy against a wide spectrum of microbes, copper oxide (CuO) and copper oxide (Cu_2O) are among the most extensively used antimicrobial agents [26–29]. CuO is less expensive than silver, can be combined easily with polymers, and has chemical and physical properties that are generally stable. Researchers are particularly interested in high-ionic-scale metal oxides with antimicrobial activity, such as CuO, because they can be produced with unusually large surface areas and distinctive crystal shapes that add to their increased potency [19].

Furthermore, copper is a necessary plant micronutrient that aids in plant growth and disease resistance. Furthermore, copper is required for the creation of key plant defense proteins, such as plastocyanin, peroxidase, and copper multiple oxidases, in response to pathogen infections [30]. These nonspecific immune responses to infection can protect against a wide range of diseases. Elmer and colleagues [31–33] were among the first to demonstrate that the presence and function of CuO NPs can affect plant nutrition and disease defense. Cu_2ONPs have also been used as a nano-fertilizer to promote disease resistance in a variety of plant/disease systems, such as asparagus/fusarium crown and root rot. The findings indicate that Cu_2ONPs can operate as a very effective Cu delivery agent, promoting disease suppression [33]. As a result, the primary purpose of this study was to look into the physiological, pharmacological, and anatomical effects of Cu_2ONPs in the control of cucumber root rot disease.

Induced resistance has been reported in viral, bacterial, and fungal diseases [18]. It offers an attractive alternative to genetic resistance, particularly for the control of diseases caused by soil-borne pathogens, which are treated with chemical pesticides. One of the new strategies for controlling plant pathogens in different crops is the use of systemic acquired resistance (SAR) on a large scale [34].

Thus, the current study aims to carry out the following objectives: (1) to fabricate Cu_2ONPs of unique size and shape, verify their direct antifungal activity against *F. solani* in vitro, (2) to evaluate their ability to induce systemic resistance against root rot disease in cucumber plants in greenhouse and field conditions, (3) to determine the mechanism of resistance induction through the expression of regulatory and defense genes and the activity of defense enzymes and finally, (4) to look into the impact of produced nanoparticles on some of the growth and production features of cucumbers that have been treated.

2. Materials and Methods

2.1. Fabrication of Cu_2ONPs

Cu_2ONPs were chemically prepared using a low-cost precipitation method, where 1 M copper sulfate ($CuSO_4$. $5H_2O$) was dissolved in 100 mL deionized water (DI), followed by the addition of 5 M sodium hydroxide (NaOH), which was also dissolved in 100 mL DI. The reaction occurs at room temperature, after which different quantities of glucose solution ($C_6H_{12}O_6H_2O$) (0.30, 0.35 M) were heated to 50 °C and dropped into the mixture to reduce CuO to Cu_2O nanoparticles. The precipitates were collected and centrifuged, then washed five times with DI and ethanol, respectively, before being dried for 24 h at 100 °C.

2.2. Characterization of Cu_2ONPs

SEM (JSM-651OLV) and X-ray diffraction (XRD, Shimadzu 6000) were used to investigate the crystal structure and morphological properties of the Cu_2ONPs, respectively. The samples' optical properties were investigated using a Kimmon He-Cd (325 nm) photoluminescence excitation laser, and the spectra were obtained using a HORIBA iHR320 spectrometer with a Synapse CCD camera. From XRD patterns, the particle size can be calculated employing Scherrer's formula [35], which is as follows:

$$D_{avg} = K\,(\lambda/\beta\,\cos\theta)$$

where K is a shape factor and is usually taken to be 0.94, λ is the wavelength of the incident X-ray (Cu Kα1, 0.15406 nm), θ is the angle of Bragg, and β is the total width at half maxima (FWHM) in radians.

2.3. Plant Materials

Cucumber (*Cucumis sativus* L.) seeds of cv. Beta alpha were obtained from the Horticulture Research Institute, Agricultural Research Center (ARC), Dokki, Giza, Egypt.

2.4. Fungal Pathogen Identification

F. solani was isolated from naturally infected cucumber plants that displayed damping-off and root rot symptoms. Cucumber samples were collected from various cultivated areas in Egypt's Giza Governorate. The isolated fungus was purified with the hyphal tip technique and identified using morphological and microscopic features [36]. The identification of fungal isolates was confirmed morphologically at the Plant Pathology Research Institute (PPRI), Agricultural Research Center (ARC), Giza, Egypt. The identified fungus was kept in a potato dextrose agar (PDA) medium at 4 ± 1 °C [37].

2.5. Pathogenicity Test

An inoculum of the pathogen *F. solani* was prepared by growing the isolate in autoclaved bottles containing (100 g sorghum, 50 g sand, and 80 mL water) and incubated at 25 ± 2 °C. The sandy loam soil was autoclaved at 121 °C for 2 h. Plastic pots (30 cm in diameter) were sterilized using 5% formalin and left for 2 days to ensure complete formalin evaporation. Soil infestation was carried out by adding the previous inoculum to each pot at the rate of 3% of the soil weight. Five surface-sterilized cucumber seeds were sown in each pot seven days after soil infestation under controlled greenhouse conditions (25 ± 2 °C, 65 ± 2% humidity). Five replicates were used for each treatment, and pots with pathogen-free sandy loam soil were used for planting a control treatment. The percentage of damping-off at 15, 30, and 45 days after sowing was recorded.

2.6. Effects of Cu_2ONP Concentrations on Fungal Growth

The fungal isolate was cultured on a PDA medium for 7 days, then plugs (5 mm) were re-cultured again onto new PDA plates (9 mm), treated with two concentrations (50 and 100 µg/L) of 0.30 and 0.35 M Cu_2O NPs, respectively. The fungicide Uniform 390 SE (azoxystrobin + mefenoxam), produced by Syngenta company, Basel, Switzerland, was used as a recommended fungicide against soil-borne disease at an application rate of 1.5 L/hectare). Five plates served as duplicates for each treatment. Plugs (5 mm) of *F. solani* grown on PDA were used as a control.

2.7. In Vivo (Greenhouse) Experiment

Cucumber seeds were planted at a rate of 5 seeds per pot. Five pots were used for each treatment. Infested soil with the pathogenic agent inocula was prepared as mentioned above in the section of the pathogenicity test. All pots were irrigated three times throughout the seven days. In this experiment, 0.30 and 0.35 M Cu_2ONPs were applied at a concentration level of 100 µg/L, and the fungicide was used at 1.5 L/hectare. All treatments were applied by immersing the seeds for 60 min in the solution before sowing. The experimental design was a randomized complete block. The greenhouse conditions were 24 °C and 60% humidity. The percentages of pre-and post-emergence damping-off and survived plants were calculated according to Shaban and El-Bramawy [38], while root rot assessment was recorded according to the scale 0-4 of Hwang and Chang [39], with minor modifications, which were as follows: 0 = healthy roots, 1 = 1–9%, 2 ≥ 9–39%, 3 ≥ 39–69% and 4 ≥ 69% and above for root discoloration. Root discoloration was recorded at the end of the experiment and calculated according to the following formula:

$$\text{Disease incidence (DI\%)} = \text{number of infected plants/total plants in the treatment} \times 100$$

$$\text{Root rot index} = (\text{total of all ratings}/(\text{total number of plants} \times 4)) \times 100$$

2.8. Laboratory Studies

2.8.1. Enzyme Activity Assay

To evaluate the effect of the tested treatments on the activity of defense enzymes (catalase, peroxidase, and polyphenol oxidase) in the treated cucumber plants, 0.5 g of freshly treated cucumber leaves were homogenized at 0–4 °C in 3 mL of 50 mM TRIS buffer (pH 7.8), containing 1 mM EDTA-Na_2 and 7.5% polyvinylpyrrolidone. The homogenates were centrifuged (12,000 rpm, 20 min, 4 °C), and the enzyme activity was estimated at 25 °C, using a typical UV-160A spectrophotometer. Catalase (CAT), peroxidase (POX), and polyphenol oxidase (PPO) activities were measured, as demonstrated by Aebi [40], Hammerschmidt et al. [41], and Malik and Singh [42], respectively.

2.8.2. RT-PCR Analysis

One week after germination, a 100 mg sample of cucumber leaf tissue was taken from the treated and control plants. With a pre-chilled mortar and pestle, the samples were immediately crushed in liquid nitrogen. Total RNA was extracted as explained by Tek and Calis [43]. The isolated RNA was utilized for qRT-PCR amplification. All qRT-PCRs were performed with real-time PCR equipment using the SYBR green method. The specificity was tested by creating a melting curve by progressively increasing the temperature to 95 °C. The gene-specific primers (*PR-1* and *LOX-1*) were used in quantitative RT-PCR (Table 1). To normalize the transcript levels for each sample, the actin gene was used as a reference gene, and the final data were calculated using the formula $2^{-\Delta\Delta CT}$ [44].

Table 1. Forward and reverse primers sequence for *PR-1*, *LOX-1* and *actin* genes.

Gene	Forward Primer ('5 ------ '3)	Reverse Primer ('5 ------ '3)
PR-1	TGCTCAACAATATGCGAACC	TCATCCACCCACAACTGAAC
LOX-1	CTCTTGGGTGGTGGTGTTTC	TGGTGGGATTGAAGTTAGCC
Actin	TGCTGGTCGTGACCTTACTG	GAATCTCTCAGCTCCGATGG

2.8.3. Microscopic Observations of Fungal Morphology

A light microscope (Leica DM1000) examination was used to study the effect of Cu_2ONPs on mycelia and spores of *F. solani*. To study the interaction between *F. solani*, the cause of damping-off and root rot of cucumber plants and copper oxide nanoparticles Cu_2ONPs (0.35) and Cu_2ONPs (0.30), small pieces of agar were cut at the parts embedded with copper oxide nanoparticles with *F. solani* growth and transferred for dehydrating and were subsequently sputter-coated with gold according to methods of Harley and Ferguson [45]. Examination and photographing were carried out using a scanning electron microscope (SEM), JEOL JSM 6510 Iv, at the Faculty of Agriculture, Mansoura University, to observe the copper oxide nanoparticles' effects through parasitism action.

2.8.4. The Anatomical Structure of Cucumber Plants

The anatomical structure of the median internode of the main roots and the grafting zone of 40-day old cucumber plants infected with *F. solani* and treated with 0.30 and 0.35 M Cu_2ONPs were investigated. The samples were sliced and fixed in a solution made up of 10 mL formalin, 5 mL glacial acetic acid, and 85 mL 70% ethyl alcohol. Following that, the samples were washed in 50% ethyl alcohol, dried in a standard butyl alcohol series, embedded in paraffin wax at 56 °C (melting point), and cut with a rotary microtome. Finally, crystal violet and erythrosine prepared in Canada balsam were used to stain the samples [46]. An optical microscope was used to examine the slides, and counts and measures (m) for various tissues were calculated.

2.9. Field Experiment

A field experiment was conducted to evaluate the efficiency of Cu_2ONPs and the recommended fungicide against the damping-off disease of cucumber. The experiment was designed in a randomized complete block with four replicates. The field in which the experiment was conducted has a previous history of the disease. Whether it was 0.30 or 0.35 M, Cu_2ONPs were used at a rate of 100 μg/L and the fungicide at 1.5 L/ha. Each replicate was $2 \times 10\ m^2$ in the area and had two rows of 1 m in width and 10 m in length. Then, the soil was irrigated for 7 days before sowing. Cucumber seeds (cv. Beta Alpha) were planted at a spacing of 50 cm at a rate of 3 seeds per hole (after soaking in the tested treatments for 1 h). The effectiveness of the treatments in lowering damping-off and disease incidence in pre-and post-emergent stages, as well as the percentages of healthy plants that survived, were recorded after 15, 30, and 45 days. The percentages of pre-and post-emergence damping-off and survived plants were estimated according to the previously described method.

2.10. Growth and Yield Parameters

Plant height, root length, wet weight and dry weight were evaluated in both treated and untreated cucumber plants to evaluate the influence of the tested treatments on some growth characteristics of cucumber plants. According to Torres-Netto et al. [47], the total chlorophyll content in fully expanded cucumber leaves was determined using a portable paper chlorophyll meter (Minolta SPAD-502, Osaka, Japan). Shoot length was measured from the base of the cucumber plants to the top in centimeters; root length was measured from the base of cucumber plants to the top root in centimeters. Fresh weight was measured in grams and the dry weight of plants (80 °C for 36 h) was measured in grams when the weight was stable. Yield parameters, such as the number of fruits/plant, fruit weight (g), fruit weight/plant (kg), and % rate of increase yield, were measured.

2.11. Statistical Analysis

All experiments were designed with a complete randomized block design. The WASP software (Web Agriculture Stat Package) was used for the analysis of variance (ANOVA) at $p \leq 0.05$. Using the SPSS v.22 software, analysis of variance and Pearson correlation tests were run to determine the relationship between all the measured parameters in this study.

3. Results

3.1. Characterization of Cu_2ONPs

The crystal structure of the fabricated nanoparticles was examined by XRD, as presented in Figure 1A,B. The strongest peak of the fabricated Cu_2O nanoparticles was formed at 2θ of around 36.43°, which is related to the (111) diffraction plane of the cubic phase of Cu_2O [32]. There are also many peaks at 2θ of 29.58°, 42.32°, 61.39°, 73.54°and 77°, which correspond to the (110), (200), (220), (311) and (222) diffraction planes for the cubic phase structure of the Cu_2ONPs, respectively [45,46]. All the observed diffraction peaks are associated with the standard polycrystalline cubic structure of Cu_2O with the Pn3m group space (JPCD NO. 05-0667) [48,49]. Only a pure Cu_2ONP phase is formed when no peaks for other phases appear. Moreover, the intensity (111) peak shows the high quality of crystallization. The calculated values for the particle size from the highest diffraction peak (111) were found to be around 30.96 and 31.91 nm for 0.30 and 0.35 Cu_2ONPs, respectively. In addition, the average particle size from all the peaks was found to be around 25.54 and 25.83 nm for 0.30 and 0.35 Cu_2O, respectively.

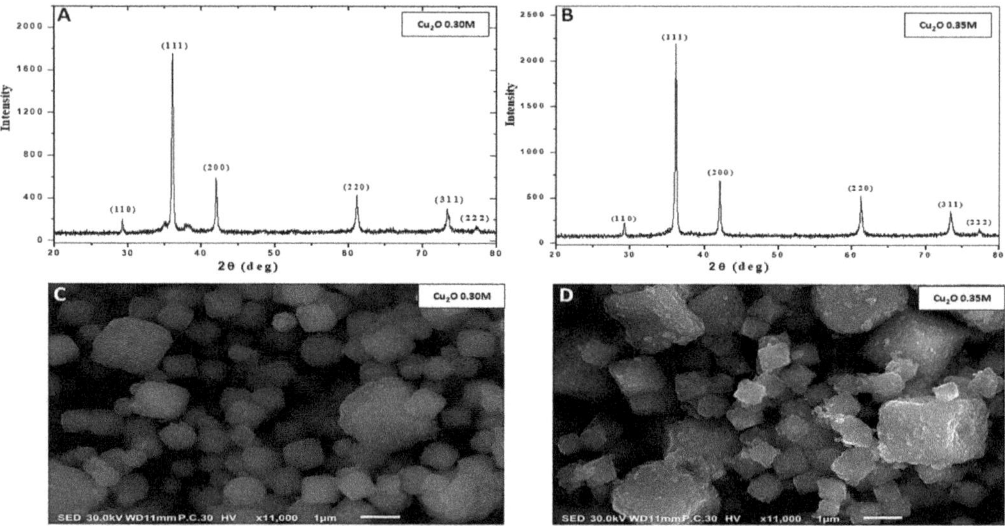

Figure 1. XRD patterns of the fabricated Cu_2O nanoparticles (**A**,**B**). Top view of SEM image of surface morphologies of fabricated Cu_2ONPs (**C**,**D**).

Figure 1C,D display the top-view SEM images of the fabricated Cu_2ONPs. It is noticeable that Cu_2ONPs demonstrate the morphology of cubic structures [35]. The results confirm and agree with the XRD measurements. From SEM images, the distributions of grain size for the fabricated samples are calculated, as shown in Figure 2A. It is clear that the grain size for the 0.30 Cu_2ONPs is smaller than the 0.35 Cu_2ONPs, where its average values were around 0.7 and 0.9 μm, respectively. Therefore, the surface area of 0.30 is higher than 0.35. Such a change in the particle size, average grain size, and surface area for the Cu_2O NPs is attributed mainly to the glucose amount during the reduction of CuO to Cu_2ONPs, where glucose acts as a reduced agent and surfactant agent at the same time. In such a case, the increase in glucose prevents the agglomeration of the Cu_2ONPs and increases the lateral and vertical growth, which produces smaller particle sizes and grain sizes for 0.30 Cu_2ONPs than 0.35 Cu_2ONPs [50,51].

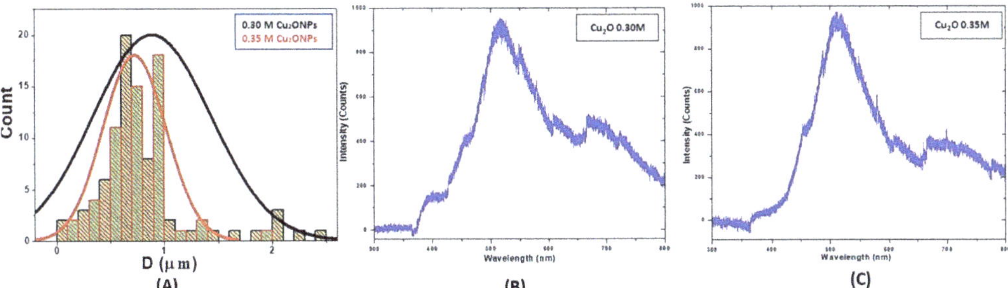

Figure 2. Average grain size distribution of 0.30 and 0.35 of Cu_2O NPs (**A**) and room temperature PL spectra of fabricated Cu_2ONPs (**B**,**C**).

Photoluminescence (PL) is a technique used to examine the recombination rate of produced electron–hole pairs and the quality of crystal for the fabricated materials [52,53]. Thus, PL spectra of the synthesized Cu_2ONPs were described, as shown in Figure 2B,C. The PL spectrum has two emission peaks, which cover up the visible light region. The first one is broader and varies from around 360 to 610 nm and is centered at about 520 nm. The other has smaller intensity and less broadening and is centered at about 705 nm. Excitonic transition series of Wannier hydrogen-like electrons associated with deep level defects are the responsible for the broad peak that can produce oxygen vacancies and/or copper interstitials [35]. This peak may be also due to the phonon-assisted excitations from the recombination process in nanoparticles [35]. The other peak resulted from the existence of other defects formed during the growth process, such as oxygen vacancies [35].

3.2. Antifungal Activity of Cu_2ONPs against *F. solani* under Laboratory Conditions

The effects of Cu_2ONPs with its two molar concentrations, Cu_2ONPs (0.35) and Cu_2ONPs (0.30), on the mycelial growth of *F. solani* compared to the fungicide are presented in Table 2 and Figure 3. The results showed that the used Cu_2ONPs and the fungicide significantly inhibited the growth of *F. solani* compared to the untreated control. The most effective treatment was the fungicide, followed by Cu_2ONPs (0.30 M) and Cu_2ONPs (0.35 M), respectively. In addition, there was a strong correlation between the inhibition percentage of the tested treatments and their concentrations.

Table 2. Radial growth and inhibition percentage of the Cu_2ONPs and fungicide against *F. solani* in vitro with regression equation and degree of correlation.

Treatment	Conc. (μg/L)	Radial Growth (mm)	Inhibition %	Regression Equation	R^2
Cu_2ONPs (0.35)	10	7.0 ^{b} ±0.54 *	22.22 ^{f} ± 0.56	Y = 1.7611x − 33.49	0.99
	25	5.8 ^{c} ± 0.66	35.56 ^{e} ± 0.76		
	50	4.6 ^{d} ±0.48	48.89 ^{d} ±0.73		
	100	2.3 ^{d} ± 0.52	74.44 ^{b}± 0.86		
Cu_2ONPs (0.30)	10	6.8 ^{c} ± 0.64	24.44 ^{d}± 0.78	Y = 1.6779x − 36.244	0.981
	25	5.4 ^{e}± 0.57	38.89 ^{c} ± 0.91		
	50	4.1 ^{f} ± 0.43	54.44 ^{b} ±0.93		

Table 2. *Cont.*

Treatment	Conc. (μg/L)	Radial Growth (mm)	Inhibition %	Regression Equation	R^2
	100	1.9 ^{f} ± 0.39	78.89 ^{a} ± 1.12		
Fungicide (Uniform 390 SE)	10	6.3 ^{d} ± 0.67	30.0 ^{d} ± 0.71	Y = 1.627x − 43.269	0.983
	25	5.1 ^{e} ± 0.55	43.33 ^{c} ± 0.89		
	50	3.5 ^{f} ± 0.45	61.11 ^{b} ± 1.07		
	100	1.3 ^{f} ± 0.41	85.56 ^{a} ± 1.14		
Control	0	9.0 ^{a} ± 0.69	0.0 ^{g} ± 0.19		

Statistical comparisons were made among treatments within a single column. * The different letters represent significant differences using Fisher's LSD test at $p \leq 0.05$. Each mean value came from three replicates.

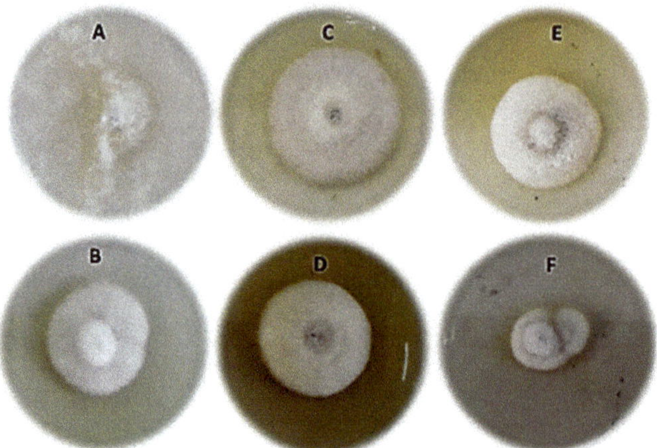

Figure 3. Effect of Cu_2ONPs and chemical fungicide on the radial growth of *Fusarium solani* **(A)** = control, **(B)** = fungicide, **(C)** = Cu_2ONPs (0.35 M) at a concentration of 50 μg/L, **(D)** = Cu_2ONPs (0.35 M) at a concentration of 100 μg/L, **(E)** = Cu_2ONPs (0.30 M) at a concentration of 50 μg/L and **(F)** = Cu_2ONPs (0.30 M) at a concentration of 100 μg/L.

3.3. *Effect of Cu_2ONPs on Disease Incidence under Greenhouse Conditions*

The effect of Cu_2ONPs compared to the recommended fungicide on the damping-off percentage and root rot disease incidence in the treated cucumber plants under greenhouse conditions is presented in Table 3. The results showed that 0.30 and 0.35 M of Cu_2ONPs and the chemical fungicide significantly reduced the damping-off percentage and disease incidence in treated cucumber plants compared to the untreated control. The highest reduction in damping-off percentage and disease incidence was recorded for the recommended fungicide, followed by 0.30 and 0.35 M of Cu_2ONPs, respectively. The reduction in the damping-off percentage and disease incidence in cucumber plants treated with 0.30 M of Cu_2ONPs was significantly higher than that of 0.35 M of Cu_2ONPs.

3.4. *Effect of Cu_2ONPs on the Activity of Defense Enzymes in Treated Cucumber Plants*

The effect of Cu_2ONPs and the recommended fungicide on the activity of defense enzymes in the treated cucumber plants is presented in Table 4. The data presented in Table 4 indicated that Cu_2ONPs and the recommended fungicide significantly increased the activities of defense-related enzymes, i.e., catalase, peroxidase and polyphenol oxidase in treated cucumber plants compared to the untreated control. The highest activity for defense enzymes was recorded for 0.30 M Cu_2ONPs, followed by 0.35 M Cu_2ONPs and fungicide, respectively.

Table 3. Effect of Cu_2ONPs compared to the chemical fungicide on the percentages of damping-off and root rot disease incidence of *F. solani* in cucumber plants under greenhouse conditions.

Treatment	Damping-Off %			Disease Incidence%	% Efficacy
	Pre-Emergence	Post-Emergence	Survival		
Cu_2ONPs (0. 30)	$13.3^{c} \pm 0.68$ *	$6.0^{c} \pm 0.26$	$80.7^{a} \pm 1.35$	$26.9^{c} \pm 0.57$	50.0
Cu_2ONPs (0.35)	$30.0^{a} \pm 0.89$	$0.0^{d} \pm 0.02$	$70.0^{c} \pm 1.23$	$28.6^{b} \pm 0.63$	46.8
Fungicide (Uniform 390 SE)	$13.3^{c} \pm 0.63$	$13.3^{b} \pm 0.46$	$73.4^{b} \pm 1.41$	$20.0^{d} \pm 0.48$	62.8
Control	$17.5^{b} \pm 0.47$	$17.5^{a} \pm 0.53$	$65.0^{d} \pm 1.12$	$53.8^{a} \pm 0.63$	0.0

Statistical comparisons were made among treatments within a single column. * The different letters represent significant differences using Fisher's LSD test at $p \leq 0.05$. Each mean value came from three replicates.

Table 4. Effect of Cu_2ONPs compared to the chemical fungicide on enzyme activities in treated cucumber plants.

Treatments	Enzyme Activity		
	CAT (Catalase) mM H_2O_2 g^{-1} FW Min^{-1}	POX (Peroxidase) mM H_2O_2 g^{-1} FW Min^{-1}	PPO (Polyphenol Oxidase) μ mol/min^{-1} g^{-1} (FW)
Cu_2ONPs (0.30)	$23.3^{a} \pm 0.63$ *	$1.397^{a} \pm 0.23$	$0.127^{a} \pm 0.52$
Cu_2ONPs (0.35)	$22.1^{b} \pm 0.61$	$1.239^{a} \pm 0.28$	$0.098^{b} \pm 0.43$
Fungicide (Uniform 390 SE)	$19.7^{c} \pm 0.46$	$0.741^{b} \pm 0.25$	$0.054^{c} \pm 0.35$
Control	$12.8^{d} \pm 0.42$	$0.329^{c} \pm 0.21$	$0.034^{d} \pm 0.22$

Statistical comparisons were made among treatments within a single column. * The different letters represent significant differences using Fisher's LSD test at $p \leq 0.05$. Each mean value came from three replicates.

3.5. Relative Expression Assay

Defense gene transcriptions were substantially up-regulated among Cu_2ONP treatments (Figure 4). Increased relative transcription levels of the *PR-1*, and *LOX-1* genes analyzed by qRT-PCR were observed in the treated cucumber plants at 7 days after treatment. It was revealed that in the Cu_2ONP-treated plants, there were more up-regulated *PR-1* and *LOX-1* than in the control group, and that the Cu_2ONP (0.30M)-treated plants had slightly higher *PR-1* and *LOX-1* transcription levels than the Cu_2ONP (0.35 M) group. The relationship between the cucumber and *F. solani* was found to be mediated by some genes.

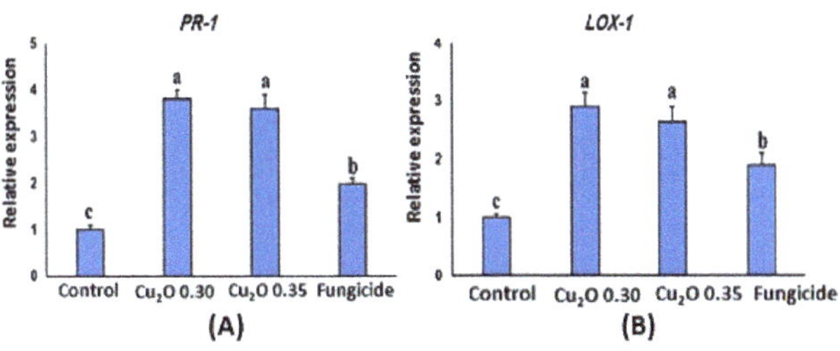

Figure 4. Expression of defense-related genes, such as *PR-1* (**A**) and *LOX-1* (**B**), in leaves of cucumber plants treated with Cu_2ONPs before challenge inoculation with *F. solani*.

3.6. Laboratory Studies

3.6.1. Microscopic Observations of Fungal Morphology

Light microscope examinations of *F. solani* mycelia and spores treated with 0.30 and 0.35 M Cu_2ONPs revealed shrinking, twisting, and collapse of treated plants (Figure 5B,C), compared to untreated *F. solani* (control) (Figure 4A).

Scanning electron microscope examination of the fungal structures taken from cucumber roots treated with 0.30 and 0.35 M Cu_2ONPs showed abnormalities and alterations in the mycelia of *F. solani* (Figure 6B), compared to the untreated control (Figure 6A). Moreover, twisting and plasmolysis of mycelial and spores and shrinking and collapsing were also observed in the roots of the treated cucumber plants (Figure 6B).

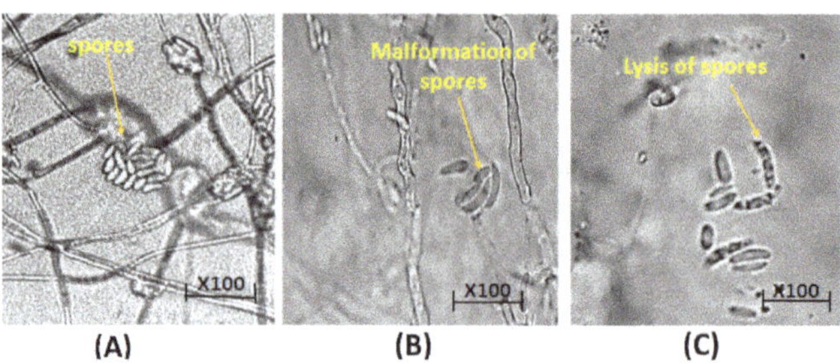

Figure 5. Light microscope observation of spores of *F. solani* showing (**A**): untreated control with normal spores and mycelium (yellow arrows). (**B**): Treated with Cu_2ONPs (0.35M), showing collapsed spores (yellow arrows) and (**C**): treated with Cu_2ONPs (0.30M), showing collapsed spores (yellow arrows).

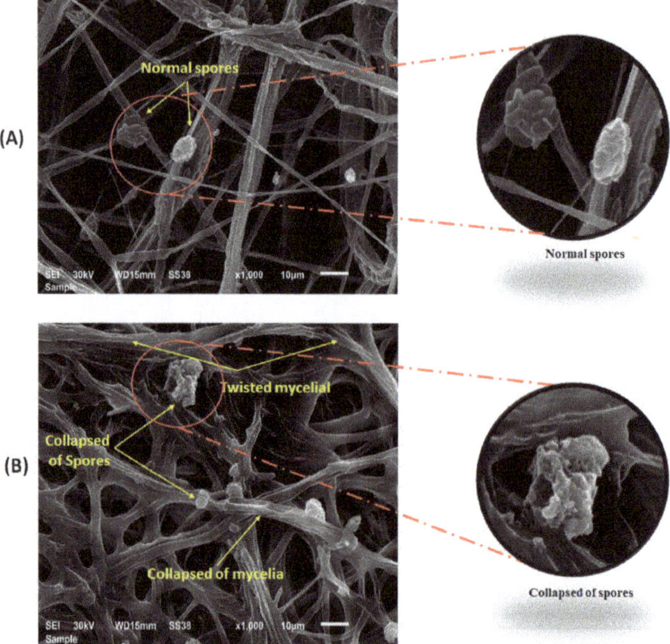

Figure 6. Scanning electron microscope observations of spores and mycelia of *F. solani* taken from growth medium on potato dextrose agar, with two sizes of nano copper showing. (**A**): Untreated control with normal spores and mycelium (yellow arrows). (**B**): Treated with Cu_2ONPs, showing collapsed mycelia and spores (yellow arrows).

3.6.2. The Anatomical Structure of Cucumber Plants

Our findings reveal that, in comparison to the untreated plants, cucumber plants treated with 0.30 and 0.35 M of Cu_2ONPs increased the anatomical characters (Figure 7B,C). When plants were treated with 0.30 and 0.35 M Cu_2ONPs, the cell wall, root cortex and mesophyll tissue (MT) thickness were all increased in comparison to the untreated plants, where the cell wall was ruptured (Figure 7A).

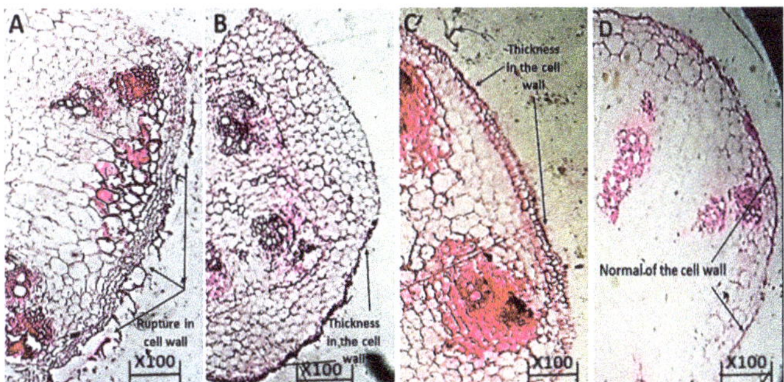

Figure 7. Effect of Cu_2ONPs on the anatomical structure of cucumber root infection with *F. solani*. (**A**): infected control with *F. solani*, (**B**): Cu_2ONPs (0.35 M), (**C**): Cu_2ONPs (0.30 M) and (**D**): uninfected.

3.7. *Effect of Cu_2ONPs on Disease Incidence in Two Locations under Field Conditions*

The effect of Cu_2ONPs compared to the recommended fungicide on damping-off percentage and root rot disease incidence in cucumber plants in Menoufia and Giza Governorates is presented in Table 5. The results showed that 0.30 and 0.35 M of Cu_2ONPs and the chemical fungicide significantly reduced the damping-off percentage and disease incidence in the treated cucumber plants compared to the untreated control in the two locations. The highest reduction in damping-off percentage and disease incidence was recorded for the recommended fungicide, followed by 0.30 and 0.35 M of Cu_2ONPs in the two locations. The efficacy of Cu_2ONPs and the recommended fungicide was higher in Giza than in Menoufia Governorate.

Table 5. Effect of Cu_2ONPs compared to the fungicide on the percentages of damping-off and root rot disease incidence of *F. solani* in cucumber plants under field conditions in Menoufia and Giza Governorates.

Treatments	Menoufia Governorate				
	Damping-Off %			Disease Incidence %	% Efficacy
	Pre-Emergence	Post-Emergence	Survival		
Cu_2ONPs (0.35)	12.0 ^{b} ± 0.65	18.7 ^{c} ± 0.68	69.3 ^{b} ± 1.23	24.7 ^{b} ± 0.54	63.8
Cu_2ONPs (0.30)	9.3 ^{c} ± 0.56	10.3 ^{b} ± 0.54	80.4 ^{a} ± 1.41	21.9 ^{c} ± 0.46	67.9
Fungicide (Uniform 390 SE)	7.3 ^{c} ± 0.51	11.3 ^{b} ± 0.57	81.4 ^{a} ± 1.46	18.0 ^{d} ± 0.44	73.6
Control	14.5a ± 0.72	29.2 ^{a} ± 0.81	56.3 ^{c} ± 1.27	68.3 ^{a} ± 1.75	0.0
			Giza Governorate		
Cu_2ONPs (0.35)	9.7 ^{b} ± 0.58	12.7 ^{c} ± 0.59	77.6 ^{c} ± 1.34	21.3 ^{b} ± 0.58	70.17
Cu_2ONPs (0.30)	7.4 ^{c} ± 0.51	11.3 ^{b} ± 0.56	81.6 ^{b} ± 1.42	20.4 ^{b} ± 0.66	71.43
Fungicide (Uniform 390 SE)	5.7 ^{c} ± 0.47	10.3 ^{b} ± 0.51	84.0 ^{a} ± 1.44	17.3 ^{c} ± 0.51	75.77
Control	13.5 ^{a} ± 0.62	31.7 ^{a} ± 0.74	54.8 ^{d} ± 1.04	71.4 ^{a} ± 1.42	0.0

Statistical comparisons were made among treatments within a single column. * The different letters represent significant differences using Fisher's LSD test at $p \leq 0.05$. Each mean value came from three replicates.

3.8. Effect of Cu_2ONPs on Total Chlorophyll and Growth Parameter of Cucumber Plants under Field Conditions

The results in Table 6 indicated that the applied Cu_2ONPs (0.30 and 0.35 M) and fungicide increased the growth parameters of treated cucumbers, such as total chlorophyll (SPAD), shoot length (cm), root length (cm), fresh and dry weight (g), under field conditions in the two locations compared to the untreated control. The highest growth parameters in the treated cucumbers were for the chemical fungicide, followed by 0.30 and 0.35 M of Cu_2ONPs, respectively. Approximately, there are no significant differences in the measured growth parameters of cucumbers between Cu_2ONPs 0.30 and 0.35 M in the two locations. The measured growth parameters of cucumber treated with Cu_2ONPs and the fungicide were significantly higher in Giza Governorate than in the Menoufia Governorate.

Table 6. Effect of Cu_2ONPs compared to the chemical fungicide on total chlorophyll and growth parameters of cucumber plants under field conditions in Menoufia and Giza Governorates.

Treatment	Total Chlorophyll (SPAD)	Shoot Length (cm)	Root Length (cm)	Fresh Weight (g)	Dry Weight (g)
			Menoufia Governorate		
Cu_2ONPs (0.35)	$32.1^{b} \pm 0.74$	$176.4^{a} \pm 1.21$	$30.8^{b} \pm 0.85$	$45.2^{a} \pm 0.67$	$4.8^{b} \pm 0.23$
Cu_2ONPs (0.30)	$33.7^{b} \pm 0.77$	$177.2^{a} \pm 1.24$	$31.1^{b} \pm 0.82$	$47.7^{b} \pm 0.86$	$5.2^{b} \pm 0.23$
Fungicide (Uniform 390 SE)	$35.5^{a} \pm 0.71$	$178.8^{a} \pm 1.32$	$32.9^{a} \pm 0.79$	$47.3^{a} \pm 0.90$	$6.7^{a} \pm 0.31$
Control	$23.4^{c} \pm 0.59$	$130.3^{b} \pm 1.21$	$23.2^{c} \pm 0.98$	$29.3^{c} \pm 0.47$	$3.3^{c} \pm 0.22$
			Giza Governorate		
Cu_2ONPs (0.35)	$35.7^{a} \pm 0.73$	$181.3^{b} \pm 1.34$	$31.2^{b} \pm 0.79$	$48.9^{a} \pm 0.76$	$6.1^{a} \pm 0.43$
Cu_2ONPs (0.30)	$36.1^{a} \pm 0.74$	$182.4^{a} \pm 1.43$	$33.2^{a} \pm 0.88$	$49.3^{a} \pm 0.56$	$6.4^{a} \pm 0.46$
Fungicide (Uniform 390 SE)	$36.5^{a} \pm 0.68$	$179.3^{b} \pm 1.44$	$33.9^{a} \pm 0.81$	$50.7^{a} \pm 0.67$	$6.3^{a} \pm 0.42$
Control	$24.4^{b} \pm 0.42$	$131.8^{c} \pm 1.07$	$25.1^{c} \pm 0.57$	$28.1^{b} \pm 0.37$	$3.5^{c} \pm 0.32$

Statistical comparisons were made among treatments within a single column. * The different letters represent significant differences using Fisher's LSD test at $p \leq 0.05$. * Each mean value came from three replicates.

3.9. Effect of Cu_2ONPs on Yield Parameter of Cucumber Plants under Field Conditions

The results in Table 7 indicated that the applied Cu_2ONPs (0.30 and 0.35 M) and fungicide increased the yield parameters of treated cucumbers, such as the number of fruits, fruits weight/plant, weight of fruits and rate of increased yield, under field conditions in the two locations compared to untreated control. The highest yield parameters in the treated cucumbers were for the chemical fungicide, followed by 0.30 and 0.35 M of Cu_2ONPs, respectively. With regard to the measured yield parameters in the two locations, there are approximately no significant differences between Cu_2ONPs (0.30) and the fungicide used. The measured yield parameters of the treated cucumbers were significantly higher in the cucumber plants treated with 0.30 M than with 0.35 M Cu_2ONPs in the two locations.

Table 7. Effect of Cu_2ONPs compared to the chemical fungicide on yield parameters of cucumber plants under field conditions in Menoufia and Giza Governorates.

Treatments	No. of Fruits	Mean Weight of Fruits (g)	Fruits Weight/Plant (kg)	% Rate of Yield Increase **
		Menoufia Governorate		
Cu_2ONPs (0.35)	$27.7^{b} \pm 0.78$ *	$73.4^{b} \pm 0.99$	$2.033^{b} \pm 0.34$	$96.4^{c} \pm 1.03$
Cu_2ONPs (0.30)	$28.3^{a} \pm 0.76$	$74.2^{a} \pm 0.95$	$2.099^{b} \pm 0.37$	$102.8^{b} \pm 1.11$
Fungicide (Uniform 390 SE)	$29.4^{a} \pm 0.79$	$74.8^{a} \pm 0.97$	$2.199^{a} \pm 0.29$	$112.5^{a} \pm 1.14$
Control	$14.3^{c} \pm 0.49$	$72.4^{c} \pm 0.93$	$1.035^{c} \pm 0.19$	$0.00^{d} \pm 0.78$

Table 7. *Cont.*

Treatments	No. of Fruits	Mean Weight of Fruits (g)	Fruits Weight/Plant (kg)	% Rate of Yield Increase **
		Giza Governorate		
Cu_2ONPs (0.35)	28.3 ^{b} ± 0.89	74.1 ^{a} ± 0.94	2.097 ^{b} ± 0.37	89.7 ^{c} ± 0.99
Cu_2ONPs (0.30)	29.4 ^{a} ± 0.87	74.9 ^{a} ± 0.98	2.202 ^{a} ± 0.43	99.3 ^{b} ± 0.94
Fungicide	29.7 ^{a} ± 0.84	75.2 ^{a} ± 1.04	2.233 ^{a} ± 0.34	102.1^a ± 1.05
Control (Uniform 390 SE)	15.4 ^{c} ± 0.53	71.8 ^{c} ± 0.97	1.105 ^{c} ± 0.29	0.0 ^{d} ± 1.01

Statistical comparisons were made among treatments within a single column. * The different letters represent significant differences using Fisher's LSD test at $p \leq 0.05$. Each mean value came from three replicates. ** Rate of yield increase = (fruit weight/plant (treatment) − fruit weight/plant (control))/(fruit weight/plant (control)) × 100.

Table 8. Pearson correlations between efficacy and total chlorophyll and growth parameters and yield parameters.

	Efficacy	Total Chlorophyll	Shoot Length	Root Length	Fresh Weight	Dry Weight	No. of Fruits	Mean Weight of Fruits	Fruits Weight/Plant	Rate of Yield Increase
Efficacy	1	0.978 **	0.995 **	0.972 **	0.989 **	0.914 **	0.998 **	0.923 **	0.998 **	0.991 **
Total chlorophyll		1	0.976 **	0.978 **	0.987 **	0.965 **	0.978 **	0.921 **	0.980 **	0.952 **
Shoot length			1	0.961 **	0.988 **	0.904 **	0.995 **	0.896 **	0.993 **	0.983 **
Root length				1	0.963 **	0.943 **	0.979 **	0.941 **	0.983 **	0.960 **
Fresh weight					1	0.917 **	0.986 **	0.911 **	0.986 **	0.968 **
Dry weight						1	0.914 **	0.922 **	0.920 **	0.888 **
No. of fruits							1	0.916 **	1.000 **	0.992 **
Mean weight of fruits								1	0.927 **	0.904 **
Fruits weight/plant									1	0.991 **
Rate of yield increase										1

** = highly significant.

3.10. Pearson Correlation

The relationship between the efficacy of the tested treatments and total chlorophyll, stem length (cm), root length (cm), fresh weight (g), dry weight (g), number of fruits, weight of fruits/plant, and rate of yield increase is indicated as shown in Table 8. The correlations between these variables were strongly positive and ranged between 0.904 and 0.998, which indicates the ability of these treatments under study to control this disease and increase growth and yield parameters.

4. Discussion

The pathogenic cucumber fungus *F. solani* is a common fungal genus that causes seed rot in cucumber seedlings, as well as pre-and post-emergency suppression of cucumber seedling production [54]. Cu_2ONPs were examined in this work to control *F. solani* in cucumbers under laboratory, greenhouse and field conditions, and the results revealed a considerable reduction in the disease's occurrence. In this investigation, copper oxide nanostructures were found to have promising antifungal action against *F. solani* under laboratory conditions. This is in agreement with [29], who found that copper oxide nanoparticles inhibited *F. solani* cultures significantly. Furthermore, the results agreed with those of Elmer and White [31], Elmer et al. [33], and Khatami et al. [29], who found that Cu_2ONPs had a high potential for controlling soil borne fungi, such as *F. solani* and *F. oxysporum*. Consolo et al. [55] showed that both Ag and Cu_2ONPs caused a significant reduction in the mycelia development of *A. alternata* and *P. oryzae* in a dose-dependent concentration. Moreover, copper compounds are still employed as fungicides to protect wood and prevent plant diseases [56,57]. From another point of view, copper oxide is a non-toxic, inorganic antimicrobial agent that inhibits the growth of a wide range of microbes [58–62]. The precise mechanism of copper oxide nanoparticles' antimicrobial activity is still unknown. Researchers have proven in some published studies that the antibacterial activity may be related to the inactivation of the DNA enzyme, and as a result, impedes replication and growth inhibition [29]. Generally, some researchers discovered that NPs operate directly as antibacterial agents, while others discovered that their main role is to change the host's nutritional status and activate defense mechanisms. Copper can be directly poisonous to microorganisms. Moreover, as fertilizer, Cu appears to contribute to host defense [33].

Accordingly, the obtained results show a significant reduction in the damping-off percentage and disease incidence in the cucumber plants treated with 0.30 and 0.35 M of Cu_2ONPs under greenhouse and field conditions (Menoufia and Giza Governorates) compared to the untreated plants. As explained by some researchers, Cu_2ONPs have high efficacy against several pathogens that cause damping-off and root rot, making them a viable alternative to fungicides in cucumber protection. It has a dual effect. The first is that it is directly toxic to diseases, and the second is that it can be used as a fertilizer, while also increasing the plant's natural defenses against pathogens. Due to their long-standing usage as contact bactericides and fungicides, CuNPs are also an essential choice in plant disease management [27,31,63].

The induction of host-plant resistance by Cu_2ONPs was confirmed by the results of enzyme activities (catalase, peroxidase, and polyphenoloxidase) and gene expression (*PR-1* and *LOX-1).* This was shown by the fact that Cu_2ONPs caused an increment in the activity of catalase, peroxidase, and polyphenol oxidase, as well as in the gene expression of *PR-1,* and *LOX-1.* Elmer [33] provided an explanation for this by demonstrating how the presence of both Cu_2ONPs and *F. oxysporum* f. sp. *niveum* strongly up-regulated the gene expression for *polyphenol oxidase* (*PPO*) and *PR1* in watermelon roots. The PPO enzyme assay results supported the gene expression findings. In order to effectively provide this micronutrient to fight disease, Cu_2ONPs may be used. According to Ashraf et al. [64], treatment with varying concentrations of CuO-CFNPs resulted in an upward trend in photosynthetic pigments, phenolic content, and stress/antioxidant enzymatic components. In addition, *PR-1* is a gene that is often expressed if SAR activity is stimulated [65]. Similarly, transcriptions were increased in Cu_2ONP treatments. The biosynthesis process of jasmonic acid, the phytohormone that controls ISR, begins with the synthesis of *LOX-1,* which is the first enzyme produced in this pathway [65].

The effect of Cu_2ONPs in controlling the disease was also confirmed using a light microscope, a scanning electron microscope, and anatomical characteristics. The significant effect of Cu_2ONPs was observed on mycelia and spores (abnormalities and alterations, twisting, plasmolysis, shrinking and collapsing) of *F. solani*. In addition, the cell wall, root cortex, and mesophyll tissue (MT) thickness were all increased with the treatment by Cu_2ONPs, in comparison to the untreated plants, where the cell wall was ruptured. According to Ashraf et al. [64], comparative exposure to larger concentrations has severe negative consequences on the mycelial surface, resulting in split, distorted, and collapsed structures with tiny vesicles, similar to polyps. It was proposed that Cu-NPs released Cu ions into the growth media, which could diffuse past the cell wall and bond with the surface of fungal cells. Our findings unmistakably demonstrate that fungal mycelia become distorted following treatment with Cu_2ONPs, which may be related to the interruption of chitin synthesis. Another interpretation revealed that the properties and chemical content of the nanoparticles, as well as their size and surface coating, when they interact with plants, cause several morphological and physiological changes [66].

These findings were in line with those of Elsharkawy et al. [67], who demonstrated that chitosan nanoparticles improved the thickness of the mesophyll tissue (MT), the thickness of the lower and upper epidermis (LE and UE), and the bundle length and width in the midrib in comparison to the control of treated wheat plants. Furthermore, Kim et al. [68] discovered that due to the disruption of the membrane integrity, *Candida albicans'* normal budding process was inhibited and the structure of the cell membrane was disrupted.

From another point of view, cucumber plants treated with Cu_2ONPs and a fungicide grew better and demonstrated a better yield in the Governorates of Menoufia and Giza, according to this study. Elmer and White [31] found that foliar spraying of CuONPs substantially enhances root content and fresh weight in eggplants, compared to untreated eggplant. Elmer [33] demonstrated that CuONP-treated plants produced and yielded 39% more fruit than the untreated controls. Furthermore, the administration of Cu-NPs aided mung bean roots and shoot growth. According to Yasmine et al. [69], when wheat was treated with 25 ppm Cu-NPs, the spike length of wheat climbed slowly or remained unchanged, while the number of grains increased dramatically.

Copper compounds have been successful in the control of crop diseases caused by certain fungi and bacteria, as their cost is low and the risks of emergence of resistant strains of pathogens to these compounds are few or low, due to these compounds having more than one site of toxic action [70]. Particularly, copper nanoparticles cause a change in protein expression, which is the key to inhibiting microbial growth [71]. Although silver compounds have the same toxicity as copper compounds on plant pathogens, Cu_2ONPs are more beneficial for use in the agricultural environment because they are less toxic than AgNPs. Cu_2ONPs also have numerous modes of inhibitory action for microbial diseases [72], allowing them to be used as an alternative to chemical fungicides to control a variety of plant pathogens that infect plants. This would be extremely useful in decreasing the negative effects of fungicides, particularly on edible plants and fresh vegetables [73]. We also hope, in the future, to increase the study of this compound in terms of toxicity and the adoption of international companies and for it to reach the final product for its application at the field level among farmers.

5. Conclusions

Cu_2ONPs inhibited the growth of *F. solani* in the laboratory, and they also reduced the disease incidence of the pathogen that causes cucumber root rot under greenhouse and field environments. They also improved cucumber growth and production characteristics. Defense enzyme activity and defense genes were expressed at higher levels in the cucumber plants treated with Cu_2ONPs than in the untreated control. In comparison to the untreated plants, scanning electron microscopy and the anatomical investigation indicated anomalies, alterations, twisting in the mycelia, diminishing spores, and the collapse of *F. solani*, as well as increased cell wall, root cortex, and mesophyll tissue (MT) thickness of the cucumber plants.

Author Contributions: Conceptualization, S.M.K., S.F.E., R.I.O., A.S.D. and M.M.E.; methodology, S.M.K., S.F.E., R.I.O., A.S.D., M.A., A.E.-S. and M.M.E.; software, S.M.K., S.F.E., R.I.O., A.S.D. and M.M.E.; validation, S.M.K., S.F.E., R.I.O., A.S.D., A.A., A.A.A.-A., K.A.A.-E., T.E., M.K. and M.M.E.; formal analysis, R.I.O., A.S.D., K.A.A.-E. and M.M.E.; investigation, S.M.K. and M.M.E.; data curation, S.M.K., S.F.E., R.I.O., A.S.D., T.E. and M.M.E.; writing—original draft preparation, S.M.K., S.F.E., R.I.O., A.S.D. and M.M.E.; writing—review and editing, S.M.K., S.F.E., R.I.O., A.S.D. and M.M.E.; visualization, S.M.K., S.F.E., R.I.O., A.S.D., A.A., A.A.A.-A., K.A.A.-E., M.K. and M.M.E., supervision, S.M.K., R.I.O., A.S.D. and M.M.E.; funding acquisition, A.A., A.A.A.-A., M.K. and M.M.E. All authors have read and agreed to the published version of the manuscript.

Funding: This research was financially supported by the Researchers Supporting Project number (RSP2022R505), King Saud University, Riyadh, Saudi Arabia.

Institutional Review Board Statement: Not applicable.

Informed Consent Statement: Not applicable.

Data Availability Statement: Not applicable.

Acknowledgments: The authors would like to extend their appreciation to the Researchers Supporting Project number (RSP2022R505), King Saud University, Riyadh, Saudi Arabia.

Conflicts of Interest: The authors declare no conflict of interest.

References

1. FAOSTAT. *Retrieved 2020-02-15. Countries-Select All; Regions-World + (Total); Elements-Production Quantity; Items-Cucumbers and Gherkins*; FAOSTAT: Rome, Italy, 2018.
2. Szalay, J. Cucumbers: Health Benefits and Nutrition Facts. 2017. Available online: https://www.livescience.com/51000-cucumber-nutrition.html (accessed on 22 January 2022).
3. Abd-El-Moneium, M.L. Integrated system to protect cucumber plants in greenhouses against diseases and pests under organic farming conditions. *Egypt. J. Agric. Res.* **2004**, *82*, 1–9.
4. Roberts, D.P.; Lohrke, S.M.; Meyer, S.L.; Chung, S. Biological agents applied individually or in combination for suppression of soil borne diseases of cucumber plants. *Crop. Prot.* **2005**, *24*, 135–141. [CrossRef]
5. Jinghua, Z.; Chang, W.; Xu, W.; Hanlian, W.; Shuge, T. Allelopathy of diseased survival on cucumber Fusarium wilt. *Acta Phytophylacica Sin.* **2008**, *35*, 317–321.
6. Elagamey, E.; Abdellatef, M.A.E.; Kamel, S.M.; Essa, T.A. *Fusarium oxysporum* isolates collected from the same geographical zone exhibited variations in disease severity and diversity in morphological and molecular characters. *Egypt. J. Phytopathol.* **2020**, *48*, 43–57. [CrossRef]
7. Cuervo-Parra, J.A.; Sánchez-López, V.; Ramírez-Suero, M.; Ramírez-Lepe, M. Morphological and molecular characterization of *Moniliophthora roreri* causal agent of frosty pod rot of cocoa tree in Tabasco, México. *Plant. Pathol. J.* **2011**, *10*, 122–127. [CrossRef]
8. Omara, R.I.; Kamel, S.M.; Hafez, Y.M.; Morsy, S.Z. Role of Non-traditional control treatments in inducing resistance against wheat leaf rust caused by *Puccinia Triticina*. *Egypt. J. Biol. Pest. Control.* **2015**, *25*, 335–344.
9. Omara, R.I.; Essa, T.A.; Khalil, A.A.; Elsharkawy, M.M. A case study of non-traditional treatments for the control of wheat stem rust disease. *Egypt. J. Biol. Pest. Control.* **2020**, *30*, 1–12. [CrossRef]
10. Krauss, U.; Hoopen, M.; Rees, R.; Stirrup, T.; Argyle, T.; George, A.; Arroyo, C.; Corrales, E.; Casanoves, F. Mycoparasitism by *Clonostachys byssicola* and *Clonostachys rosea* on *Trichoderma* spp. from cocoa (*Theobroma cacao*) and implication for the design of mixed biocontrol agents. *Biol. Control.* **2013**, *67*, 317–327. [CrossRef]
11. Mbarga, J.B.; Begoude, B.A.d.; Ambang, Z.; Meboma, M.; Kuate, J.; Schiffers, B.; Wwbank, W.; Dedieu, I.; Hoopen, G.M. A new oil-based formulation of *Trichoderma asperellum* for the biological control of cacao black pod disease caused by Phytophthora megakarya. *Biol. Control.* **2014**, *77*, 15–22. [CrossRef]
12. Kazemi, M.; Akbari, A.; Sabouri, Z.; Soleimanpour, S.; Zarrinfar, H.; Khatami, M.; Darroudi, M. Green synthesis of colloidal selenium nanoparticles in starch solutions and investigation of their photocatalytic, antimicrobial, and cytotoxicity effects. *Bioprocess Biosyst. Eng.* **2021**, *44*, 1215–1225. [CrossRef]
13. Kazemi, M.; Akbari, A.; Zarrinfar, H.; Soleimanpour, S.; Sabouri, Z.; Khatami, M.; Darroudi, M. Evaluation of antifungal and photocatalytic activities of gelatin-stabilized selenium oxide nanoparticles. *J. Inorg. Organomet. Polym.* **2020**, *30*, 3036–3044. [CrossRef]
14. Rostami, N.; Alidadi, H.; Zarrinfar, H.; Ketabi, D.; Tabesh, H. Interventional Effect of nanosilver paint on fungal load of indoor air in a hospital ward. *Can. J. Infect. Dis. Med. Microbiol.* **2021**, *20*, 8658600. [CrossRef] [PubMed]
15. Alghuthaymi, M.A.; Kalia, A.; Bhardwaj, K.; Bhardwaj, P.; Abd-Elsalam, K.A.; Valis, M.; Kuca, K. Nanohybrid antifungals for control of plant diseases: Current status and future perspectives. *J. Fungi.* **2021**, *7*, 48. [CrossRef] [PubMed]
16. Shenashen, M.; Derbalah, A.; Hamza, A.; Mohamed, A.; El Safty, S. Recent trend in controlling root rot disease of tomato caused by *Fusarium solani* using aluminasilica nanoparticles. *Int. J. Adv. Res. Biol. Sci.* **2017**, *4*, 105–119.
17. Derbalah, A.; Elsharkawy, M.M.; Hamza, A.; El-Shaer, A. Resistance induction in cucumber and direct antifungal activity of zirconium oxide nanoparticles against *Rhizoctonia solani*. *Pestic. Biochem. Physiol.* **2019**, *157*, 230–236. [CrossRef] [PubMed]
18. Derbalah, A.S.H.; Elsharkawy, M.M. A new strategy to control Cucumber mosaic virus using fabricated NiO-nanostructures. *J. Biotechnol.* **2019**, *306*, 134–141. [CrossRef] [PubMed]
19. Stoimenov, P.K.; Klinger, R.L.; Marchin, G.L.; Klabunde, K.J. Metal oxide nanoparticles as bactericidal agents. *Langmuir* **2002**, *18*, 6679–6686. [CrossRef]
20. Theivasanthi, T.; Alagar, M. Studies of copper nanoparticles effects on micro-Organisms. *Ann. Bio. Res.* **2011**, *2*, 368–373.
21. Chou, W.L.; Yu, D.G.; Yang, M.C. The preparation and characterization of silver-loading cellulose acetate hollow fiber membrane for water treatment. *Poly. Adv. Technol.* **2005**, *16*, 600–607. [CrossRef]
22. Sambhy, V.; MacBride, M.M.; Peterson, B.R.; Sen, A. Silver bromidenanoparticle/polymer composites: Dual action tunable antimicrobial materials. *J. Am. Chem. Soc.* **2006**, *128*, 9798–9808. [CrossRef]
23. Lee, C.; Kim, Y.; Lee, W.I.; Nelson, K.L.; Yoon, J.; Sedlak, D.L. Bactericidal effect of zero-valence iron nanoparticles on Escherichia coli. *Environ. Sci. Technol.* **2008**, *42*, 4927–4933. [CrossRef] [PubMed]
24. Gondal, M.A.; Dastageer, M.A.; Khalil, A.; Hayat, K.; Yamani, Z.H. Nanostructured ZnO synthesis and its application for effective disinfection Escherichia coli microorganism in water. *J. Nanopart. Res.* **2011**, *133*, 423–430.

25. Kavitha, T.; Yuvaraj, H. A facile approach to the synthesis of high-quality NiO nanorods: Electrochemical and antibacterial properties. *J. Mater. Chem.* **2011**, *21*, 15686–15691. [CrossRef]
26. Xiong, L.; Tong, Z.H.; Chen, J.J.; Li, L.L.; Yu, H.Q. Morphology-dependent antimicrobial activity of Cu/CuxO nanoparticles. *Ecotoxicology* **2015**, *24*, 2067–2072. [CrossRef] [PubMed]
27. Giannousi, K.; Sarafidis, S.; Mourdikoudis, G.; Pantazaki, A.; Dendrinou-Samara, C. Selective synthesis of Cu_2O and Cu/Cu_2O NPs: Antifungal activity to yeast saccharomyces cerevisiae and DNA interaction. *Inorg. Chem.* **2014**, *53*, 9657–9666. [CrossRef] [PubMed]
28. Huang, S.; Wang, L.; Liu, L.; Hou, Y. Nanotechnology in agriculture, livestock, and aquaculture in China. "A review". *Agron. Sustain. Dev.* **2015**, *35*, 369–400. [CrossRef]
29. Khatami, M.; Varma, R.; Heydari, M.; Peydayesh, M.; Sedighi, A.; Askari, H.A.; Rohani, M.; Baniasadi, M.; Arkia, S.; Seyedi, F.; et al. Copper oxide nanoparticles greener aynthesis using tea and its antifungal efficiency on *Fusarium solani*. *Geomicrobiol. J.* **2019**, *36*, 777–781. [CrossRef]
30. Evans, I.; Solberg, E.; Huber, D.M. *Mineral Nutrition and Plant*; Datnoff, L.E., Elmer, W.H., Huber, D.N., Eds.; APS Press: Paul, MN, USA, 2007.
31. Elmer, W.H.; White, J. Nanoparticles of CuO improves growth of eggplant and tomato in disease infested soils. *Environ. Sci. Nano* **2016**, *3*, 1072–1079. [CrossRef]
32. Borgatta, J.; Ma, C.; Hudson-Smith, N.; Elmer, W.; Pérez, C.D.P.; De La Torre-Roche, R.; Zuverza, M.N.; Haynes, C.L.; White, J.C.; Hamers, R.L. Copper nanomaterials suppress root fungal disease in watermelon (*Citrullus lanatus*): Role of particle morphology, composition, and dissolution behavior. *CS Sustain. Chem. Eng.* **2018**, *6*, 14847–14856. [CrossRef]
33. Elmer, W.H.; De La Torre-Roche, R.; Pagano, L.; Majumdar, S.; Zuverza-Mena, N.; Dimpka, C.; Gardea-Torresdey, J.; White, W. Effect of metalloid and metallic oxide nanoparticles on Fusarium wilt of watermelon. *Plant. Dis.* **2018**, *102*, 1394–1401. [CrossRef]
34. Nelson, E.B.; Boehm, M.J. Compost-induced suppression of turf grass diseases. *Bio. Cycle* **2002**, *43*, 51.
35. Ismail, W.; El-Shafai, N.M.; El-Shaer, A.; Abdelfatah, M. Impact of substrate type on the surface and properties of electrodeposited Cu_2O nanostructure films as an absorber layer for solar cell applications. *Mater. Sci Semicond. Process.* **2020**, *120*, 105335. [CrossRef]
36. Booth, C. *The Genus Fusarium*, 2nd ed.; Commonwealth Mycological Institute: Surrey, UK, 1971; p. 237.
37. Allen, O.N. *Experiments in Soil Bacteriology*; Burgess Publishing Co.: Minneapolis, MN, USA, 1950.
38. Shaban, W.I.; El-Bramawy, M.A. Impact of dual inoculation with *Rhizobium* and *Trichoderma* on damping-off, root rot diseases and plant growth parameters of some legumes field crop under greenhouse conditions. *Int. Res. J. Agric. Sci. Soil. Sci.* **2011**, *1*, 98–108.
39. Hwang, S.F.; Chang, K.F. Incidence and severity of root rot disease complex of field pea in northern Alberta. *Can. Plant Dis Surv* **1989**, *69*, 139–141.
40. Aebi, H. Catalase in vitro. *Methods Enzym.* **1984**, *105*, 121–126.
41. Hammerschmidt, R.; Nuckles, E.; Kuć, J. Association of enhanced peroxidase activity with induced systemic resistance of cucumber to *Colletotrichum Lagenarium*. *Physiol. Plant Pathol.* **1982**, *20*, 73–82. [CrossRef]
42. Malik, C.P.; Singh, M.B. Plant Emynology and Histoenzymology, Kalyani Publishers. *Indian Print. Navin. Shanndara Delhi* **1980**, 54–56.
43. Tek, M.I.; Calis, O. Mechanisms of resistance to powdery mildew in cucumber. *Phytopathol. Mediterr.* **2022**, *61*, 119–127. [CrossRef]
44. Livak, K.J.; Schmittgen, T.D. Analysis of relative gene expression data using real-time quantitative PCR and the 2−CT method. *Methods* **2001**, *25*, 402–408. [CrossRef]
45. Harley, M.M.; Ferguson, I.K. The role of the SEM in pollen morphology and plant systematics. In *Scanning Electron Microscopy in Taxonomy and Functional Morphology*; Oxford University Press: Oxford, England, 1990; pp. 45–68.
46. Nassar, M.A.; El-Sahar, K.F. *Botanical Preparations and Microscopy (Micro Technique)*; Academic Bookshop: Giza Governorate, Egypt, 1998; p. 219.
47. Torres-Netto, A.; Campostrini, E.; Oliveira, J.G.; Smith, R.E.B. Photosynthetic pigments, nitrogen, chlorophyll a fluorescence and SPAD-502 readings in coffee leaves. *Sci. Hortic.* **2005**, *104*, 199–209. [CrossRef]
48. El-Shaer, A.; Ismail, W.; Abdelfatah, M. Towards low cost fabrication of inorganic white light emitting diode based on electrodeposited Cu_2O thin film/TiO_2 nanorods heterojunction. *Mater. Res. Bull.* **2019**, *116*, 111–116. [CrossRef]
49. Abdelfatah, M.; Ismail, W.; El-Shaer, A. Low cost inorganic white light emitting diode based on submicron ZnO rod arrays and electrodeposited Cu_2O thin film. *Mater. Sci. Semicond. Process.* **2018**, *81*, 44–47. [CrossRef]
50. El-Shafai, N.M.; Abdelfatah, M.; El-Mehasseb, I.M.; Ramadan, M.S.; Ibrahim, M.M.; El-Shaer, A.; El-Kemary, M.A.; Masoud, M.S. Enhancement of electrochemical properties and photocurrent of copper oxide by heterojunction process as a novel hybrid nanocomposite for photocatalytic anti-fouling and solar cell applications. *Sep. Purif. Technol.* **2021**, *267*, 118631. [CrossRef]
51. Abdelfatah, M.; Salah, H.; El-Henawey, M.; Oraby, A.; El-Shaer, A.; Ismail, W. Insight into co concentrations effect on the structural, optical, and photoelectrochemical properties of ZnO rod arrays for optoelectronic applications. *J. Alloy Compd.* **2021**, *873*, 159875. [CrossRef]
52. El-Shaer, A.; Abdelfatah, K.R.; Mahmoud, S.; Eraky, M. Correlation between photoluminescence and positron annihilation lifetime spectroscopy to characterize defects in calcined MgO nanoparticles as a first step to explain antibacterial activity. *J. Alloy Compd.* **2020**, *817*, 152799. [CrossRef]

53. El-Shafai, N.; Shukry, M.M.; El-Mehasseb, I.M.; Abdelfatah, M.; Ramadan, M.S.; El-Shaer, A.; El-Kemary, M. Electrochemical property, antioxidant activities, water treatment and solar cell applications of titanium dioxide-zinc oxide hybrid nanocomposite based on graphene oxide nanosheet. *Mater. Sci. Eng. B-Adv.* **2020**, *259*, 114596. [CrossRef]
54. Ziedan, E.H.E.; Moataza, S. Efficacy of nanoparticles on seed borne fungi and their pathological potential of cucumber. *Inter. J. PharmTech. Res.* **2016**, *9*, 16–24.
55. Consolo, V.F.; AndrésTorres-Nicolini, A.T.; Alvarez, V.A. Mycosinthetized Ag, CuO and ZnO nanoparticles from a promising *Trichoderma harzianum* strain and their antifungal potential against important phytopathogens. *Nat. Res.* **2020**, *10*, 1–10. [CrossRef] [PubMed]
56. Franich, R.A. Chemistry of weathering and solubilisation of copper fungicide and the effect of copper on germination, growth, metabolism, and reproduction of *Dothistroma pini*. *New Zealand J. For. Sci.* **1988**, *18*, 318–328.
57. Freeman, M.H.; McIntyre, C.R. A comprehensive review of cooper-based wood preservatives with a focus on new micronized or dispersed copper systems. *For. Prod. J.* **2008**, *58*, 6–27.
58. Jamdagni, P.; Khatri, P.; Rana, J.S. Green synthesis of zinc oxide nanoparticles using flower extract of *Nyctanthes arbor-tristis* and their antifungal activity. *J. King Saud. Univ. Sci.* **2018**, *30*, 168–175. [CrossRef]
59. Babaei, S.; Bajelani, F.; Mansourizaveleh, O.; Abbasi, A.; Oubari, F. A study of the bactericidal effect of copper oxide nanoparticles on *Shigella sonnei* and *Salmonella typhimurium*. *J. Babol. Univ. Med. Sci.* **2017**, *19*, 76–81.
60. Hamedi, S.; Shojaosadati, S.A.; Mohammadi, A. Evaluation of the catalytic, antibacterial and anti-biofilm activities of the Convolvulus arvensis extract functionalized silver nanoparticles. *J. Photochem. Photobiol. B* **2017**, *167*, 36–44. [CrossRef] [PubMed]
61. Karthik, K.; Dhanuskodi, S.; Gobinath, C.; Prabukumar, S.; Sivaramakrishnan, S. Photocatalytic and antibacterial activities of hydrothermally prepared cdo nanoparticles. *J. Mater. Sci.* **2017**, *28*, 11420–11429. [CrossRef]
62. Khorrami, M.B.; Sadeghnia, H.R.; Pasdar, A.; Ghayour-Mobarhan, M.; RiahiZanjani, B.; Darroudi, M. Role of pullulan in preparation of ceria nanoparticles and investigation of their biological activities. *J. Mol. Struct.* **2018**, *1157*, 127–131. [CrossRef]
63. Li, Y.; Yang, D.; Cui, J. Graphene oxide loaded with copper oxide nanoparticles as an antibacterial agent against *Pseudomonas syringae* pv. *tomato*. *RSC Adv.* **2017**, *7*, 38853–38860. [CrossRef]
64. Ashraf, H.; Anjum, T.; Riaz, S.; Ahmad, I.S.; Irudayaraj, J.; Javed, S.; Qaiserf, U.; Shahzad Naseem, S. Inhibition mechanism of green-synthesized copper oxide nanoparticles from *Cassia fistula* towards *Fusarium oxysporum* by boosting growth and defense response in tomatoes. *Env. Sci. Nano* **2021**, *8*, 1729–1748. [CrossRef]
65. Spoel, S.H.; Dong, X. How do plants achieve immunity? Defence without specialized immune cells. *Nat. Rev. Immunol.* **2012**, *12*, 89–100. [CrossRef]
66. Khodakovskaya, M.V.; de Silva, K.; Biris, A.S.; Dervishi, E.; Villagarcia, H. Carbon nanotubes induce growth enhancement of tobacco cells. *Am. Chem. Soc. Nano* **2012**, *3*, 2128–2135. [CrossRef]
67. Elsharkawy, M.M.; Omara, R.I.; Mostafa, Y.S.; Alamri, S.A.; Hashem, A.; Alrumman, S.; Ahmed, A.A. Mechanism of wheat leaf rust control using chitosan nanoparticles and salicylic acid. *J. Fungi* **2022**, *8*, 304. [CrossRef] [PubMed]
68. Kim, S.W.; Kim, K.S.; Lamsal, K.; Kim, Y.J.; Kim, S.B.; Jung, M.; Sim, S.J.; Kim, H.S.; Chang, S.J.; Kim, H.K.; et al. An in vitro study of the antifungal effect of silver nanoparticles on oak wilt pathogen Raffaelea sp. *J. Microbio.l Biotechnol.* **2009**, *19*, 760–764.
69. Yasmeen, F.; Raja, N.I.; Razzaq, A.; Komatsu, S. Proteomic and physiological analyses of wheat seeds exposed to copper and iron nanoparticles. *Biochim. Biophys. Acta* **2017**, *1865*, 28–42. [CrossRef] [PubMed]
70. Keller, A.A.; Huang, Y.; Nelson, J. Detection of nanoparticles in edible plant tissues exposed to nano-copper using single-particle ICP-MS. *J. Nanopart. Res.* **2018**, *20*, 1–13. [CrossRef]
71. Su, Y.; Zheng, X.; Chen, Y.; Li, M.; Liu, K. Alteration of intracellular protein expressions as a key mechanism of the deterioration of bacterial denitrification caused by copper oxide nanoparticles. *Sci. Rep.* **2015**, *5*, 15824. [CrossRef]
72. Goyal, P.; Chakraborty, S.; Misra, S.K. Multifunctional Fe_3O_4-ZnO nanocomposites for environmental remediation applications. *Environ. Nanotechnol. Monit. Manag.* **2018**, *10*, 28–35. [CrossRef]
73. Dreyer, D.R.; Park, S.; Bielawski, C.W.; Ruoff, R.S. The chemistry of graphene oxide. *Chem. Soc. Rev.* **2010**, *39*, 228–240. [CrossRef]

Journal of
Fungi

Article

Polyphenol-Capped Biogenic Synthesis of Noble Metallic Silver Nanoparticles for Antifungal Activity against *Candida auris*

Maqsood Ahmad Malik [1,*], Maha G. Batterjee [1], Majid Rasool Kamli [2,3], Khalid Ahmed Alzahrani [1], Ekram Y. Danish [1] and Arshid Nabi [4]

1 Chemistry Department, Faculty of Science, King Abdulaziz University, P.O. Box 80203, Jeddah 21589, Saudi Arabia; mbatterjee@kau.ed.sa (M.G.B.); kaalzahrani1@kau.edu.sa (K.A.A.); eydanish@kau.edu.sa (E.Y.D.)
2 Department of Biological Sciences, Faculty of Sciences, King Abdulaziz University, P.O. Box 80203, Jeddah 21589, Saudi Arabia; mkamli@kau.edu.sa
3 Center of Excellence in Bionanoscience Research, King Abdulaziz University, Jeddah 21589, Saudi Arabia
4 Department of Chemistry, University of Malaya, Kuala Lumpur 50603, Malaysia; arshidpharmachem@gmail.com
* Correspondence: mamalik@kau.edu.sa

Abstract: In terms of reduced toxicity, the biologically inspired green synthesis of nanoparticles has emerged as a promising alternative to chemically fabricated nanoparticles. The use of a highly stable, biocompatible, and environmentally friendly aqueous extract of *Cynara cardunculus* as a reducing and capping agent in this study demonstrated the possibility of green manufacturing of silver nanoparticles (CC-AgNPs). UV–visible spectroscopy validated the development of CC-AgNPs, indicating the surface plasmon resonance (SPR) λ_{max} band at 438 nm. The band gap of CC-AgNPs was found to be 2.26 eV. SEM and TEM analysis examined the surface morphology of CC-AgNPs, and micrographs revealed that the nanoparticles were spherical. The crystallinity, crystallite size, and phase purity of as-prepared nanoparticles were confirmed using XRD analysis, and it was confirmed that the CC-AgNPs were a face-centered cubic (fcc) crystalline-structured material. Furthermore, the role of active functional groups involved in the reduction and surface capping of CC-AgNPs was revealed using the Fourier transform infrared (FTIR) spectroscopic technique. CC-AgNPs were mostly spherical and monodispersed, with an average size of 26.89 nm, and were shown to be stable for a longer period without any noticeable change at room temperature. Further, we checked the antifungal mechanism of CC-AgNPs against *C. auris* MRL6057. The minimum inhibitory concentrations (MIC) and minimum fungicidal concentrations (MFC) were 50.0 µg/mL and 100.0 µg/mL respectively. The cell count and viability assay confirmed the fungicidal potential of CC-AgNPs. Further, the analysis showed that CC-AgNPs could induce apoptosis and G2/M phase cell cycle arrest in *C. auris* MRL6057. Our results also suggest that the CC-AgNPs were responsible for the induction of mitochondrial toxicity. TUNEL assay results revealed that higher concentrations of CC-AgNPs could cause DNA fragmentation. Therefore, the present study suggested that CC-AgNPs hold the capacity for antifungal drug development against *C. auris* infections.

Keywords: green synthesis; polyphenols; cell cycle; *Candida auris*

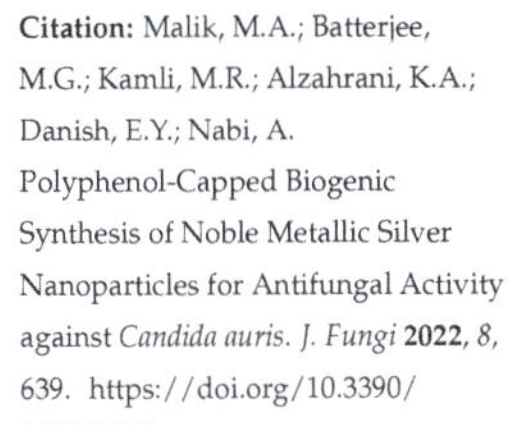

Citation: Malik, M.A.; Batterjee, M.G.; Kamli, M.R.; Alzahrani, K.A.; Danish, E.Y.; Nabi, A. Polyphenol-Capped Biogenic Synthesis of Noble Metallic Silver Nanoparticles for Antifungal Activity against *Candida auris*. *J. Fungi* **2022**, *8*, 639. https://doi.org/10.3390/jof8060639

Academic Editor: Kamel A. Abd-Elsalam

Received: 20 April 2022
Accepted: 7 June 2022
Published: 16 June 2022

1. Introduction

Green nanotechnology is a fast-emerging science with potential applications in the pharmaceutical, healthcare, biomedical, and drug delivery fields [1–3]. It was reported that a variety of metallic nanoparticles, including gold and silver nanomaterials, are being developed for use in a wide range of scientific applications [4]. Because of their excellent antioxidant and antibacterial capabilities, plasmonic silver nanoparticles (AgNPs) have recently received a lot of attention [5,6]. The surface plasmon resonance (SPR) of metal nanoparticles makes them interesting because of their applications in photocatalysis, sensors, biodevices, drug storage and loading, antimicrobial activity, and spectroscopic applications [7–11]. The SPR of the metal nanoparticles depends on the shape, size, and surrounding dielectric medium, as SPR is the resonant oscillation of conduction electrons under appropriate light illumination [12–14]. Silver nanoparticles are well-known noble metallic materials with strong antimicrobial and photocatalytic properties because of their high sensitivity, chemical stability, and better light absorption and optical properties [15–17].

The synthesis of metal nanoparticles involves various physical and chemical routes, which are quite expensive, require high energy, and have various toxicity issues associated with these approaches [18], therefore new cost-efficient, non-toxic, and eco-friendly synthesis techniques were adopted [19]. Bioactive substances, such as plant materials and microbes, and biowastes, such as vegetable waste, fruit peel trash, eggshell, and agricultural waste; can be used to synthesize different metal nanoparticles [20]. The restrictions of synthetic approaches are overwhelmed using green chemistry methods, which are economical and require less time to synthesize nanoparticles. Hence, many researchers performed the synthesis of nanoparticles via a green chemistry approach [21–23]. The metabolites present in the plant extract play a significant role in the reduction, nucleation, growth, stability, and capping of the silver nanoparticles [24]. The reducing capacity of plant extracts depends on water-soluble phenolic compounds, which have a key function in the reduction of Ag ions [24]. The method used to prepare metallic nanoparticles, nature of the solvent, mixing ratio, concentration, pH, temperature of the reaction mixture, and strength of the reducing agent are all key factors that influence the size, morphology, and stability of the nanoparticles [25–27]. Furthermore, Kim et al. deduced the concentration-dependent inhibitory cytotoxicity against Escherichia coli and Staphylococcus aureus using silver nanoparticles within a range of 13.5 nm [28]. Pauksch et al. studied cell proliferation, viability, and bone-forming cells upon incubation with AgNPs over time and in a dose-dependent manner [29]. Further, emphasis on AgNPs in biomaterials may lead to decreased cytotoxicity due to the possible reduced chance of AgNP cellular uptake; meanwhile, a window may open for future AgNP clinical and pharmaceutical applications in real-time medicinal practice. Along similar lines, the biocompatibility of biogenic AgNPs was investigated by P. Kumar Panda et al. in zebrafish embryos. However, both computational and experimental analysis was utilized in a concentration-dependent manner, the AgNPs enhanced oxidative stress accumulation and internalization depending on an intrinsic atomic interaction with the proteins, including sod1, tp53, and apoa1-mttp. In addition, it was ascertained that the biogenic AgNPs developed from silver grass were significantly biocompatible and eco-compatible and could be used for biomedical and ecological applications [30]. Silver nanoparticles, those developed via biogenic essence are proven to be biocompatible, such as oligodynamic characteristics of such biogenic NPs have been explored for thousands of years ago. In particular, cups made up of silver were used as a therapeutic agent in the Roman Empire [31]. It is worth mentioning that based on the inherent microbial inhibition against fungi and bacteria on the surface of AgNPs, this makes them a comparatively efficient antimicrobial candidate relative to other biogenic metal nanoparticles [32–34]. The biocompatibility of AgNPs was further ascertained after the continuous release of small amounts of silver ions from the surface of AgNPs, which was responsible for the inhibition of bacterial growth on the surface of nanoparticles, as well as on the metal surface. In real life medical applications, AgNPs are being used in medical operations, including impregnated catheters and in wound dressings [35]. Moreover, the AgNPs are being used as highly antibacterial agents nowadays and show potent efficacy as antimicrobial agents at concentrations $\leq$10 µg/g and retain an efficient potency against biofilm formation, as reported in previous studies [28,36–40]. Keeping the biogenic and biocompatible yield of polyphenol-capped silver nanoparticles, along with their inherent antimicrobial properties, in mind, here we aimed to investigate the antifungal activities against *C. auris* strains.

Silver nanoparticles synthesized using green chemistry approaches show antioxidant [41], antibacterial, and anti-inflammatory properties [42]. Recently, bloodstream infections caused by *Candida auris* have been spreading and widely reported in different parts of the globe [43]. This species of Candida was initially reported in 2009 in Japan [44] as an evolving multidrug-resistant (MDR) yeast pathogen and is mainly responsible for septicemia, resulting in a high rate of mortality. The spread of *C. auris* is identified as a risk in healthcare units and leads to outbreaks. Additionally, unlike other species of *Candida*, this pathogen can persist and flourish for a long time on both dry and moist surfaces in clinics and hospitals [45,46]. The MDR property of *C. auris* was described [47] and the scenario becomes further complicated by the formation of biofilms [48] and active efflux pumps. Therefore, there is a need to look for new and efficient antifungal strategies to combat this evolving yeast pathogen and prevent nosocomial outbreaks. Considering the importance of silver nanoparticles, the present work deals with the *Cynara cardunculus* extract assisted preparation of silver nanoparticles using a simple, nontoxic, and economical approach. The structural properties of CC-AgNPs were investigated using different spectroscopic and microscopic techniques to determine the surface morphology, elemental composition, crystallinity, and optical properties. Further, the present study aimed to investigate the antifungal activities of CC-AgNPs against *C. auris* strains.

2. Materials and Methodology

2.1. Materials

Cynara cardunculus, commonly known as artichoke, was collected from a local market in Jeddah, Saudi Arabia. Silver nitrate ($AgNO_3$, ≥99.0%) and ethanol (CH_3CH_2OH, 95.0%) were purchased from Sigma–Aldrich, St. Louis, MO, USA. All the chemicals used in this study were of analytical grade and were used without additional treatment. Highly pure double-distilled water (DDW) was utilized for the preparation of silver precursor and *Cynara cardunculus* solution.

2.2. Preparation of *Cynara cardunculus* Extract

The collected *Cynara cardunculus* were washed several times with distilled water, dried until the moisture was completely removed, and then ground into a fine powder. The *Cynara cardunculus* powder (10 g) was dispersed in an Erlenmeyer flask containing 250 mL distilled water and further heated at 80 °C for 60 min to achieve the completed extraction of biomolecules. The unfiltered solution was kept at room temperature for 12 h, and after that, the resulting extract was filtered through Whatman filter paper No. 1 using vacuum filtration apparatus. The filtered aqueous solution of *Cynara cardunculus was* stored in a refrigerator at 4 °C for further experimental use. It is recommended that fresh *Cynara cardunculus* extract (no more than 5 days after extraction) is used for synthesizing CC-AgNPs.

2.3. Preparation and Physicochemical Characterization of CC-AgNPs

The preparation of the silver nanoparticles was initiated by optimizing the amount of *Cynara cardunculus* extract required for the synthesis of CC-AgNPs. After several optimizing experiments, 14 mL of *Cynara cardunculus* extract was added to 20 mL of 1.4×10^{-4} M silver nitrate solution under continuous stirring using a magnetic stirrer. The color of the reaction mixture changed after just 5 min of reaction time and was analyzed using a double-beam Thermo Scientific Evolution 300 UV–visible spectrophotometer. The *Cynara-cardunculus*-mediated CC-AgNPs were purified, and the precipitated pellets were collected by using a BIOBASE centrifuge at a centrifugation speed of 5000 rpm for 20 min. The acquired pellets were then dispersed in distilled water and successively washed several times to completely remove the unbound compounds from the surface of the CC-AgNPs. Furthermore, the obtained material was subsequently dried at 90 °C for 5 h and then calcined at 500 °C for 3 h in a muffle furnace to remove all surface impurities and increase the crystallinity.

The successful reduction of the silver metal ions using *Cynara cardunculus* extract was initially validated by recording the absorbance of the reaction mixture in the wavelength range of 200–800 nm using a double-beam Thermo Scientific Evolution 300 UV–visible spectrophotometer (Thermo Fisher Scientific, Waltham, MA, USA). All spectra were recorded at room temperature, in a quartz cuvette cell (path length 1 cm). A powder X-ray diffractometer (XRD) (D8 Advance, Bruker, Karlsruhe, Germany) set to 40 kV and 40 mA with 1.54 Å CuKα radiation was used to acquire the XRD pattern of the as-prepared CC-AgNPs in the scan range of 20–80 θ. Fourier transform infrared spectroscopy (FTIR) analysis of CC-AgNPs was performed on a Bruker ALPHA II FT-IR (Bruker Optics GmbH & Co., Rosenheim, Germany) spectrometer to assess the possible involvement of the functional groups in the *Cynara cardunculus* extract in the reduction and stabilization/capping of the CC-AgNPs. Transmission electron microscopy (TEM) (JOEL, JEM-2100F, Tokyo, Japan; accelerating voltage of 200 kV) measurements were performed to analyze the morphology and the particle size distribution of the CC-AgNPs. Scanning electron microscopy (SEM) (ZEISS-SEM, Oberkochen, Germany) equipped with an energy dispersive spectroscopy (EDS) was used to investigate the surface morphology and the elemental composition of the CC-AgNPs. Malvern Zetasizer (Malvern Panalytical Ltd., Enigma Business Park, Malvern, UK) examined the zeta potential and the particle size distribution of the CC-AgNPs. The thermal stability of the as-prepared CC-AgNPs was analyzed using thermogravimetric analysis (TGA) in the temperature range of 30–800 °C under a N2 atmosphere with a heating rate of 10 °C/min using a Perkin-Elmer Pyris Diamond thermogravimetric analyzer (PerkinElmer LAS (UK)Ltd., Llantrisant, UK).

2.4. Antifungal Activity of CC-AgNPs

In the present study, the *C. auris* clinical strain MRL6057 was used. The strain was obtained from the National Institute of Communicable Diseases (NICD), South Africa, and preserved in the department as a glycerol stock. The antifungal action of CC-AgNPs was evaluated against *C. auris* MRL6057 by using a broth microdilution assay recommended in the standard M27 document (fourth ed.) [49]. The concentrations used for the test NPs and positive control/amphotericin B (AmB; Sigma-Aldrich, St. Louis, MO, USA) were 200–0.19 μg/mL and 16–0.031 μg/mL, respectively. Before reading the MIC values, which are the minimum concentration of the compound/drug that inhibited the yeast growth, all the plates were kept at 37 °C for 48 h. Later, the minimum fungicidal concentration (MFC) was estimated by further growing the cells from each well at 37 °C for 24 h on Sabouraud dextrose agar (SDA; Merck, Darmstadt, Germany). Again, the lowest concentration with less than five colonies on the agar plate was recorded as the MFC.

2.5. Effect on Cell Viability and Count

The candidacidal phenomenon of CC-AgNPs was quantified using the count and viability kit provided by Muse^{TM}. Briefly, yeast cells were grown for 8–10 h in Sabouraud dextrose broth (SDB) (Merck (Pty) Ltd., Johannesburg, South Africa) followed by centrifugation (3000 rpm, 5 min) and resuspension in fresh growth media. The yeast cells were adjusted to a density of 1×10^6 CFU/mL and subjected to various strengths of test NPs (0.5 MIC, MIC, and 2 MIC) for 4 h. Afterward, the yeast cells were washed and mixed with a Muse^{TM} kit reagent (20 μL of yeast cells + 380 μL reagent), then incubated for 5 min at room temperature. The viability and cell count were estimated using a Muse^{TM} cell analyzer. The experiment included a negative and positive control (H_2O_2, 10 mM; Merck, Darmstadt, Germany).

2.6. Effect on Cell Cycle

The impact of NPs on the yeast cell cycle was investigated using a Muse^{TM} cell cycle kit following the steps given by the manufacturer. Briefly, the cells were propagated for 8–10 h in a fresh medium (SDB) and then centrifuged at 3000 rpm for 4 min. Then, the cells were re-suspended in a fresh medium and the turbidity was adjusted to 1×10^6 CFU/mL. Later, the cells were subjected to various strengths of CC-AgNPs (0.5 MIC, MIC, and 2 MIC) for 4 h. In post-incubation, the cells were washed and fixed in chilled 70% ethyl alcohol (Sigma Aldrich Co., St. Louis, MO, USA), mixed with cell cycle reagent in equal proportions, and incubated in the dark for 30 min. The experiment included both negative and positive control.

2.7. Effect on Mitochondrial Membrane Potential

The impact of test NPs on the mitochondrial membrane potential of *C. auris* was measured using a JC-10 assay kit (Abcam, Cambridge, UK). The experiment was done using the steps given by the manufacturer. The cells (mid-log phase) were subjected to various concentrations of test NPs (0.5 MIC, MIC, and 2 MIC) for 4 h. They were subjected to protoplast preparation, as described previously by Lone et al. [50]. Then, 90 μL of *C. auris* protoplasts were mixed with 50 μL JC-10 dye and distributed in different wells of a 96-well microtiter plate (clear bottom-black walled; Thermo Fisher Scientific, Dreieich, Germany) for 1 h in the dark. After that, 50 μL of buffer-B was added to the plate and centrifuged for 2 min at 800 rpm. The readings were captured at Ex/Em = 490/530 nm (X) and 540/590 nm (Y) in microplate readers (Molecular Devices, San Jose, CA, USA). The variation was measured in terms of the Y/X ratio. A decreased ratio confirmed the depolarization of the mitochondrial membrane. Moreover, the experiment included both negative and positive control.

2.8. Effect on DNA Fragmentation

The DNA fragmentation and condensation in *C. auris* MRL6057 due to CC-AgNPs were examined using a terminal deoxynucleotidyl transferase dUTP nick-end labeling (TUNEL) assay. The protocol used the In Situ Cell Death Detection Kit, Fluorescein (Roche Diagnostics, Mannheim, Germany), and the instructions provided by the manufacturer were followed. Briefly, *C. auris* cells were subjected to various concentrations of NPs (0.5 MIC, MIC, and 2 MIC) for 4 h and were subjected to protoplast preparation. Later, Triton X-100 (0.25%) was used to permeabilize the protoplasts, followed by incubation at 37 °C for 20 min. Later, the cells were mixed with TUNEL reagent and incubated at 37 °C for 1 h in a dark humidified box. Subsequently, samples were examined with fluorescence microscopy at Ex/Em = 495/519 nm (Carl Zeiss Microscopy, Jena, Germany). The experiment included both negative and positive control.

2.9. Haemolytic Activity

The cytotoxic potential of various concentrations of given NPs (0.5MIC, MIC, and 2MIC) was evaluated on horse RBCs (NHLS, Johannesburg, South Africa) and stated as percent hemolysis. The method was adopted from a method reported elsewhere [50]. One percent Triton X-100 was considered as the positive control, and sterile PBS solution was considered the negative control. The calculation for percent hemolysis was as follows:

$$\%\ Hemolysis = \frac{[(A450\ of\ treated\ sample) - (A450\ of\ negative\ control)]}{[(A450\ of\ positive\ control) - (A450\ of\ negative\ control)]} \times 100 \quad (1)$$

Experiments were repeated thrice, and a two-way ANOVA test was used for determining the statistical significance of the results. Additionally, p-values ≤ 0.05 were measured statistically.

3. Results and Discussion

The green and sustainable fabrication of metal nanomaterials using polyphenolic compounds has attracted huge attention because it has more advantages over the chemical route synthesis of nanomaterials [51,52]. Here, we demonstrated a simple polyphenol-capped biogenic reduction method to prepare silver nanoparticles using *Cynara cardunculus* extract as a reducing and capping/stabilizing agent. The incorporation of *Cynara cardunculus* extract in silver nitrate aqueous solution resulted in a gradual color change of the reaction mixture from pale yellow to brown and finally to a deep brown, which implied the formation of stable CC-AgNPs. The existing water-soluble polyphenolic molecules in *Cynara carduncculus* extract successfully reduced the available silver metal ion (Ag^{+}) to metallic silver nanoparticles (Ag^{0}) [53]. The bioreduction of metal ions to metallic nanoparticles takes place through an activation step, which involves the reduction of available silver metal ions followed by the nucleation and growth steps in which metal nanoparticles of definite shapes and sizes are formed [54]. Finally, the stabilization step takes place, in which the *Cynara cardunculus* extract metabolites similarly play the role of surface-capping agents to prevent the nanoparticles from agglomerating [54]. The proposed mechanism of CC-AgNPs formation via *Cynara cardunculus* extract is depicted in systematic Figure 1.

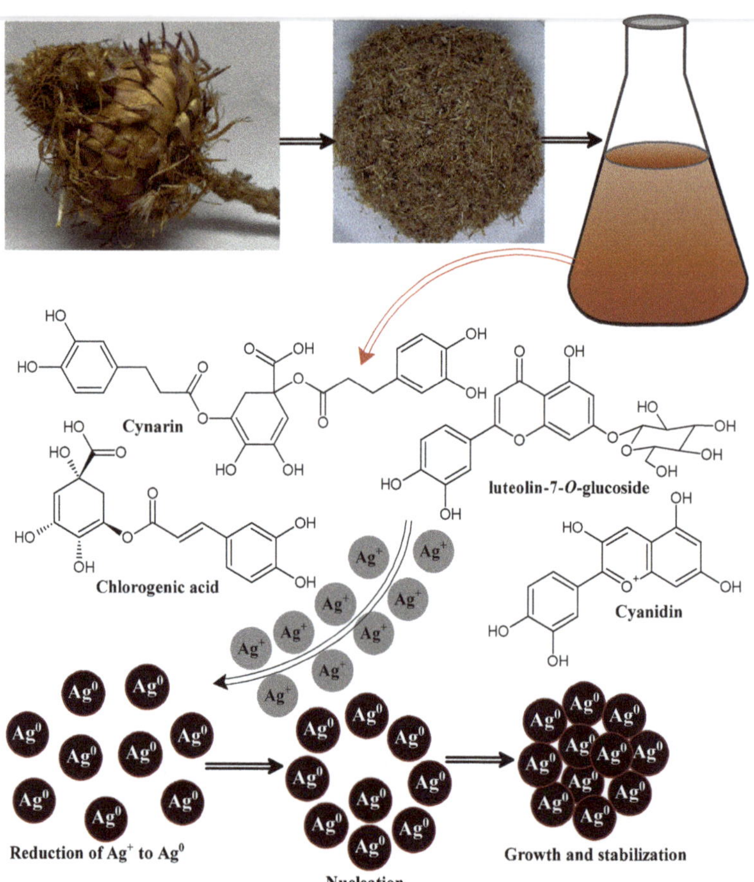

Figure 1. Schematic diagram showing the possible mechanism after the biosynthesis of CC-AgNPs using *Cynara cardunculus* extract.

Cynara cardunculus, commonly named the artichoke plant, originated in southern Europe, and belongs to the family Asteraceae, which includes daisies and sunflowers, and is mostly cultivated as a horticultural crop in Italy [55–57]. The different parts of this plant possess the phytochemical composition, including (i) phenolic acid derivatives: mono and di-caffeoylquinic acid compounds [58,59] and neochlorogenic and chlorogenic acids [60,61]; (ii) flavonoids: luteolin, luteolin-7-O-glycoside, and luteolin-7-O-rutinoside [62]; (iii) sesquiterpene glycosides: cynarascolo-side A/B and cynarascoloside C; (iv) sesquiterpene lactones: cynaropicrin and grossheimin [63]; (v) triterpene saponins, including cynarasaponin E, J, C, A/H, and F/I; and (vi) the presence of amino and fatty acids was reported by Farag et al. in *Cynara cardunculus* extract, mostly concentrated in the roots, including hydroxyoctadecatrienoic acid, hydroxy-oxo-octadecatrienoic acid, tyrosyl-l-leucin, and dihydroxy-octadecatrienoic acid [64].

The UV-visible spectra were obtained as the initial analysis to record the formation of *Cynara-cardunculus*-extract-mediated CC-AgNPs. In general, UV-vis spectra were analyzed to infer valuable information about the shape, size, and distribution of the biosynthesized nanoparticles established using surface plasmon resonance (SPR) bands [65]. The UV-vis spectrum is important for deducing the role played by plant extract with inherited phytochemical bioactive compounds being involved in the biosynthesis of CC-AgNPs. UV-visible spectroscopy is a useful technique to determine the SPR band of the noble metal silver nanoparticles (plasmonic) due to the free electron excitation [65]. The plasmonic nanoparticles are quite distinguished from other magnetic, polymeric, and semiconductor nanoparticles because of their unique surface plasmon resonance [65]. The position of the SPR peak generally depends on the shape, state of aggregation, and particle size of the nanoparticles [65–68]. In this study, the initial physical observation of color change in colorless silver nitrate solution with the addition of *Cynara cardunculus* extract confirmed the biosynthesis of polyphenol capped AgNPs. Meanwhile, with the addition of *Cynara cardunculus* extract to colorless silver nitrate solution, the color of the reaction mixture changed from light yellow to brown within 5 to 10 min of incubation. Furthermore, the reaction mixture turned dark brown after 30 min of incubation, demonstrating the reduction of silver metal ions (Ag^+) to silver nanoparticles (Ag^0). The *Cynara cardunculus* extract phytochemicals acted as reducing and stabilizing agents in the biogenic synthesis of stable CC-AgNPs. The present study involved further exploration using different $AgNO_3$ concentrations and different volumes of *Cynara cardunculus* aqueous extract, and the stability of the as-prepared CC-AgNPs was also investigated by recording the UV-visible spectra at different times from 30 min to 360 min, as shown in Figure 2a–c. Figure 2a clearly shows a sharp peak maximum at ca. 438 nm; meanwhile, with an increase in time and concentration of $AgNO_3$, the increased absorption intensities were predominantly intensified. The fact of increased intensity with increased concentration of $AgNO_3$ also emphasized the biogenesis of CC-AgNPs from the *Cynara cardunculus* extract increases. It is worth mentioning that the observed increase in the absorption intensity with the increase in time was due to the reduction of silver ions (Ag^+) to elemental silver (Ag^0). To determine the role of increased plant extract on the biosynthesis of CC-AgNPs, different plant extract concentrations ranging from 2 mL to 14 mL were explored under UV-vis spectra, as depicted in Figure 2b. The observed data revealed that with an increase in the concentration of plant extract, an increased absorbance intensity was obtained, with a maximum peak intensity at ca. 438 nm, further demonstrating the biogenesis of many spherical CC-AgNPs at higher concentrations of *Cynara cardunculus* extract. The study was further extended to establish the role of different time intervals ranging from 30 min–24 h under optimal reaction conditions, as depicted in Figure 2c. Figure 2d shows the UV-vis spectrum of the silver nitrate solution, *Cynara cardunculus* extract, and CC-AgNPs under optimal reaction conditions. Furthermore, the UV-visible spectra data revealed that the absorbance intensity of the reaction mixture increased with time, and the solution remained steady after more than 24 h of incubation, indicating that stable nanoparticle formation in the solution was successfully completed [69,70]. The optical images of silver nanoparticle formation shown in Figure 2c,d (inset) also confirmed the formation of CC-AgNPs with an increase in color intensity with incubation time from 30 min to 24 h. Subsequently, the CC-AgNPs were centrifuged, washed, dried, and calcined before being used for further studies.

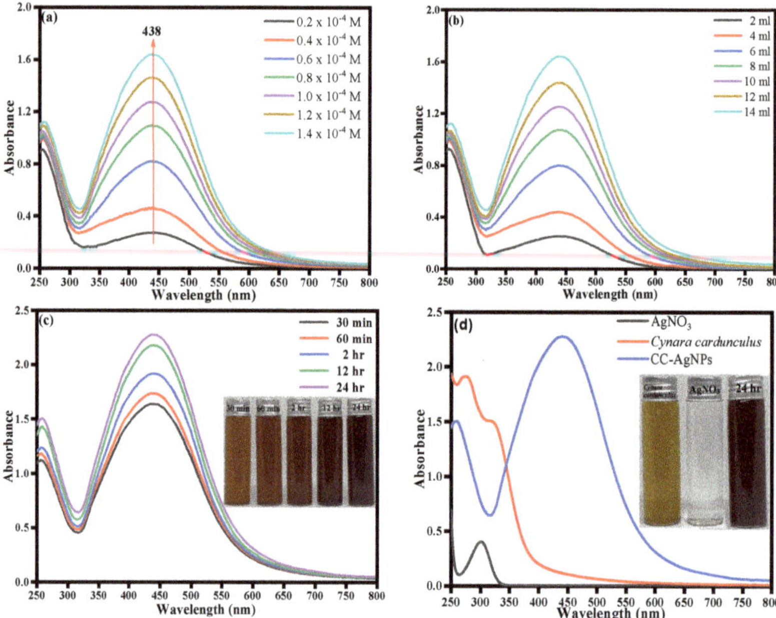

Figure 2. UV-vis absorption spectra of CC-AgNPs recorded as a function of different (**a**) $AgNO_3$ concentrations and (**b**) plant extract concentrations; (**c**) UV-visible spectra at different times from 30 min to 360 min (inset: optical images of CC-AgNPs at different time intervals); and (**d**) UV-visible spectra of silver nitrate solution, *Cynara cardunculus* extract, and CC-AgNPs (inset: optical images of *Cynara cardunculus* extract, silver nitrate solution, and CC-AgNPs).

The FTIR analysis was used to deduce the role of different functional groups present in the phytochemical composition of *Cynara cardunculus* extract as reducing/stabilizing agents in the biogenesis of CC-AgNPs, as depicted in Figure 3a. However, the standard peaks of CC-AgNPs were compared with *Cynara cardunculus* extract to analyze the biosynthesis of CC-AgNPs. Figure 3 reveals the presence of absorption peaks in CC-AgNPs at 3298.1 cm^{-1}, 2860.8 cm^{-1}, 1616.6 cm^{-1}, 1489.7 cm^{-1}, 1261.1 cm^{-1}, 1218.8 cm^{-1}, 1187.8 cm^{-1}, 1080.6 cm^{-1}, 998.7 cm^{-1}, 632.0 cm^{-1}, and 530.4 cm^{-1}. However, the *Cynara cardunculus* extract possessed similar peaks as observed in standard CC-AgNPs, which demonstrated the biogenesis of CC-AgNPs upon phytochemical reduction. The peak intensity band that appeared at approximately 3300 cm^{-1}, i.e., 3312.2 cm^{-1} and 3298.1 cm^{-1}, were attributed to the presence of polyphenolic or polysaccharide –OH stretching vibrations. However, it was reported that the enol forms of such polyphenols or polysaccharides are reduced to quinone in extract mixture and are usually interpreted as a peak shift of –OH groups toward a higher frequency ranging between 3400 cm^{-1} and 3456 cm^{-1} [71]. A protruding peak intensity appeared at 2860.8 cm^{-1}, which was assigned to C–H typical stretching vibrations from the CH_2 groups of aliphatic compounds and are believed to have occurred after the reduction of $AgNO_3$ [72]. The existence of sharp peak intensity in the range between 2820 cm^{-1} and 2760 cm^{-1}, i.e., 2790.3 cm^{-1}, in the *Cynara cardunculus* extract was attributed to the presence of N-CH_3 and C-H stretching vibrations corresponding to methyl-amino substituted groups [73]. In addition, the appeared peaks at 1619.4 cm^{-1} and 1616.6 cm^{-1} lying between 1650 cm^{-1} and 1600 cm^{-1} corresponded to stretching vibrations due to conjugated ketones and aromatic ring stretching (-C=C-C) vibrations. The presence of peaks between 1550 cm^{-1} and 1400 cm^{-1}, for example, 1534.8 cm^{-1} and 1489.7 cm^{-1}, is believed to be due to the N-O stretching vibration of aromatic nitro compounds. The existence of an intensity peak in the range between 1410 cm^{-1} and 1310 cm^{-1}, such as the 1388.1 cm^{-1} peak, in the *Cynara cardunculus* extract was assigned to –OH bending in phenolic and tertiary alcoholic groups. Similarly, the peak at 1252.7 cm^{-1} in *Cynara cardunculus* extract and the peak at 1261.1 cm^{-1} in CC-AgNPs were due to the presence of C-N stretching vibrations of primary aromatic amines. In addition, the occurrence of sharp peak intensities at 1083.4 cm^{-1} and 1080.6 cm^{-1} were the corresponding C-N stretching vibration bands of aliphatic amines, in the *Cynara cardunculus* extract and CC AgNPs, respectively. Moreover, the peaks between 1320 cm^{-1} and 1210 cm^{-1}, viz., 1227.3 cm^{-1} of the *Cynara cardunculus* extract and 1218.8 cm^{-1} of the CC-AgNPs, were attributed to C-O stretching vibrations. Furthermore, the peaks between 1190 cm^{-1} and 1130 cm^{-1}, i.e., 1185.0 cm^{-1} of the *Cynara cardunculus* L. extract and 1187.8 cm^{-1} of the CC-AgNPs, signified the presence of C-N stretching vibrations of secondary amines. The observed sharp intensity peak in the range 1055 cm^{-1}–1000 cm^{-1}, i.e., 1007.2 cm^{-1} of the *Cynara cardunculus* extract, was due to the corresponding cyclohexane ring vibrations. The observed peak intensities at relatively lower frequencies between 850 cm^{-1} to 1000 cm^{-1}, i.e., 900.0 cm^{-1} of the *Cynara cardunculus* extract and 998.7 cm^{-1} of the CC-AgNPs, were attributed to the hydrogen-bonded OH out-of-plane bending vibration modes. Meanwhile, the observed peaks between 660 cm^{-1} and 630 cm^{-1}, such as 637.6 cm^{-1} of the *Cynara cardunculus* L. extract and 632.0 cm^{-1} of the CC-AgNPs, were believed to occur due to the presence of C-S thio-substituted compounds. The important observed peak intensity in both *Cynara cardunculus L.* extract and CC-AgNPs at 530.4 cm^{-1} was caused by the reduction of Ag^+ to Ag^0 in the biosynthesis of CC-AgNPs from *Cynara cardunculus* aqueous extract. In conclusion, the presence of polyphenolic vibrations, along with other functional group stretching vibrations, of *Cynara cardunculus* extract demonstrated their role as reducing and stabilizing agents in the biosynthesis of CC-AgNPs.

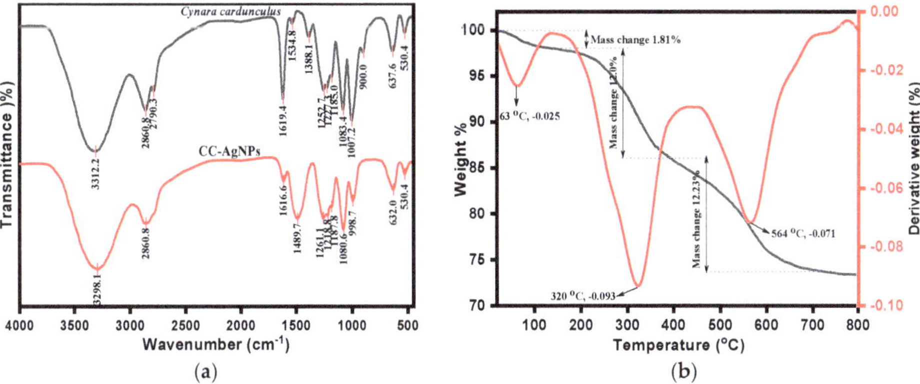

Figure 3. (**a**) FTIR spectra and (**b**) TGA/DTG analysis of biogenic CC-AgNPs.

The phytochemicals play essential roles as stabilizing and capping agents in the biogenesis of nanoparticles. Thus, the weight loss and thermal stability of such were investigated, which depended upon the adsorption of phytochemicals onto the surface of biogenic CC-AgNPs. Although the thermogravimetric analysis was operated at a heating rate of 10 °C/min under a nitrogen atmosphere, we inferred the weight loss in a stepwise manner for biogenic CC-AgNPs, as depicted in Figure 3b. The overall weight loss in this study of thermal decomposition associated with CC-AgNPs from *Cynara cardunculus* extract was approximately 26.64% by weight. This observed weight loss in the first region between 0–220 °C was 1.81% by weight occurred after the loss of the adsorbed surrounding moisture and volatile residues of phytochemicals onto the surface of CC-AgNPs. Furthermore, the observed TGA peak at 63 °C for the derivative weight (by %) of −0.025 corresponded to the degradation of volatile phytocompounds onto the surface of the biosynthesized CC-AgNPs. The second weight loss between 150–300 °C was 12.6% by weight, which was observed after the thermal degradation of heterocyclic volatile phytochemical compounds that thermally decomposed onto the adsorbed surface of CC-AgNPs. The third region of thermal decomposition after 300 °C, with the DTG peak at 320 °C for the derivative weight (by %) of −0.093 and the DTG peak at 564 °C for the derivative weight (by %) of −0.071 were attributed to the thermal decomposition with a weight loss (by % weight) of about 12.23%, which was believed to occur after the thermal decomposition of phytochemical constituents, including polyphenolic compounds, flavonoids, and polysaccharides. These phytocompounds have an important appealing role in the capping and stabilization of the biosynthesized surface morphology of CC-AgNPs. The overall results of the TGA/DTG demonstrated the thermally stable biosynthesis of CC-AgNPs from *Cynara cardunculus* extract.

The as-synthesized CC-AgNPs were subjected to XRD analysis to deduce the crystalline nature of the particles. The XRD pattern of the green-synthesized CC-AgNPs from *Cynara cardunculus* extract was obtained and is depicted in Figure 4. It is clear from Figure 4 that the diffraction peaks observed at 2θ angles of 38.16° (111), 44.40° (200), 64.58° (220) and 77.38° (311) corresponded to the face-centered cubic (fcc) structure of metallic silver. Moreover, the obtained XRD results were a good approximation to the JCPDS Card NO. 04-0783 results [74]. However, the presence of a prominent peak at 38.16° was due to the crystalline Ag, showing that the biosynthesized CC-AgNPs were encompassed with crystalline Ag lattice sites. The average particle size found via XRD analysis was calculated using the Scherrer equation ($d = K\lambda/\beta\cos\theta$), where d is the crystallite size, K represents the Scherer constant equal to 0.9, λ is the wavelength of the X-ray source (typically 1.5406 Å), β is the FWHM in radians, and θ is the peak position (Bragg angle), as tabulated in Table 1. The average crystalline size was found to be 17.26 nm by using the Scherrer calculation; meanwhile, similar numerical values of crystalline size were found in the SEM analysis, as discussed in another section in detail. No additional reflection other than Ag-lattices was observed, demonstrating the purity of the biosynthesized CC-AgNPs from the *Cynara cardunculus* extract.

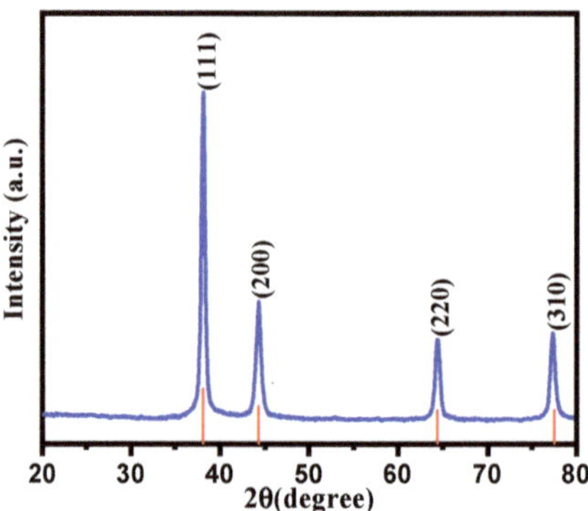

Figure 4. XRD spectrum of CC-AgNPs from the *Cynara cardunculus* extract.

Table 1. Lattice parameters and crystalline sizes of CC-AgNPs from the XRD patterns.

2θ (°)	FWHM	Miller Indices (hkl)	d_{hkl} d-Spacing (Å)	Crystal Size d (nm)	d (Average)
38.13	0.4353	(111)	2.358	19.30	17.26 nm
44.27	0.6005	(200)	2.044	14.28	
64.42	0.5394	(220)	1.445	17.40	
77.312	0.5636	(310)	1.233	18.04	

The transmission electron microscopy (TEM) and scanning electron microscopy–energy-dispersive X-ray spectroscopy (SEM-EDX) analysis of the as-synthesized CC-AgNPs were used to perform the study of the exterior surface morphology, structural characterization, and elemental composition, as shown in Figure 5. The TEM analysis of the green-synthesized CC-AgNPs from *Cynara cardunculus aqueous* extract showed them to be well defined with a homogenous distribution of nearly spherical nanoparticles with an average particle size ranging between 14 to 43 nm. The larger particle sizes were observed due to the formation of agglomerates of the small particles. The average (23.74 nm), minimum (9.07 nm), and maximum (59.28 nm) particle sizes of the as-prepared CC-AgNPs were found from the particle size histogram using ImageJ software, as shown in Figure 5b. During the SEM analysis, the surface morphology showed agglomeration of spherical CC-AgNPs, as shown in Figure 5c. Besides the SEM analysis, EDX analysis was performed for the elemental analysis and purity of the as-prepared CC-AgNPs, as shown in Figure 5d. The observed intensity ranges emphasized the strong spectral signals corresponding to the silver region (Ag). The observed sharp intensity signal at 3 KeV was found due to the adsorption of the metallic silver region and emphasized the presence of biosynthesized nanocrystallites of CC-AgNPs. The other signals found in the 0–0.5 KeV range were attributed to the presence of adsorbed oxygen and carbon atoms. The overall results of surface morphology and elemental analysis indicated that the biosynthesized CC-AgNPs using *Cynara cardunculus aqueous* extract were of high purity.

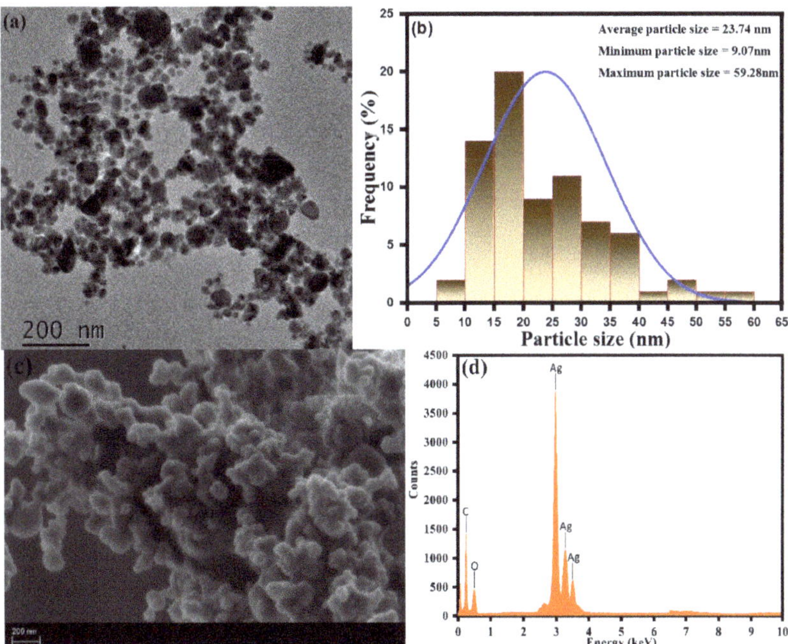

Figure 5. (**a**) TEM, (**b**) particle size histogram, (**c**) SEM, and (**d**) EDX spectra of the CC-AgNPs.

A zeta potential analysis was undertaken to analyze the stability of the biosynthesized CC-AgNPs from *Cynara cardunculus* extract in their colloidal state. The results obtained using zeta potential analysis are depicted in Figure 6a. Values of zeta = −26.2 mV and −6.38 mV was observed, showing a negative charge on the surface of the CC-AgNPs, further emphasizing their stability in the colloidal state and the role of phytochemicals as surface-capping agents. The observed negative values of the zeta potential were due to from the absorbed phytochemicals with negative surface charge onto the surface, possibly because of the presence of functional groups, such as OH-, CO-, and COO-. However, the detailed mechanism and possibility of occurrence of such functional groups are discussed in another section related to characterization using FTIR analysis. In general, the presence of the negative surface charge on the surface of CC-AgNPs resulted in preventing the aggregation and stabilizing the CC-AgNPs upon electrostatic repulsion among the negative charges. Moreover, the size distribution of the as-synthesized CC-AgNPs was analyzed using dynamic light scattering (DLS) analysis, as shown in Figure 6b. The DLS results of the CC-AgNPs shown in Figure 6b revealed that the average size of the particles in the optimal condition was 127 nm with a polydispersity (PDI) of 0.515. The bigger hydrodynamic diameter shown in the DLS results as compared with the TEM and XRD results was because of the presence of the capping agents and some agglomerated CC-AgNPs in the reaction mixture.

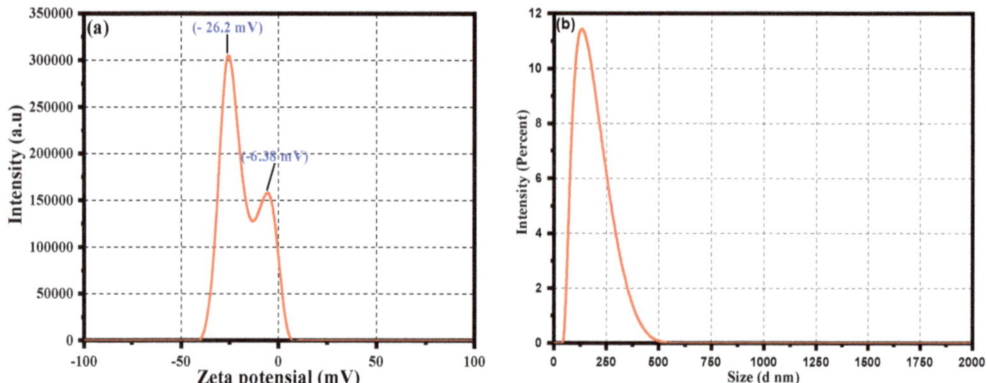

Figure 6. (**a**) Zeta potential and (**b**) particle size distribution analysis of the CC-AgNPs.

3.1. Anti-Candida Activity of CC-AgNPs

The CC-AgNPs were found to be active against *C. auris* MRL6057, and the MIC values were reported as 50.0 µg/mL, whereas the MFC was found to be 100.0 µg/mL. In comparison, the MIC and MFC values of AmB against *C. auris* MRL6057 were found to be 4.0 and 8.0 µg/mL, respectively. According to published MIC breakpoints for *C. auris* [75], the clinical strains of *C. auris* with MIC ≥2 µg/mL are considered resistant to AmB. Therefore, *C. auris* MRL6057 was deemed to be resistant to AmB.

C. auris displays high resistance to commonly used drugs [76]. In vitro examinations demonstrated the reduced susceptibility of ≥90% of *C. auris* isolates to fluconazole [77]. In comparison, 3–73% and 13–35% of clinical isolates of *C. auris* were found to be resistant to voriconazole and AmB, respectively [78,79]. Furthermore, in the USA, ≥99% of these isolates were reported to be less susceptible to fluconazole, around two-thirds were resistant to AmB, and approximately 4% were found to be resistant to echinocandins class antifungals [80]. The global emergence of pan-resistant strains of *C. auris* and their ability to persist in healthcare settings has redrawn the attention of researchers and healthcare experts to this pathogenic yeast [79].

Researchers investigated the candidacidal potential of AgNPs against *C. albicans*, and AgNPs were found to be potential inhibitors of growth and viability, both alone and in combination; for instance, a low strength of 1.8 mg/mL AgNPs in combination to cationic carboxilane inhibited the growth of *C. albicans* [81]. Additionally, L-3,4-dihydroxyphenylalanine capped with AgNPs showed fungicidal activity at a concentration of 31.2–62.5 µg/mL [82]. These findings support our investigation and that AgNPs can be a potential candidate for drug development against *C. auris*. Therefore, further research that analyzed the in-depth mechanism of antifungal action of CC-AgNPs was required.

3.2. CC-AgNPs Impede Cell Count and Viability in C. auris

A susceptibility assay of *C. auris* was performed against CC-AgNPs to measure the growth and viability of these cells (Figure 7). The unexposed cells had 98.5% live cells, whereas, after exposure to H_2O_2, the percentage of live cells was 8.4%. Moreover, the reduction in the number of live cells was dependent on the concentration of the NPs. Therefore, the percentage of live cells decreased with increasing concentration of NPs, where 41.6%, 26.6%, and 2.2% live cells were recorded at values of 0.5MIC, MIC, and 2MIC, respectively. The results confirmed that these NPs entirely stopped the growth and survival of *C. auris*, and thus, corroborated the anti-*Candida* potency of AgNPs.

The antimicrobial activity of AgNPs was well studied by various researchers [81,82]. The current findings were in accord with the results obtained by Wani and co-workers 2013, where they showed the potent anti-*Candida* effect of metallic NPs [83]. The antifungal activity of metallic NPs is attributed to their ability to disrupt the membrane porosity and induce cellular damage, ROS production, damage of nucleic acid, and disruption of important biological enzymes [84].

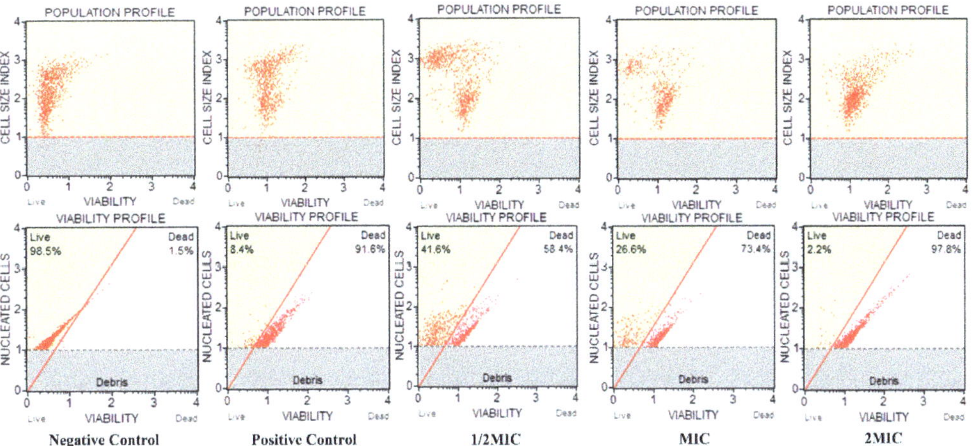

Figure 7. CC-AgNPs affected *C. auris* cell numbers and viability. The figure shows the *C. auris* viability and population profile. Negative control: unexposed *C. auris* cells; positive control: H_2O_2 exposed cells; *C. auris* exposed to various MIC values of the test NPs.

3.3. CC-AgNPs Arrested the Cell Cycle in C. auris

The CC-AgNPs may result in the induction of cellular apoptosis in *C. auris*, and thus, the potency of these NPs on the cell cycle was investigated. Consequently, the number of cells dispersed in various phases of the cell cycle must be different from that present in the healthy cells, representing cell cycle blockage. Hence, the change in DNA content was evaluated quantitatively with the help of fluorescence generated using DNA tagged with PI throughout the cellular growth.

The results are summarized in Table 2, where the unexposed experiment had the maximum number of cells in the G0/G1 phase, followed by the S phase and G2/M phase, whereas, the cells of the positive control were mostly accumulated in the S phase, followed by the G0/G1 and G2/M phase. However, after exposure to the CC-AgNPs, the cell cycle was blocked at two different phases: the S phase and G2/M phase (Figure 8a,b). At 0.5 MIC and MIC, the cells were arrested in the S phase of the cell cycle, with the distribution percentages of 44.5% and 60.2%, respectively. Furthermore, at 2MIC, the cells were arrested in the G2/M phase (58.7%). Altogether, the present findings discovered that the CC-AgNPs at a lower concentration arrested the cells in the S phase, whereas, at a higher concentration, the G2/M phase is blocked, and therefore, had a conspicuous role in blocking cell cycle advancement in *Candida*.

Table 2. Cell cycle in *C. auris*.

Experiment	Phases of Cell Cycle	Cell Percentage (%)
Negative control	G0/G1	67
	S	26.8
	G2/M	6.2
Positive control	G0/G1	23.8
	S	66.7
	G2/M	9.3
0.5MIC (25 μg/mL)	G0/G1	48.2
	S	44.5
	G2/M	7.2
MIC (50 μg/mL)	G0/G1	33.2
	S	60.2
	G2/M	6.1
2MIC (100 μg/mL)	G0/G1	0.6
	S	40.3
	G2/M	58.7

These findings were in good agreement with previous results where researchers showed the inhibitory effect of various compounds on the cell cycle of *Candida* species. For instance, clioquinol, crambescidin-816, and crambescidin-089 were found to block the G2/M phase in *Candida* and Saccharomyces cerevisiae [85–87]. Furthermore, the damaged cell cycle alters the cellular morphology, which increases the chance of the host's immune system recognizing the yeast cells [88]. Thus, CC-AgNPs can impede the cell cycle in *C. auris* and boost its identification by the host's immune system.

3.4. CC-AgNPs Depolarized the Mitochondrial Membrane Potential in C. auris

The mitochondrial membrane potential is one of the primary steps in fungal apoptosis owing to the mitochondria's crucial role in cell survival. The results obtained in this study revealed the potential of the test CC-AgNPs to cause mitochondrial membrane disintegration in *C. auris* (Figure 9). Mitochondrial membrane disruption results in pore formation, which leads to the depolarization and movement of different apoptotic factors. Mitochondrial depolarization was also related to the cytochrome c release and was often observed during early apoptosis. The test CC-AgNPs caused mitochondrial toxicity induction, which was followed by the loss of membrane potential, oxidative phosphorylation inhibition, and changes in calcium sequestration [89]. Other studies reported the potential of silver nanoparticles to depolarize mitochondrial membranes and cause apoptosis in *C. albicans*, which was related to an increase in hydroxyl radicals [90]. A study by Zhu and co-workers reported the impact of iron nanoparticles in causing apoptosis in human umbilical endothelial cells by causing membrane depolarization [90]. These studies further supported our claims and are congruent with our findings that metal nanoparticles can cause mitochondrial membrane depolarization and with our conclusions that metal nanoparticles can cause mitochondrial membrane depolarization and apoptosis in fungal cells. However, to further verify these claims, more studies involving other metal nanoparticles, including mono-, bi-, and/or tri- metallic nanoparticles, are needed to compare their effects in different fungal cells in terms of causing apoptosis related to mitochondrial membrane depolarization.

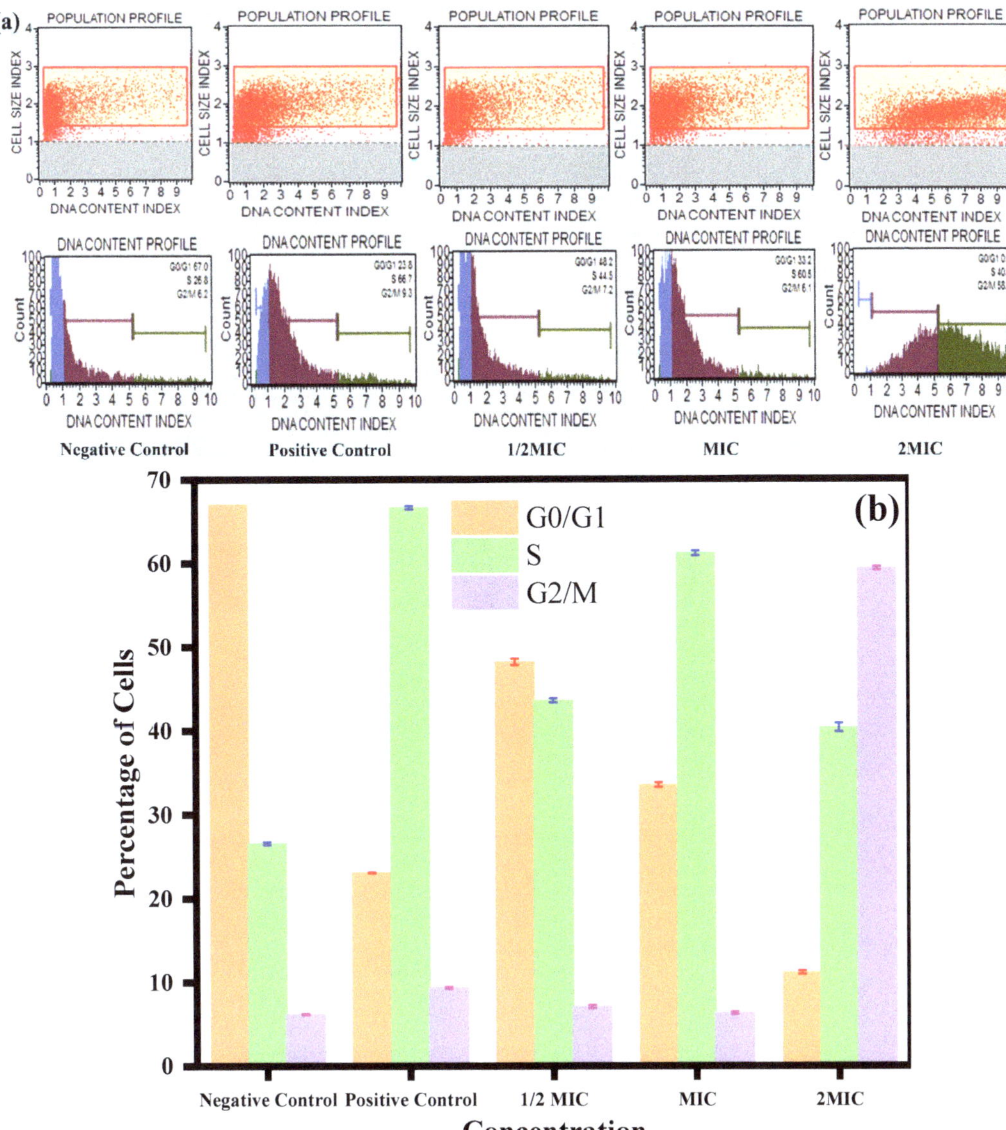

Figure 8. Cell cycle analysis of *C. auris*. (**a**) Effect of CC-AgNPs at various concentrations on cell cycle progression in *C. auris*. (**b**) Representative histograms of the *C. auris* cell cycle at various CC-AgNP concentrations. Positive controls were cells treated with H_2O_2 (10 mM) and negative controls were untreated cells.

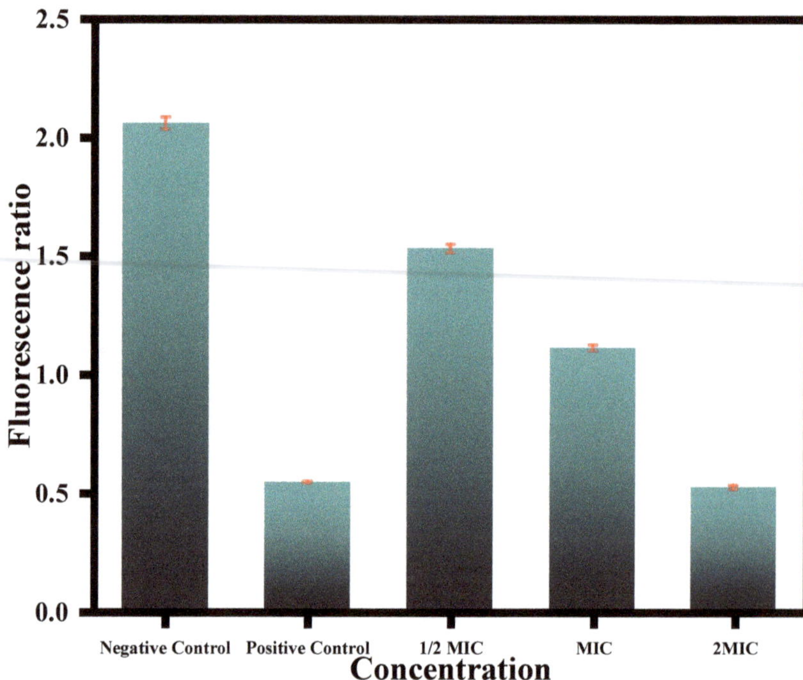

Figure 9. Effect of CC-AgNPs at varying concentrations on mitochondrial membrane depolarization in *C. auris* cells. Positive and negative controls were represented by cells treated with 10 mM H_2O_2 and untreated cells, respectively.

3.5. The CC-AgNPs Elicited DNA Fragmentation in C. auris

The microscopic analysis for the TUNEL assay results revealed that at higher concentrations of CC-AgNPs, green solid fluorescence suggested DNA breakage, as observed for cells treated with 10 mM H_2O_2 (Figure 10). Therefore, these results demonstrated the potential of the test NPs to cause DNA fragmentation in *C. auris*. At a lower concentration (25 µg/mL), the degree of DNA fragmentation was not much, which was depicted by fewer TUNEL-positive yeast cells. However, with increasing concentration (50–100 µg/mL), the degree of DNA damage also increased, which was reflected in the higher number of TUNEL-positive yeast cells.

DNA fragmentation is one of the significant markers related to morphological changes to identify the late apoptosis in yeast cells. The DNA fragmentation can be visualized using a TUNEL assay, which labels the free 3′-OH termini with modified nucleotides catalyzed by terminal deoxynucleotidyl transferase. The TUNEL assay was observed as the most dependable method to study late apoptosis [91]. The test nanoparticles' cell cycle arrest and DNA fragmentation suggested that CC-AgNPs can damage nucleic acids in *C. auris* and other pathogenic yeasts. It was also predicted that these nanoparticles, besides damaging nucleic acids in *C. auris*, can also damage antioxidant enzymes and cause lipid peroxidation in fungal cells. Overall, the results from this study suggested that CC-AgNPs led to nucleic acid fragmentation and mitochondrial membrane depolarization, which are the characteristic markers of apoptosis, thus, validating the idea that AgNPs induce late apoptosis in yeast cells and have dual antifungal action modes, including membrane disruption. The overall result suggested that CC-AgNPs led to mitochondrial membrane depolarization and DNA fragmentation, which are crucial apoptosis characteristics.

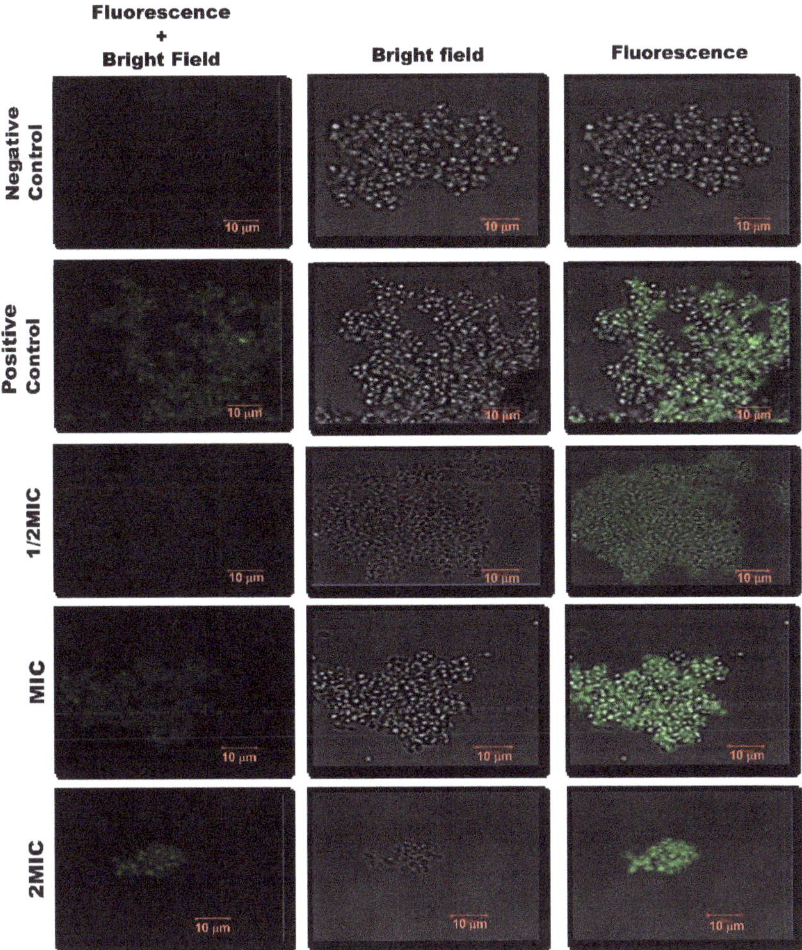

Figure 10. Confocal scanning fluorescence images of *C. auris* when treated with different concentrations of CC-AgNPs. Untreated cells were the negative control, whereas cells treated with H_2O_2 (10 mM) were the positive control.

The CC-AgNPs in this study showed potent antifungal activity with a dual antifungal mode of action by causing cell cycle arrest and cellular apoptosis. However, to further escalate these nanoparticles to the next steps of drug development, it is essential to check their toxicity on host cells. To this end, the CC-AgNPs were tested for hemolytic activity against horse blood cells. The results obtained in this study revealed only 3–7% cell lysis when treated with CC-AgNPs at $\frac{1}{2}$ half of MIC and MIC respectively. Figure 11, thus confirming that CC-AgNPs are safe for in vivo animal experiments. Even at higher concentrations (2MIC), only 13% hemolysis was observed; however, this concentration is not considered safe for testing in animal models. Furthermore, our results also reported no lysis observed in untreated control cells, whereas 100% cell lysis was observed with Triton X, which served as the positive control.

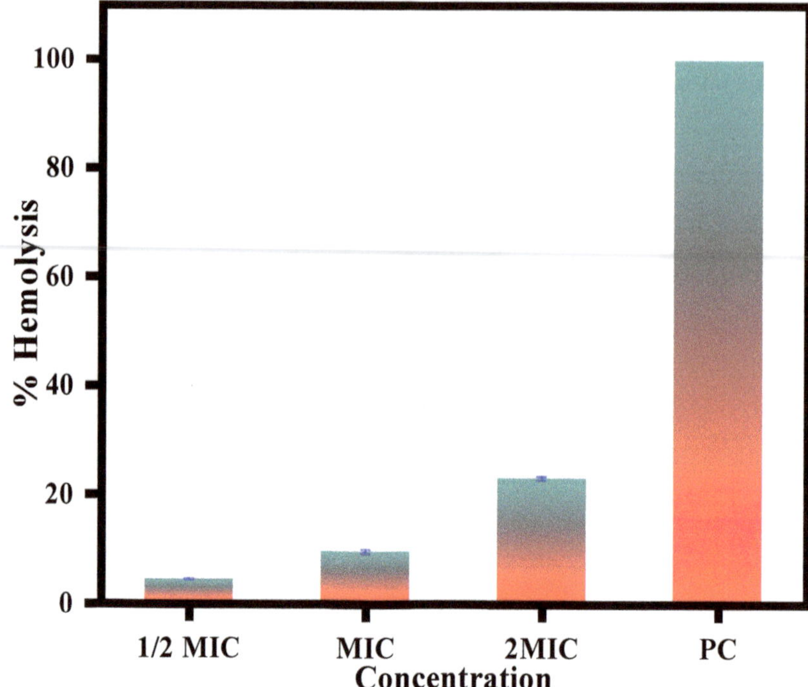

Figure 11. Hemolytic activity of CC-AgNPs using horse erythrocytes showed no lysis in untreated cells, whereas 100% hemolysis was caused by Triton X, which was the positive control.

4. Conclusions

In this work, chemically stable silver nanoparticles (CC-AgNPs) were prepared via a phytochemically induced synthesis process using *Cynara cardunculus* extract as a reducing and capping agent. The present work was facile, cost-effective, and ecofriendly and did not require any solvent except water, which made this process highly advantageous. The formation of CC-AgNPs was confirmed by various microscopic and spectroscopic techniques before utilizing them for antifungal activities against *C. auris*. The experimental conditions for the preparation of CC-AgNPs were optimized via a surface plasmon (SPR) peak at 438 nm using UV-visible spectroscopy. Furthermore, the CC-AgNPs directly inhibited the cell cycle and arrested cells in the G2/M phase and could be a potential lead for antifungal drug development. Our results demonstrated that the as-prepared silver nanoparticles had good antifungal performance against *C. auris* and could be further explored for exceptional and enhanced biomedical applications.

Author Contributions: Conceptualization, M.A.M., M.R.K., M.G.B. and A.N.; Data curation, M.G.B. and K.A.A. and E.Y.D.; Formal analysis, M.A.M., M.R.K., M.G.B. and A.N.; Investigation, K.A.A. and E.Y.D.; Methodology, M.A.M., M.R.K., M.G.B. and A.N.; Project administration, M.A.M.; Software, K.A.A. and E.Y.D.; Supervision, M.A.M., M.R.K. and E.Y.D.; Validation, M.A.M., K.A.A. and E.Y.D.; Writing–original draft, M.A.M., M.R.K., M.G.B. and A.N.; Writing–review & editing, M.A.M., M.R.K., M.G.B., A.N., K.A.A. and E.Y.D. All authors have read and agreed to the published version of the manuscript.

Funding: This research received funding from the Institutional Fund Projects under grant no. (IFPHI-136-130-2021) supported by the Ministry of Education and King Abdulaziz University, Deanship of Scientific Research (DSR), Jeddah, Saudi Arabia.

Institutional Review Board Statement: Not applicable.

Informed Consent Statement: Not applicable.

Data Availability Statement: All data created is provided in this study.

Acknowledgments: The authors extend their appreciation to the Deputyship for Research & Innovation, Ministry of Education in Saudi Arabia for funding this research work through the project number IFPHI-136-130-2021 and King Abdulaziz University, DSR, Jeddah, Saudi Arabia.

Conflicts of Interest: The authors declare that they have no known competing financial interests or personal relationships that could have appeared to influence the work reported in this paper.

References

1. Medina-Cruz, D.; Mostafavi, E.; Vernet-Crua, A.; Cheng, J.; Shah, V.; Cholula-Diaz, J.L.; Guisbiers, G.; Tao, J.; García-Martín, J.M.; Webster, T.J. Green nanotechnology-based drug delivery systems for osteogenic disorders. *Expert Opin. Drug Deliv.* **2020**, *17*, 341–356. [CrossRef] [PubMed]
2. Xia, Q.H.; Ma, Y.J.; Wang, J.W. Biosynthesis of silver nanoparticles using Taxus yunnanensis callus and their antibacterial activity and cytotoxicity in human cancer cells. *Nanomaterials* **2016**, *6*, 160. [CrossRef] [PubMed]
3. Pugazhendhi, A.; Edison, T.N.J.I.; Karuppusamy, I.; Kathirvel, B. Inorganic nanoparticles: A potential cancer therapy for human welfare. *Int. J. Pharm.* **2018**, *539*, 104–111. [CrossRef]
4. Islam, M.A.; Jacob, M.V.; Antunes, E. A critical review on silver nanoparticles: From synthesis and applications to its mitigation through low-cost adsorption by biochar. *J. Environ. Manag.* **2021**, *281*, 111918. [CrossRef] [PubMed]
5. Alsammarraie, F.K.; Wang, W.; Zhou, P.; Mustapha, A.; Lin, M. Green synthesis of silver nanoparticles using turmeric extracts and investigation of their antibacterial activities. *Colloids Surf. B Biointerfaces* **2018**, *171*, 398–405. [CrossRef]
6. Kumari, M.M.; Jacob, J.; Philip, D. Green synthesis and applications of Au–Ag bimetallic nanoparticles. *Spectrochim. Acta Part A Mol. Biomol. Spectrosc.* **2015**, *137*, 185–192. [CrossRef]
7. Jana, J.; Ganguly, M.; Pal, T. Enlightening surface plasmon resonance effect of metal nanoparticles for practical spectroscopic application. *RSC Adv.* **2016**, *6*, 86174–86211. [CrossRef]
8. Kamal, E.F.B.; Fen, Y.W. The Principle of Nanomaterials Based Surface Plasmon Resonance Biosensors and Its Potential for Dopamine Detection. *Molecules* **2020**, *25*, 2769. [CrossRef]
9. Amirjani, A.; Haghshenas, D.F. Ag nanostructures as the surface plasmon resonance (SPR)-based sensors: A mechanistic study with an emphasis on heavy metallic ions detection. *Sens. Actuators B Chem.* **2018**, *273*, 1768–1779. [CrossRef]
10. Bindhu, M.R.; Umadevi, M. Surface plasmon resonance optical sensor and antibacterial activities of biosynthesized silver nanoparticles. *Spectrochim. Acta Part A Mol. Biomol. Spectrosc.* **2014**, *121*, 596–604. [CrossRef]
11. Yang, X.; Du, Y.; Li, D.; Lv, Z.; Wang, E. One-step synthesized silver micro-dendrites used as novel separation mediums and their applications in multi-DNA analysis. *Chem. Commun.* **2011**, *47*, 10581–10583. [CrossRef] [PubMed]
12. Kim, D.M.; Park, J.S.; Jung, S.-W.; Yeom, J.; Yoo, S.M. Biosensing Applications Using Nanostructure-Based Localized Surface Plasmon Resonance Sensors. *Sensors* **2021**, *21*, 3191. [CrossRef] [PubMed]
13. Alex, K.V.; Pavai, P.T.; Rugmini, R.; Prasad, M.S.; Kamakshi, K.; Sekhar, K.C. Green Synthesized Ag Nanoparticles for Bio-Sensing and Photocatalytic Applications. *ACS Omega* **2020**, *5*, 13123–13129. [CrossRef] [PubMed]
14. Kim, M.; Lee, J.-H.; Nam, J.-M. Plasmonic Photothermal Nanoparticles for Biomedical Applications. *Adv. Sci.* **2019**, *6*, 1900471. [CrossRef]
15. Ravichandran, V.; Vasanthi, S.; Shalini, S.; Shah, S.A.A.; Tripathy, M.; Paliwal, N. Green synthesis, characterization, antibacterial, antioxidant and photocatalytic activity of Parkia speciosa leaves extract mediated silver nanoparticles. *Results Phys.* **2019**, *15*, 102565. [CrossRef]
16. Alshehri, A.A.; Malik, M.A. Phytomediated photo-induced green synthesis of silver nanoparticles using Matricaria chamomilla L. and its catalytic activity against rhodamine B. *Biomolecules* **2020**, *10*, 1604. [CrossRef]
17. Naaz, F.; Farooq, U.; Khan, M.A.M.; Ahmad, T. Multifunctional Efficacy of Environmentally Benign Silver Nanospheres for Organic Transformation, Photocatalysis, and Water Remediation. *ACS Omega* **2020**, *5*, 26063–26076. [CrossRef]
18. Duan, H.; Wang, D.; Li, Y. Green chemistry for nanoparticle synthesis. *Chem. Soc. Rev.* **2015**, *44*, 5778–5792. [CrossRef]
19. Kurhade, P.; Kodape, S.; Choudhury, R. Overview on green synthesis of metallic nanoparticles. *Chem. Pap.* **2021**, *75*, 5187–5222. [CrossRef]
20. Saratale, R.G.; Karuppusamy, I.; Saratale, G.D.; Pugazhendhi, A.; Kumar, G.; Park, Y.; Ghodake, G.S.; Bharagava, R.N.; Banu, J.R.; Shin, H.S. A comprehensive review on green nanomaterials using biological systems: Recent perception and their future applications. *Colloids Surf. B Biointerfaces* **2018**, *170*, 20–35. [CrossRef]
21. Abdullah, N.I.S.B.; Ahmad, M.B.; Shameli, K. Biosynthesis of silver nanoparticles using Artocarpus elasticus stem bark extract. *Chem. Cent. J.* **2015**, *9*, 61. [CrossRef] [PubMed]
22. Ajith, P.; Murali, A.S.; Sreehari, H.; Vinod, B.S.; Anil, A.; Smitha, C.S. Green synthesis of silver nanoparticles using Calotropis gigantea extract and its applications in antimicrobial and larvicidal activity. *Mater. Today Proc.* **2019**, *18*, 4987–4991. [CrossRef]
23. Sowmya, C.; Lavakumar, V.; Venkateshan, N.; Ravichandiran, V.; Saigopal, D. Exploration of Phyllanthus acidus mediated silver nanoparticles and its activity against infectious bacterial pathogen. *Chem. Cent. J.* **2018**, *12*, 42. [CrossRef] [PubMed]

24. Ahn, E.-Y.; Jin, H.; Park, Y. Assessing the antioxidant, cytotoxic, apoptotic and wound healing properties of silver nanoparticles green-synthesized by plant extracts. *Mater. Sci. Eng. C* **2019**, *101*, 204–216. [CrossRef] [PubMed]
25. Subramanian, R.; Subbramaniyan, P.; Raj, V. Antioxidant activity of the stem bark of Shorea roxburghii and its silver reducing power. *SpringerPlus* **2013**, *2*, 28. [CrossRef]
26. Jalani, N.S.; Zati-Hanani, S.; Teoh, Y.P.; Abdullah, R. Short review: The effect of reaction conditions on plant-mediated synthesis of silver nanoparticles. *Mater. Sci. Forum Trans. Tech. Publ.* **2018**, *917*, 145–151. [CrossRef]
27. Kim, H.-S.; Seo, Y.S.; Kim, K.; Han, J.W.; Park, Y.; Cho, S. Concentration effect of reducing agents on green synthesis of gold nanoparticles: Size, morphology, and growth mechanism. *Nanoscale Res. Lett.* **2016**, *11*, 230. [CrossRef]
28. Kim, J.S.; Kuk, E.; Yu, K.N.; Kim, J.-H.; Park, S.J.; Lee, H.J.; Kim, S.H.; Park, Y.K.; Park, Y.H.; Hwang, C.-Y. Antimicrobial effects of silver nanoparticles. *Nanomed. Nanotechnol. Biol. Med.* **2007**, *3*, 95–101. [CrossRef]
29. Pauksch, L.; Hartmann, S.; Rohnke, M.; Szalay, G.; Alt, V.; Schnettler, R.; Lips, K.S. Biocompatibility of silver nanoparticles and silver ions in primary human mesenchymal stem cells and osteoblasts. *Acta Biomater.* **2014**, *10*, 439–449. [CrossRef]
30. Panda, P.K.; Kumari, P.; Patel, P.; Samal, S.K.; Mishra, S.; Tambuwala, M.M.; Dutt, A.; Hilscherová, K.; Mishra, Y.K.; Varma, R.S. Molecular nanoinformatics approach assessing the biocompatibility of biogenic silver nanoparticles with channelized intrinsic steatosis and apoptosis. *Green Chem.* **2022**, *24*, 1190–1210. [CrossRef]
31. Silver, S.; Phung, L.T.; Silver, G. Silver as biocides in burn and wound dressings and bacterial resistance to silver compounds. *J. Ind. Microbiol. Biotechnol.* **2006**, *33*, 627–634. [CrossRef] [PubMed]
32. Bahrulolum, H.; Nooraei, S.; Javanshir, N.; Tarrahimofrad, H.; Mirbagheri, V.S.; Easton, A.J.; Ahmadian, G. Green synthesis of metal nanoparticles using microorganisms and their application in the agrifood sector. *J. Nanobiotechnol.* **2021**, *19*, 86. [CrossRef] [PubMed]
33. Peiris, M.; Fernando, S.; Jayaweera, P.; Arachchi, N.; Guansekara, T. Comparison of antimicrobial properties of silver nanoparticles synthesized from selected bacteria. *Indian J. Microbiol.* **2018**, *58*, 301–311. [CrossRef] [PubMed]
34. Bedlovičová, Z. Green synthesis of silver nanoparticles using actinomycetes. In *Green Synthesis of Silver Nanomaterials*; Elsevier: Amsterdam, The Netherlands, 2022; pp. 547–569.
35. Wijnhoven, S.W.; Peijnenburg, W.J.; Herberts, C.A.; Hagens, W.I.; Oomen, A.G.; Heugens, E.H.; Roszek, B.; Bisschops, J.; Gosens, I.; Van De Meent, D. Nano-silver–a review of available data and knowledge gaps in human and environmental risk assessment. *Nanotoxicology* **2009**, *3*, 109–138. [CrossRef]
36. Cho, K.-H.; Park, J.-E.; Osaka, T.; Park, S.-G. The study of antimicrobial activity and preservative effects of nanosilver ingredient. *Electrochim. Acta* **2005**, *51*, 956–960. [CrossRef]
37. Choi, O.; Deng, K.K.; Kim, N.-J.; Ross Jr, L.; Surampalli, R.Y.; Hu, Z. The inhibitory effects of silver nanoparticles, silver ions, and silver chloride colloids on microbial growth. *Water Res.* **2008**, *42*, 3066–3074. [CrossRef]
38. Hernández-Sierra, J.F.; Ruiz, F.; Pena, D.C.C.; Martínez-Gutiérrez, F.; Martínez, A.E.; Guillén, A.d.J.P.; Tapia-Pérez, H.; Castañón, G.M. The antimicrobial sensitivity of Streptococcus mutans to nanoparticles of silver, zinc oxide, and gold. *Nanomed. Nanotechnol. Biol. Med.* **2008**, *4*, 237–240. [CrossRef]
39. Kalishwaralal, K.; BarathManiKanth, S.; Pandian, S.R.K.; Deepak, V.; Gurunathan, S. Silver nanoparticles impede the biofilm formation by Pseudomonas aeruginosa and Staphylococcus epidermidis. *Colloids Surf. B Biointerfaces* **2010**, *79*, 340–344. [CrossRef]
40. Mohanty, S.; Mishra, S.; Jena, P.; Jacob, B.; Sarkar, B.; Sonawane, A. An investigation on the antibacterial, cytotoxic, and antibiofilm efficacy of starch-stabilized silver nanoparticles. *Nanomed. Nanotechnol. Biol. Med.* **2012**, *8*, 916–924. [CrossRef]
41. Mohan, S.C.; Sasikala, K.; Anand, T.; Vengaiah, P.; Krishnaraj, S. Green synthesis, antimicrobial and antioxidant effects of silver nanoparticles using Canthium coromandelicum leaves extract. *Res. J. Microbiol.* **2014**, *9*, 142. [CrossRef]
42. Chartarrayawadee, W.; Charoensin, P.; Saenma, J.; Rin, T.; Khamai, P.; Nasomjai, P.; Too, C.O. Green synthesis and stabilization of silver nanoparticles using *Lysimachia foenum-graecum* Hance extract and their antibacterial activity. *Green Process. Synth.* **2020**, *9*, 107–118. [CrossRef]
43. Lara, H.; Ixtepan-Turrent, L.; Jose Yacaman, M.; Lopez-Ribot, J. Inhibition of Candida auris biofilm formation on medical and environmental surfaces by silver nanoparticles. *ACS Appl. Mater. Interfaces* **2020**, *12*, 1021. [CrossRef] [PubMed]
44. Satoh, K.; Makimura, K.; Hasumi, Y.; Nishiyama, Y.; Uchida, K.; Yamaguchi, H. Candida auris sp. nov., a novel ascomycetous yeast isolated from the external ear canal of an inpatient in a Japanese hospital. *Microbiol. Immunol.* **2009**, *53*, 41–44. [CrossRef] [PubMed]
45. Osei Sekyere, J. Candida auris: A systematic review and meta-analysis of current updates on an emerging multidrug-resistant pathogen. *Microbiologyopen* **2018**, *7*, e00578. [CrossRef] [PubMed]
46. Fijan, S.; Šostar-Turk, S.; Cencič, A. Implementing hygiene monitoring systems in hospital laundries in order to reduce microbial contamination of hospital textiles. *J. Hosp. Infect.* **2005**, *61*, 30–38. [CrossRef] [PubMed]
47. Welsh, R.M.; Bentz, M.L.; Shams, A.; Houston, H.; Lyons, A.; Rose, L.J.; Litvintseva, A.P. Survival, persistence, and isolation of the emerging multidrug-resistant pathogenic yeast Candida auris on a plastic health care surface. *J. Clin. Microbiol.* **2017**, *55*, 2996–3005. [CrossRef]
48. Sarma, S.; Upadhyay, S. Current perspective on emergence, diagnosis and drug resistance in Candida auris. *Infect. Drug Resist.* **2017**, *10*, 155. [CrossRef]
49. *CLSI document M38–A2*; Reference Method for Broth Dilution Antifungal Susceptibility Testing of Filamentous Fungi. Approved Standard. CLSI: Wayne, PA, USA, 2008.

50. Lone, S.A.; Wani, M.Y.; Fru, P.; Ahmad, A. Cellular apoptosis and necrosis as therapeutic targets for novel eugenol tosylate congeners against Candida albicans. *Sci. Rep.* **2020**, *10*, 1191. [CrossRef]
51. Sharma, D.; Kanchi, S.; Bisetty, K. Biogenic synthesis of nanoparticles: A review. *Arab. J. Chem.* **2019**, *12*, 3576–3600. [CrossRef]
52. Vijayaraghavan, K.; Ashokkumar, T. Plant-mediated biosynthesis of metallic nanoparticles: A review of literature, factors affecting synthesis, characterization techniques and applications. *J. Environ. Chem. Eng.* **2017**, *5*, 4866–4883. [CrossRef]
53. Makarov, V.; Love, A.; Sinitsyna, O.; Makarova, S.; Yaminsky, I.; Taliansky, M.; Kalinina, N. "Green" nanotechnologies: Synthesis of metal nanoparticles using plants. *Acta Nat. (A a B)* **2014**, *6*, 35–44. [CrossRef]
54. Talabani, R.F.; Hamad, S.M.; Barzinjy, A.A.; Demir, U. Biosynthesis of Silver Nanoparticles and Their Applications in Harvesting Sunlight for Solar Thermal Generation. *Nanomaterials* **2021**, *11*, 2421. [CrossRef] [PubMed]
55. Fratianni, F.; Tucci, M.; De Palma, M.; Pepe, R.; Nazzaro, F. Polyphenolic composition in different parts of some cultivars of globe artichoke (*Cynara cardunculus* L. var. *scolymus* (L.) Fiori). *Food Chem.* **2007**, *104*, 1282–1286. [CrossRef]
56. Lombardo, S.; Pandino, G.; Mauromicale, G.; Knödler, M.; Carle, R.; Schieber, A. Influence of genotype, harvest time and plant part on polyphenolic composition of globe artichoke [*Cynara cardunculus* L. var. *scolymus* (L.) Fiori]. *Food Chem.* **2010**, *119*, 1175–1181. [CrossRef]
57. Mandim, F.; Petropoulos, S.A.; Giannoulis, K.D.; Santos-Buelga, C.; Ferreira, I.C.F.R.; Barros, L. Chemical Composition of *Cynara cardunculus* L. var. *altilis* Bracts Cultivated in Central Greece: The Impact of Harvesting Time. *Agronomy* **2020**, *10*, 1976. [CrossRef]
58. Jun, N.-J.; Jang, K.-C.; Kim, S.-C.; Moon, D.-Y.; Seong, K.-C.; Kang, K.-H.; Tandang, L.; Kim, P.-H.; Cho, S.-M.K.; Park, K.-H. Radical scavenging activity and content of cynarin (1, 3-dicaffeoylquinic acid) in Artichoke (*Cynara scolymus* L.). *J. Appl. Biol. Chem.* **2007**, *50*, 244–248.
59. Pinelli, P.; Agostini, F.; Comino, C.; Lanteri, S.; Portis, E.; Romani, A. Simultaneous quantification of caffeoyl esters and flavonoids in wild and cultivated cardoon leaves. *Food Chem.* **2007**, *105*, 1695–1701. [CrossRef]
60. El Sayed, A.M.; Hussein, R.; Motaal, A.A.; Fouad, M.A.; Aziz, M.A.; El-Sayed, A. Artichoke edible parts are hepatoprotective as commercial leaf preparation. *Rev. Bras. Farmacogn.* **2018**, *28*, 165–178. [CrossRef]
61. Moglia, A.; Lanteri, S.; Comino, C.; Acquadro, A.; de Vos, R.; Beekwilder, J. Stress-induced biosynthesis of dicaffeoylquinic acids in globe artichoke. *J. Agric. Food Chem.* **2008**, *56*, 8641–8649. [CrossRef]
62. Schütz, K.; Persike, M.; Carle, R.; Schieber, A. Characterization and quantification of anthocyanins in selected artichoke (*Cynara scolymus* L.) cultivars by HPLC–DAD–ESI–MS n. *Anal. Bioanal. Chem.* **2006**, *384*, 1511–1517. [CrossRef]
63. Fritsche, J.; Beindorff, C.M.; Dachtler, M.; Zhang, H.; Lammers, J.G. Isolation, characterization and determination of minor artichoke (*Cynara scolymus* L.) leaf extract compounds. *Eur. Food Res. Technol.* **2002**, *215*, 149–157. [CrossRef]
64. Farag, M.A.; Gad, H.A.; Heiss, A.G.; Wessjohann, L.A. Metabolomics driven analysis of six Nigella species seeds via UPLC-qTOF-MS and GC–MS coupled to chemometrics. *Food Chem.* **2014**, *151*, 333–342. [CrossRef] [PubMed]
65. Albeladi, S.S.R.; Malik, M.A.; Al-thabaiti, S.A. Facile biofabrication of silver nanoparticles using Salvia officinalis leaf extract and its catalytic activity towards Congo red dye degradation. *J. Mater. Res. Technol.* **2020**, *9*, 10031–10044. [CrossRef]
66. Desai, R.; Mankad, V.; Gupta, S.K.; Jha, P.K. Size distribution of silver nanoparticles: UV-visible spectroscopic assessment. *Nanosci. Nanotechnol. Lett.* **2012**, *4*, 30–34. [CrossRef]
67. Durán, N.; Marcato, P.D.; Alves, O.L.; De Souza, G.I.; Esposito, E. Mechanistic aspects of biosynthesis of silver nanoparticles by several Fusarium oxysporum strains. *J. Nanobiotechnol.* **2005**, *3*, 8. [CrossRef]
68. Rani, P.; Trivedi, L.; Gaurav, S.S.; Singh, A.; Shukla, G. Green synthesis of silver nanoparticles by Cassytha filiformis L. extract and its characterization. *Mater. Today Proc.* **2022**, *49*, 3510–3516. [CrossRef]
69. Shameli, K.; Ahmad, M.B.; Zargar, M.; Yunus, W.M.Z.W.; Rustaiyan, A.; Ibrahim, N.A. Synthesis of silver nanoparticles in montmorillonite and their antibacterial behavior. *Int. J. Nanomed.* **2011**, *6*, 581. [CrossRef]
70. Mistry, H.; Thakor, R.; Patil, C.; Trivedi, J.; Bariya, H. Biogenically proficient synthesis and characterization of silver nanoparticles employing marine procured fungi Aspergillus brunneoviolaceus along with their antibacterial and antioxidative potency. *Biotechnol. Lett.* **2021**, *43*, 307–316. [CrossRef]
71. Kathiresan, K.; Manivannan, S.; Nabeel, M.; Dhivya, B. Studies on silver nanoparticles synthesized by a marine fungus, Penicillium fellutanum isolated from coastal mangrove sediment. *Colloids Surf. B Biointerfaces* **2009**, *71*, 133–137. [CrossRef]
72. Li, R.; Pan, Y.; Li, N.; Wang, Q.; Chen, Y.; Phisalaphong, M.; Chen, H. Antibacterial and cytotoxic activities of a green synthesized silver nanoparticles using corn silk aqueous extract. *Colloids Surf. A: Physicochem. Eng. Asp.* **2020**, *598*, 124827. [CrossRef]
73. Farvin, K.S.; Jacobsen, C. Phenolic compounds and antioxidant activities of selected species of seaweeds from Danish coast. *Food Chem.* **2013**, *138*, 1670–1681. [CrossRef] [PubMed]
74. Karthik, R.; Govindasamy, M.; Chen, S.-M.; Cheng, Y.-H.; Muthukrishnan, P.; Padmavathy, S.; Elangovan, A. Biosynthesis of silver nanoparticles by using Camellia japonica leaf extract for the electrocatalytic reduction of nitrobenzene and photocatalytic degradation of Eosin-Y. *J. Photochem. Photobiol. B Biol.* **2017**, *170*, 164–172. [CrossRef] [PubMed]
75. Centers for Disease Control and Prevention. Antifungal Susceptibility Testing and Interpretation. 2020. Available online: https://www.cdc.gov/fungal/candida-auris/c-auris-antifungal.html (accessed on 10 April 2022).
76. Navalkele, B.D.; Revankar, S.; Chandrasekar, P. Candida auris: A worrisome, globally emerging pathogen. *Expert Rev. Anti-Infect. Ther.* **2017**, *15*, 819–827. [CrossRef] [PubMed]
77. Alfouzan, W.; Dhar, R.; Albarrag, A.; Al-Abdely, H. The emerging pathogen Candida auris: A focus on the Middle-Eastern countries. *J. Infect. Public Health* **2019**, *12*, 451–459. [CrossRef]

78. Chowdhary, A.; Anil Kumar, V.; Sharma, C.; Prakash, A.; Agarwal, K.; Babu, R.; Dinesh, K.; Karim, S.; Singh, S.; Hagen, F. Multidrug-resistant endemic clonal strain of Candida auris in India. *Eur. J. Clin. Microbiol. Infect. Dis.* **2014**, *33*, 919–926. [CrossRef]
79. Lockhart, S.R.; Etienne, K.A.; Vallabhaneni, S.; Farooqi, J.; Chowdhary, A.; Govender, N.P.; Colombo, A.L.; Calvo, B.; Cuomo, C.A.; Desjardins, C.A. Simultaneous emergence of multidrug-resistant Candida auris on 3 continents confirmed by whole-genome sequencing and epidemiological analyses. *Clin. Infect. Dis.* **2017**, *64*, 134–140. [CrossRef]
80. Ostrowsky, B.; Greenko, J.; Adams, E.; Quinn, M.; O'Brien, B.; Chaturvedi, V.; Berkow, E.; Vallabhaneni, S.; Forsberg, K.; Chaturvedi, S. Candida auris isolates resistant to three classes of antifungal medications—New York, 2019. *Morb. Mortal. Wkly. Rep.* **2020**, *69*, 6. [CrossRef]
81. Peña-González, C.E.; Pedziwiatr-Werbicka, E.; Martín-Pérez, T.; Szewczyk, E.M.; Copa-Patiño, J.L.; Soliveri, J.; Pérez-Serrano, J.; Gómez, R.; Bryszewska, M.; Sánchez-Nieves, J. Antibacterial and antifungal properties of dendronized silver and gold nanoparticles with cationic carbosilane dendrons. *Int. J. Pharm.* **2017**, *528*, 55–61. [CrossRef]
82. Al-Nbaheen, M.; Ali, D.; Bouslimi, A.; Al-Jassir, F.; Megges, M.; Prigione, A.; Adjaye, J.; Kassem, M.; Aldahmash, A. Human stromal (mesenchymal) stem cells from bone marrow, adipose tissue and skin exhibit differences in molecular phenotype and differentiation potential. *Stem Cell Rev. Rep.* **2013**, *9*, 32–43. [CrossRef]
83. Wani, I.A.; Ahmad, T. Size and shape dependant antifungal activity of gold nanoparticles: A case study of Candida. *Colloids Surf. B Biointerfaces* **2013**, *101*, 162–170. [CrossRef]
84. Dlugaszewska, J.; Dobrucka, R. Effectiveness of biosynthesized trimetallic Au/Pt/Ag nanoparticles on planktonic and biofilm Enterococcus faecalis and Enterococcus faecium forms. *J. Clust. Sci.* **2019**, *30*, 1091–1101. [CrossRef]
85. Rubiolo, J.A.; Ternon, E.; López-Alonso, H.; Thomas, O.P.; Vega, F.V.; Vieytes, M.R.; Botana, L.M. Crambescidin-816 acts as a fungicidal with more potency than crambescidin-800 and-830, inducing cell cycle arrest, increased cell size and apoptosis in Saccharomyces cerevisiae. *Mar. Drugs* **2013**, *11*, 4419–4434. [CrossRef] [PubMed]
86. Stefanini, I.; Rizzetto, L.; Rivero, D.; Carbonell, S.; Gut, M.; Heath, S.; Gut, I.G.; Trabocchi, A.; Guarna, A.; Ben Ghazzi, N. Deciphering the mechanism of action of 089, a compound impairing the fungal cell cycle. *Sci. Rep.* **2018**, *8*, 5964. [CrossRef] [PubMed]
87. Yan, C.; Wang, S.; Wang, J.; Li, H.; Huang, Z.; Sun, J.; Peng, M.; Liu, W.; Shi, P. Clioquinol induces G2/M cell cycle arrest through the up-regulation of TDH3 in Saccharomyces cerevisiae. *Microbiol. Res.* **2018**, *214*, 1–7. [CrossRef]
88. Akter, M.; Sikder, M.T.; Rahman, M.M.; Ullah, A.A.; Hossain, K.F.B.; Banik, S.; Hosokawa, T.; Saito, T.; Kurasaki, M. A systematic review on silver nanoparticles-induced cytotoxicity: Physicochemical properties and perspectives. *J. Adv. Res.* **2018**, *9*, 1–16. [CrossRef]
89. Hwang, I.s.; Lee, J.; Hwang, J.H.; Kim, K.J.; Lee, D.G. Silver nanoparticles induce apoptotic cell death in Candida albicans through the increase of hydroxyl radicals. *FEBS J.* **2012**, *279*, 1327–1338. [CrossRef]
90. Zhu, M.-T.; Wang, Y.; Feng, W.-Y.; Wang, B.; Wang, M.; Ouyang, H.; Chai, Z.-F. Oxidative stress and apoptosis induced by iron oxide nanoparticles in cultured human umbilical endothelial cells. *J. Nanosci. Nanotechnol.* **2010**, *10*, 8584–8590. [CrossRef]
91. Daniel, B.; DeCoster, M.A. Quantification of sPLA2-induced early and late apoptosis changes in neuronal cell cultures using combined TUNEL and DAPI staining. *Brain Res. Protoc.* **2004**, *13*, 144–150. [CrossRef]

Journal of *Fungi*

Article

Multifunctional Silver Nanoparticles Based on Chitosan: Antibacterial, Antibiofilm, Antifungal, Antioxidant, and Wound-Healing Activities

Amr M. Shehabeldine [1], Salem S. Salem [1,*], Omar M. Ali [2], Kamel A. Abd-Elsalam [3], Fathy M. Elkady [4] and Amr H. Hashem [1,*]

1 Botany and Microbiology Department, Faculty of Science, Al-Azhar University, Nasr City, Cairo 11884, Egypt; dramrshehab@azhar.edu.eg
2 Department of Chemistry, Turabah University College, Turabah Branch, Taif University, Taif 21944, Saudi Arabia; om.ali@tu.edu.sa
3 Plant Pathology Research Institute, Agricultural Research Centre, Giza 12619, Egypt; kamelabdelsalam@gmail.com
4 Microbiology and Immunology Department, Faculty of Pharmacy (Boys), Al-Azhar University, Nasr City, Cairo 11884, Egypt; fathyelkady2426.el@azhar.edu.eg
* Correspondence: salemsalahsalem@azhar.edu.eg (S.S.S.); amr.hosny86@azhar.edu.eg (A.H.H.)

Abstract: The purpose of this study is to create chitosan-stabilized silver nanoparticles (Chi/Ag-NPs) and determine whether they were cytotoxic and also to determine their characteristic antibacterial, antibiofilm, and wound healing activities. Recently, the development of an efficient and environmentally friendly method for synthesizing metal nanoparticles based on polysaccharides has attracted a lot of interest in the field of nanotechnology. Colloidal Chi/Ag-NPs are prepared by chemical reduction of silver ions in the presence of Chi, giving Chi/Ag-NPs. Physiochemical properties are determined by Fourier-transform infrared spectroscopy (FTIR), X-ray diffraction (XRD), transmission electron microscopy (TEM), dynamic light scattering (DLS), and scanning electron microscopy with energy-dispersive X-ray spectroscopy (SEM-EDX) analyses. TEM pictures indicate that the generated Chi/Ag-NPs are nearly spherical in shape with a thin chitosan covering around the Ag core and had sizes in the range of 9–65 nm. In vitro antibacterial activity was evaluated against *Staphylococcus aureus* and *Pseudomonas aeruginosa* by a resazurin-mediated microtiter plate assay. The highest activity was observed with the lowest concentration of Chi/Ag-NPs, which was 12.5 μg/mL for both bacterial strains. Additionally, Chi/Ag-NPs showed promising antifungal features against *Candida albicans*, *Aspergillus fumigatus*, *Aspergillus terreus*, and *Aspergillus niger*, where inhibition zones were 22, 29, 20, and 17 mm, respectively. Likewise, Chi/Ag-NPs revealed potential antioxidant activity is 92, 90, and 75% at concentrations of 4000, 2000, and 1000 μg/mL, where the IC_{50} of Chi/Ag-NPs was 261 μg/mL. Wound healing results illustrated that fibroblasts advanced toward the opening to close the scratch wound by roughly 50.5% after a 24-h exposure to Chi/Ag-NPs, greatly accelerating the wound healing process. In conclusion, a nanocomposite based on AgNPs and chitosan was successfully prepared and exhibited antibacterial, antibiofilm, antifungal, antioxidant, and wound healing activities that can be used in the medical field.

Keywords: chitosan; chitosan/silver nanoparticles; antimicrobial; antioxidant; wound healing

Citation: Shehabeldine, A.M.; Salem, S.S.; Ali, O.M.; Abd-Elsalam, K.A.; Elkady, F.M.; Hashem, A.H. Multifunctional Silver Nanoparticles Based on Chitosan: Antibacterial, Antibiofilm, Antifungal, Antioxidant, and Wound-Healing Activities. *J. Fungi* **2022**, *8*, 612. https://doi.org/10.3390/jof8060612

Academic Editor: David S. Perlin

Received: 23 April 2022
Accepted: 6 June 2022
Published: 8 June 2022

1. Introduction

Antimicrobial resistance occurs when bacteria, viruses, fungi, and parasites evolve over time and become less drug-responsive, making infections more difficult to treat and raising the risk of sickness, severe illness, and death. Antibiotics and other antimicrobial drugs are rendered ineffective by drug resistance, and infections are becoming increasingly difficult or impossible to treat. Therefore, it is necessary to design and develop new compounds that overcome these limitations. Recently, nanoparticles (NPs) have been successfully used to deliver therapeutic agents [1–4] in the diagnosis of chronic diseases, to reduce bacterial infections, and in the food and clothing industries as antimicrobial agents [5–9]. Because of their antimicrobial properties and unique mode of action, specialized NPs provide an attractive alternative to conventional antibiotics in the development of a new generation of antibiotics. Green synthesis of metallic NPs is an economical, easy, and environmentally friendly approach [10–17]. Due to their various applications in the medical field, silver NPs (AgNPs) have been proposed as treatment agents to overcome the problem of drug resistance caused by the abuse of antibiotics [18–20]. The mechanism by which silver exerts its biological activity is still poorly understood. The bacteria that makeup biofilms are a primary concern of medical microbiology. These membranes are formed by secreting a layer of polymer consisting of sugars, nucleic acids, and proteins, which are formed on biological and non-biological membranes and surfaces [21,22]. This layer plays an important role by restricting the diffusion of antibiotics into biofilm-producing cells and making them resistant to antibiotics. It has been found that bacterial cells growing within biofilms secrete different surface molecules and virulence factors and exhibit low growth rates, which enhance their pathogenicity by several folds [23]. Chitosan is a linear polysaccharide that is obtained from the deacetylation of chitin, a naturally occurring polymer present in the shells of prawns and other crustaceans [24]. It is one of the most commonly used biopolymers in a wide range of applications, including fabrics, cosmetics, water treatment, and food processing [25–27]. Previous studies confirmed that chitosan has multiple roles in nanoparticle synthesis, stabilization, and their applications [28,29]. The wound healing process is described by a series of responses including inflammatory, tissue regenerating, and tissue-remodeling processes [30]. Wound dressings are biomaterials of synthetic or natural origin that support wound healing by providing suitable micro-environments capable of attracting the cells to the wounded area [31]. In recent years, among the various wound dressing options, hydrogels have been highlighted for their unique properties, thus making them ideal for promoting an environment conducive to tissue regeneration [32]. Herein, AgNPs have been synthesized in the presence of chitosan using a new environmentally friendly synthesis process, and their biological activities, namely, antibacterial, antibiofilm, antifungal, antioxidant, and wound healing promoting skills, have been evidenced.

2. Materials and Methods

2.1. Synthesis of Chitosan/Silver Nanoparticles (Chi/Ag-NPs)

Chitosan/silver nanoparticles (Chi/Ag-NPs) were synthesized utilizing a chemical reduction process with chitosan as a reducing and stabilizing agent. With minor changes, the synthesis of Chitosan/Silver nanoparticles (Chi/Ag-NPs) was carried out according to the technique reported by [33]. Chitosan (0.2%) was produced by mixing in 0.5% acetic acid. After that, the chitosan solution was filtered to create a homogeneous solution. NaOH and $AgNO_3$ had just been made. To the chitosan solution, an aliquot of 2 mL of 2 mM $AgNO_3$ and 100 µL of 0.5 M NaOH (pH 10) was added. At 85 °C, the chitosan solution was agitated for 4 h. The colorless chitosan solution became yellow, then brown, indicating that Chi/Ag-NPs had been synthesized. Finally, the Chi/Ag-NPs solution was stored, and the precipitate was filtered, washed with distilled water, and dried in an oven at 90 °C for 4 h. Chitosan purchased from Sigma-Aldrich (St. Louis, MO, USA), and Silver nitrate ($AgNO_3$) was obtained from Fisher Scientific (Mumbai, India). Other chemicals and reagents used in this study were purchased from Modern Lab Co., Madhya Pradesh, India, in analytical grade without any purification required.

2.2. Characterization of Chi/Ag-NPs

A variety of instrumental analytical methods were used to characterize the Chi/Ag-NPs. Using a Spectrum Two IR Spectrometer (PerkinElmer Inc., Shelton, WA, USA) and these techniques, the total internal reflectance/Fourier-transform Infrared (ATR-FTIR) spectra was used to semi-quantitatively measure the observable IR spectrum of the Chi/Ag-NPs by evaluating the transmittance over a spectral region of 4000 to 400 cm^{-1}. To achieve a suitable signal quality, all spectra were collected at a 4 cm^{-1} resolution by collecting 32 scans [34]. A Diano X-ray diffractometer (Philips) with a CuK radiation source (λ = 0.15418 nm) activated at 45 kV, as well as a generator (PW, 1930) and a goniometer (PW, 1820), was used to study the XRD pattern of the produced Chi/Ag-NPs. The shape and size of the prepared Chi/Ag-NPs were observed using the TEM method. The Ultra-High Resolution transmission electron microscope (JEOL-2010, Tokyo, Japan) with a voltage of 200 kV was employed. A drop of the particle solution was placed on a carbon-coated copper grid and dried under a light to create TEM grids. The NicompTM-380 ZLS size analyzer from the United States (USA) was used to calculate the pore size distribution and zeta potential of the prepared Chi/Ag-NPs using dynamic light scattering (DLS). For particle size detection, laser beam scattering at 170° was utilized, with the zeta potential recorded at 18°. A field emission scanning electron microscope (SEM) installed with a Field Emission-Gun (Quanta, 250-FEG) and connected with an energy-dispersive X-ray analyzer (EDX, Unit) with an excitation source of 30 kV for energy-dispersive X-ray evaluation (EDX) and mapping were used to examine the surfaces of the prepared Chi/Ag-NPs.

2.3. Microbial Strains and Reagent

Staphylococcus aureus ATCC® 25923™ and *Pseudomonas aeruginosa* MTCC1034 were cultivated in Luria broth medium and incubated at 37 °C, for 16–18 h. Fungal strains used are *Candida albicans* ATCC 90028, *Aspergillus niger* RCMB 02724, *A. terreus* RCMB 02574, and *A. fumigatus* RCMB 02568. These four fungal strains were inoculated on malt extract agar (MEA) (Oxoid) plates; then incubated for 3–5 days at 28 ± 2 °C; then kept at 4 °C for further use [35–38]. Chitosan, low molecular weight, and crystal violet were purchased from Sigma-Aldrich (St Louis, MO, USA). MTT [3-(4,5-dimethylt hiazol-2-yl)-2,5-diphenyltetrazolium bromide]. Silver nitrate (AgNO3) was obtained from Fisher Scientific (Mumbai, India). Normal human skin cell line (BJ-1) was cultured in RPMI 1640 medium supplemented with 10% heated fetal bovine serum, 1% of 2 mM l-glutamine, 50 IU/mL penicillin, and 50 µg/mL streptomycin.

2.4. Broth Microdilution Assay

Briefly, a fresh culture on LB broth media at a turbidity equivalent to that of 0.5 McFarland standard, 500 µL of each bacterial culture were added into a 96-well polystyrene flat-bottomed microtiter plate [39]. Tested sample was added to bacterial suspension in each well at a final concentration ranging from 0 to 1000 µg/mL. Growth control wells contained only bacteria in LB media. Two-fold serial dilutions of the tested samples Chi/Ag-NPs, were made starting with the first well by adding 50 µL of the tested sample, dissolved at a concentration of 1000 µg/mL. To each of the wells, 10 µL of the diluted culture (0.5 McFarland standards) was added. After incubation at 37 °C for 24 h, 5 µL resazurin indicator (prepared by dissolving 0.016 g in 100 mL of sterile distilled water) was also added to all 96 wells. Then the microtiter plate was incubated in the dark. We observed with the naked eye any change in color from purple to pink as a positive result. The lowest concentration of the tested sample in which discoloration occurred was recorded as the MIC value. All experiments were performed in triplicate [40]. The MBC for each sample was calculated by plating the contents of the first three wells that showed no visible bacteria growth onto LB plates and incubating for 24 h [41].

2.5. Evaluation of Anti-Biofilm Activity

Biofilm experiments were performed using static biofilm model. Effect of tested sample Chi/Ag-NPs on *S. aureus* ATCC® 25923™ and *P. aeruginosa* MTCC1034 biofilm formation were determined by the crystal violet staining method [42,43]. Prepared compound was diluted into 96-well plates as described above; six concentrations diluted from 0.5xMIC. The plates were incubated under aerobic conditions at 37 °C for 48 h. discarding the liquid mixture; the wells were stained with 0.1 mL 0.4% crystal violet for 15 min after being washed with sterile water twice. Then, samples were rinsed with distilled water twice and the dye bound to biofilm was solubilized by adding ethanol (95%). Absorbance of the isolated dye was measured quantitatively at 540 nm. The biofilm inhibition percentage was calculated using the following formula [44]:

$$[(\text{OD growth control} - \text{OD sample})/\text{OD growth control}] \times 100$$

OD: optical density

2.6. *Antifungal Activity*

The test of diffusion in agar was performed in accordance with the document M51-A2 of the Clinical Laboratory Standard Institute [45] with minor adaptations. Fungal strains were initially grown on MEA plates and incubated at 30 °C for 3–5 days [46–48]. The fungal suspension was prepared in sterilized phosphate buffer solution (PBS) pH 7.0, and then the inoculum was adjusted to 10^7 spores/mL after counting in a cell counter chamber. One ml was uniformly distributed on agar MEA Plates. In total, 100 µL of each Chi/Ag-NPs, Ag+, Chi, and nystatin were put in Agar wells (7 mm) then incubated at 30 °C. After 72 h of incubation, the inhibition zone diameter was measured [49,50].

2.7. *Antioxidant Activity*

Different concentrations Chi/Ag-NPs, Ag+, Chi, and Ascorbic acid rangeing from 3.9 to 4000 µg/mL were used to determine the ability to scavenge DPPH (2,2-diphenyl-1-picrylhydrazyl) radicals. DPPH solution (800 µL) was mixed with 200 µL of the specific concentration and incubated for 30 min at 25 °C in darkness. After this time, centrifugation was performed at 13,000 rpm for 5 min, then absorbance was measured at 517 nm [46]. Antioxidant activity was calculated by the following equation:

$$\text{Antioxidant activity } (\%) = \frac{\text{Control absorbance} - \text{Sample absorbance}}{\text{Control absorbance}} \times 100$$

2.8. *Determination of Safe Dose on the Proliferation of Normal Human Skin Fibroblast Cell Line by Sulphorhodamine B (SRB) Assay*

Cytotoxicity was investigated using normal human skin cell line (BJ-1), Cell viability was assessed by SRB assay [51]. Aliquots of 100 µL cell suspension (5×10^3 cells) were in 96-well plates and incubated in complete media for 24 h. Before addition to the culture medium, tested substance Chi/Ag-NPs and standard substance drugs doxorubicin (DOX) were dissolved in DMSO and followed by serial dilution for 6 points ranging from 200 µg/mL to 1.56 µg/mL. After 72 h of exposure, cells were fixed by replacing media with 150 µL of 10% TCA and incubated at 4 °C for 1 h. Aliquots of 70 µL SRB solution (0.4% *w*/*v*) were added and incubated in a dark place at room temperature for 10 min. Plates were washed 3 times with 1% acetic acid and allowed to air-dry overnight. Then, 150 µL of TRIS (10 mM) was added to dissolve protein-bound SRB stain [52]. The effective safe concentration (EC100) value (at 100% cell viability) of each tested extract was estimated by GraphPad Instat software (version 6.01), California, USA. GraphPad Instat software was used to evaluate the cytotoxicity activity of the tested material, which was expressed as an IC_{50} value. The experiments were carried out three times. As a positive control, doxorubicin was used. The levels of cytotoxic effects were categorized as cytotoxic (IC_{50} 2.00 µg/mL), moderately cytotoxic (IC_{50} 2.00–89.00 µg/mL), and non-toxic (IC_{50} > 90.00 µg/mL) according to the Special Programme for Research and Training in Tropical Diseases (WHO—Tropical Diseases) [53]. The rate of the cytotoxicity (CT%) was estimated by the following expression:

$$\text{CT\%} = \text{Ac} - \text{At}/\text{Ac} \times 100$$

In which, Ac and At are the absorbance of the control sample and the test sample, respectively.

2.9. In Vitro Wound-Healing Assay

The wound-healing potential of the final formulations was assessed by in vitro wound-healing assay [54]. To this aim, Human Skin Fibroblast cell line were seeded at a density of 3×10^5/well onto a coated 6-well plate in 5% FBS-DMEM at 37 °C and 5% CO_2 to obtain a monolayer of cells [55] then, a scratch was made across the middle of each well using a sterile 1000 µL pipet tip. The plate was washed thoroughly with PBS. Control wells were replenished with fresh medium while drug wells were treated with fresh media containing drug. Images were taken using an inverted microscope at the indicated time intervals. The plate was incubated at 37 °C and 5% CO_2 in-between time points. The migration rate can be expressed as the percentage of area reduction of wound closure, which increases as cells migrate over time [56].

$$\text{Wound closure\%: } (A_0 - A_t/A_0) \times 100,$$

where A_0 = 0 h is the average area of the wound measured immediately after scratching (time zero), and A_t = Δh is the average area of the wound measured h hours after the scratch is performed.

2.10. Statistical Analysis

Data are presented as means ± SD of at least three independent experiments. Comparisons are made by the Student's *t*-test or by ANOVA when appropriate. Differences are considered statistically significant at $p < 0.05$. Statistical analysis was carried out estimated by GraphPad Instat software, (version 6.01), San Diego, CA, USA.

3. Results and Discussion

3.1. Characterization of Chi/Ag-NPs

Nanostructures have piqued curiosity as a fast-evolving class of materials with a wide range of uses. Nanomaterials have been described using a number of methodologies to explain their size, crystalline structure, elemental content, and a range of other physical features. There are various physical qualities that may be examined using many methods. The many strengths and limits of each methodology make it difficult to choose the best method, and an aggregate characterization approach is frequently necessary [57]. FTIR analysis of Chi/Ag-NPs was carried out to validate the decrease in molecular interaction and capping agent, which is important for the nanoparticles' synthesis and stabilization. This method is commonly used to analyze nanostructures qualitatively. The FTIR spectra of Ag NPs revealed absorption peaks at 3435 cm^{-1}, 2940 cm^{-1}, 2869 cm^{-1}, 1725 cm^{-1}, 1366 cm^{-1}, 1251 cm^{-1}, 1132 cm^{-1}, 948 cm^{-1}, and 731 cm^{-1}, which correspond to linkage groups (Figure 1A). Moreover, the peaks at 3435 cm^{-1} matched to -OH-group stretch-vibrations. Asymmetric and symmetric -CH_2 groups may be ascribed to the bands at 2940 cm^{-1} and 2869 cm^{-1}, respectively. Carbonyl expands vibrations in aldehydes, ketones, and carboxylic acids, which correlate to the peak at 1725 cm^{-1}. The existence of a strong 1725 cm^{-1} signal in Chi/Ag-NPs shows that silver ion (Ag+) reduction is accompanied by hydroxyl group oxidation in chitosan structures. C-N- stretching is shown by the peak seen at 1366 cm^{-1}. The -C–O–C stretching is responsible for the peak at 1251 cm^{-1}. The C-O stretching vibration is shown by the peak at 1132 cm^{-1}. The absorption peak at 948 cm^{-1} conforms to the β-d-glucose unit's typical absorption. One of the most extensively used methods for the characterization of NPs is X-ray diffraction (XRD). The crystalline nature, phase behavior, lattice constants, and particle sizes are commonly determined using XRD. The significant peaks appeared at 38.8°, 47.7°, 63.2°, and 76.9°, correlating to lattice-planes (111), (200), (220), and (311) are shown in XRD diffraction of synthesized Chi/Ag-NPs. The Chi/Ag-NPs XRD results were found to be highly similar to JCPDS card no. 04-0783. As a result, XRD revealed that AgNPs were generated by reducing $AgNO_3$ with chitosan (Figure 1B), and the crystal lattice complemented previous studies. Furthermore, the most prominent diffraction peaks at 16.2° and 22.5° confirmed the crystalline form of chitosan [58].

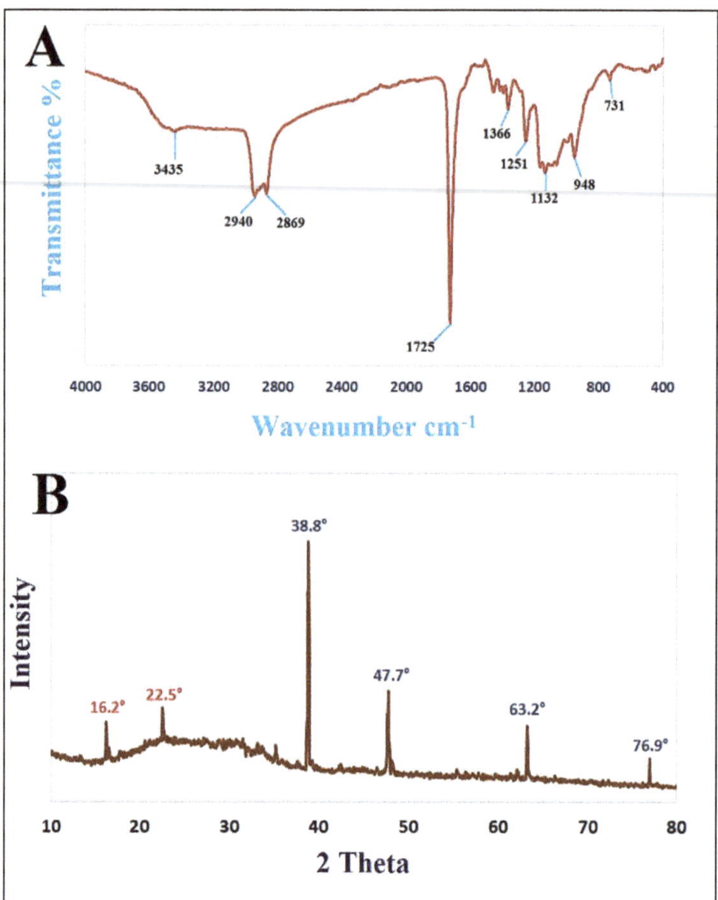

Figure 1. FTIR spectrum (**A**) and XRD pattern (**B**) of Chi/Ag NPs.

The TEM pictures indicated that the generated Chi/Ag-NPs were nearly spherical, polydisperse, and had sizes in the range of 9–65 nm Figure 2A. The particles had a spherical shape with a thin chitosan covering around the Ag core (Figure 2B). Furthermore, TEM micrographs revealed a uniform dispersion of the chitosan covering surrounding the Ag-nanostructures. When Chi/Ag-NPs were tested, no aggregation was found, indicating that the nanostructures were entirely covered with a polymer. Ag-nanoparticles coated with chitosan did, in fact, have a transparent coating surrounding their core. In another paper, it was discovered that the size of Chi/Ag-NPs ranges around 10–230 nm [33]. Figure 2C shows the Chi/Ag-NPs' areas selected electron diffraction (SAED) pattern, which shows excellent sharp rings and confirms the Ag-nanostructures' crystalline character. DLS is among the most often used techniques to detect the distribution of particle size in a colloidal mixture based on intensity. The Chi/Ag-NPs produced were a polydispersed combination of spherical Chi/Ag-NPs with an average diameter of 117.6 nm (Figure 2D). Because the acquired size utilizing DLS is not only connected to the metallic core of particles but also impacted by capping proteins around the particles, the resulting particle sizes were over expressed when compared to those observed utilizing TEM. The Chi/Ag-NPs had a zeta potential of −48.76 mV (Figure 2E), showing that Chi/Ag-NPs in suspension were dispersed optimally. The surface charge of the Chi/Ag-NPs is represented by the negative charge of the zeta potential value.

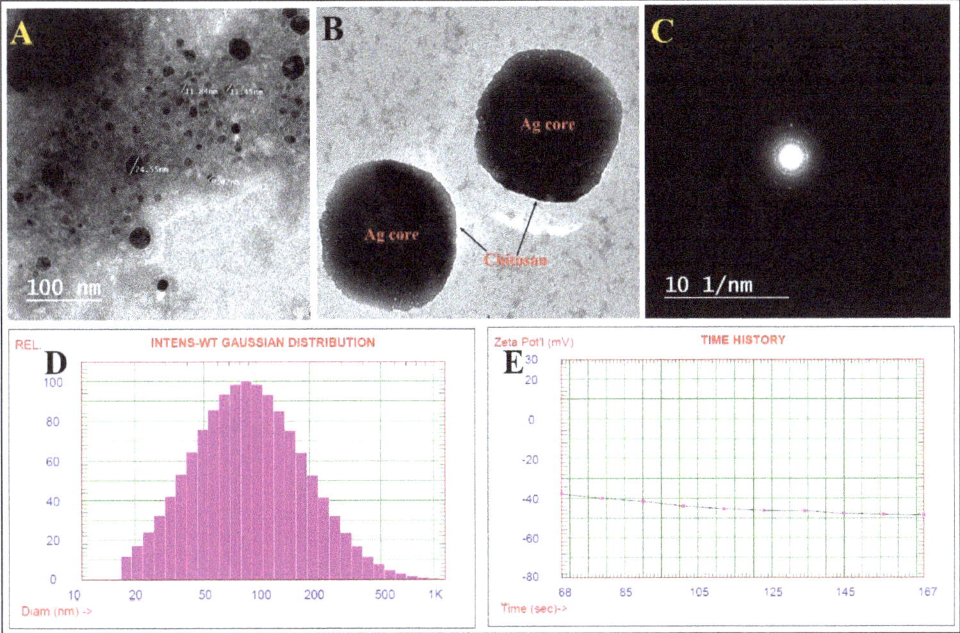

Figure 2. TEM images (**A**,**B**), SAED pattern (**C**), DLS (**D**),and zeta potential (**E**) of Chi-Ag-NPs.

As shown in Figure 3A, the SEM was used to evaluate the surface morphology and particle size of Chi/Ag-NPs. Chi/Ag-NPs had a form that was virtually spherical. The particle size varied from 24 to 80 nm on average. EDX analysis was used to determine the elemental composition of the Chi/Ag-NPs powder. In the Chi/Ag-NPs, the EDX spectra revealed the existence of several well-defined bands associated with silver [Ag], oxygen [O], and carbon [C] components (Figure 3B). The carbon [C] and oxygen [O] signals come from the chitosan, whereas the silver [Ag] peak indicates the creation of Ag-nanostructures. Furthermore, EDX spectra revealed the generation of very pure Chi/Ag-NPs with no additional impurity-related peaks.

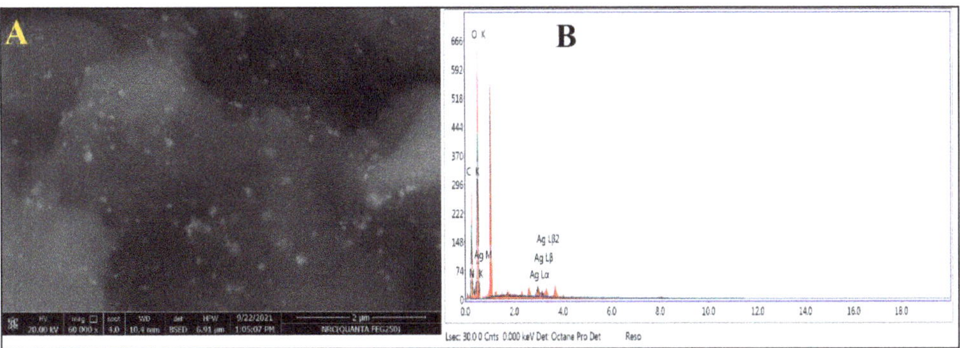

Figure 3. (**A**) SEM image and (**B**) EDX spectrum of prepared Chi/Ag-NPs.

3.2. Determination of Minimum Inhibitory Concentration (MIC) by Resazurin Stain

In this study, Chi/Ag-NPs were tested as an antimicrobial against selected gram-positive and gram-negative bacteria. Using a resazurin-mediated microtiter plate assay, the antibacterial activity of Chi, Ag+, and Chi/Ag-NPs was investigated against *P. aeruginosa* and *S. auerus* bacterial strains. The color shift of the resazurin indicator was used to visually assess the inhibitory action of Chi, Ag+, and Chi/Ag-NPs. Figure 4 depicts the color shift seen at various concentrations of Chi, Ag+, and Chi/Ag-NPs. Chitosan and Ag+ exhibited moderate antibacterial activity against *P. aeruginosa* and *S. auerus*, with MICs ranging from 25 to 100 µg/mL, while Chi/Ag-NPs had increased antibacterial activity against both bacterial strains. Figure 4 shows the MIC and MBC values for Chi, Ag+, and Chi/Ag-NPs. The highest activity was observed with the lowest concentration of Chi/Ag-NPs, which was 12.5 µg/mL for both *P. aeruginosa and S. auerus*. Table 1 summarizes the results, which show the mean MIC and MBC values for each antibacterial agent tested. The increased antimicrobial activity of Ag-incorporated chitosan materials due to a high infiltration of the silver component results in a high bactericidal activity, which is consistent with our experimental findings results. These findings also show that chitosan-based antibacterial silver nano may have a dual mechanism of action. Chitosan is the enhanced activity as a result of Chi/Ag-NPs, which acts as a stabilizing agent as well as a carrier for Ag-NPs. This could also be due to the increased surface area and positive surface density, which allow for better interaction with negatively charged bacterial cell membranes, enhancing alteration in cell permeability and penetration of nano-sized particles into the bacterial cell, resulting in its death [59,60]. As expected, the mode of action could also be due to Ag+ interactions from the AgNPs and Chi/Ag-NPs conjugates with DNA or available proteins in the bacterial cell wall, which ultimately lead to cell death [61].

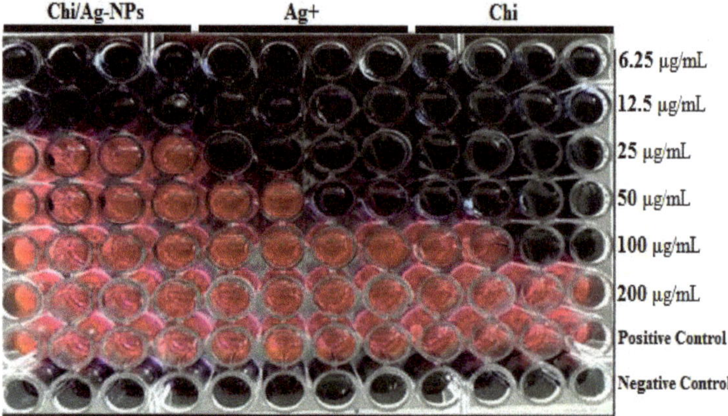

Figure 4. Resazurin dye test for determining minimum inhibitory concentration against *S. aureus* ATCC® 25923™ and *P. aeruginosa* MTCC1034.

Table 1. Minimum inhibition concentrations and minimum bactericidal concentrations of Chi, Ag+, and Chi/Ag-NPs samples against gram-negative *P. aeruginosa* and gram-positive *S. aureus*.

	MIC µg/mL		MBC µg/mL	
Ag+	25.0	50.0	100.0	ND
Chi	50.0	100.0	100.0	ND
Chi/AgNPs	12.5	12.5	25.0	50.0

ND: not detected; MIC: minimum inhibitory concentration, MBC: minimum bactericidal concentration.

3.3. Anti-Biofilm Evaluation

Biofilm inhibition was measured using a standard crystal violet assay, and the results were expressed as a percentage. Figure 5 shows the anti-biofilm activity of the Chi/Ag-NPs used in their synthesis at different concentrations. Bacterial biofilm inhibition of Chi/Ag-NPs demonstrated significant biofilm inhibiting activity ($p < 0.05$) against *P. aeruginosa* but not against *S. auerus*. In a biofilm quantification assay, test bacterial pathogens showed a concentration-dependent decrease in biofilm formation (Figure 5). The antibiofilm activity of Chi/Ag-NPs against *P. aeruginosa* at sub-MIC levels reduced biofilm formation by 78 and 69% at 0.5 × MIC and 0.25 × MIC, respectively. Furthermore, the antibiofilm activity of *S. auerus* by Chi/Ag-NPs at sub-MIC levels reduced biofilm formation by 48% and 36% at 0.5 × MIC, and 0.25 × MIC, respectively. At the sub-MIC levels tested, this compound had the highest inhibitory potential for biofilm formation against *P. aeruginosa* and *S. auerus* biofilm formation without inhibiting planktonic growth. One of the most important reasons for the effect of silver nanocomposites on biofilm inhibition activity is their particle size, as smaller particles have a larger surface area for interaction with microorganisms when compared to the bacterial control [62]. Chi/Ag-NPs hydrogel showed reduced biomass biofilm formation when compared to the bacterial control. Higher levels of biofilm reductase have been reported when AgNPs particles are less than 100 nm in size, which inhibits the synthesis and secretion of extracellular polysaccharides [63]. In the production and secretion of exopolysaccharides (EPSs) by bacterial cells, environmental signals trigger the production of EPS in bacteria. As a result, biofilm formation will be limited if EPS formation can be suppressed [64]. Our results indicated that Ag-NPs with a size of 9–65 nm in the Chi/Ag-NPs hydrogel help prevent biofilm formation. Biosynthesized AgNPs have previously been shown to have anti-biofilm activity against *P. aeruginosa* and *S. epidermidis* by AgNPs with a diameter of 100 nm, which reduced biofilm formation by 95–98% [62]. Another study reported on biofilm inhibition at 15 mg/mL from silver nanoparticles resulted in an 89% inhibition of biofilm in *S. aureus* and 75% in *E. coli.* The new findings also demonstrated that the bacteria studied are extremely sensitive to Chi/Ag-NPs, implying that the complicated biofilm signaling system is linked to cell viability. Recently, research has been undertaken on conjugating produced nanoparticles with polymers for use in combating biofilm development [65].

3.4. Antifungal Activity

In this study, the antifungal activity of Chi/Ag-NPs and starting materials was evaluated as shown in Figure 6 and Table 2. Results revealed that Chi/Ag-NPs exhibited promising antifungal activity against tested fungal strains compared to starting materials and nystatin as a standard antifungal. Inhibition zones of Chi/Ag-NPs towards *C. albicans*, *A. fumigatus*, *A. terreus*, and *A. niger* were 22, 29, 20, and 17 mm, respectively. On the other hand, start materials (Ag+ and Chi) exhibited very weak antifungal activity on some tested fungal strains. Nystatin exhibited promising antifungal activity against *C. albican* but gave weak antifungal against *A. fumigatus*, *A. terreus*, and *A. niger*. Therefore, the prepared Chi/Ag-NPs are promising as antifungal agents compared to nystatin. Furthermore, the MIC of Chi/Ag-NPs and starting materials were detected as shown in Table 2. Results illustrated that MIC_s of Chi/Ag-NPs against *C. albicans*, *A. fumigatus*, *A. terreus*, and *A. niger* were 250, 62.5, 500, and 1000 µg/mL, respectively. A Chi/Ag-NPs solution can be used as a bamboo-straw-coating material to inhibit the growth of *A. flavus* [66]. The outstanding antifungal activity of Chi/Ag-NPs is attributed to the positive charge of the amino group in chitosan is combined with negative charge components of the fungal cell, thus Chi/Ag-NPs may suppress the fungal growth by chelating various transitional metal ions, inhibiting enzymes and by impairing the exchange with the medium. The increase in the antimicrobial activity is due to the greater stability of Chi/Ag-NPs in an aqueous medium because chitosan protects them from aggregation [67].

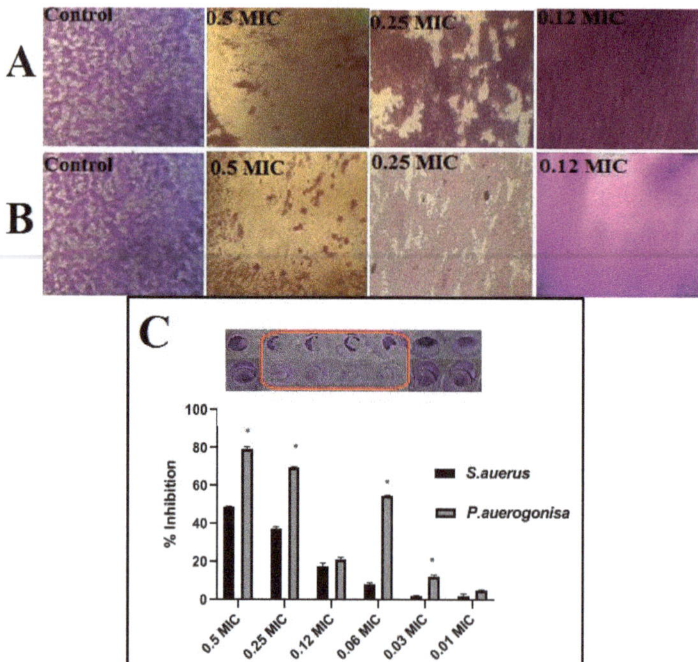

Figure 5. Light-inverted microscopic images of *S. aureus* (**A**) and *P. aeruginosa* (**B**) biofilms grown with various concentrations of Chi/Ag-NPs. At concentrations above 0.12xMIC bacteria have appeared as aggregated together to perform normal biofilm. *P. aeruginosa* and *S. auerus* biofilm inhibition in the presence of Chi/Ag-NPs at Sub. MIC (**C**). The absorbance of the control was considered to represent 100% of biofilm (results were considered significant when compared to control; * $p < 0.05$. Data are presented as mean ± SD, $n = 4$).

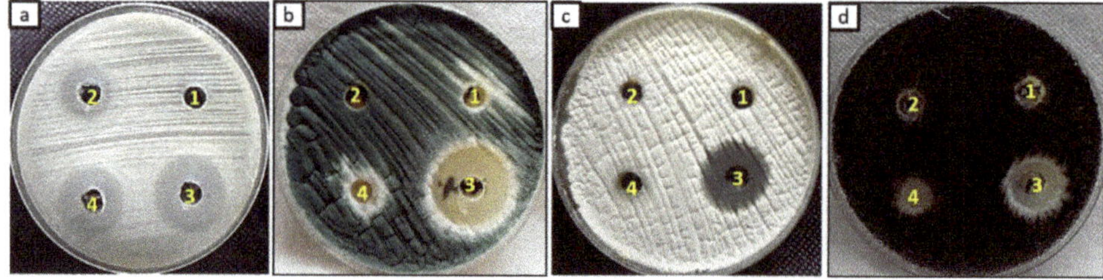

Figure 6. Antifungal activity of Chi (1), Ag+ (2), Chi/Ag-NPs (3), and Nystatin (4) toward *C. albicans* (**a**), *A. fumgatus* (**b**), *A. terreus* (**c**), and *A. niger* (**d**) using agar-well diffusion method.

Table 2. Inhibition zones and MIC of Chi/Ag-NPs compared to start materials.

Organisms	*C. albicans*		*A. fumgatus*		*A. terreus*		*A. niger*	
Materials	**IZ/mm (4000 µg/mL)**	**MIC µg/mL**	**IZ/mm (4000 µg/mL)**	**MIC µg/mL**	**IZ/mm (4000 µg/mL)**	**MIC µg/mL**	**IZ/mm (4000 µg/mL)**	**MIC µg/mL**
Chi	Nil	ND	8.00	4000	Nil	ND	Nil	ND
Ag+	16.00	1000	Nil	ND	Nil	ND	8.00	4000
Chi/Ag-NPs	22.00	250	29.00	62.5	20.00	500	17.00	1000
NS	21.00	250	10.00	2000	8.00	4000	9.00	4000

3.5. Antioxidant Activity

In biological systems, free radicals are generated as a result of the interaction of biomolecules with molecular oxygen [68]. Therefore, antioxidant compounds are used to resist the ROS effect. Antioxidant activity of Chi/Ag-NPs was evaluated compared to starting materials and ascorbic acid, as shown in Figure 7. The result showed that Chi/Ag-NPs revealed antioxidant activity were 92, 90, and 75% at concentrations of 4000, 2000, and 1000 μg/mL. Moreover, the IC_{50} of Chi/Ag-NPs was 261 μg/mL but was 3.9 μg/mL for ascorbic acid. The mechanism of action of the antioxidant activity of Chi/Ag-NPs is attributed to the binding of transition metal ion catalysts, decomposition of peroxides, inhibition of chain reaction, and inhibition of continued hydrogen abstraction [69]. The highest antioxidant activity is attributed to the presence of various bio-reductive groups of the phytochemicals present on the surface of the Ag-NPs [70].

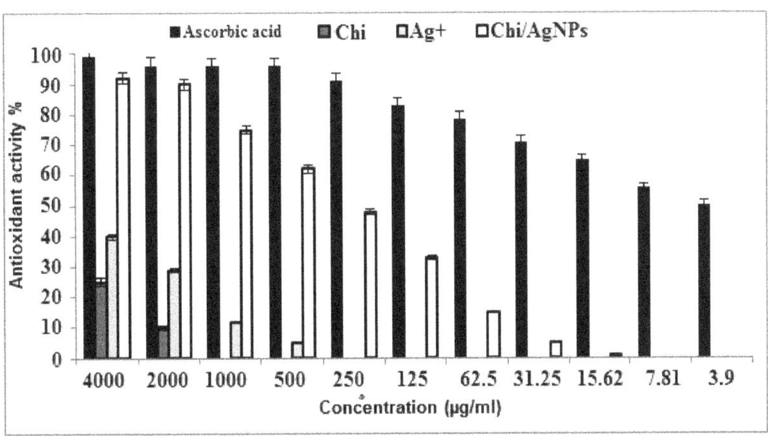

Figure 7. Antioxidant activity of Chi, Ag+, and Chi/Ag-NPs.

3.6. Cytotoxicity on Normal Human Skin Cell Line (BJ-1)

Cytotoxic studies are needed to avoid toxicity on normal cells. The in vitro cytotoxicity effects of Chi/Ag-NPs against normal human skin cell line (BJ-1) significantly inhibited the proliferation of (BJ-1) human skin cell line in a concentration-dependent manner (0–200 μg/mL). The half-maximal inhibitory concentrations (IC_{50}) of Chi/Ag-NPs and Dox were calculated to be 119.2 and 3.9 μg/mL cells, respectively. So, according to the Special Programme for Research and Training in Tropical Diseases (WHO—Tropical Diseases), Chi/Ag-NPs were considered non-toxic (IC50 > 90.00 μg/mL) Figure 8. The toxicity of silver nanoparticles loaded with chitosan depends on concentration, species, and particle size [71]. In the present study, As previously reported by researchers [72], an increase in nanoparticle concentration increases the cytotoxic effect on normal cell lines in a dose-dependent manner. The half-maximal inhibitory concentrations (IC_{50}) of Chi/Ag-NPs were calculated to be 119.2 μg/mL. Our results demonstrate that Chi/Ag-NPs are less toxic to normal human skin cell line (BJ-1) cells, whereas doxorubicin is more toxic.

3.7. Cell Migration Assay (Wound Scratch Assay)

In this study, concentrations of Chi/Ag-NPs less than 100 μg/mL were used to investigate wound healing activities in human skin fibroblasts. The effects of Chi/Ag-NPs on the healing process were studied using an in vitro scratch wound healing assay. In Figure 9, fibroblasts advanced toward the opening to close the scratch wound by roughly 50.5% after a 24-h exposure to Chi/Ag-NPs, greatly accelerating the wound healing process compared to the control 17.5%. Because of its non-toxicity, anti-inflammatory impact, biocompatibility, retention of fibroblast growth factors, and stimulation of human skin fibroblast activities, chitosan has been widely employed as a wound dressing material [73]. It stimulates cell adhesion and proliferation and helps in the organization of the extracellular matrix. The above results are in line with those of Hajji et al. [74], which showed that silver nanoparticles prepared with chitosan promote wound healing, reduce infection, and reduce the risk of silver absorption. Based on these findings, it can be concluded that Chi/Ag-NPs can significantly speed up wound healing. The findings are consistent with those published by Souto et al. [75], who found that a spongy bilayer dressing containing CS–Ag nanoparticles dramatically expedited the healing of cutaneous wounds.

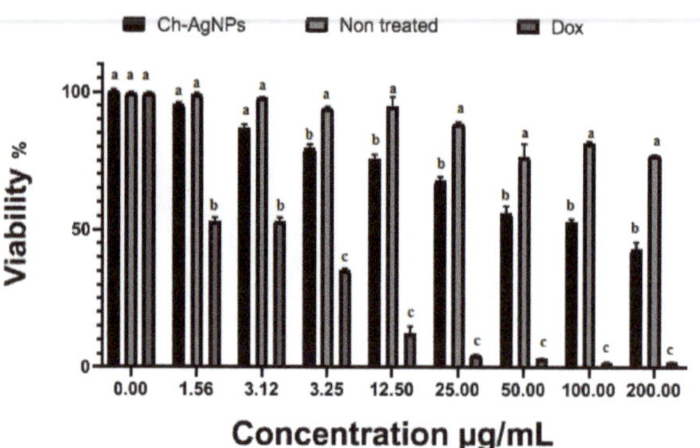

Figure 8. In vitro cytotoxicity effects on Chi/Ag-NPs and doxorubicin against normal human skin cell line (BJ-1) was assessed by SRB colorimetric assay. Within each column, different letters indicate significant differences among values ($p < 0.05$) based on one-way ANOVA estimated by GraphPad Instat software, (version 6.01), San Diego, CA, USA.

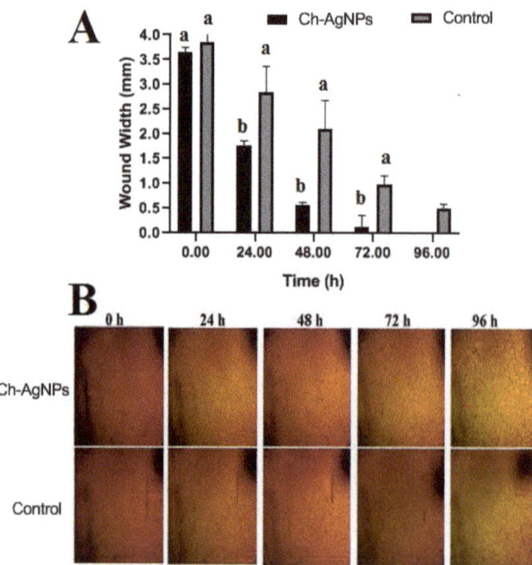

Figure 9. (**A**) Effects of different treatments on the wound area contraction (0–96 h). Values are given as mean ± SD (n = 3/group). Different letters indicate significant differences ($p < 0.05$). (**B**) Representative phase contrast micrographs of cells treated with 100 µg/mL Chi/Ag-NPs at 0 and 24 h. Wound closure rates are expressed as percentage of scratch closure from after 0 to 96 h compared to initial area. Red and black lines mean center and wide of the wound.

This section may be divided by subheadings. It should provide a concise and precise description of the experimental results, their interpretation, as well as the experimental conclusions that can be drawn.

4. Conclusions

In the current study, Chi/Ag-NPs with multifunctional biological purposes were prepared. Chi/Ag-NPs had highly antibacterial activity against gram-positive and gram-negative bacteria. They also reported antibiofilm activity as well as antioxidant activity. Furthermore, Chi/Ag-NPs exhibited promising antifungal activity towards unicellular and multicellular fungi. Chi/Ag-NPs were seen to significantly speed up wound healing at non-toxic concentrations due to their biocompatibility and good absorption of wound exudates. Data demonstrated the potentialities of Chi/Ag-NPs to be used as an alternative to antimicrobial drugs.

Author Contributions: Conceptualization, A.H.H., A.M.S. and S.S.S.; methodology, A.H.H., A.M.S., F.M.E. and S.S.S.; software, A.H.H., A.M.S. and S.S.S.; validation, A.H.H., A.M.S. and S.S.S.; formal analysis, A.H.H., A.M.S. and S.S.S.; investigation, A.H.H., A.M.S., S.S.S., F.M.E. and O.M.A.; resources, A.H.H., A.M.S. and S.S.S.; K.A.A.-E. data curation, A.H.H., A.M.S. and S.S.S.; writing—original draft preparation, A.H.H., A.M.S. and S.S.S.; writing—review and editing A.H.H., A.M.S., S.S.S., K.A.A.-E., F.M.E. and O.M.A.; visualization, A.H.H., A.M.S., S.S.S., K.A.A.-E., F.M.E. and O.M.A. All authors have read and agreed to the published version of the manuscript.

Funding: This research received no external funding.

Institutional Review Board Statement: Not applicable.

Informed Consent Statement: Not applicable.

Data Availability Statement: Not applicable.

Acknowledgments: The authors extend their appreciation to the Botany and Microbiology Department, Faculty of Science, Al-Azhar University and the Taif University Researchers Supporting Project (TURSP-2020/81) at Taif University in Taif, Saudi Arabia for providing the necessary research facilities.

Conflicts of Interest: The authors declare no conflict of interest.

References

1. Mitchell, M.J.; Billingsley, M.M.; Haley, R.M.; Wechsler, M.E.; Peppas, N.A.; Langer, R. Engineering precision nanoparticles for drug delivery. *Nat. Rev. Drug Discov.* **2021**, *20*, 101–124. [CrossRef] [PubMed]
2. Shehabeldine, A.M.; Hashem, A.H.; Wassel, A.R.; Hasanin, M. Antimicrobial and antiviral activities of durable cotton fabrics treated with nanocomposite based on zinc oxide nanoparticles, acyclovir, nanochitosan, and clove oil. *Appl. Biochem. Biotechnol.* **2021**, *194*, 783–800. [CrossRef] [PubMed]
3. Salem, S.S.; Hammad, E.N.; Mohamed, A.A.; El-Dougdoug, W. A Comprehensive Review of Nanomaterials: Types, Synthesis, Characterization, and Applications. *Biointerface Res. Appl. Chem.* **2023**, *13*, 14. [CrossRef]
4. Aref, M.S.; Salem, S.S. Bio-callus synthesis of silver nanoparticles, characterization, and antibacterial activities via Cinnamomum camphora callus culture. *Biocatal. Agric. Biotechnol.* **2020**, *27*, 101689. [CrossRef]
5. Salem, S.S.; Ali, O.M.; Reyad, A.M.; Abd-Elsalam, K.A.; Hashem, A.H. Pseudomonas indica-Mediated Silver Nanoparticles: Antifungal and Antioxidant Biogenic Tool for Suppressing Mucormycosis Fungi. *J. Fungi* **2022**, *8*, 126. [CrossRef]
6. Salem, S.S. Bio-fabrication of Selenium Nanoparticles Using Baker's Yeast Extract and Its Antimicrobial Efficacy on Food Borne Pathogens. *Appl. Biochem. Biotechnol.* **2022**, *194*, 1898–1910. [CrossRef]
7. Salem, S.S.; Fouda, A. Green Synthesis of Metallic Nanoparticles and Their Prospective Biotechnological Applications: An Overview. *Biol. Trace Elem. Res.* **2021**, *199*, 344–370. [CrossRef]
8. Sharaf, M.H.; Nagiub, A.M.; Salem, S.S.; Kalaba, M.H.; El Fakharany, E.M.; Abd El-Wahab, H. A new strategy to integrate silver nanowires with waterborne coating to improve their antimicrobial and antiviral properties. *Pigment. Resin Technol.* **2022**. [CrossRef]
9. Eid, A.M.; Fouda, A.; Niedbała, G.; Hassan, S.E.D.; Salem, S.S.; Abdo, A.M.; Hetta, H.F.; Shaheen, T.I. Endophytic streptomyces laurentii mediated green synthesis of Ag-NPs with antibacterial and anticancer properties for developing functional textile fabric properties. *Antibiotics* **2020**, *9*, 641. [CrossRef]

10. Okba, M.M.; Baki, P.M.A.; Abu-Elghait, M.; Shehabeldine, A.M.; El-Sherei, M.M.; Khaleel, A.E.; Salem, M.A. UPLC-ESI-MS/MS profiling of the underground parts of common Iris species in relation to their anti-virulence activities against Staphylococcus aureus. *J. Ethnopharmacol.* **2022**, *282*, 114658. [CrossRef]
11. Hasanin, M.; Al Abboud, M.A.; Alawlaqi, M.M.; Abdelghany, T.M.; Hashem, A.H. Ecofriendly Synthesis of Biosynthesized Copper Nanoparticles with Starch-Based Nanocomposite: Antimicrobial, Antioxidant, and Anticancer Activities. *Biol. Trace Elem. Res.* **2021**, *200*, 2099–2112. [CrossRef] [PubMed]
12. Hasanin, M.; Hashem, A.H.; Lashin, I.; Hassan, S.A.M. In vitro improvement and rooting of banana plantlets using antifungal nanocomposite based on myco-synthesized copper oxide nanoparticles and starch. *Biomass Convers. Biorefinery* **2021**. [CrossRef]
13. Hashem, A.H.; Abdelaziz, A.M.; Askar, A.A.; Fouda, H.M.; Khalil, A.M.A.; Abd-Elsalam, K.A.; Khaleil, M.M. Bacillus megaterium-Mediated Synthesis of Selenium Nanoparticles and Their Antifungal Activity against Rhizoctonia solani in Faba Bean Plants. *J. Fungi* **2021**, *7*, 195. [CrossRef] [PubMed]
14. Hashem, A.H.; Al Abboud, M.A.; Alawlaqi, M.M.; Abdelghany, T.M.; Hasanin, M. Synthesis of Nanocapsules Based on Biosynthesized Nickel Nanoparticles and Potato Starch: Antimicrobial, Antioxidant and Anticancer Activity. *Starch-Stärke* **2022**, *74*, e2100165. [CrossRef]
15. Hashem, A.H.; Khalil, A.M.A.; Reyad, A.M.; Salem, S.S. Biomedical applications of mycosynthesized selenium nanoparticles using Penicillium expansum ATTC 36200. *Biol. Trace Elem. Res.* **2021**, *199*, 3998–4008. [CrossRef]
16. Lashin, I.; Hasanin, M.; Hassan, S.A.M.; Hashem, A.H. Green biosynthesis of zinc and selenium oxide nanoparticles using callus extract of Ziziphus spina-christi: Characterization, antimicrobial, and antioxidant activity. *Biomass Convers. Biorefinery* **2021**. [CrossRef]
17. Hashem, A.H.; Salem, S.S. Green and ecofriendly biosynthesis of selenium nanoparticles using Urtica dioica (stinging nettle) leaf extract: Antimicrobial and anticancer activity. *Biotechnol. J.* **2022**, *17*, 2100432. [CrossRef]
18. Singh, A.; Gautam, P.K.; Verma, A.; Singh, V.; Shivapriya, P.M.; Shivalkar, S.; Sahoo, A.K.; Samanta, S.K. Green synthesis of metallic nanoparticles as effective alternatives to treat antibiotics resistant bacterial infections: A review. *Biotechnol. Rep.* **2020**, *25*, e00427. [CrossRef]
19. Al-Rajhi, A.M.H.; Salem, S.S.; Alharbi, A.A.; Abdelghany, T.M. Ecofriendly synthesis of silver nanoparticles using Kei-apple (Dovyalis caffra) fruit and their efficacy against cancer cells and clinical pathogenic microorganisms. *Arab. J. Chem.* **2022**, *15*, 103927. [CrossRef]
20. Salem, S.S.; El-Belely, E.F.; Niedbała, G.; Alnoman, M.M.; Hassan, S.E.D.; Eid, A.M.; Shaheen, T.I.; Elkelish, A.; Fouda, A. Bactericidal and in-vitro cytotoxic efficacy of silver nanoparticles (Ag-NPs) fabricated by endophytic actinomycetes and their use as coating for the textile fabrics. *Nanomaterials* **2020**, *10*, 2082. [CrossRef]
21. Kim, H.S.; Sun, X.; Lee, J.-H.; Kim, H.-W.; Fu, X.; Leong, K.W. Advanced drug delivery systems and artificial skin grafts for skin wound healing. *Adv. Drug Deliv. Rev.* **2019**, *146*, 209–239. [CrossRef] [PubMed]
22. Tavares, T.D.; Antunes, J.C.; Padrão, J.; Ribeiro, A.I.; Zille, A.; Amorim, M.T.P.; Ferreira, F.; Felgueiras, H.P. Activity of specialized biomolecules against gram-positive and gram-negative bacteria. *Antibiotics* **2020**, *9*, 314. [CrossRef] [PubMed]
23. Dumitrache, A. Understanding Biofilms of Anaerobic, Thermophilic and Cellulolytic Bacteria: A Study towards the Advancement of Consolidated Bioprocessing Strategies. Doctoral Dissertation, University of Toronto, Toronto, ON, Canada, 2014.
24. Li, J.; Fu, J.; Tian, X.; Hua, T.; Poon, T.; Koo, M.; Chan, W. Characteristics of chitosan fiber and their effects towards improvement of antibacterial activity. *Carbohydr. Polym.* **2022**, *280*, 119031. [CrossRef] [PubMed]
25. Wang, W.; Meng, Q.; Li, Q.; Liu, J.; Zhou, M.; Jin, Z.; Zhao, K. Chitosan derivatives and their application in biomedicine. *Int. J. Mol. Sci.* **2020**, *21*, 487. [CrossRef]
26. Wei, S.; Ching, Y.C.; Chuah, C.H. Synthesis of chitosan aerogels as promising carriers for drug delivery: A review. *Carbohydr. Polym.* **2020**, *231*, 115744. [CrossRef]
27. Saad, E.M.; Elshaarawy, R.F.; Mahmoud, S.A.; El-Moselhy, K.M. New Ulva lactuca Algae Based Chitosan Bio-composites for Bioremediation of Cd(II) Ions. *J. Bioresour. Bioprod.* **2021**, *6*, 223–242. [CrossRef]
28. Almalik, A.; Benabdelkamel, H.; Masood, A.; Alanazi, I.O.; Alradwan, I.; Majrashi, M.A.; Alfadda, A.A.; Alghamdi, W.M.; Alrabiah, H.; Tirelli, N. Hyaluronic acid coated chitosan nanoparticles reduced the immunogenicity of the formed protein corona. *Sci. Rep.* **2017**, *7*, 10542. [CrossRef]
29. Tekie, F.S.M.; Hajiramezanali, M.; Geramifar, P.; Raoufi, M.; Dinarvand, R.; Soleimani, M.; Atyabi, F. Controlling evolution of protein corona: A prosperous approach to improve chitosan-based nanoparticle biodistribution and half-life. *Sci. Rep.* **2020**, *10*, 9664. [CrossRef]
30. Gonzalez, A.C.d.O.; Costa, T.F.; Andrade, Z.d.A.; Medrado, A.R.A.P. Wound healing-A literature review. *An. Bras. De Dermatol.* **2016**, *91*, 614–620. [CrossRef]
31. Verma, J.; Kanoujia, J.; Parashar, P.; Tripathi, C.B.; Saraf, S.A. Wound healing applications of sericin/chitosan-capped silver nanoparticles incorporated hydrogel. *Drug Deliv. Transl. Res.* **2017**, *7*, 77–88. [CrossRef]
32. Zhao, Y.; Li, Z.; Song, S.; Yang, K.; Liu, H.; Yang, Z.; Wang, J.; Yang, B.; Lin, Q. Skin-inspired antibacterial conductive hydrogels for epidermal sensors and diabetic foot wound dressings. *Adv. Funct. Mater.* **2019**, *29*, 1901474. [CrossRef]
33. Dara, P.K.; Mahadevan, R.; Digita, P.A.; Visnuvinayagam, S.; Kumar, L.R.G.; Mathew, S.; Ravishankar, C.N.; Anandan, R. Synthesis and biochemical characterization of silver nanoparticles grafted chitosan (Chi-Ag-NPs): In vitro studies on antioxidant and antibacterial applications. *SN Appl. Sci.* **2020**, *2*, 665. [CrossRef]

34. Shehabeldine, A.M.; Elbahnasawy, M.A.; Hasaballah, A.I. Green Phytosynthesis of Silver Nanoparticles Using Echinochloa stagnina Extract with Reference to Their Antibacterial, Cytotoxic, and Larvicidal Activities. *BioNanoScience* **2021**, *11*, 526–538. [CrossRef]
35. Fouda, A.; Khalil, A.; El-Sheikh, H.; Abdel-Rhaman, E.; Hashem, A. Biodegradation and detoxification of bisphenol-A by filamentous fungi screened from nature. *J. Adv. Biol. Biotechnol.* **2015**, *2*, 123–132. [CrossRef]
36. Hashem, A.H.; Hasanin, M.S.; Khalil, A.M.A.; Suleiman, W.B. Eco-green conversion of watermelon peels to single cell oils using a unique oleaginous fungus: Lichtheimia corymbifera AH13. *Waste Biomass Valorization* **2019**, *11*, 5721–5732. [CrossRef]
37. Suleiman, W.; El-Sheikh, H.; Abu-Elreesh, G.; Hashem, A. Recruitment of Cunninghamella echinulata as an Egyptian isolate to produce unsaturated fatty acids. *Res. J. Pharm. Biol. Chem. Sci.* **2018**, *9*, 764–774.
38. Suleiman, W.; El-Skeikh, H.; Abu-Elreesh, G.; Hashem, A. Isolation and screening of promising oleaginous Rhizopus sp and designing of Taguchi method for increasing lipid production. *J. Innov. Pharm. Biol. Sci.* **2018**, *5*, 8–15.
39. Elshikh, M.; Ahmed, S.; Funston, S.; Dunlop, P.; McGaw, M.; Marchant, R.; Banat, I.M. Resazurin-based 96-well plate microdilution method for the determination of minimum inhibitory concentration of biosurfactants. *Biotechnol. Lett.* **2016**, *38*, 1015–1019. [CrossRef]
40. Chakansin, C.; Yostaworakul, J.; Warin, C.; Kulthong, K.; Boonrungsiman, S. Resazurin rapid screening for antibacterial activities of organic and inorganic nanoparticles: Potential, limitations and precautions. *Anal. Biochem.* **2022**, *637*, 114449. [CrossRef]
41. Sawasdidoln, C.; Taweechaisupapong, S.; Sermswan, R.W.; Tattawasart, U.; Tungpradabkul, S.; Wongratanacheewin, S. Growing Burkholderia pseudomallei in biofilm stimulating conditions significantly induces antimicrobial resistance. *PLoS ONE* **2010**, *5*, e9196. [CrossRef]
42. Tofiño-Rivera, A.; Ortega-Cuadros, M.; Galvis-Pareja, D.; Jiménez-Rios, H.; Merini, L.J.; Martínez-Pabón, M.C. Effect of Lippia alba and Cymbopogon citratus essential oils on biofilms of Streptococcus mutans and cytotoxicity in CHO cells. *J. Ethnopharmacol.* **2016**, *194*, 749–754. [CrossRef] [PubMed]
43. Shehabeldine, A.M.; Ashour, R.M.; Okba, M.M.; Saber, F.R. Callistemon citrinus bioactive metabolites as new inhibitors of methicillin-resistant Staphylococcus aureus biofilm formation. *J. Ethnopharmacol.* **2020**, *254*, 112669. [CrossRef] [PubMed]
44. Chaieb, K.; Kouidhi, B.; Jrah, H.; Mahdouani, K.; Bakhrouf, A. Antibacterial activity of Thymoquinone, an active principle of Nigella sativa and its potency to prevent bacterial biofilm formation. *BMC Complementary Altern. Med.* **2011**, *11*, 29. [CrossRef] [PubMed]
45. Clinical and Laboratory Standards Institute. *Reference Method for Broth Dilution Antifungal Susceptibility Testing of Yeasts*; National Committee for Clinical Laboratory Standards: Wayne, PA, USA, 2002.
46. Khalil, A.M.A.; Abdelaziz, A.M.; Khaleil, M.M.; Hashem, A.H. Fungal endophytes from leaves of Avicennia marina growing in semi-arid environment as a promising source for bioactive compounds. *Lett. Appl. Microbiol.* **2020**, *72*, 263–274. [CrossRef]
47. Khalil, A.M.A.; Hashem, A.H. Morphological changes of conidiogenesis in two aspergillus species. *J. Pure Appl. Microbiol.* **2018**, *12*, 2041–2048. [CrossRef]
48. Khalil, A.M.A.; Hashem, A.H.; Abdelaziz, A.M. Occurrence of toxigenic Penicillium polonicum in retail green table olives from the Saudi Arabia market. *Biocatal. Agric. Biotechnol.* **2019**, *21*, 101314. [CrossRef]
49. Shehabeldine, A.; El-Hamshary, H.; Hasanin, M.; El-Faham, A.; Al-Sahly, M. Enhancing the Antifungal Activity of Griseofulvin by Incorporation a Green Biopolymer-Based Nanocomposite. *Polymers* **2021**, *13*, 542. [CrossRef]
50. Dacrory, S.; Hashem, A.H.; Hasanin, M. Synthesis of cellulose based amino acid functionalized nano-biocomplex: Characterization, antifungal activity, molecular docking and hemocompatibility. *Environ. Nanotechnol. Monit. Manag.* **2021**, *15*, 100453. [CrossRef]
51. Vajrabhaya, L.-O.; Korsuwannawong, S. Cytotoxicity evaluation of a Thai herb using tetrazolium (MTT) and sulforhodamine B (SRB) assays. *J. Anal. Sci. Technol.* **2018**, *9*, 15. [CrossRef]
52. Orellana, E.A.; Kasinski, A.L. Sulforhodamine B (SRB) assay in cell culture to investigate cell proliferation. *Bio-Protocol* **2016**, *6*, e1984. [CrossRef]
53. Houdkova, M.; Urbanova, K.; Doskocil, I.; Soon, J.W.; Foliga, T.; Novy, P.; Kokoska, L. Anti-staphylococcal activity, cytotoxicity, and chemical composition of hexane extracts from arils and seeds of two Samoan *Myristica* spp. *South Afr. J. Bot.* **2021**, *139*, 1–5. [CrossRef]
54. Rameshk, M.; Sharififar, F.; Mehrabani, M.; Pardakhty, A.; Farsinejad, A.; Mehrabani, M. Proliferation and in vitro wound healing effects of the Microniosomes containing *Narcissus tazetta* L. bulb extract on primary human fibroblasts (HDFs). *DARU J. Pharm. Sci.* **2018**, *26*, 31–42. [CrossRef] [PubMed]
55. Alshalal, I.A.R. Development of In Vitro Models to Investigate the Role of Decidualisation, PDGF, and ho-1 on Stromal–Trophoblast Interaction. Doctoral Dissertation, University of Birmingham, Birmingham, UK, 2020.
56. Grada, A.; Otero-Vinas, M.; Prieto-Castrillo, F.; Obagi, Z.; Falanga, V. Research techniques made simple: Analysis of collective cell migration using the wound healing assay. *J. Investig. Dermatol.* **2017**, *137*, e11–e16. [CrossRef] [PubMed]
57. Mourdikoudis, S.; Pallares, R.M.; Thanh, N.T. Characterization techniques for nanoparticles: Comparison and complementarity upon studying nanoparticle properties. *Nanoscale* **2018**, *10*, 12871–12934. [CrossRef]
58. Alshehri, M.A.; Trivedi, S.; Panneerselvam, C. Efficacy of chitosan silver nanoparticles from shrimp-shell wastes against major mosquito vectors of public health importance. *Green Processing Synth.* **2020**, *9*, 675–684. [CrossRef]
59. Shehabeldine, A.; Hasanin, M. Green synthesis of hydrolyzed starch–chitosan nano-composite as drug delivery system to gram negative bacteria. *Environ. Nanotechnol. Monit. Manag.* **2019**, *12*, 100252. [CrossRef]

60. Venault, A.; Chen, S.-J.; Lin, H.-T.; Maggay, I.; Chang, Y. Bi-continuous positively-charged PVDF membranes formed by a dual-bath procedure with bacteria killing/release ability. *Chem. Eng. J.* **2021**, *417*, 128910. [CrossRef]
61. Damm, C.; Münstedt, H.; Rösch, A. The antimicrobial efficacy of polyamide 6/silver-nano-and microcomposites. *Mater. Chem. Phys.* **2008**, *108*, 61–66. [CrossRef]
62. Gurunathan, S.; Han, J.W.; Kwon, D.-N.; Kim, J.-H. Enhanced antibacterial and anti-biofilm activities of silver nanoparticles against Gram-negative and Gram-positive bacteria. *Nanoscale Res. Lett.* **2014**, *9*, 373. [CrossRef]
63. Barabadi, H.; Mohammadzadeh, A.; Vahidi, H.; Rashedi, M.; Saravanan, M.; Talank, N.; Alizadeh, A. Penicillium chrysogenum-derived silver nanoparticles: Exploration of their antibacterial and biofilm inhibitory activity against the standard and pathogenic Acinetobacter baumannii compared to tetracycline. *J. Clust. Sci.* **2021**, 1–14. [CrossRef]
64. Ferreiraa, M.R.; Gomesa, S.C.; Moreiraa, L.M. Mucoid switch in Burkholderia cepacia complex bacteria: Triggers, molecular mechanisms and implications in pathogenesis. *Adv. Appl. Microbiol.* **2019**, *107*, 113–140.
65. Rajivgandhi, G.N.; Maruthupandy, M.; Li, J.-L.; Dong, L.; Alharbi, N.S.; Kadaikunnan, S.; Khaled, J.M.; Alanzi, K.F.; Li, W.-J. Photocatalytic reduction and anti-bacterial activity of biosynthesized silver nanoparticles against multi drug resistant Staphylococcus saprophyticus BDUMS 5 (MN310601). *Mater. Sci. Eng. C* **2020**, *114*, 111024. [CrossRef] [PubMed]
66. Bao, V.-V.Q.; Vuong, L.D.; Waché, Y. Synthesis of chitosan–silver nanoparticles with antifungal properties on bamboo straws. *Nanomater. Energy* **2021**, *10*, 111–117. [CrossRef]
67. Shrifian-Esfahni, A.; Salehi, M.T.; Nasr-Esfahni, M.; Ekramian, E. Chitosan-modified superparamgnetic iron oxide nanoparticles: Design, fabrication, characterization and antibacterial activity. *Chemik* **2015**, *69*, 19–32.
68. Phaniendra, A.; Jestadi, D.B.; Periyasamy, L. Free radicals: Properties, sources, targets, and their implication in various diseases. *Indian J. Clin. Biochem.* **2015**, *30*, 11–26. [CrossRef]
69. Rehana, D.; Mahendiran, D.; Kumar, R.S.; Rahiman, A.K. Evaluation of antioxidant and anticancer activity of copper oxide nanoparticles synthesized using medicinally important plant extracts. *Biomed. Pharmacother.* **2017**, *89*, 1067–1077. [CrossRef]
70. Nunes, M.R.; de Souza Maguerroski Castilho, M.; de Lima Veeck, A.P.; da Rosa, C.G.; Noronha, C.M.; Maciel, M.V.O.B.; Barreto, P.M. Antioxidant and antimicrobial methylcellulose films containing Lippia alba extract and silver nanoparticles. *Carbohydr. Polym.* **2018**, *192*, 37–43. [CrossRef]
71. Hu, Y.-L.; Qi, W.; Han, F.; Shao, J.-Z.; Gao, J.-Q. Toxicity evaluation of biodegradable chitosan nanoparticles using a zebrafish embryo model. *Int. J. Nanomed.* **2011**, *6*, 3351.
72. Mishra, V.; Nayak, P.; Singh, M.; Tambuwala, M.M.; Aljabali, A.A.; Chellappan, D.K.; Dua, K. Pharmaceutical aspects of green synthesized silver nanoparticles: A boon to cancer treatment. *Anti-Cancer Agents Med. Chem.* **2021**, *21*, 1490–1509. [CrossRef]
73. Croisier, F.; Jérôme, C. Chitosan-based biomaterials for tissue engineering. *Eur. Polym. J.* **2013**, *49*, 780–792. [CrossRef]
74. Hajji, S.; Khedir, S.B.; Hamza-Mnif, I.; Hamdi, M.; Jedidi, I.; Kallel, R.; Boufi, S.; Nasri, M. Biomedical potential of chitosan-silver nanoparticles with special reference to antioxidant, antibacterial, hemolytic and in vivo cutaneous wound healing effects. *Biochim. Biophys. Acta (BBA)-Gen. Subj.* **2019**, *1863*, 241–254. [CrossRef] [PubMed]
75. Souto, E.B.; Ribeiro, A.F.; Ferreira, M.I.; Teixeira, M.C.; Shimojo, A.A.; Soriano, J.L.; Naveros, B.C.; Durazzo, A.; Lucarini, M.; Souto, S.B. New nanotechnologies for the treatment and repair of skin burns infections. *Int. J. Mol. Sci.* **2020**, *21*, 393. [CrossRef] [PubMed]

Journal of
Fungi

Article

Biological Synthesis of Low Cytotoxicity Silver Nanoparticles (AgNPs) by the Fungus *Chaetomium thermophilum*—Sustainable Nanotechnology

Mariana Fuinhas Alves [1,*], Ariane Caroline Campos Paschoal [2], Tabata D'Maiella Freitas Klimeck [3], Crisciele Kuligovski [2], Bruna Hilzendeger Marcon [2,3], Alessandra Melo de Aguiar [2,4,*] and Patrick G. Murray [1,*]

[1] Shannon Applied Biotechnology Centre, Department of Applied Science, Faculty of Applied Sciences and Technology, Moylish Campus, Technological University of the Shannon: Midlands Midwest, Moylish, V94 EC5T Limerick, Ireland
[2] Laboratório de Biologia Básica de Células-Tronco, Instituto Carlos Chagas, FIOCRUZ Paraná, Curitiba 81350-010, PR, Brazil; arianepaschoal@hotmail.com (A.C.C.P.); crisciele.kuligovski@fiocruz.br (C.K.); bruna.marcon@fiocruz.br (B.H.M.)
[3] Rede de Plataformas Tecnológicas FIOCRUZ-Plataforma de Microscopia, Instituto Carlos Chagas, FIOCRUZ Paraná, Curitiba 81350-010, PR, Brazil; tabata.klimeck@gmail.com
[4] Rede de Plataformas Tecnológicas FIOCRUZ-Bioensaios em Métodos Alternativos em Citotoxicidade, Instituto Carlos Chagas, FIOCRUZ Paraná, Curitiba 81350-010, PR, Brazil
* Correspondence: mariana.alves@lit.ie (M.F.A.); alessandra.aguiar@fiocruz.br (A.M.d.A.); patrick.murray@lit.ie (P.G.M.)

Citation: Alves, M.F.; Paschoal, A.C.C.; Klimeck, T.D.F.; Kuligovski, C.; Marcon, B.H.; de Aguiar, A.M.; Murray, P.G. Biological Synthesis of Low Cytotoxicity Silver Nanoparticles (AgNPs) by the Fungus *Chaetomium thermophilum*—Sustainable Nanotechnology. *J. Fungi* **2022**, *8*, 605. https://doi.org/10.3390/jof8060605

Academic Editor: Kamel A. Abd-Elsalam

Received: 22 May 2022
Accepted: 2 June 2022
Published: 4 June 2022

Abstract: Fungal biotechnology research has rapidly increased as a result of the growing awareness of sustainable development and the pressing need to explore eco-friendly options. In the nanotechnology field, silver nanoparticles (AgNPs) are currently being studied for application in cancer therapy, tumour detection, drug delivery, and elsewhere. Therefore, synthesising nanoparticles (NPs) with low toxicity has become essential in the biomedical area. The fungus *Chaetomium thermophilum* (*C. thermophilum*) was here investigated—to the best of our knowledge, for the first time—for application in the production of AgNPs. Transmission electronic microscopy (TEM) images demonstrated a spherical AgNP shape, with an average size of 8.93 nm. Energy-dispersive X-ray spectrometry (EDX) confirmed the presence of elemental silver. A neutral red uptake (NRU) test evaluated the cytotoxicity of the AgNPs at different inhibitory concentrations (ICs). A half-maximal concentration (IC_{50} = 119.69 µg/mL) was used to predict a half-maximal lethal dose (LD_{50} = 624.31 mg/kg), indicating a Global Harmonized System of Classification and Labelling of Chemicals (GHS) acute toxicity estimate (ATE) classification category of 4. The fungus extract showed a non-toxic profile at the IC tested. Additionally, the interaction between the AgNPs and the Balb/c 3T3 NIH cells at an ultrastructural level resulted in preserved cells structures at non-toxic concentrations (IC_{20} = 91.77 µg/mL), demonstrating their potential as sustainable substitutes for physical and chemically made AgNPs. Nonetheless, at the IC_{50}, the cytoplasm of the cells was damaged and mitochondrial morphological alteration was evident. This fact highlights the fact that dose-dependent phenomena are involved, as well as emphasising the importance of investigating NPs' effects on mitochondria, as disruption to this organelle can impact health.

Keywords: biosynthesis; *Chaetomium thermophilum*; cytotoxicity; fungus; silver nanoparticles

1. Introduction

Over-exploitation of natural resources and exponential human growth are the roots of modern social concerns [1]. In this context, sustainable consumption and production patterns, extensively described by the United Nations in the Sustainable Development Agenda as Goal 12, have inevitably become needed [2]; hence, the growing attention given to sustainable development and the pressing need to explore sustainable options [3]. As a result, physical and chemical processes are gradually being replaced by biological ones. This is not different in the nanotechnology field, as the biological route to synthesising nanoparticles (NPs) offers the benefits of environmental compatibility, scalability, and low or reduced production costs [4,5]. Also, the use of biological organisms as NP biofactories minimizes the use of hazardous chemicals, generating fewer or non-toxic end-products and, consequently fewer unwanted byproducts [6].

Silver nanoparticles (AgNPs) have been used in different fields of biotechnology; for example, to enhance the effectiveness of antibiotics and increase antibacterial activity by killing pathogenic and multiple-drug-resistant bacteria [7]. Additionally, they have been used to inhibit the viability of cancer cell lines [8,9]. Moreover, they have wide applications due to their antibacterial, antifungal, antiviral, anti-inflammatory, anti-angiogenic, and anticancer properties [10–12]. The advantages of using fungi rather than other microorganisms in NP synthesis include their metal tolerance and bioaccumulation capacity, their economic viability, and their suitability for handling biomass during downstream of processing and large-scale production [4].

Different fungi species, such as *Aspergillus fumigatus* (*A. fumigatus*), *Cladosporium halotolerans*, *Fusarium oxysporum* (*F. oxysporum*), *Penicillium italicum*, and *Trichoderma longibrachiatum* have been successfully used to synthesise AgNPs [8,10,13–15]. However, to the best of our knowledge, the species *Chaetomium thermophilum* (*C. thermophilum*) has not yet been investigated for NP synthesis. Hence, this study aimed to investigate the utilisation of their fungal metabolites in the extracellular synthesis of AgNPs.

Regardless of the benefits of nanotechnology advances, nanomaterials' physicochemical properties are still a source of concern with respect to the risks related to the production process, safety, and other environmental issues [16,17]. Furthermore, as human exposure to NPs is inevitable, it is crucial to understand their interactions with cellular systems and their toxicological impact [18]. Hence, toxicology research has been gaining significant attention.

Several toxicity tests are available nowadays, such as cytotoxicity, neurotoxicity, genotoxicity, and ecotoxicity tests. Nevertheless, the Interagency Coordinating Committee on the Validation of Alternative Methods (ICCVAM) states that acute oral toxicity is usually the first tested in order to assess chemical hazards regarding classification, labelling, risk assessment, diagnosis, treatment, and prognosis toward chemical exposure [19].

Acute oral toxicity tests can be investigated in vivo or in vitro. However, for the implementation of alternative methodologies to use with animals, in vitro tests are essential, especially for stages of initial refinement of new substances with promising applications [19,20]. In this context, Balb/c 3T3 NIH (murine fibroblast) cells are recommended substrates for in vitro acute cytotoxicity testing [19,20]. This cell line has already been used to predict the cytotoxicity of AgNPs, and they have been proven to be more sensitive and accurate in toxicological evaluation than in vivo studies [21,22].

In vitro studies have reported that the interaction between AgNPs and cell cultures can cause diverse cytotoxicity outcomes, depending on the physical and chemical nature of the AgNPs and the cell lineage [12]. Toxicology studies have reported cytotoxicity effects, such as damage to the cell membrane and, consequently, alteration in the cell permeability, as well as severe morphological changes, especially in the mitochondria, leading to the impairment of this organelle [23–25]. Thus, there is a pressing need to predict nanomaterials' toxicological impacts and to establish use of efficient, safe, reliable, and non-toxic NPs. The purpose of this study was, therefore, to: (1) investigate the ability of metabolites of the fungus *C. thermophilum* to synthesise AgNPs, (2) estimate AgNPs' cytotoxicity using the neutral red uptake (NRU) assay, and (3) evaluate the interaction of AgNPs with the Balb/c 3T3 NIH cell line and further investigate potential safety hazards associated with biotechnological applications. The results of this study demonstrate the successful biological synthesis of AgNPs. The NRU cytotoxicity test predicted a lethal dose (LD_{50}) value that indicated a Global Harmonized System of Classification and Labelling of

Chemicals (GHS) category of 4 (300 mg/kg to 2000 mg/kg). Furthermore, preserved cell structures were observed following the interaction of AgNPs and Balb/c 3T3 NIH cells at an inhibitory concentration (IC) of IC_{20} and with an evident mitochondrial morphological alteration at IC_{50}. These results highlight the applicability of the fungi system as a source of bio-nanomaterials with low cytotoxicity, low cost, and less impact on the environment that may eventually lead to sustainable development in the green nanotechnology field.

2. Materials and Methods

2.1. Production of *Chaetomium thermophilum* Cell-Free Extract

The production of cell-free extract of the fungus *Chaetomium thermophilum var. thermophilum* (*C. thermophilum*), Centraal Bureau voor Schimmelcultures (CBS) collection number 143.50, was adapted from AbdelRahim et al., Hamedi et al., and Katapodis et al. [26–28]. The fungus was cultivated, for three days, at 45 °C in sterile *C. thermophilum* minimal agar (CTMA), composed of 0.10 g/L calcium chloride dihydrate ($CaCl_2 \cdot 2H_2O$, Honeywell, Seelze, Germany), 15.00 g/L magnesium sulphate heptahydrate ($MgSO_4 \cdot 7H_2O$, VWR, Dublin, Ireland), following chemicals were bought from Sigma-Aldrich (St. Louis, MO, USA): 30 mL/L corn steep liquor, 1.00 g/L yeast extract, 3.00 g/L potassium phosphate monobasic (KH_2PO_4), 2.00 g/L potassium phosphate dibasic (K_2HPO_4), 0.70 g/L noble agar. Trace mineral salts chemicals were bought from Sigma-Aldrich (St. Louis, MO, USA): 5.00 mg/L iron (II) sulphate heptahydrate ($FeSO_4 \cdot 7H_2O$), 1.40 mg/L zinc sulfate heptahydrate ($ZnSO_4 \cdot 7H_2O$), 1.60 mg/L manganese (II) sulfate tetrahydrate ($MnSO_4 \cdot 4H_2O$), 0.20 mg/L cobalt (III) chloride hexahydrate ($CoCl_3 \cdot 6H_2O$). In aseptic conditions, three samples (taken from the mycelia mat borders to guarantee actively growing fungal cells) were inoculated in 250 mL Erlenmeyer flasks with sterile *C. thermophilum* minimal medium (CTMM, without agar) at 45 °C and 120 rpm for five days. Fungal cells were filtered, washed thoroughly with sterile deionised water, dried, and weighed accurately. In order to induce secretion of secondary metabolites, the cells were transferred to sterile deionised water (stress liquid medium) with a ratio of 1 g of cells to 10 mL and incubated in the shaker for three days. The secreted fungal extract was separated from the fungus cells by muslin filtration, followed by centrifugation at 5000 rpm for 20 min at 25 °C and subsequent 0.22 µm membrane filtration. The *C. thermophilum* cell-free extract was stored at 4 °C until further use.

2.2. Biological Synthesis and Characterisation of AgNPs

2.2.1. AgNP Synthesis

The AgNP synthesis activity level of the *C. thermophilum* cell-free extract was tested following the study by Alves and Murray [3]. Briefly, the cell-free extract was heated for 15 min at 90 °C; then, 127.40 mg/L silver nitrate ($AgNO_3$, Sigma-Aldrich, St. Louis, MO, USA) was added for a total volume of 500 µL, and heating was continued for one hour at 90 °C. In the same reaction conditions, a reaction synthesis control was produced using ultrapure water and $AgNO_3$.

2.2.2. AgNP Characterization

The resulting AgNPs were characterised based following the study by Alves and Murray [3]. Ultraviolet–visible spectrophotometry (UV-Vis, BioTek Synergy 4 Microplate Reader, Bad Friedrichshall, Germany) was first used to measure the localised surface plasmon resonance (LSPR) absorbance. AgNPs were scanned between 300 nm and 1000 nm, with 2 nm steps. Afterwards, transmission electron microscopy (TEM, JEOL JEM 2100 Field Emission Electron Microscope, Tokyo, Japan) was used to evaluate the shape and size distribution of the AgNPs. Samples were drop-coated onto Formvar carbon-coated copper grids with a 200 µm mesh size and dried over 24 h. Images were obtained using a JEOL JEM 2100 Field Emission Electron Microscope operated at 200 kV with a field emission electron gun equipped with a Gatan Ultrascan digital camera. An average of 200 NPs was recorded from several TEM images, with AxioVision Rel 4.8 software used to evaluate size distribution. Furthermore, the elemental composition was analysed using energy dispersive X-ray spectrometry (EDX, Hitachi 3000, Tokyo, Japan). Five microliters of the samples was drop-coated onto polished aluminium slides and dried in the oven at 60 °C for 1 h, thrice. A Hitachi 3000 electron microscope with EDX capability at 15 kV accelerating voltage and a working distance of 2 mm was used to obtain the ED spectra of the samples.

2.3. Mammalian Cell Culture

Skin fibroblasts from murine embryo Balb/c mice 3T3 NIH (clone A31) cells (Balb/c 3T3 NIH cells, Rio de Janeiro Cell Bank, Rio de Janeiro, Brazil), the cells recommended by ICCVAM [29], were cultivated in routine medium containing high-glucose (4.5 g/L) Dulbecco's Modification of Eagle's Medium (DMEM, Gibco Invitrogen, Carlsbad, CA, USA) supplemented with non-heat-inactivated 10% fetal bovine serum (FBS, Gibco Invitrogen, Carlsbad, CA, USA) and 4 mM L-Glutamine (Gibco Invitrogen, Carlsbad, CA, USA) at 37 °C, 90% humidity, and 5.0% CO_2/air [29,30].

2.4. Cytotoxicity Evaluation Profile in Mammalian Cell

The cytotoxicity evaluation of the *C. thermophilum* AgNPs and of the *C. thermophilum* cell-free extract was based on the Organisation for Economic Co-operation and Development (OECD) guidelines from the Environment, Health and Safety Publications Series on Testing and Assessment No. 129 guidance document on using cytotoxicity tests to estimate starting doses for acute oral systemic toxicity tests [20]. In brief, the cells were plated into the inner wells of 96-well tissue culture microtiter plates at a density of 2.5×10^3 cells (100 µL/well); the outer wells were filled with culture medium and then cultivated for 24 h at 37 °C, 90% humidity, and 5.0% CO_2/air. After 24 h of incubation, the culture medium was removed. The AgNPs were diluted immediately prior to use with a solution of 4.5 g/L DMEM, 4 mM L-Glutamine (Gibco Invitrogen, Carlsbad, CA, USA), 100 IU/mL penicillin (Sigma-Aldrich, St. Louis, MO, USA), and 100 µg/mL streptomycin (Sigma-Aldrich, St. Louis, MO, USA), according to ICCVAM recommendations [31]. Eight serial dilutions of AgNPs ranging from 31.03 µg/mL to 460.32 µg/mL were prepared. For the AgNP serial dilution, the log-factor of 1.47 was used. Eight serial dilutions of the *C. thermophilum* cell-free extract ranging from 0.38 µg/mL to 1200 µg/mL were also prepared, with a serial dilution log factor of 3.16. Plates were incubated for 48 h at 37 °C, 90% humidity, and 5.0% CO_2/air. After this period, Balb/c 3T3 NIH cells were stained with a neutral red (NR) medium composed of 25 µg/mL NR dye (Sigma-Aldrich, St. Louis, MO, USA). NR was extracted from cells using 250 µL/well of NR desorb solution (freshly prepared with 49 parts water, 50 parts ethanol, and 1 part glacial acetic acid, all bought from Sigma-Aldrich, St. Louis, MO, USA) over a period of 20 min in a shaker protected from light [32,33]. The optical density of the samples was measured at a wavelength of 540 nm using a Multi-Modo Synergy H1 (Biotek, Winooski, Vermont, EUA) spectrophotometer reader.

2.5. Evaluation of Balb/c 3T3 NIH Cells and AgNP Interaction by TEM

Balb/c 3T3 NIH were plated into 6-well tissue culture microtiter plates at a density of 7.5 × 104 cells/well in 3 mL of the routine medium and cultivated for 24 h at 37 °C, 90% humidity, and 5.0% CO_2/air. After this period, the routine medium was discharged and 1.5 mL of a fresh one was added. The cells were submitted to three treatments: a control (cells without AgNPs), a non-toxic concentration (IC_{20}), and the IC_{50}. Plates were incubated for 6 h and rinsed with pre-warmed phosphate buffered saline (PBS, Sigma-Aldrich, St. Louis, MO, USA). The cells were kept in a fixing solution composed of 2.5% glutaraldehyde (Sigma-Aldrich, St. Louis, MO, USA) and diluted in 0.1 M sodium cacodylate buffer solution added over a period of 24 h. The cells were washed with 0.1 M sodium cacodylate buffer (Electron Microscopy Sciences) and post-fixed with 1% osmium tetroxide (Electron Microscopy Sciences, Hatfield, PA, USA), 0.8% potassium ferricyanide (Electron Microscopy Sciences, Hatfield, PA, USA), 5 mM calcium chloride, and 0.1 M sodium cacodylate buffer. After washing, the samples were dehydrated using a graded acetone series (30%, 50%,

70%, 90%, and 100%) and embedded using EMBed 812 resin (Electron Microscopy Sciences, Hatfield, PA, USA). Ultrathin sections of each sample were obtained using a Leica EM UC6 ultramicrotome (Leica, Wetzlar, Germany). The samples were contrast-stained with 5% uranyl acetate (Sigma-Aldrich, St. Louis, MO, USA) for 30 min and with lead citrate (Sigma-Aldrich, St. Louis, MO, USA) for 5 min, then analysed using a JEOL JEM1400-Plus TEM (JEOL, Tokyo, Japan) [30,33].

2.6. Statistical Analysis

All assays were run in triplicate. Data were expressed as means ± standard deviation (St. Dev). The software packages used to analyse the data generated in the characterisation process were Gen5, Microsoft Office Excel, and Quantax 70 Microanalysis. Cell viability based on optical density data was analysed in Microsoft Office Excel. Outliers were analysed with the statistics Grubbs test (available online at www.graphpad.com/quickcalcs/Grubbs1.cfm, accessed on 12 June 2021). GraphPad Prism® 5.0 was used to create a sigmoidal dose–response (variable slope) with four parameters, rearranged in the Hill function. IC_{20}, IC_{50}, and IC_{80} were expressed graphically with mean and standard deviation. IC_{50} was used to predict LD_{50} using the formula: $\log LD_{50}$ (mg/kg) = 0.372 $\log IC_{50}$ (μg/mL) + 2.024 (R^2 = 0.325) [30–33]. Once the LD_{50} was predicted, it was possible to classify the AgNPs according to the Globally Harmonized System of Classification and Labelling of Chemicals (GHS) classification [34]. The acceptance criteria of the assay followed ICCVAM guidelines [31]. Sodium dodecyl sulphate (SDS, Sigma-Aldrich, St. Louis, MO, USA) was used as the control drug, and the acceptance criteria of the assay also followed the ICCVAM guidelines [31,35].

3. Results

3.1. Production of *Chaetomium thermophilum* Cell-Free Extract

The fungus *C. thermophilum* grows with septate hyphae that develop initially with a lightly hyaline colour (glass-like) and change over time to olivaceous (brownish olive) with thicker septate walls [36]. Photos of the fungus growth in CTMA plates and of the cell-free extract were taken, as shown in Figure 1A,B, and the mycelia cells were analysed using a microscope (VWR, Dublin, Ireland) with 100× magnification (Figure 1C).

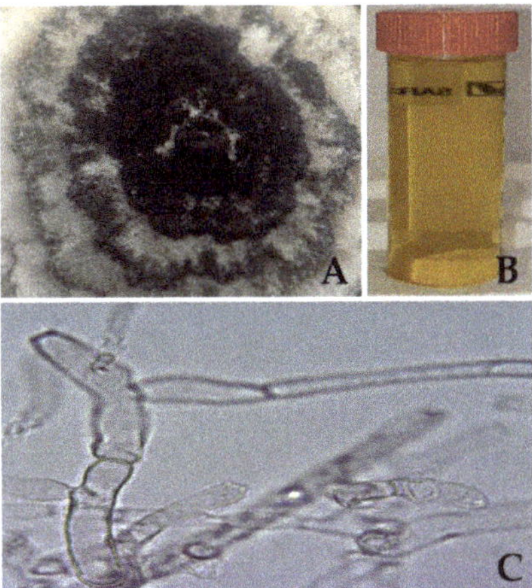

Figure 1. *C. thermophilum* biological characterisation. (**A**) Fungal growth in CTMA medium. (**B**) Fungal cell-free extract ready to be used in nanoparticles synthesis. (**C**) Microscope image of mycelia cells with 100× magnification.

3.2. AgNP Characterisation

The LSPR absorbance in the wavelength region of 380 nm to 435 nm indicates AgNP synthesis. The UV-Vis spectrophotometry spectrum of the AgNPs displayed a maximum wavelength of 405 ± 1.15 nm, with a correspondent maximum absorbance value of 1.250 ± 0.01, demonstrating a different pattern from the spectrum of the synthesis reaction control (Figure 2). The $AgNO_3$ control solution did not present an absorbance peak in the relevant UV-Vis region.

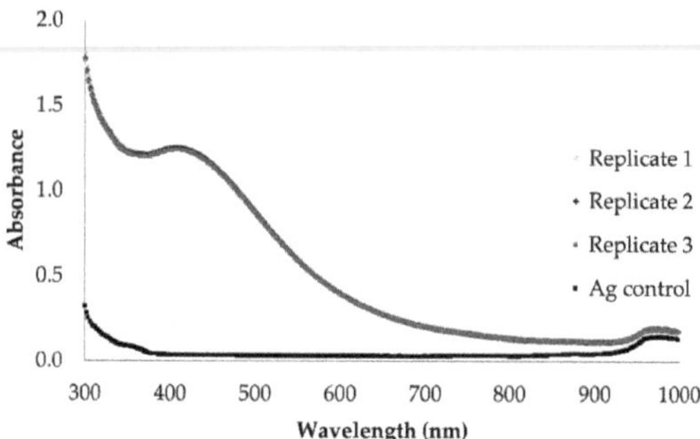

Figure 2. Biological AgNPs physicochemical characterisation. UV/Vis spectrophotometry analysis of the biologically synthesised AgNPs by the fungus *C. thermophilum*.

Figure 3 shows a 60,000× magnification TEM image of the spherical-shaped AgNPs synthesised. Statistical analysis revealed that the AgNPs' size distribution ranged from 4.72 nm to 30.73 nm, with an average size of 8.93 ± 2.29 nm. Furthermore, EDX analysis (data not shown) was carried out on the samples, and the presence of elemental silver was confirmed at 3 keV where AgNPs were identified in the TEM.

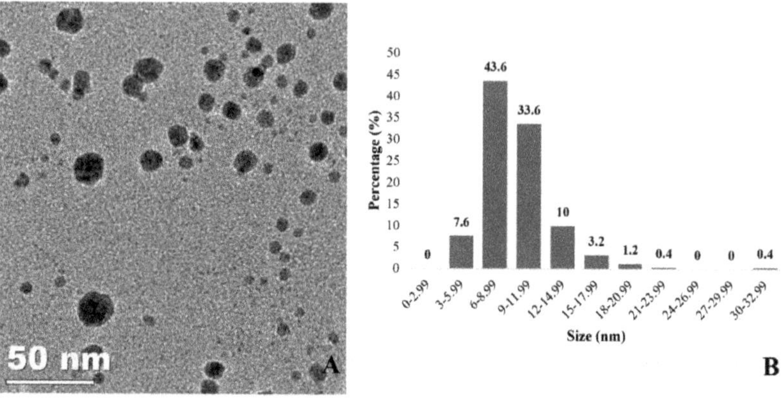

Figure 3. Biological AgNPs physicochemical characterisation. (**A**) TEM image (60000× magnification, 50 nm scale) showing spherical shaped AgNPs synthesised by the fungus *C. thermophilum*. (**B**) Size distribution, with the measurement obtained using ImageJ software, with 8.93 ± 2.29 nm average size.

3.3. NRU Cytotoxicity Evaluation

The cytotoxicity of the AgNPs synthesised using *C. thermophilum* cell-free extract was evaluated using the NRU following OECD guidelines. AgNP concentrations ranging from 31.03 μg/mL to 460.32 μg/mL (dilution log factor: 1.47) were tested. Dose–response curves were obtained using GraphPad Prism® and Excel software (Figure 4). The average IC values were as follows: IC_{20} = 91.77 ± 24.24 μg/mL, IC_{50} = 119.69 ± 21.15 μg/mL, and IC_{80} = 144.92 ± 23.22 μg/mL. The IC_{50} was used to predict an LD_{50} value of 624.31 ± 41.87 mg/kg, suggesting that the AgNPs synthesised might belong to GHS category 4 with regard to inducing acute toxicity (Figure 5 and Table 1).

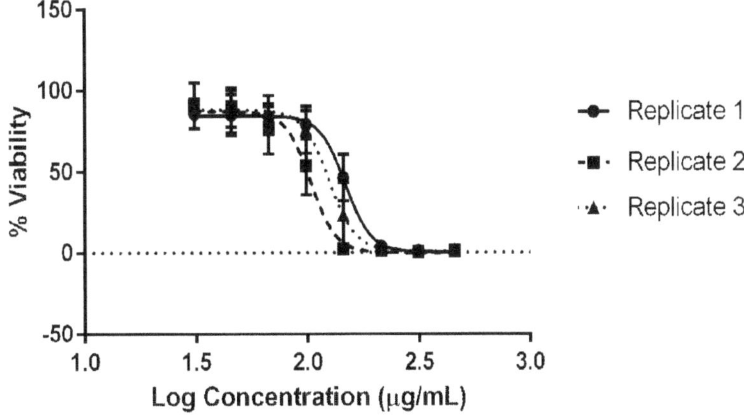

Figure 4. Cytotoxicity evaluation of the AgNPs biologically synthesised by the fungus *C. thermophilum*: the AgNP dose–response curves (Hill function fit) of the NRU assay using the skin fibroblasts from the murine embryo Balb/c 3T3 NIH cell line.

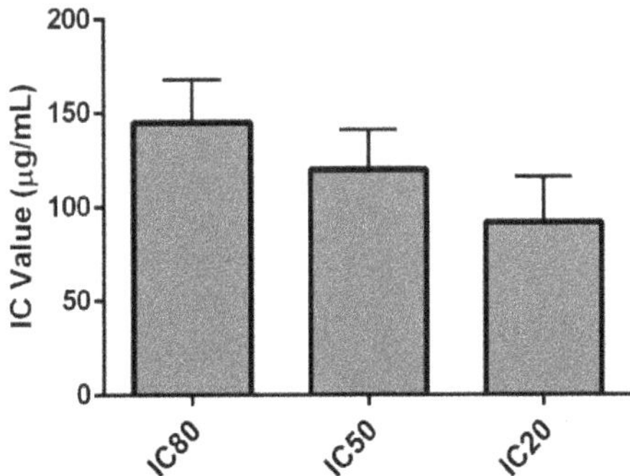

Figure 5. Cytotoxicity evaluation of the AgNPs biologically synthesised by the fungus *C. thermophilum* using the NRU assay with the skin fibroblasts from the murine embryo Balb/c 3T3 NIH cell line: IC values.

Table 1. Cytotoxicity evaluation of the AgNPs biologically synthesised by the fungus *C. thermophilum* using the NRU assay with the skin fibroblasts from the murine embryo Balb/c 3T3 NIH cell line: IC values, the predicted LD_{50} (which, given all at once, could cause the death of 50% of a group of test animals) and the GHS for the AgNPs' cytotoxic effects.

IC	μg/mL	Predicted LD_{50} mg/kg	GHS
IC_{80}	144.92 ± 23.33	624.31 ± 41.87	4
IC_{50}	119.69 ± 21.15		
IC_{20}	91.77 ± 24.24		

Additionally, the cytotoxicity of the *C. thermophilum* cell-free extract was analysed to evaluate potential toxicity derived from fungal metabolites. Concentrations ranging from 0.38 μg/mL to 1200 μg/mL (dilution log factor: 3.16) were tested. At 1200 μg/mL, the highest IC tested for the extract (2.5 times higher than the maximum AgNP concentration tested), a maximum of 50% of the cells were killed (Figure 6.). Thus, the IC_{50} and, consequently, the LD_{50} were not statistically calculated, nor was the GHS acute toxicity classification, as this was based on the LD_{50} values. Hence, the extract cytotoxicity was beyond the evaluated cytotoxicity range of this test. Therefore, the *C. thermophilum* cell-free extract was effectively non-toxic at the IC tested.

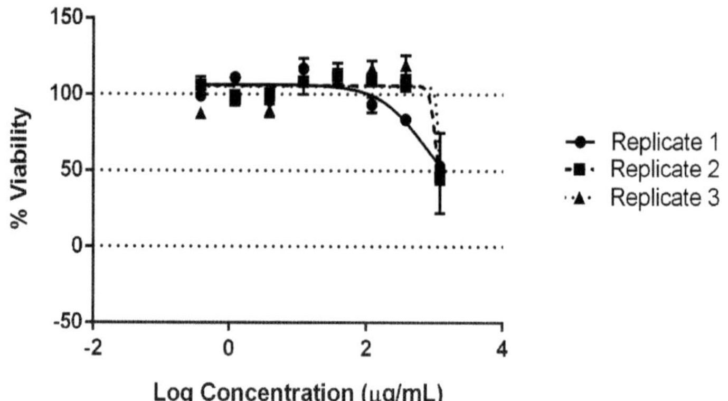

Figure 6. Cytotoxicity evaluation of the *C. thermophilum* cell-free-extract fungal metabolites: dose–response curves (Hill function fit) of the NRU assay using skin fibroblasts from the murine embryo Balb/c 3T3 NIH cell line.

3.4. AgNPs-Balb/c 3T3 NIH Cells Interaction

The interaction between the AgNPs and the Balb/c 3T3 NIH cells was evaluated at an ultrastructural level in three different scenarios: (1) a control in which the cells were not exposed to the AgNPs, (2) a non-cytotoxic AgNP concentration (IC_{20}), and (3) the IC_{50}. Compared to the control (Figure 7A,B), most cells had a preserved ultrastructure at the IC_{20} (Figure 7C,D). Hence, no evidence of cell alteration at the ultrastructural level was detected. However, at the IC_{50} (Figure 7E,F), the cell cytoplasm was damaged, and mitochondrial morphological alteration was evident.

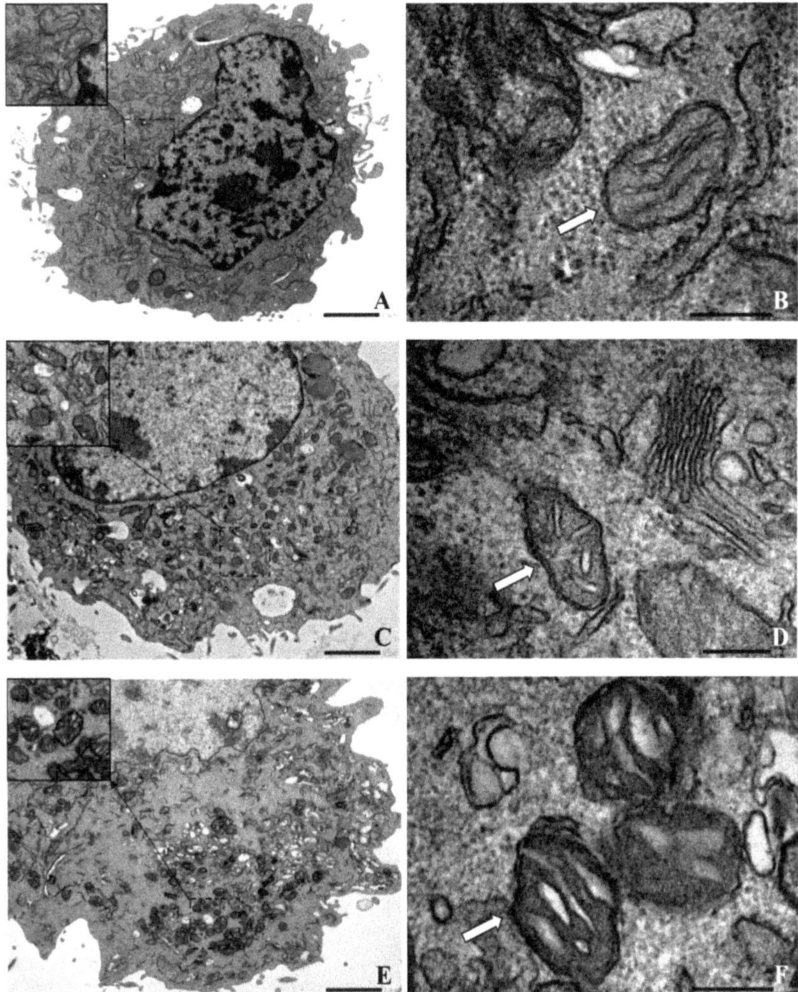

Figure 7. TEM ultrastructural evaluation of the interaction between the AgNPs biologically synthesised by the fungus *C. thermophilum* cell-free extract and skin fibroblasts from the murine embryo Balb/c 3T3 NIH cell line. (**A**) Negative control (Balb/c 3T3 NIH cells without treatment). (**B**) Mitochondria organelle of Balb/c 3T3 NIH cells without treatment. (**C**) Balb/c 3T3 NIH cells exposed to a non-toxic AgNP concentration (IC_{20} = 91.77 ± 24.24 µg/mL), demonstrating no significant cell damage or alteration at the ultrastructural level. (**D**) Mitochondria organelle of Balb/c 3T3 NIH cells exposed to a non-toxic AgNP concentration (IC_{20}). (**E**) Balb/c 3T3 NIH cells exposed to an IC_{50} (119.69 ± 21.15 µg/mL), showing that the cell cytoplasm was damaged. (**F**) Mitochondria organelle of Balb/c 3T3 NIH cells exposed to an IC_{50} with an evident morphological alteration.

4. Discussion

Research involving silver nanostructures has expanded rapidly due to their promising applications within the biomedicine and biotechnology fields. However, despite their exceptional physicochemical properties, concerns related to environmental toxicity and health-related hazards exist [16,17]. Therefore, efforts have been made in order to develop eco-friendly approaches to synthesising AgNPs. Overall, the advantage of the biological route involves the process simplicity, the use of non-hazardous reducing and stabilising reagents, and comparatively low production costs [4,5].

This study showed that the fungus *C. thermophilum* cell-free extract can biologically synthesise spherically shaped AgNPs with an average size of 8.93 ± 2.29 nm. The genus *Chaetomium* has been previously investigated with regard to NP synthesis. The *Chaetomium* species *C. globosum* and *C. cupreum* were used to produce copolymer NPs with bioactive compounds from their crude methanol extract [37]. Furthermore, the species *C. globosum* was used in iron NP production [38]. To the extent of our knowledge, the species *C. thermophilum* has not been yet investigated for NP synthesis.

C. thermophilum is a moderately thermophilic fungus with slow growth at 35 °C but with optimum growth temperatures in the 40 to 55 °C range [36]. Its genome sequence was first described by Amlacher et al. [39]. It has since been used as a model organism system for biophysical research. Recently, this fungus was involved in high-temperature protein adaptation studies that demonstrated that protein glycosylation and deglycosylation are the mechanisms that allow their thermophily [40]. Furthermore, different authors have investigated this species' enzymes in the breakdown of lignocellulose for biofuel-renewable biomass, as well as for their potential to degrade cellulosic waste [18,41].

In vitro cytotoxicity tests are used to determine the toxicity of substances in cell lines as an alternative to direct animal testing. In vitro systems are relevant models when investigating the common toxicity mechanisms of AgNPs because they are cost-effective and allow direct assessment of NPs, providing valuable data for the screening of toxicity [12]. The advantages relate to the costs, the time required to obtain the final results, and ethical issues [42]. The NRU assay is an example of an in vitro cytotoxicity test. The OECD published a guideline document (No. 129 from the Environment, Health and Safety Publications Series on Testing and Assessment) for this test in 2010. The cell lines Balb/c 3T3 (murine fibroblast) or NHK (normal human keratinocytes) are used to measure toxicity as a concentration-dependent reduction of the chemical NR cell uptake after the substance's test exposure [20]. Obtaining cytotoxicity results with a sensitive cell line, such as Balb/c 3T3, is a valuable screening approach. However, additional tests must be taken, not only to assess toxicity mechanisms but also to assess the effectiveness in other models of biomedical relevance.

It is important to highlight that the cytotoxicity effects depend on the AgNPs' physical and chemical nature, concentration, and incubation time; the presence of serum and, hence, protein coronas; the presence of ion release, agglomeration in the cell medium, and intracellular localization; the cell lineage; and other factors [12]. In this context, colourimetry tests, such as tests involving NRU, 3-(4,5-dimethylthiazol-2-yl)-2,5-diphenyl-2H-tetrazolium bromide (MTT), and (3-(4,5-dimethylthiazol-2-yl)-5-(3-carboxymethoxyphenyl)-2-(4-sulfophenyl)-2H-tetrazolium) (MTS), are often used in the cytotoxicity evaluation of AgNPs.

The GHS establishes that the acute toxic estimate (ATE) via the oral exposure route can be used to classify substances into five categories based on the LD_{50} values. A lower LD_{50} (mg/kg) is an indicator of greater toxicity [34]. This study successfully employed the OECD-based methodology (Figure 4). The NRU cytotoxicity assay was used to calculate the value of the IC_{50} (119.69 ± 21.15 µg/mL) of the biological AgNPs produced by the fungus *C. thermophilum* and to predict an LD_{50} value of 624.31 ± 41.87 mg/kg. Hence, this indicates a GHS ATE classification as category 4 (Figure 5). For contextual reference, the NRU assay control drug, sodium dodecyl sulphate (SDS), which is used in cleaning and hygiene products, has a rat oral LD_{50} of 977 mg/kg, also fitting into class 4 (300 mg/kg to 2000 mg/kg) [43].

The AgNPs biologically synthesised by the fungus *C. thermophilum* were significantly less toxic than chemically synthesised AgNPs, which have reported IC_{50} values ranging from 2.20 µg/mL to 10 µg/mL, indicating higher toxicity [22,23,44,45]. Importantly, they were also found to be 4 to 30 times less toxic than other AgNPs synthesised with different species (Table 2) [45–47]. It is essential to highlight that AgNPs are currently being explored for applications in cancer therapy, tumour detection, drug delivery, wound dressing, and elsewhere [46,48–52]. Therefore, synthesising NPs with lower toxicity, as demonstrated in this study, may lead to an expansion in the range of applications in the biomedical field.

Table 2. Comparative IC_{50} of AgNPs synthesised in different ways.

AgNP SYNTHESIS	IC_{50} µg/mL	CYTOTOXIC TEST	CELL LINEAGE	REF.
Biological: *Fusarium semitectum*	260.00	MTT	HGF human fibroblast	[45]
Biological: *Gloeophyllum striatum*	28.76	MTT	L929 mouse fibroblasts	[47]
Biological: *Streptomyces* sp.	64.50	MTT	L929 mouse fibroblasts	[53]
Biological: *Streptomyces xinghaiensis*	4.0	MTT	BALB/c 3T3 fibroblasts	[54]
Biological: *Canna edulis*	18.00	NRU/MTT	L929 mouse fibroblasts	[6]
Chemical: PVP-AgNP	2.80	NRU	BALB/c 3T3 fibroblasts	[22]
Chemical: PVP-AgNP	2.80	NRU	BALB/c 3T3 fibroblasts	[21]
Chemical: $Na_3C_6H_5O_7$-AgNP	10.00 *	MTS	BALB/c 3T3 fibroblasts	[55]
Chemical: $Na_3C_6H_5O_7$-AgNP	7.00	NRU	NCTC 929 fibroblast	[44]
Biological: *Chaetomium thermophilum*	119.69	NRU	Balb/c 3T3 fibroblast	Present study

* Substantial numbers of dead cells (56.8%).

Although multiple species of fungi have been claimed to synthesise AgNPs, some of these species are pathogenic. For example, *F. oxysporum* was reported to synthesise well-dispersed, spherically shaped AgNPs with sizes ranging from 5 nm to 13 nm. However, this species is responsible for soil-borne diseases that are extremely difficult to control, affecting food plants such as tomatoes, bananas, and onions [8,46]. Another example is the fungus *A. fumigatus*, described as responsible for the extracellular synthesis of extremely small (0.68 nm) cube-shaped AgNPs [15]. Nonetheless, this genus is also responsible for a parenchymal lung disease called aspergillosis [56].

While fungi produce a large variety of secondary metabolites, which may be critical for NP synthesis and stabilisation, they are also known to produce harmful self-preservation chemicals, including mycotoxins. In small concentrations, these low-molecular-weight compounds can be toxic to vertebrates and other animal groups [57]. Hence, fungal extracts also require toxicity evaluation for any proposed biological applications of NPs synthesised using fungi.

In this study, it was demonstrated that the *C. thermophilum* cell-free extract used to synthesise the AgNPs was effectively non-toxic at the maximum IC tested (1200 µg/mL, Figure 6). Similar results were obtained in a cytotoxicity study of culinary–medicinal mushroom aqueous extract using the NRU assay with Balb/c 3T3 NIH cells. The non-toxic fungal extracts of *Ganoderma lucidum* (IC_{50} = 1350 µg/mL), *Ganoderma neo-japonicum* (IC_{50} = 1780 µg/mL), *Hericium erinaceus* (IC_{50} = 3530 µg/mL), *Lignosus rhinocerotis* (IC_{50} = 5600 µg/mL), and others were reported [58].

Furthermore, this study evaluated the interaction between the AgNPs and the Balb/c 3T3 NIH cells at an ultrastructural level (Figure 7). The different mechanisms by which AgNPs induce cell death include Ag ion release, disruption of cell membrane integrity, oxidative stress, protein or deoxyribonucleic acid (DNA) damage, generation of reactive oxygen species, and apoptotic cell death [59,60]. The ultrastructural damage caused by NP–cell interaction is often identified using TEM.

The TEM study demonstrated that most cells had preserved ultrastructures at the IC_{20} (91.77 ± 24.24 μg/mL, Figure 7C,D). However, at the IC_{50} (119.69 ± 21.15 μg/mL), the cell cytoplasm was damaged (Figure 7E). The mitochondrial morphological alteration was evident in the swelling of the inner membrane (Figure 7F). Similar results were found after analysing the interaction between mitochondria and chemically synthesised AgNPs. For example, a significant decrease in the mitochondrial membrane potential, adenosine diphosphate (ADP)-induced depolarisation, and respiratory control ratio were reported in rat liver mitochondria exposed to 40 nm and 80 nm of AgNPs. The function impairment was mainly attributed to changes in the membrane permeability [25]. Furthermore, exposure of AgNPs and titanium NPs to rat liver mitochondria was also demonstrated to lower the respiratory control ratio and induce mitochondrial swelling [61]. Additionally, the interaction of adult Wistar rats with chemically produced AgNPs (10 ± 4 nm) at a low dosage (0.2 mg/kg b.w.) resulted in mitochondrial swelling and cristolysis (damage of cristae) caused by silver nano-granules in the brain [24].

Mitochondria are cytoplasmic, double- membrane-bound organelles known to play an essential role in cellular energy production and to participate in calcium signalling, cell growth, differentiation, and death [62]. Moreover, mitochondria dysfunction has been related to many diseases, including neurodegenerative disorders; Huntington's, Parkinson's, and Alzheimer's diseases; epilepsy; schizophrenia; and Leigh syndrome [63–65]. Thus, it is essential to investigate the effects of NPs on mitochondria at a structural level, as disruption of this organelle can result in health effects.

5. Conclusions

As sustainable development awareness has grown, there has been a shift in focus to the use of biological sources. In the nanotechnology field, fungi have been investigated for the production of AgNPs in an environmentally friendly manner. This study reported the successful biological synthesis of AgNPs using the fungus C. *thermophilum* and is, to the best of our knowledge, the first time this fungus has been studied with regard to NPs production. Furthermore, it demonstrated low AgNP cytotoxicity towards a reference mammalian cell line, Balb/c 3T3 NIH. Additionally, the interaction of the AgNPs with this cell line showed preserved cell structures at non-toxic IC_{20}. Therefore, the biologically synthesised AgNPs described in this study have potential as sustainable substitutes for physically and chemically made AgNPs; hence, they are a step in the right direction in achieving sustainable development in the nanotechnology field.

Author Contributions: Conceptualization, A.M.d.A. and P.G.M.; formal analysis, M.F.A.; investigation, M.F.A., A.C.C.P., T.D.F.K. and C.K.; resources, A.M.d.A. and P.G.M.; writing—original draft preparation, M.F.A., A.C.C.P., B.H.M., A.M.d.A. and P.G.M.; writing—review and editing, M.F.A., A.M.d.A. and P.G.M.; supervision, C.K., B.H.M., A.M.d.A. and P.G.M. All authors have read and agreed to the published version of the manuscript.

Funding: This research was performed with financial support from the Technological University of Shannon: Midlands Midwest (TUS), the Instituto Carlos Chagas (ICC)-Fiocruz Paraná, Erasmus+ (KA107-2015), and Fundação Araucária (20/2017, Protocol 48122).

Institutional Review Board Statement: Not applicable.

Informed Consent Statement: Not applicable.

Acknowledgments: All authors wish to acknowledge Shannon Applied Biotechnology Centre (ShannonABC) and Rede de Plataformas Tecnológicas da FIOCRUZ for the use of their facilities to conduct this research.

Conflicts of Interest: The authors declare no conflict of interest. The funders had no role in the design of the study; in the collection, analyses, or interpretation of data; in the writing of the manuscript, or in the decision to publish the results.

References

1. Mukherjee, D.; Singh, S.; Kumar, M.; Kumar, V.; Datta, S.; Dhanjal, D.S. Fungal Biotechnology: Role and Aspects. In *Fungi and Their Role in Sustainable Development: Current Perspectives*; Gehlot, P., Singh, J., Eds.; Springer: Singapore, 2018; pp. 91–103. [CrossRef]
2. United Nations. Transforming Our World, the 2030 Agenda for Sustainable Development, D.o.E.a.S. Affairs, Editor. 2020. Available online: https://sustainabledevelopment.un.org/content/documents/21252030%20Agenda%20for%20Sustainable%20Development%20web.pdf (accessed on 22 May 2022).
3. Alves, M.F.; Murray, P.G. Biological Synthesis of Monodisperse Uniform-Size Silver Nanoparticles (AgNPs) by Fungal Cell-Free Extracts at Elevated Temperature and pH. *J. Fungi* **2022**, *8*, 439. [CrossRef] [PubMed]
4. Alghuthaymi, M.A.; Almoammar, H.; Rai, M.; Said-Galiev, E.; Abd-Elsalam, K.A. Myconanoparticles: Synthesis and their role in phytopathogens management. *Biotechnol. Biotechnol. Equip.* **2015**, *29*, 221–236. [CrossRef] [PubMed]
5. Kaabipour, S.; Hemmati, S. A review on the green and sustainable synthesis of silver nanoparticles and one-dimensional silver nanostructures. *Beilstein J. Nanotechnol.* **2021**, *12*, 102–136. [CrossRef] [PubMed]
6. Otari, S.V.; Pawar, S.H.; Patel, S.K.S.; Singh, R.K.; Kim, S.Y.; Lee, J.H.; Zhang, L.; Lee, J.K. *Canna edulis* Leaf Extract-Mediated Preparation of Stabilized Silver Nanoparticles: Characterization, Antimicrobial Activity, and Toxicity Studies. *J. Microbiol. Biotechnol.* **2017**, *27*, 731–738. [CrossRef]
7. Brown, A.N.; Smith, K.; Samuels, T.A.; Obare, S.O.; Scott, M.E. Nanoparticles functionalized with ampicillin destroy multiple-antibiotic-resistant isolates of *Pseudomonas aeruginosa* and *Enterobacter aerogenes* and methicillin-resistant *Staphylococcus aureus*. *Appl. Environ. Microbiol.* **2012**, *78*, 2768–2774. [CrossRef]
8. Husseiny, S.M.; Salah, T.A.; Anter, H.A. Biosynthesis of size controlled silver nanoparticles by Fusarium oxysporum, their an-tibacterial and antitumor activities. *Beni-Suef Univ. J. Basic Appl. Sci.* **2015**, *4*, 225–231. [CrossRef]
9. Rajeshkumar, S.; Malarkodi, C.; Vanaja, M.; Annadurai, G. Anticancer and enhanced antimicrobial activity of biosynthesizd silver nanoparticles against clinical pathogens. *J. Mol. Struct.* **2016**, *1116*, 165–173. [CrossRef]
10. Ameen, F.; Al-Homaidan, A.A.; Al-Sabri, A.; Almansob, A.; AlNadhari, S. Anti-oxidant, anti-fungal and cytotoxic effects of silver nanoparticles synthesized using marine fungus Cladosporium halotolerans. *Appl. Nanosci.* 2021. [CrossRef]
11. Firdhouse, M.J.; Lalitha, P. Biosynthesis of silver nanoparticles and its applications. *J. Nanotechnol.* **2015**, *2015*, 1–18. [CrossRef]
12. Zhang, X.F.; Liu, Z.G.; Shen, W.; Gurunathan, S. Silver Nanoparticles: Synthesis, Characterization, Properties, Applications, and Therapeutic Approaches. *Int. J. Mol. Sci.* **2016**, *17*, 1534. [CrossRef]
13. Elamawi, R.M.; Al-Harbi, R.E.; Hendi, A.A. Biosynthesis and characterization of silver nanoparticles using Trichoderma longibrachiatum and their effect on phytopathogenic fungi. *Egypt. J. Biol. Pest. Control.* **2018**, *28*, 28. [CrossRef]
14. Nayak, B.K.; Nanda, A.; Prabhakar, V. Biogenic synthesis of silver nanoparticle from wasp nest soil fungus, *Penicillium italicum* and its analysis against multi drug resistance pathogens. *Biocatal. Agric. Biotechnol.* **2018**, *16*, 412–418. [CrossRef]
15. Shahzad, A.; Saeed, H.; Iqtedar, M.; Hussain, S.Z.; Kaleem, A.; Abdullah, R.; Sharif, S.; Naz, S.; Saleem, F.; Aihetasham, A.; et al. Size Controlled Production of Silver Nanoparticles by *Aspergillus fumigatus* BTCB10: Likely Anti-bacterial and Cytotoxic Effects. *J. Nanomater.* **2019**, *2019*, 1–14. [CrossRef]
16. He, X.; Hwang, H.M. Nanotechnology in food science: Functionality, applicability, and safety assessment. *J. Food Drug Anal.* **2016**, *24*, 671–681. [CrossRef]
17. Krug, H. Nanosafety Research-Are We on the Right Track? *Angew. Chem. Int. Ed. Engl.* **2014**, *53*, 12304–12319. [CrossRef]
18. Li, J.; Zhang, B.; Chang, X.; Gan, J.; Li, W.; Niu, S.; Wu, T.; Zhang, T.; Tang, M.; Xue, Y. Silver nanoparticles modulate mitochondrial dynamics and biogenesis in HepG2 cells. *Environ. Pollut.* **2020**, *256*, 113430. [CrossRef]
19. ICCVAM. Guidance Document on Using In Vitro Data to Estimate In Vivo Starting Doses for Acute Toxicity, H.a.H. Services, Editor. US Public Health Service. 2001. Available online: https://ntp.niehs.nih.gov/iccvam/docs/acutetox_docs/guidance0801/iv_guide.pdf (accessed on 22 May 2022).
20. Organisation for Economic Co-Operation and Development. Guidance Document on Using Cytotoxicity Tests to Estimate Starting Doses for Acute Oral Systemic Toxicity Tests. 2010. Available online: https://www.oecd.org/officialdocuments/publicdisplaydocumentpdf/?cote=env/jm/mono%282010%2920&doclanguage=en (accessed on 22 May 2022).
21. Mannerström, M.; Zou, J.; Toimela, T.; Pyykkö, I.; Heinonen, T. The applicability of conventional cytotoxicity assays to predict safety/toxicity of mesoporous silica nanoparticles, silver and gold nanoparticles and multi-walled carbon nanotubes. *Toxicol. Vitr.* **2016**, *37*, 113–120. [CrossRef]
22. Zou, J.; Feng, H.; Mannerström, M.; Heinonen, T.; Pyykkö, I. Toxicity of silver nanoparticle in rat ear and BALB/c 3T3 cell line. *J. Nanobiotechnol.* **2014**, *12*, 52. [CrossRef]
23. McShan, D.; Ray, P.C.; Yu, H. Molecular toxicity mechanism of nanosilver. *J. Food Drug Anal.* **2014**, *22*, 116–127. [CrossRef]
24. Skalska, J.; Dąbrowska-Bouta, B.; Frontczak-Baniewicz, M. Sulkowski, G.; Strużyńska, L. A low dose of nanoparticulate silver induces mitochondrial dysfunction and autophagy in adult rat brain. *Neurotox. Res.* **2020**, *38*, 650–664. [CrossRef]
25. Teodoro, J.S.; Simões, A.M.; Duarte, F.V.; Rolo, A.P.; Murdoch, R.C.; Hussain, S.M.; Palmeira, C.M. Assessment of the toxicity of silver nanoparticles in vitro: A mitochondrial perspective. *Toxicol. Vitr.* **2011**, *25*, 664–670. [CrossRef] [PubMed]
26. AbdelRahim, K.; Mahmoud, S.Y.; Ali, A.M.; Almaary, K.S.; Mustafa, A.E.-Z.M.A.; Husseiny, S.M. Extracellular biosynthesis of silver nanoparticles using *Rhizopus stolonifer*. *Saudi J. Biol. Sci.* **2017**, *24*, 208–216. [CrossRef] [PubMed]

27. Hamedi, S.; Ghaseminezhad, M.; Shokrollahzadeh, S.; Shojaosadati, S.A. Controlled biosynthesis of silver nanoparticles using nitrate reductase enzyme induction of filamentous fungus and their antibacterial evaluation. *Artif. Cells Nanomed. Biotechnol.* **2017**, *45*, 1588–1596. [CrossRef] [PubMed]
28. Katapodis, P.; Christakopoulou, V.; Kekos, D.; Christakopoulos, P. Optimization of xylanase production by *Chaetomium thermophilum* in wheat straw using response surface methodology. *Biochem. Eng. J.* **2007**, *35*, 136–141. [CrossRef]
29. ICCVAM. Recommended Test Method Protocol Normal Human Keratinocyte NRU Cytotoxicity Test Method. 2006. Available online: https://ntp.niehs.nih.gov/iccvam/docs/protocols/ivcyto-nhk.pdf (accessed on 22 May 2022).
30. Reus, T.L.; Machado, T.N.; Bezerra, A.G., Jr.; Marcon, B.H.; Paschoal, A.C.C.; Kuligovski, C.; Aguiar, A.M.; Dallagiovanna, B. Dose-dependent cytotoxicity of bismuth nanoparticles produced by LASiS in a reference mammalian cell line BALB/c 3T3. *Toxicol. In Vitro* **2018**, *53*, 99–106. [CrossRef]
31. ICCVAM. In Vitro Cytotoxicity Test Methods for Estimating Acute Oral Systemic Toxicity—Background Review Document. 2006. Available online: https://ntp.niehs.nih.gov/iccvam/docs/acutetox_docs/brd_tmer/at-tmer-complete.pdf (accessed on 22 May 2022).
32. Abud, A.P.; Zych, J.; Reus, T.L.; Kuligovski Moraes, E.; Dallagiovanna, B.; Aguiar, A.M. The use of human adipose-derived stem cells based cytotoxicity assay for acute toxicity test. *Regul. Toxicol. Pharmacol.* **2015**, *73*, 992–998. [CrossRef]
33. Reus, T.L.; Machado, T.N.; Marcon, B.H.; Paschoal, A.C.C.; Ribeiro, I.R.S.; Cardoso, M.B.; Dallagiovanna, B.; Aguiar, A.M. Dose-dependent cell necrosis induced by silica nanoparticles. *Toxicol. In Vitro* **2020**, *63*, 104723. [CrossRef]
34. United Nations. Global Harmonized System of Classification and Labelling of Chemicals (GHS). 2019. Available online: https://unece.org/fileadmin/DAM/trans/danger/publi/ghs/ghs_rev08/ST-SG-AC10-30-Rev8e.pdf (accessed on 22 May 2022).
35. ICCVAM. In Vitro Cytotoxicity Test Methods for Estimating Starting Doses for Acute Oral Systemic Toxicity Testing. 2006. Available online: https://ntp.niehs.nih.gov/iccvam/docs/acutetox_docs/brd_tmer/brdvol2_nov2006.pdf (accessed on 22 May 2022).
36. La-Touche, C.J. On a thermophile species of *Chaetomium. Trans. Br. Mycol. Soc.* **1950**, *33*, 94–104. [CrossRef]
37. Dar, J.; Soytong, K. Construction and characterization of copolymer nanomaterials loaded with bioactive compounds from *Chaetomium* species. *J. Agric. Technol.* **2014**, *10*, 823–831. Available online: https://www.thaiscience.info/journals/Article/IJAT/10934631.pdf (accessed on 22 May 2022).
38. Kaul, R.K.; Kumar, P.; Burman, U.; Joshi, P.; Agrawal, A.; Raliya, R.; Tarafdar, J.C. Magnesium and iron nanoparticles production using microorganisms and various salts. *Mater. Sci.-Pol.* **2012**, *30*, 254–258. [CrossRef]
39. Amlacher, S.; Sarges, P.; Flemming, D.; van Noort, V.; Kunze, R.; Devos Damien, P.; Arumugam, M.; Bork, P.; Hurt, E. Insight into Structure and Assembly of the Nuclear Pore Complex by Utilizing the Genome of a Eukaryotic Thermophile. *Cell* **2011**, *146*, 277–289. [CrossRef] [PubMed]
40. Gao, J.; Li, Q.; Li, D. Novel Proteome and N-Glycoproteome of the Thermophilic Fungus *Chaetomium thermophilum* in Response to High Temperature. *Front. Microbiol.* **2021**, *12*, 1200. [CrossRef] [PubMed]
41. Darwish, A.M.G.; Abdel-Azeem, A.M. *Chaetomium Enzymes and Their Applications. Recent Developments on Genus Chaetomium*; Springer International Publishing: Cham, Switzerland, 2020; pp. 241–249. [CrossRef]
42. Fard, J.K.; Jafari, S.; Eghbal, M.A. A Review of Molecular Mechanisms Involved in Toxicity of Nanoparticles. *Adv. Pharm. Bull.* **2015**, *5*, 447–454. [CrossRef] [PubMed]
43. MERCK. Sodium Dodecyl Sulfate—Safety Data Sheet. 2021. Available online: https://www.sigmaaldrich.com/IE/en/sds/mm/8.17034 (accessed on 22 May 2022).
44. Salomoni, R.; Leo, P.; Montemor, A.F.; Rinaldi, B.G.; Rodrigues, M.F.A. Antibacterial effect of silver nanoparticles in *Pseudomonas aeruginosa. Nanotechnol. Sci. Appl.* **2017**, *10*, 115. [CrossRef]
45. Halkai, K.; Mudda, J.A.; Shivanna, V.; Patil, V.; Rathod, V.; Halkai, R. Cytotoxicity evaluation of fungal-derived silver nanoparticles on human gingival fibroblast cell line: An In Vitro study. *J. Conserv. Dent.* **2019**, *22*, 160–163. [CrossRef]
46. Rahimi, M.; Noruzi, E.B.; Sheykhsaran, E.; Ebadi, B.; Kariminezhad, Z.; Molaparast, M.; Mehrabani, M.G.; Mehramouz, B.; Yousefi, M.; Ahmadi, R.; et al. Carbohydrate polymer-based silver nanocomposites: Recent progress in the antimicrobial wound dressings. *Carbohydr. Polym.* **2020**, *231*, 115696. [CrossRef]
47. Zawadzka, K.; Felczak, A.; Nowak, M.; Kowalczyk, A.; Piwonski, I.; Lisowska, K. Antimicrobial activity and toxicological risk assessment of silver nanoparticles synthesized using an eco-friendly method with *Gloeophyllum striatum. J. Hazard. Mater.* **2021**, *418*, 126316. [CrossRef]
48. Hasanzadeh, M.; Feyziazar, M.; Solhi, E.; Mokhtarzadeh, A.; Soleymani, J.; Shadjou, N.; Jouyban, A.; Mahboob, S. Ultrasensitive immunoassay of breast cancer type 1 susceptibility protein (BRCA1) using poly (dopamine-beta cyclodextrine-Cetyl trimethylammonium bromide) doped with silver nanoparticles: A new platform in early stage diagnosis of breast cancer and efficient management. *Microchem. J.* **2019**, *145*, 778–783. [CrossRef]
49. Karuppaiah, A.; Siram, K.; Selvaraj, D.; Ramasamy, M.; Babu, D.; Sankar, V. Synergistic and enhanced anticancer effect of a facile surface modified non-cytotoxic silver nanoparticle conjugated with gemcitabine in metastatic breast cancer cells. *Mater. Today Commun.* **2020**, *23*, 100884. [CrossRef]
50. Khansa, I.; Schoenbrunner, A.R.; Kraft, C.T.; Janis, J.E. Silver in Wound Care—Friend or Foe?: A Comprehensive Review. *Plast. Reconstr. Surg.-Glob. Open* **2019**, *7*, e2390. [CrossRef]
51. Nigam Joshi, P.; Agawane, S. Athalye, M.C.; Jadhav, V.; Sarkar, D.; Prakash, R. Multifunctional inulin tethered silver-graphene quantum dots nanotheranostic module for pancreatic cancer therapy. *Mater. Sci. Eng. C* **2017**, *78*, 1203–1211. [CrossRef] [PubMed]

52. Pothipor, C.; Wiriyakun, N.; Putnin, T.; Ngamaroonchote, A.; Jakmunee, J.; Ounnunkad, K.; Laocharoensuk, R.; Aroonyadet, N. Highly sensitive biosensor based on graphene–poly (3-aminobenzoic acid) modified electrodes and porous-hollowed-silver-gold nanoparticle labelling for prostate cancer detection. *Sens. Actuators B Chem.* **2019**, *296*, 126657. [CrossRef]
53. Składanowski, M.; Golinska, P.; Rudnicka, K.; Dahm, H.; Rai, M. Evaluation of cytotoxicity, immune compatibility and antibacterial activity of biogenic silver nanoparticles. *Med. Microbiol. Immunol.* **2016**, *205*, 603–613. [CrossRef] [PubMed]
54. Wypij, M.; Czarnecka, J.; Swiecimska, M.; Dahm, H.; Rai, M.; Golinska, P. Synthesis, characterization and evaluation of antimicrobial and cytotoxic activities of biogenic silver nanoparticles synthesized from *Streptomyces xinghaiensis* OF1 strain. *World J. Microbiol. Biotechnol.* **2018**, *34*, 23. [CrossRef]
55. Lee, Y.-H.; Cheng, F.-Y.; Chiu, H.-W.; Tsai, J.-C.; Fang, C.-Y.; Chen, C.-W.; Wang, Y.-J. Cytotoxicity, oxidative stress, apoptosis and the autophagic effects of silver nanoparticles in mouse embryonic fibroblasts. *Biomaterials* **2014**, *35*, 4706–4715. [CrossRef]
56. Baluku, J.B.; Nuwagira, E.; Bongomin, F.; Denning, D.W. Pulmonary TB and chronic pulmonary aspergillosis: Clinical differences and similarities. *Int. J. Tuberc. Lung Dis.* **2021**, *25*, 537–546. [CrossRef]
57. Bennett, J.W.; Klinch, M. Mycotoxins. *Clin. Microbiol. Rev.* **2003**, *16*, 497–516. [CrossRef]
58. Phan, C.-W.; David, P.; Naidu, M.; Wong, K.-H.; Sabaratnam, V. Neurite outgrowth stimulatory effects of culinary-medicinal mushrooms and their toxicity assessment using differentiating Neuro-2a and embryonic fibroblast BALB/3T3. *BMC Complement. Altern. Med.* **2013**, *13*, 261. [CrossRef]
59. Ali, H.; Aboud, M.; Alwan, S. Biological Synthesis of Silver Nanoparticles from *Saprolegnia parasitica*. *J. Phys. Conf. Ser.* **2019**, *1294*, 062090. [CrossRef]
60. Bressan, E.; Ferroni, L.; Gardin, C.; Rigo, C.; Stocchero, M.; Vindigni, V.; Cairns, W.; Zavan, B. Silver Nanoparticles and Mitochondrial Interaction. *Int. J. Dent.* **2013**, *2013*, 312747. [CrossRef]
61. Pereira, L.C.; Pazin, M.; Franco-Bernardes, M.F.; Martins, A.D.C., Jr.; Barcelos, G.R.M.; Pereira, M.C.; Mesquita, J.P.; Rodrigues, J.L.; Barbosa, F.; Dorta, D.J. A perspective of mitochondrial dysfunction in rats treated with silver and titanium nanoparticles (AgNPs and TiNPs). *J. Trace Elem. Med. Biol.* **2018**, *47*, 63–69. [CrossRef] [PubMed]
62. Osellame, L.D.; Blacker, T.S.; Duchen, M.R. Cellular and molecular mechanisms of mitochondrial function. *Best Pract. Res. Clin. Endocrinol. Metab.* **2012**, *26*, 711–723. [CrossRef] [PubMed]
63. Farshbaf, M.J.; Ghaedi, K. Huntington's Disease and Mitochondria. *Neurotox. Res.* **2017**, *32*, 518–529. [CrossRef] [PubMed]
64. Ruhoy, I.S.; Saneto, R.P. The genetics of Leigh syndrome and its implications for clinical practice and risk management. *Appl. Clin. Genet.* **2014**, *7*, 221–234. [CrossRef]
65. Wu, Y.; Chen, M.; Jiang, J. Mitochondrial dysfunction in neurodegenerative diseases and drug targets via apoptotic signaling. *Mitochondrion* **2019**, *49*, 35–45. [CrossRef] [PubMed]

Journal of *Fungi*

Article

Biological Synthesis of Monodisperse Uniform-Size Silver Nanoparticles (AgNPs) by Fungal Cell-Free Extracts at Elevated Temperature and pH

Mariana Fuinhas Alves * and Patrick G. Murray 

Shannon Applied Biotechnology Centre, Department of Applied Science, Faculty of Applied Sciences and Technology, Moylish Campus, Technological University of the Shannon, Midlands Midwest, Moylish, V94 EC5T Limerick, Ireland; patrick.murray@lit.ie
* Correspondence: mariana.alves@lit.ie

Abstract: Fungi's ability to convert organic materials into bioactive products offers environmentally friendly solutions for diverse industries. In the nanotechnology field, fungi metabolites have been explored for green nanoparticle synthesis. Silver nanoparticle (AgNP) research has grown rapidly over recent years mainly due to the enhanced optical, antimicrobial and anticancer properties of AgNPs, which make them extremely useful in the biomedicine and biotechnology field. However, the biological synthesis mechanism is still not fully established. Therefore, this study aimed to evaluate the combined effect of time, temperature and pH variation in AgNP synthesis using three different fungi phyla (Ascomycota, Basidiomycota and Zygomycota) represented by six different fungi species: *Cladophialophora bantiana* (*C. bantiana*), *Penicillium antarcticum* (*P. antarcticum*), *Trametes versicolor* (*T. versicolor*), *Trichoderma martiale* (*T. martiale*), *Umbelopsis isabellina* (*U. isabellina*) and *Bjerkandera adusta* (*B. adusta*). Ultraviolet–visible (UV-Vis) spectrophotometry and transmission electron microscopy (TEM) results demonstrated the synthesis of AgNPs of different sizes (3 to 17 nm) and dispersity percentages (25 to 95%, within the same size range) using fungi extracts by changing physicochemical reaction parameters. It was observed that higher temperatures (90 °C) associated with basic pH (9 and 12) favoured the synthesis of monodisperse small AgNPs. Previous studies demonstrated enhanced antibacterial and anticancer properties correlated with smaller nanoparticle sizes. Therefore, the biologically synthesised AgNPs shown in this study have potential as sustainable substitutes for chemically made antibacterial and anticancer products. It was also shown that not all fungi species (*B. adusta*) secrete metabolites capable of reducing silver nitrate ($AgNO_3$) precursors into AgNPs, demonstrating the importance of fungal screening studies.

Keywords: biosynthesis; fungus; nanotechnology; reaction optimisation; silver nanoparticles

Citation: Alves, M.F.; Murray, P.G. Biological Synthesis of Monodisperse Uniform-Size Silver Nanoparticles (AgNPs) by Fungal Cell-Free Extracts at Elevated Temperature and pH. *J. Fungi* **2022**, *8*, 439. https://doi.org/10.3390/jof8050439

Academic Editor: Kamel A. Abd-Elsalam

Received: 25 March 2022
Accepted: 20 April 2022
Published: 23 April 2022

1. Introduction

The conversion of organic materials into bioactive products by fungi offers sustainable solutions, especially as substitutes for toxic chemicals, which is extremely important for a bio-based circular economy [1,2]. Metabolites secreted by fungi in cell-free extracts are bioactive compounds with various applications. Hence, there is an increase in fungal biotechnology research, particularly focusing on fungal growth optimisation and potential applications of the metabolites secreted by fungi [1,2]. One application of fungi that has been explored is in the nanotechnology field in green bio-based nanoparticle synthesis.

The resistance to toxic heavy metals displayed by microorganisms, such as the bacteria *Pseudomonas aeruginosa* and the fungus *Aspergillus niger*, both used in chemical detoxification, inspired the development of greener routes to synthesise nanomaterials [3,4]. Nowadays, the use of hazardous chemicals and toxic reducing and stabilising agents in nanoparticle synthesis are gradually being replaced with more sustainable, safe and cost-effective synthesis routes [3,5]. As a result, these agents are being substituted by biological agents such as viruses, bacteria, fungi, yeasts, algae and plants or bio-extracted compounds such as proteins/peptides, carbohydrates and vitamins [6–8].

The benefits of eco-friendly biological synthesis include enormous biodiversity, environmental compatibility, scalability, energy-efficient methodologies, low or reduced production cost, and less toxic or nontoxic end-products [8–14]. However, biological methods mainly produce nanoparticle solutions with heterogeneous morphologies, and the synthesis mechanism is still not fully established [3,15–17].

According to the International Organization for Standardization (ISO), a nanoparticle is a "nano-object, [discrete piece of material with one, two or three external dimensions in the nanoscale (length range approximately from 1 nm to 100 nm)], with all external dimensions in the nanoscale where the lengths of the longest and the shortest axes of the nano-object do not differ significantly" [18]. The physicochemical properties of nanoparticles are determined by their morphology. In other words, their properties are size- and shape-dependent. It has been previously demonstrated that nanoparticle size can be tuned by changing reaction parameters such as salt precursor concentration, stabilising agents, time, temperature and pH [19–21]. However, in the biological green myco-synthesis field, the metabolites involved in reducing and stabilising the nanoparticles can be distinctly different from species to species, resulting in nanoparticles with diverse morphologies [22]. Therefore, it is essential to optimise parameters and to develop controlled procedures to synthesise stable and monodisperse nanoparticles.

AgNP research has grown rapidly over the years due to the optical, antimicrobial and anticancer properties of AgNPs, which make them extremely useful in the biomedicine and biotechnology field [4–7,23,24]. For example, the inhibitory growth effects of AgNPs have been reported against bacteria such as *Pseudomonas aeruginosa*, *Escherichia coli* (*E. coli*) and *Staphylococcus aureus* (*S. aureus*) [25–27]. Furthermore, AgNPs' anticancer properties were also demonstrated by their cytotoxic effect on human liver cancer cells (HepG2), breast cancer cells (MCF-7) and colorectal cancer cells (HCT116) [28–30]. In this scenario, the size of the nanoparticle plays an essential role in its application, as the enhanced antibacterial and anticancer properties were correlated with smaller nanoparticle sizes [25–27,31]. Nevertheless, the biological nanoparticle synthesis's optimum reaction parameters are still not fully elucidated. Therefore, the purpose of this study was (1) to investigate the ability of metabolites secreted by six fungi to synthesise AgNPs and (2) to evaluate the effects of time, temperature and pH variation in combination on the synthesis of AgNPs. This study demonstrated that not all fungi can produce metabolites that can reduce $AgNO_3$ into AgNPs. In addition, physicochemical parameters have a significant influence on the stability, size and dispersity of AgNPs. These results highlight that fungal screening is important to determine metabolite functionality. The results also determined optimal parameters for nanoparticle product development.

2. Materials and Methods

2.1. Microorganisms

The six microorganisms utilised in this were study obtained from the Technological University of Shannon: Midland Midwest (TUS) biobank: *C. bantiana*, *P. antarcticum* (*Centraal Bureau voor Schimmelcultures* CBS 100491), *T. versicolor*, *T. martiale*, *U. isabellina* and *B. adusta*.

2.2. Production of Fungal Cell-Free (FCF) Extract

The production of the FCF extract was adapted from AbdelRahim et al., Hamedi et al. and Katapodis et al. [32–34]. Fungi were cultivated at room temperature in an MYGP agar medium composed of agar (15 g/L, Formedium, Hunstanton, UK), malt extract (3 g/L, OXOID, Hampshire, UK), yeast extract (3 g/L, Formedium), glucose (10 g/L, Formedium) and peptone (5 g/L Formedium). In aseptic conditions, fungal cells taken from the mycelia mat borders to guarantee active growth were inoculated in 250 mL Erlenmeyer flasks with MYGP medium (previous composition without agar) and placed in an incubator shaker at room temperature at 120 rpm for five days. Cells were washed thoroughly with sterile deionised water, filtered, dried and then accurately weighed. To induce the secretion of secondary metabolites, the cells were transferred to a stress medium, composed of sterile deionised water, with a ratio of 1 g of cells to 10 mL and incubated in the shaker for three days. The fungal secreted extract was separated from the fungus cells by filtration, followed by centrifugation at 5000 rpm for 20 min, at 25 °C, with subsequent 0.22 μm membrane filtration. FCF extracts were stored at 4 °C until further use.

Prior to use, all media were autoclaved at a temperature of 121 °C and pressure of 15 psi for 15 min.

2.3. Biological Synthesis of AgNPs

The activity level of the fungal cell-free (FCF) extracts in the synthesis of AgNPs was tested based on the studies of Al-Khuzai et al. [19]. $AgNO_3$ at a concentration of 0.5 mM ($AgNO_3$, Sigma-Aldrich St. Louis, MO, USA) was used, varying the reaction time (1, 3 and 6 h), temperature (20, 45 and 90 °C) and pH (6, 9 and 12), in the water bath (Julabo TW12, Seelbach, Germany). All reactions were conducted in triplicate.

The pH of the FCF extracts was adjusted using a pH meter (Hach, Cork, Ireland), sodium hydroxide (NaOH) and hydrochloric acid (HCl), both chemicals bought from Sigma-Aldrich (St. Louis, MO, USA). Ultrapure water and $AgNO_3$ were used as synthesis reaction control in the same reaction conditions.

2.4. AgNPs Characterisation

AgNPs were first characterised by ultraviolet–visible spectrophotometry (UV-Vis, BioTek Synergy 4 Microplate Reader, Bad Friedrichshall, Germany) through localised surface plasmon resonance (LSPR) absorbance measurement, scanned between 300 nm and 1000 nm, with 2 nm steps. Selected AgNPs were then characterised using high-resolution TEM (FEI Titan 80) to evaluate the shape and size distribution of the resulting AgNPs. Samples were drop-coated onto the Formvar carbon-coated copper grids with a 200 μm mesh size (Agar Scientific, Essex, UK) and dried over 24 h. Images were obtained using the FEI Titan 80 operated at 300 kV using a field emission electron gun equipped with a Gatan Ultrascan digital camera. ImageJ software was utilised to measure an average of 200 NPs to evaluate size distribution. Finally, energy-dispersive X-ray spectrometry (EDS) was used to analyse the elemental composition of the synthesised NPs. For this method, 5 μL of the samples was drop-coated trice onto polished aluminium slides and dried in the oven at 60 °C for 1 h. An electron microscope (Hitachi 3000, Tokyo, Japan) at 15 kV accelerating voltage and a working distance of 2 mm was used to obtain the EDS spectra of the samples.

2.5. Statistical Analysis

All assays were set in triplicate. Data were expressed as the mean ± standard deviation (St. Dev). The software packages used to analyse the data generated in the characterisation process were Gen5, Microsoft Office Excel and Quantax 70 Microanalysis.

3. Results

3.1. Production of Fungal Cell-Free Extract

Fungi were initially cultured on MYPG agar plates and are displayed in Figure 1.

Figure 1. Fungus growth in MYPG agar plates (first column), MYGP medium (second column), stress medium (third column) and the fungal cell-free extract (fourth column): *Cladophialophora bantiana* (**A–D**), *Penicillium antarcticum* (**E–H**), *Trametes versicolor* (**I–L**), *Trichoderma martiale* (**M–P**), *Umbelopsis isabellina* (**Q–T**) and *Bjerkandera adusta* (**U–Y**).

The mycelia weight of each fungus after fungal growth in MYGP medium and the pH of the FCF extract after the growth phase (MYGP medium) and after the stress phase (sterile deionised water) are displayed in Table 1. *P. antarcticum* was revealed to have the fastest growth in MYGP medium, which correlates with greater volume production of the FCF stress extract that can lead to future scale-up evaluation. An increase in the pH of the extracts, from growth to stress phase, was observed for the majority of the fungi except for the *U. isabellina* FCF extract, which exhibited no change in the pH between the two phases.

Table 1. Mycelia weight and FCF extract pH of each fungus in the growth phase (MYGP medium) and FCF extract pH after the stress phase (sterile deionised water).

Species	Growth Phase		Stress Phase
	Mycelia (g)	FCF Extract pH	FCF Extract pH
C. bantiana	7.0	5.1	7.9
P. antarcticum	10.4	6.5	7.1
T. versicolor	9.3	7.2	7.9
T. martiale	8.8	5.2	7.7
U. isabellina	7.4	7.8	7.8
B. adusta	4.5	5.3	8.3

3.2. Biological Synthesis of AgNPs

The activity level of the FCF stress extracts was tested by the addition of 0.5 mM $AgNO_3$, varying the reaction time (1, 3 and 6 h), temperature (20, 45 and 90 °C) and pH (6, 9 and 12). All reactions were conducted in triplicate. The presence of a clear and defined LSPR band in the region of 400 nm in the UV-Vis spectrophotometric analysis indicates AgNP synthesis.

Figures 2–7, respectively, show the UV-Vis spectrum results of the 27 reactions in the following order: *C. bantiana, P. antarcticum, T. versicolor, T. martiale, U. isabellina* and *B. adusta* (Figures 2–7). *P. antarticum* and *T. versicolor* FCF stress extracts were the only ones capable of synthesising AgNPs at all the pH values tested with a defined LSPR band in the region of 400 nm (Figures 3 and 4). On the other hand, *B. adusta* FCF stress extract could not synthesise AgNPs in any conditions analysed (Figure 7).

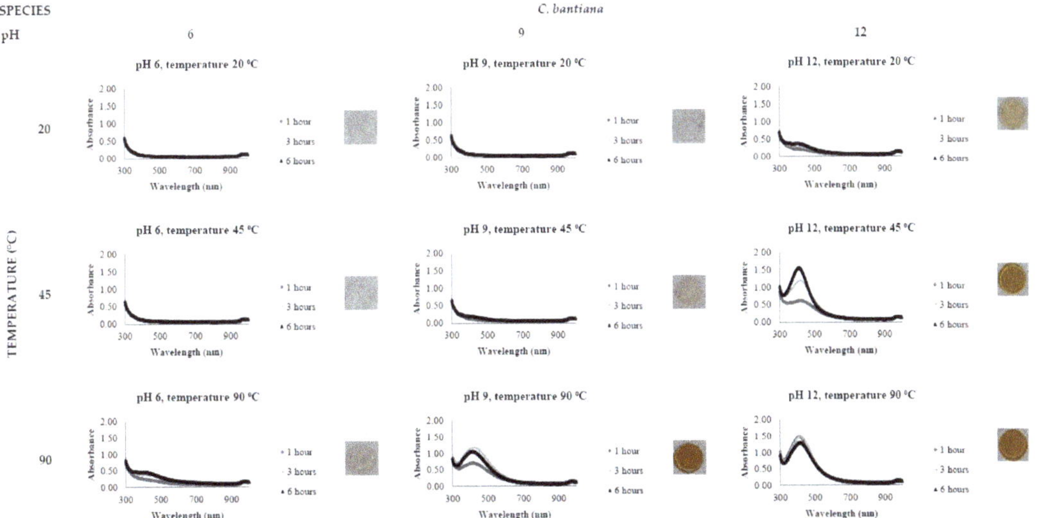

Figure 2. LSPR spectra of *C. bantiana* cell-free extract in the synthesis of AgNPs, by the addition of 0.5 mM $AgNO_3$, varying the reaction time (1, 3 and 6 h), temperature (20, 45 and 90 °C) and pH (6, 9 and 12), through UV-Vis spectrophotometry analysis.

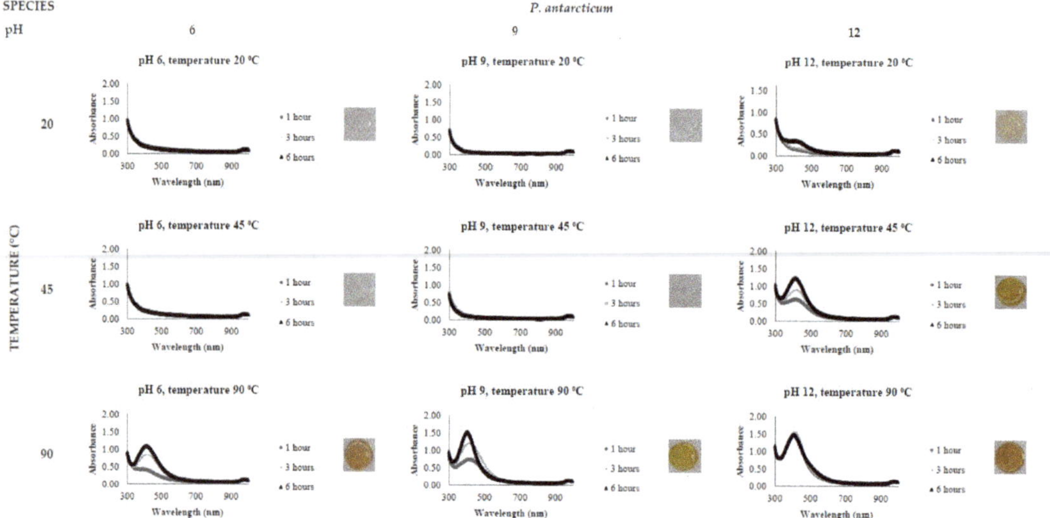

Figure 3. LSPR spectra of *P. antarcticum* cell-free extract in the synthesis of AgNPs, by the addition of 0.5 mM $AgNO_3$, varying the reaction time (1, 3 and 6 h), temperature (20, 45 and 90 °C) and pH (6, 9 and 12), through UV-Vis spectrophotometry analysis.

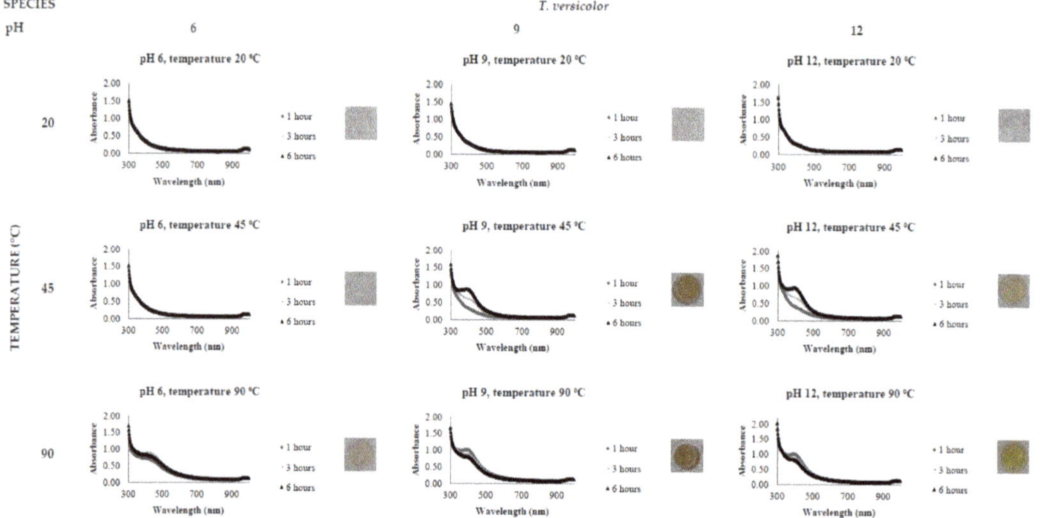

Figure 4. LSPR spectra of *T. versicolor* cell-free extract in the synthesis of AgNPs, by the addition of 0.5 mM $AgNO_3$, varying the reaction time (1, 3 and 6 h), temperature (20, 45 and 90 °C) and pH (6, 9 and 12), through UV-Vis spectrophotometry analysis.

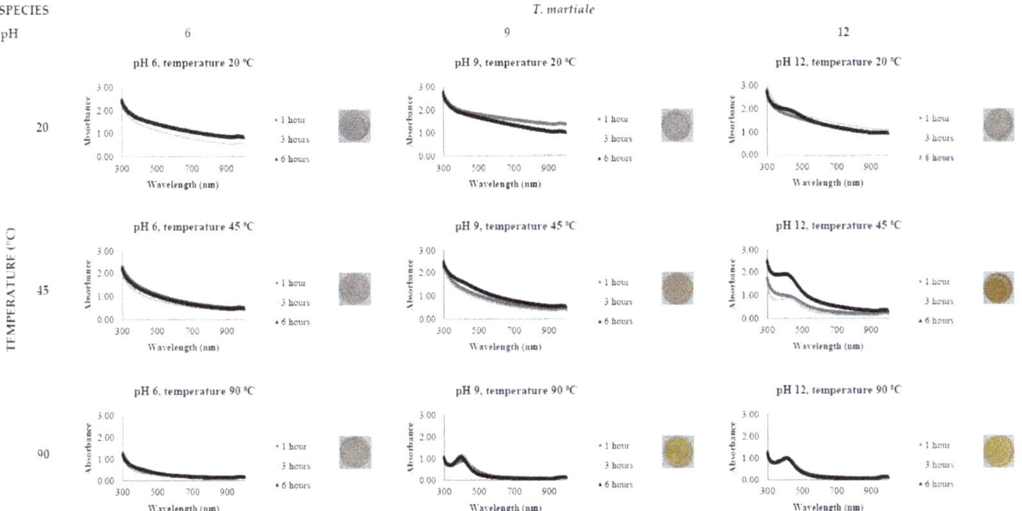

Figure 5. LSPR spectra of *T. martiale* cell-free extract in the synthesis of AgNPs, by the addition of 0.5 mM $AgNO_3$, varying the reaction time (1, 3 and 6 h), temperature (20, 45 and 90 °C) and pH (6, 9 and 12), through UV-Vis spectrophotometry analysis.

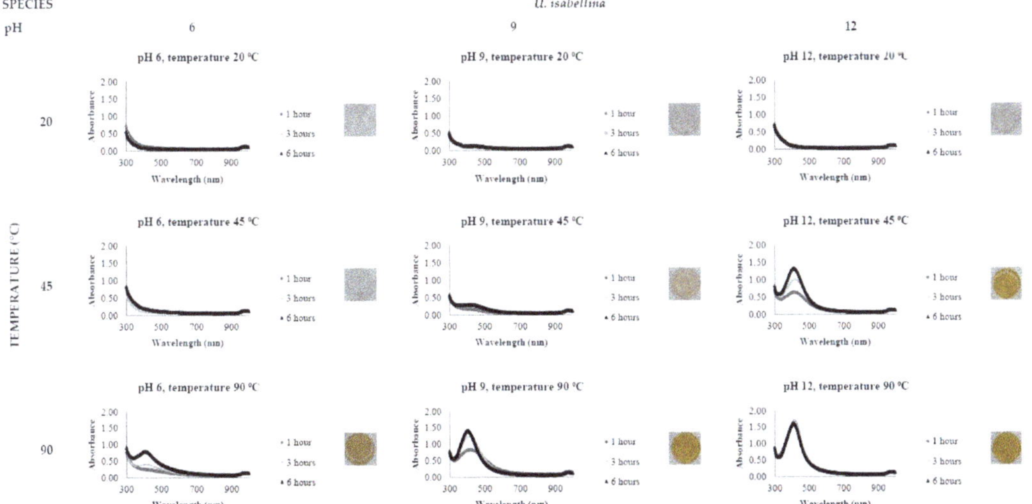

Figure 6. LSPR spectra of *U. isabellina* cell-free extract in the synthesis of AgNPs, by the addition of 0.5 mM $AgNO_3$, varying the reaction time (1, 3 and 6 h), temperature (20, 45 and 90 °C) and pH (6, 9 and 12), through UV-Vis spectrophotometry analysis.

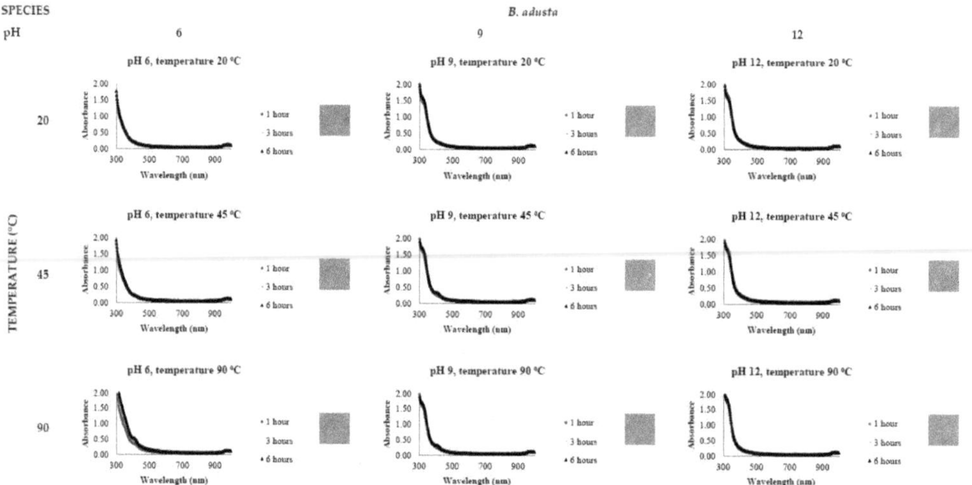

Figure 7. LSPR spectra of *B. adusta* cell-free extract in the synthesis of AgNPs, by the addition of 0.5 mM $AgNO_3$, varying the reaction time (1, 3 and 6 h), temperature (20, 45 and 90 °C) and pH (6, 9 and 12), through UV-Vis spectrophotometry analysis.

It was possible to notice that higher temperatures associated with a basic pH led to a quicker reaction time, a narrower spectrum (indicative of a monodisperse nanoparticle size distribution) and higher absorbance values (indicative of a greater $AgNO_3$ to AgNP conversion rate). In addition, it was shown that the pH significantly affects AgNP synthesis at higher temperatures, which can be visually observed by examining the LSPR absorbance spectrum. Therefore, the reaction conditions of 0.5 mM $AgNO_3$, 90 °C, different pH values (6, 9 and 12) and one hour reaction time were selected for UV-Vis spectrophotometric statistical analysis (Table 2).

Table 2. UV-Vis spectrophotometric statistical analysis of the wavelength associated with a maximum absorbance of the AgNPs synthesised using FCF extract, 0.5 mM $AgNO_3$, 1 h reaction time, 90 °C and different pH values (6, 9 and 12).

Species	pH	AgNPs	
		Wavelength (nm) Mean ± St. Dev	Absorbance Mean ± St. Dev
C. bantiana	6	Nd [1]	Nd [1]
	9	413 ± 1.018	0.706 ± 0.013
	12	411 ± 1.018	1.479 ± 0.019
P. antarcticum	6	390 ± 2.828	0.449 ± 0.013
	9	415 ± 1.414	0.746 ± 0.020
	12	410 ± 0.000	1.523 ± 0.028
T. versicolor	6	390 ± 4.000	0.729 ± 0.036
	9	392 ± 0.000	1.019 ± 0.018
	12	394 ± 0.000	1.016 ± 0.046
T. martiale	6	Nd [1]	Nd [1]
	9	404 ± 0.000	0.892 ± 0.028
	12	405 ± 1.414	1.010 ± 0.029
U. isabellina	6	Nd [1]	Nd [1]
	9	415 ± 1.155	0.833 ± 0.018
	12	410 ± 0.000	1.599 ± 0.087
B. adusta	6	Nd [1]	Nd [1]
	9	Nd [1]	Nd [1]
	12	Nd [1]	Nd [1]

[1] Nd, not determined.

3.3. AgNP Characterisation

The AgNPs synthesised using fungal stress extract and 0.5 mM $AgNO_3$ at 90 °C and different pH values (6, 9 and 12) for one hour reaction time were selected for further TEM characterisation and size distribution analysis, as shown in Figure 8. TEM images, on a 20 nm scale, indicated that spherical-shaped AgNPs were synthesised. EDS analysis (data not shown) was carried out on the samples, and the presence of elemental silver was confirmed at 3 keV where AgNPs were identified in TEM.

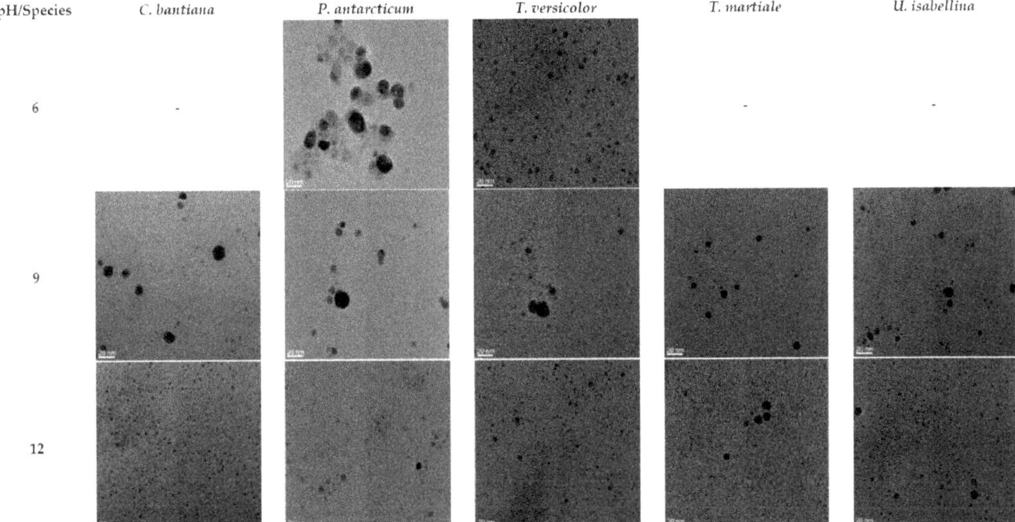

Figure 8. TEM (20 nm scale) characterisation of the biological AgNPs. AgNPs were synthesised by adding 0.5 mM $AgNO_3$, with 1 h reaction time, at a temperature of 90 °C, varying the pH. Images of the AgNPs synthesised at pH 6 are shown in the second row, pH 9 in the third row and pH 12 in the last row. AgNPs synthesised using the fungal cell-free extracts of *C. bantiana* are shown in the second column, *P. antarcticum* in the third column, *T. versicolor* in the fourth column, *T. martiale* in the fifth column and *U. isabellina* in the last column.

Furthermore, it was possible to note that smaller nanoparticles were synthesised when changing the pH from acid to basic. In addition, the percentage of the nanoparticles in the same size range increased, decreasing the polydispersity of the AgNPs solutions. The statistical analysis regarding the AgNPs' size distribution is shown in Table 3.

Table 3. Size statistical analysis of the AgNPs synthesised using FCF extract, 0.5 mM $AgNO_3$, 1 h reaction time, 90 °C and different pH values (6, 9 and 12), calculated using ImageJ.

Species	pH	AgNP Size (nm) Mean ± St. Dev	Highest Percentage within the Size Range
C. bantiana	6	Nm [1]	Nm [1]
	9	7.019 ± 4.494	41.0% (4–7.99 nm)
	12	3.062 ± 1.423	94.5% (0–3.99 nm)
P. antarcticum	6	16.811 ± 8.580	25.0% (16–19.99 nm)
	9	7.884 ± 4.183	50.5% (4–7.99 nm)
	12	5.943 ± 2.364	76.5% (4–7.99 nm)

Table 3. *Cont.*

Species	pH	AgNP Size (nm) Mean ± St. Dev	Highest Percentage within the Size Range
T. versicolor	6	6.526 ± 1.459	78.0% (4–7.99 nm)
	9	7.336 ± 6.707	48.0% (0–3.99 nm)
	12	4.816 ± 3.503	65.0% (0–3.99 nm)
T. martiale	6	Nm [1]	Nm [1]
	9	3.214 ± 2.654	70.0% (0–3.99 nm)
	12	4.051 ± 2.640	86.5% (0–3.99 nm)
U. isabellina	6	Nm [1]	Nm [1]
	9	4.239 ± 2.920	67.5% (0–3.99 nm)
	12	3.676 ± 1.818	78.5% (0–3.99 nm)
B. adusta	6	Nm[1]	Nm [1]
	9	Nm[1]	Nm [1]
	12	Nm[1]	Nm [1]

[1] Nm, not measured.

4. Discussion

Research into AgNP synthesis using eco-friendly approaches has grown rapidly because of the promising applications of these nanoparticles within the biomedicine and biotechnology field. In this context, myco-synthesis has been investigated thoroughly due to the tolerance and metal bioaccumulation capacity of fungi. Other reasons include economic viability, ease in handling biomass during downstream processing and large-scale production, which are advantages of using fungi rather than other microorganisms [14].

Fungi species such as *Aspergillus fumigatus*, *Cladosporium halotolerans*, *Fusarium oxysporum* and *Trichoderma longibrachiatum* have been previously shown to synthesise AgNPs [35–38]. However, despite efforts to establish controlled procedures to synthesise AgNPs biologically, the optimum reaction parameters display contradicting results and variability. Fungi are living organisms with different capabilities; therefore, optimum parameters for one species can be completely different from those for another. This could be due to the difference in the metabolic pathway response under stress conditions and the biosynthesis mechanism for each organism [19].

Several studies explore the effect of one parameter individually instead of analysing the combined impact of multiple variables [21,22,39,40]. Therefore, this study explored the ability of six different fungi, *C. bantiana*, *P. antarcticum*, *T. versicolor*, *T. martiale*, *U. isabellina* and *B. adusta*, to synthesise AgNPs, evaluating the combined effect of time, temperature and pH.

The results of this study demonstrate the following: (1) Nanoparticles produced by the same fungi metabolites can vary in both size and dispersion. This variance is correlated with physiochemical reaction parameters. (2) Optimal conditions to synthesise stable, monodisperse and smaller NPs were found at a higher pH of 12 and at a higher temperature of 90 °C. (3) Not all fungi (*B. adusta*) could reduce $AgNO_3$ precursors into AgNPs, demonstrating the importance of fungal screening studies.

The influence of pH (3 to 11) on the biological synthesis of AgNPs has been previously explored [19–21]. Studies have shown that basic pH is optimal for biological AgNP synthesis. For example, studies using extracts of the fungi *Saprolegnia parasitica*, *Neurospora crassa* and *Sclerotinia sclerotiorum* demonstrated alkali conditions as optimum in the myco-synthesis of AgNPs [19–21]. Our UV-Vis spectrophotometer study (Figures 2–7) demonstrated that FCF extracts of only two out of the six species (*P. antarcticum* and *T. versicolor*) were capable of reducing $AgNO_3$ into AgNPs in acidic pH (6) at an elevated temperature (90 °C). This may be caused by the formation of silver chloride (AgCl) when HCl is used to change the pH of the solution, which prevents the formation of AgNPs. Furthermore, it was also shown that basic pH (9 and 12) associated with high temperatures (45 and 90 °C) led to a quicker reaction time, a higher absorbance yield and a narrower spectrum, indicative of a monodisperse nanoparticle size distribution. Our findings could be explained by the fact that basic pH benefits the reduction of Ag+ ions to produce AgNPs by providing electrons for the reaction. As a result, more hydroxide (OH-) anions are available to participate in the reduction reaction, increasing the reduction strength [23,33].

The influence of temperature on the biological synthesis of nanoparticles is a cause for contradicting results. However, the results obtained in this research are similar to those found by Saxena et al., who reported maximum synthesis achieved at the highest temperature studied [21], and by Al-Khuzai et al., who reported a reduction in the nanoparticle size when increasing the temperature from 25 to 90 °C [19]. Previous biological AgNP synthesis studies analysed the reaction between 3 and 48 h [22,32]. Our UV-Vis spectrophotometer study (Figures 2–7) demonstrated that AgNPs completed synthesis within the first hour at basic pH and elevated temperature (90 °C). In this context, it is essential to notice that for rapid synthesis, the reaction rate increases with the increase in the temperature. However, the denaturation or inactivation of potential enzymes and other active molecules responsible for the synthesis can also occur at elevated or low temperatures [8,32]. As higher temperatures were used for synthesis, reducing and stabilising capacity was not correlated with enzyme activity as denaturation occurs at this temperature.

It is known that the composition, shape and size of the nanoparticles determine their physical and chemical properties, such as reactivity, biological interactions and optical properties [41,42]. Therefore, establishing an accurate relationship between the reaction parameter conditions and their effect on the morphology is extremely important for their future efficacy and performance applications. Several studies demonstrated the influence of the size and shape on the AgNPs' properties, revealing enhanced antibacterial and anticancer properties correlated with smaller nanoparticle sizes [25–30]. For example, the antibacterial effect of AgNPs in the size range of 5 to 100 nm was tested against *E. coli*, *Bacillus subtilis* and *S. aureus*, showing the fastest bactericidal activity with the use of the smallest nanoparticle size [25]. Moreover, it was demonstrated that chemically made small spherical-shaped AgNPs had enhanced antibacterial activity against *Pseudomonas aeruginosa* and *E. coli* compared to larger ones and also compared to triangular-shaped AgNPs [26]. The microscopy study revealed not only that we established defined parameters to synthesise nanoparticles of different sizes using the same fungal cell-free stress extract but also demonstrated that at a high temperature, when increasing the reaction pH, smaller nanoparticles were synthesised and the percentage of the nanoparticles in the same size range increased, hence positively increasing the monodispersity of the AgNP solutions (Figure 8 and Table 3). Table 3 displays the sizes of the AgNPs synthesised using 0.5 mM $AgNO_3$ for 1 h reaction time at 90 °C and basic pH (9 and 12), which were less than 10 nm. Furthermore, Figure 8 displays spherical-shaped AgNPs. Therefore, the biologically synthesised AgNPs shown in this study have potential as sustainable substitutes for chemically made antibacterial and anticancer products. Hence, they represent a step in the right direction towards achieving a bio-based circular economy.

5. Conclusions

As sustainability concerns grow, there is a shift in focus to a bio-based circular economy. Fungi have the ability to convert organic materials into bioactive products in an environmentally friendly manner. AgNPs have huge potential in biomedicine and biotechnology applications, especially due to their enhanced antibacterial and anticancer properties. However, chemically made AgNPs can be environmentally harsh. On the other hand, fungi biological synthesis, as we describe, can be used as a sustainable route to greener NP synthesis. However, the biological synthesis mechanism is still not fully established. Hence, this work evaluated the combined effect of pH, time and temperature in the mycosynthesis of AgNPs. Five fungi out of six tested had the ability to produce metabolites that can synthesise spherical AgNPs, albeit with different sizes (3 to 17 nm) and dispersity percentages (25 to 95%, within the same size range). We conclude that smaller and monodisperse AgNPs were favourably synthesised at elevated temperatures (90 °C) associated with basic pH (9 and 12).

Author Contributions: Investigation, M.F.A.; Supervision, P.G.M.; Writing—original draft, M.F.A.; Writing—review & editing, P.G.M. All authors have read and agreed to the published version of the manuscript.

Funding: This research was performed with financial support from the Technological University of the Shannon: Midlands Midwest (TUS).

Institutional Review Board Statement: Not applicable.

Informed Consent Statement: Not applicable.

Data Availability Statement: Not applicable.

Conflicts of Interest: The authors declare no conflict of interest. The funders had no role in the design of the study; in the collection, analysis or interpretation of data; in the writing of the manuscript; or in the decision to publish the results.

References

1. Meyer, V.; Basenko, E.Y.; Benz, J.P.; Braus, G.H.; Caddick, M.X.; Csukai, M.; De Vries, R.P.; Endy, D.; Frisvad, J.C.; Gunde-Cimerman, N.; et al. Growing a circular economy with fungal biotechnology: A white paper. *Fungal Biol. Biotechnol.* **2020**, *7*, 5. [CrossRef] [PubMed]
2. Takahashi, J.A.; Barbosa, B.V.R.; Martins, B.D.A.; Guirlanda, C.P.; Moura, M.A.F. Use of the Versatility of Fungal Metabolism to Meet Modern Demands for Healthy Aging, Functional Foods, and Sustainability. *J. Fungi* **2020**, *6*, 223. [CrossRef] [PubMed]
3. Duan, H.; Wang, D.; Li, Y. Green chemistry for nanoparticle synthesis. *Chem. Soc. Rev.* **2015**, *44*, 5778–5792. [CrossRef] [PubMed]
4. Yin, K.; Wang, Q.; Lv, M.; Chen, L. Microorganism remediation strategies towards heavy metals. *Chem. Eng. J.* **2019**, *360*, 1553–1563. [CrossRef]
5. Balu, S.; Andra, S.; Kannan, S.; Muthalagu, M. Facile synthesis of silver nanoparticles with medicinal grass and its biological assessment. *Mater. Lett.* **2020**, *259*, 126900. [CrossRef]
6. Cele, T. Preparation of Nanoparticles. In *Engineered Nanomaterials—Health and Safety*; IntechOpen: London, UK, 2020.
7. Iravani, S.; Korbekandi, H.; Mirmohammadi, S.V.; Zolfaghari, B. Synthesis of silver nanoparticles: Chemical, physical and biological methods. *Res. Pharm. Sci.* **2014**, *9*, 385–406.
8. Kaabipour, S.; Hemmati, S. A review on the green and sustainable synthesis of silver nanoparticles and one-dimensional silver nanostructures. *Beilstein J. Nanotechnol.* **2021**, *12*, 102–136. [CrossRef]
9. Correa-Llantén, D.N.; A Muñoz-Ibacache, S.; E Castro, M.; A Muñoz, P.; Blamey, J.M. Gold nanoparticles synthesized by *Geobacillus* sp. strain ID17 a thermophilic bacterium isolated from Deception Island, Antarctica. *Microb. Cell Factories* **2013**, *12*, 75. [CrossRef]
10. Dudhane, A.A.; Waghmode, S.R.; Dama, L.B.; Mhaindarkar, V.P.; Sonawane, A.; Katariya, S. Synthesis and Characterisation of Gold Nanoparticles using Plant Extract of *Terminalia arjuna* with Anti-bacterial Activity. *Int. J. Nanosci. Nanotechnol.* **2019**, *15*, 75–82.
11. Kitching, M.; Ramani, M.; Marsili, E. Fungal biosynthesis of gold nanoparticles: Mechanism and scale up. *Microb. Biotechnol.* **2014**, *8*, 904–917. [CrossRef]
12. Kuppusamy, P.; Yusoff, M.M.; Maniam, G.P.; Govindan, N. Biosynthesis of metallic nanoparticles using plant derivatives and their new avenues in pharmacological applications–An updated report. *Saudi Pharm. J.* **2016**, *24*, 473–484. [CrossRef] [PubMed]
13. Nindawat, S.; Agrawal, V. Fabrication of silver nanoparticles using *Arnebia hispidissima* (Lehm.) A. DC. root extract and unravelling their potential biomedical applications. *Artif. Cells Nanomed. Biotechnol.* **2019**, *47*, 166–180. [CrossRef] [PubMed]
14. Thakkar, K.N.; Mhatre, S.S.; Parikh, R.Y. Biological synthesis of metallic nanoparticles. *Nanomed. Nanotechnol. Biol. Med.* **2010**, *6*, 257–262. [CrossRef] [PubMed]
15. Das, S.; Marsili, E. A green chemical approach for the synthesis of gold nanoparticles: Characterisation and mechanistic aspect. *Rev. Environ. Sci. Biotechnol.* **2010**, *9*, 199–204. [CrossRef]
16. Madden, O.; Naughton, M.D.; Moane, S.; Murray, P.G. Mycofabrication of common plasmonic colloids, theoretical considerations, mechanism and potential applications. *Adv. Colloid Interface Sci.* **2015**, *225*, 37–52. [CrossRef]
17. Shah, M.; Fawcett, D.; Sharma, S.; Tripathy, S.K.; Poinern, G.E.J. Green Synthesis of Metallic Nanoparticles via Biological Entities. *Materials* **2015**, *8*, 7278–7308. [CrossRef]
18. ISO. ISO/TR 18401, in Nanotechnologies—Plain language explanation of selected terms from the ISO/IEC 80004 Series. 2017. Available online: https://www.iso.org/obp/ui/#iso:std:iso:tr:18401:ed-1:v1:en (accessed on 24 March 2022).

19. Al-Khuzai, R.; Aboud, M.; Alwan, S. Biological Synthesis of Silver Nanoparticles from *Sprolegnia parasitica*. *J. Phys. Conf. Ser.* **2019**, *1294*, 062090. [CrossRef]
20. Princy, K.F.; Gopinath, A. Optimisation of physicochemical parameters in the biofabrication of gold nanoparticles using marine macroalgae *Padina tetrastromatica* and its catalytic efficacy in the degradation of organic dyes. *J. Nanostruct. Chem.* **2018**, *8*, 333–342. [CrossRef]
21. Saxena, J.; Sharma, P.K.; Sharma, M.M.; Singh, A. Process optimisation for green synthesis of silver nanoparticles by *Sclerotinia sclerotiorum* MTCC 8785 and evaluation of its antibacterial properties. *SpringerPlus* **2016**, *5*, 861. [CrossRef]
22. Quester, K.; Avalos-Borja, M.; Castro-Longoria, E. Controllable Biosynthesis of Small Silver Nanoparticles Using Fungal Extract. *J. Biomater. Nanobiotechnol.* **2016**, *7*, 118–125. [CrossRef]
23. Salem, S.S.; Ali, O.M.; Reyad, A.M.; Abd-Elsalam, K.A.; Hashem, A.H. *Pseudomonas indica*-Mediated Silver Nanoparticles: Antifungal and Antioxidant Biogenic Tool for Suppressing Mucormycosis Fungi. *J. Fungi* **2022**, *8*, 126. [CrossRef] [PubMed]
24. Zaki, S.A.; Ouf, S.A.; Albarakaty, F.M.; Habeb, M.M.; Aly, A.A.; Abd-Elsalam, K.A. *Trichoderma harzianum*-Mediated ZnO Nanoparticles: A Green Tool for Controlling Soil-Borne Pathogens in Cotton. *J. Fungi* **2021**, *7*, 952. [CrossRef] [PubMed]
25. Agnihotri, S.; Mukherji, S.; Mukherji, S. Size-controlled silver nanoparticles synthesised over the range 5–100 Nm using the same protocol and their antibacterial efficacy. *RSC Adv.* **2013**, *4*, 3974–3983. [CrossRef]
26. Raza, M.A.; Kanwal, Z.; Rauf, A.; Sabri, A.N.; Riaz, S.; Naseem, S. Size- and Shape-Dependent Antibacterial Studies of Silver Nanoparticles Synthesised by Wet Chemical Routes. *Nanomaterials* **2016**, *6*, 74. [CrossRef] [PubMed]
27. Salomoni, R.; Léo, P.; Montemor, A.; Rinaldi, B.; Rodrigues, M. Antibacterial effect of silver nanoparticles in *Pseudomonas aeruginosa*. *Nanotechnol. Sci. Appl.* **2017**, *10*, 115–121. [CrossRef]
28. El-Naggar, N.E.-A.; Hussein, M.H.; El-Sawah, A.A. Bio-fabrication of silver nanoparticles by phycocyanin, characterization, in vitro anticancer activity against breast cancer cell line and in vivo cytotoxicity. *Sci. Rep.* **2017**, *7*, 10844. [CrossRef]
29. Gurunathan, S.; Qasim, M.; Park, C.; Yoo, H.; Kim, J.-H.; Hong, K. Cytotoxic Potential and Molecular Pathway Analysis of Silver Nanoparticles in Human Colon Cancer Cells HCT. *Int. J. Mol. Sci.* **2018**, *19*, 2269. [CrossRef]
30. Saratale, R.G.; Benelli, G.; Kumar, G.; Kim, D.S.; Saratale, G.D. Bio-fabrication of silver nanoparticles using the leaf extract of an ancient herbal medicine, dandelion (*Taraxacum officinale*), evaluation of their antioxidant, anticancer potential, and antimicrobial activity against phytopathogens. *Environ. Sci. Pollut. Res.* **2017**, *25*, 10392–10406. [CrossRef]
31. Lee, S.H.; Jun, B.-H. Silver Nanoparticles: Synthesis and Application for Nanomedicine. *Int. J. Mol. Sci.* **2019**, *20*, 865. [CrossRef]
32. AbdelRahim, K.; Mahmoud, S.Y.; Ali, A.M.; Almaary, K.S.; Mustafa, A.E.-Z.M.; Husseiny, S.M. Extracellular biosynthesis of silver nanoparticles using *Rhizopus stolonifer*. *Saudi J. Biol. Sci.* **2017**, *24*, 208–216. [CrossRef]
33. Hamedi, S.; Ghaseminezhad, M.; Shokrollahzadeh, S.; Shojaosadati, S.A. Controlled biosynthesis of silver nanoparticles using nitrate reductase enzyme induction of filamentous fungus and their antibacterial evaluation. *Artif. Cells Nanomed. Biotechnol.* **2016**, *45*, 1588–1596. [CrossRef]
34. Katapodis, P.; Christakopoulou, V.; Kekos, D.; Christakopoulos, P. Optimisation of xylanase production by *Chaetomium thermophilum* in wheat straw using response surface methodology. *Biochem. Eng. J.* **2007**, *35*, 136–141. [CrossRef]
35. Ameen, F.; Al-Homaidan, A.A.; Al-Sabri, A.; Almansob, A.; AlNAdhari, S. Anti-oxidant, anti-fungal and cytotoxic effects of silver nanoparticles synthesised using marine fungus *Cladosporium halotolerans*. *Appl. Nanosci.* **2021**. [CrossRef]
36. Elamawi, R.M.; Al-Harbi, R.E.; Hendi, A.A. Biosynthesis and characterisation of silver nanoparticles using *Trichoderma longibrachiatum* and their effect on phytopathogenic fungi. *Egypt. J. Biol. Pest Control.* **2018**, *28*, 28. [CrossRef]
37. Husseiny, S.M.; Salah, T.A.; Anter, H.A. Biosynthesis of size controlled silver nanoparticles by *Fusarium oxysporum*, their antibacterial and antitumor activities. *Beni-Suef Univ. J. Basic Appl. Sci.* **2015**, *4*, 225–231. [CrossRef]
38. Shahzad, A.; Saeed, H.; Iqtedar, M.; Hussain, S.Z.; Kaleem, A.; Abdullah, R.; Sharif, S.; Naz, S.; Saleem, F.; Aihetasham, A.; et al. Size-Controlled Production of Silver Nanoparticles by *Aspergillus fumigatus* BTCB10: Likely Antibacterial and Cytotoxic Effects. *J. Nanomater.* **2019**, *2019*, 5168698. [CrossRef]
39. Dada, A.O.; Adekola, F.A.; Dada, F.E.; Adelani-Akande, A.T.; Bello, M.O.; Okonkwo, C.R.; Inyinbor, A.A.; Oluyori, A.P.; Olayanju, A.; Ajanaku, K.O.; et al. Silver nanoparticle synthesis by *Acalypha wilkesiana* extract: Phytochemical screening, characterization, influence of operational parameters, and preliminary antibacterial testing. *Heliyon* **2019**, *5*, e02517. [CrossRef]
40. Marciniak, L.; Nowak, M.; Trojanowska, A.; Tylkowski, B.; Jastrzab, R. The Effect of pH on the Size of Silver Nanoparticles Obtained in the Reduction Reaction with Citric and Malic Acids. *Materials* **2020**, *13*, 5444. [CrossRef]
41. Khan, I.; Saeed, K.; Khan, I. Nanoparticles: Properties, applications and toxicities. *Arab. J. Chem.* **2019**, *12*, 908–931. [CrossRef]
42. Ridolfo, R.; Tavakoli, S.; Junnuthula, V.; Williams, D.S.; Urtti, A.; Van Hest, J.C.M. Exploring the Impact of Morphology on the Properties of Biodegradable Nanoparticles and Their Diffusion in Complex Biological Medium. *Biomacromolecules* **2020**, *22*, 126–133. [CrossRef]

Journal of
Fungi

Article

Mycosynthesis, Characterization, and Mosquitocidal Activity of Silver Nanoparticles Fabricated by *Aspergillus niger* Strain

Mohamed A. Awad [1], Ahmed M. Eid [2,*], Tarek M. Y. Elsheikh [1], Zarraq E. Al-Faifi [3], Nadia Saad [4], Mahmoud H. Sultan [2], Samy Selim [5], Areej A. Al-Khalaf [6] and Amr Fouda [2,*]

[1] Department of Zoology and Entomology, Faculty of Science, Al-Azhar University, Nasr City, Cairo 11884, Egypt; mohamed_awad@azhar.edu.eg (M.A.A.); telsheikh64@yahoo.com (T.M.Y.E.)
[2] Department of Botany and Microbiology, Faculty of Science, Al-Azhar University, Nasr City, Cairo 11884, Egypt; prof.mahmoud@azhar.edu.eg
[3] Center for Environmental Research and Studies, Jazan University, P.O. Box 2097, Jazan 42145, Saudi Arabia; zalfifi@jazanu.edu.sa
[4] Department of Mathematics, Faculty of Science, Helwan University, Cairo 11795, Egypt; nadia_saad@science.helwan.edu.eg
[5] Department of Clinical Laboratory Sciences, College of Applied Medical Sciences, Jouf University, P.O. Box 72388, Sakaka 72341, Saudi Arabia; sabdulsalam@ju.edu.sa
[6] Biology Department, College of Science, Princess Nourah Bint Abdulrahman University, P.O. Box 84428, Riyadh 11671, Saudi Arabia; aaalkhalaf@pnu.edu.sa
* Correspondence: aeidmicrobiology@azhar.edu.eg (A.M.E.); amr_fh83@azhar.edu.eg (A.F.); Tel.: +20-100-015-4414 (A.M.E.); +20-111-335-1244 (A.F.)

Abstract: Herein, silver nanoparticles (Ag-NPs) were synthesized using an environmentally friendly approach by harnessing the metabolites of *Aspergillus niger* F2. The successful formation of Ag-NPs was checked by a color change to yellowish-brown, followed by UV-Vis spectroscopy, Fourier transforms infrared (FT-IR), Transmission electron microscopy (TEM), and X-ray diffraction (XRD). Data showed the successful formation of crystalline Ag-NPs with a spherical shape at the maximum surface plasmon resonance of 420 nm with a size range of 3–13 nm. The Ag-NPs showed high toxicity against I, II, III, and IV instar larvae and pupae of *Aedes aegypti* with LC50 and LC90 values of 12.4–22.9 ppm and 22.4–41.4 ppm, respectively under laboratory conditions. The field assay exhibited the highest reduction in larval density due to treatment with Ag-NPs (10× LC50) with values of 59.6%, 74.7%, and 100% after 24, 48, and 72 h, respectively. The exposure of *A. aegypti* adults to the vapor of burning Ag-NPs-based coils caused a reduction of unfed individuals with a percentage of 81.6 ± 0.5% compared with the positive control, pyrethrin-based coils (86.1 ± 1.1%). The ovicidal activity of biosynthesized Ag-NPs caused the hatching of the eggs with percentages of 50.1 ± 0.9, 33.5 ± 1.1, 22.9 ± 1.1, and 13.7 ± 1.2% for concentrations of 5, 10, 15, and 20 ppm, whereas Ag-NPs at a concentration of 25 and 30 ppm caused complete egg mortality (100%). The obtained data confirmed the applicability of biosynthesized Ag-NPs to the biocontrol of *A. aegypti* at low concentrations.

Keywords: biosynthesis; silver nanoparticles; biocontrol; larvicidal; smoke toxicity; ovicidal

Citation: Awad, M.A.; Eid, A.M.; Elsheikh, T.M.Y.; Al-Faifi, Z.E.; Saad, N.; Sultan, M.H.; Selim, S.; Al-Khalaf, A.A.; Fouda, A. Mycosynthesis, Characterization, and Mosquitocidal Activity of Silver Nanoparticles Fabricated by *Aspergillus niger* Strain. *J. Fungi* **2022**, *8*, 396. https://doi.org/10.3390/jof8040396

Academic Editor: Kamel A. Abd-Elsalam

Received: 17 March 2022
Accepted: 8 April 2022
Published: 13 April 2022

1. Introduction

Nanotechnology deals with the production of new particles at a nano-scale (1–100 nm) [1]. Nanotechnology science paved the way for discovering active compounds that can be incorporated into various fields such as sensors, magnetic devices, the textile industry, wastewater treatment, heavy metal removal, drug delivery, optoelectronics, the agricultural sector, biomedical (antimicrobial, antitumor, cytotoxicity, and cosmetics), and parasitology [2–5]. These activities are due to the unique physical, chemical, and structural nanoparticle (NPs) properties such as sizes, shapes, the charge of the surface, stability, compatibles, and the proportion between small particle size to a large surface area [6,7]. Nanoparticles are synthesized by chemical, physical and biological methods. The main disadvantages of the chemical and physical methods are expensive, producing hazardous substances, and requiring harsh conditions (such as temperature and pressure) during synthesis [8,9]. These disadvantages paved the way to green or biological synthesis using different biological beings such as plants and microorganisms (bacteria, fungi, actinomycetes, yeast, and algae) [10–12]. Fungi are gaining more attention than other microorganisms for fabricating different metal and metal oxide NPs [13,14]. This activity can be attributed to the ability of fungi to have high heavy metal tolerance, easy scale-up, high biomass production, easy handling, low toxicity, and production of various quantities of secondary metabolites that increase the stability of NPs [15,16]. Various metals, metal oxide NPs, and nanocomposites were fabricated by fungi such as Ag, Au, Se, ZnO, CuO, MgO, CuO/ZnO nanocomposites, etc. [3,8,17,18].

Mosquitoes are one of the most abundant vectors for various deadly human and animal pathogens such as viruses, protozoa, and bacteria [19]. A variety of deadly diseases such as dengue, yellow fever, filariasis, chikungunya, malaria, Zika virus, and West Nile virus are considered mosquito-borne diseases [20]. According to the database of the World Health Organization, approximately 50 to 100 million dengue fever cases appear every year worldwide [21]. *Aedes aegypti* is a common mosquito species in the tropic as well as subtropic countries and is characterized by its ability to transmit several virus-causing diseases such as Zika, dengue, chikungunya, and yellow fever [22]. Interestingly, the viral diseases caused by *A. aegypti* do not have effective vaccines except for yellow fever, which has had an effective vaccine since the 1940s [23]. Therefore, successful protection from *A. aegypti*-caused diseases is accomplished by preventing the spread of the vector.

The *Aedes* spp. control was achieved by synthetic pesticides such as organophosphates, organochlorines, dichloro-diphenyl-trichloroethane (DDT), carbamates, and pyrethroid [24,25]. The continuous usage of these chemicals leads to increased mosquito resistance, negative impacts on soil fertility, toxic groundwater, and adverse effects on the surrounding ecosystem [2,26]. Therefore, the main target for researchers is to discover alternative eco-friendly, cost-effective, and safe tools to control and prevent the spread of insect vectors to overcome the negative impacts of chemical insecticides.

Concerning the control of mosquito vectors, green synthesized NPs are used due to their eco-friendliness, rapid effects, cost-effectiveness, high stability, and absence of negative impacts on public health compared to chemical insecticides [27]. Recently, selenium nanoparticles (Se-NPs) synthesized by *Penicillium chrysogenum* showed significant molluscicide toxicity against *Biomphlaria alexandrina* snails at a concnetration of 5.9 mg L^{-1} after 96 h. Additionally, it showed cercaricidal and miracidicidal effect on *Schistosoma mansoni* [28]. In our recent study, magnesium oxide NPs (MgO-NPs) fabricated by *Cystoseira crinita* showed high efficacy as larvicidal, pupicidal, and repellence activity for *Musca domestica* [29]. Silver nanoparticles (Ag-NPs) are considered one of the most critical metal NPs which can be integrated into different biomedical and biotechnological applications. Recently, Ag-NPs were used to control mosquito vectors such as *Culex quinquefasciatus*, *Anopheles stephensi*, and *Aedes aegypti* [2,30].

Therefore, the efficiency of metabolites secreted by fungal strains for the synthesis of Ag-NPs and the use of the final product to control the *Aedes aegypti* mosquito was investigated. The biomass filtrate of *Aspergillus niger* F2 that contains a variety of metabolites was used as a biocatalyst to form Ag-NPs as an eco-friendly approach. The physicochemical characterizations of biosynthesized NP were accomplished by the color change of fungal biomass filtrate, UV-Vis spectrophotometry, Fourier transforms infrared (FT-IR), transmission electron microscopy (TEM), and X-ray diffraction (XRD). The mechanisms of synthesized nano-silver to control *A. aegypti* including larvicidal activity against different instar larvae under laboratory and field conditions, pupicidal, smoke toxicity, and ovicidal activity were investigated.

2. Materials and Methods

2.1. Fungal Strain

The biological synthesis of Ag-NPs was done using the fungal strain *Aspergillus niger* F2 which was previously isolated from the archeological manuscript [31]. This manuscript was dated back to 1677 A.D. and collected from Al-Azhar Library, Cairo, Egypt. The identification of the selected fungal strain was achieved by cultural characteristics, microscopic examination, and molecular identification by sequencing of the internal transcribed spacer (ITS) gene [32]. The obtained gene sequence was deposited in GenBank under accession number MK452259.

2.2. Mycosynthesis of Ag-NPs

Two disks (1.0 cm in diameter) of three-day-old *A. niger* F2 culture were inoculated in potato dextrose (PD) broth media and incubated for 120 h at 30 ± 2 °C. At the end of the incubation period, the inoculated PD was filtered with Whatman (No. 1) filter paper to collect the fungal biomass which was washed thrice with sterilized distilled H_2O to remove any adhering medium components. After that, approximately 10 g of collected fungal biomass was mixed with 100 mL of distilled H_2O and incubated for 24 h under shaking conditions (150 rpm). The previous mixture was centrifuged at 5000 rpm for 10 min to collect the supernatant (fungal biomass filtrate) which was used as a biocatalyst for eco-friendly synthesis of Ag-NPs as follows: 16.9 µg of $AgNO_3$ (metal precursor) was mixed with 100 mL of fungal biomass filtrate to get a final concentration of 1 mM, adjust the pH of the mixture at 8 using 1 M NaOH which added drop-wisely under stirring condition, and incubated for 24 h at 35 ± 2 °C. The successful synthesis of Ag-NPs was checked through a color change of fungal biomass filtrate from colorless to yellowish-brown. The resultant was collected, washed thrice with distilled H_2O, and subjected to oven-dry at 150 °C overnight [33]. The negative control with either $AgNO_3$ aqueous solution or fungal biomass filtrate without $AgNO_3$ was running with the experiment under the same condition.

2.3. Ag-NPs Characterization

The synthesis of Ag-NPs was checked by measuring the absorbance properties of a synthesized aqueous solution by UV-Vis spectroscopy (JENWAY 6305, Spectrophotometer, 230 V/50 Hz, Staffordshire, UK) at a different wavelength in the range of 300 to 600 nm with 10 nm intervals [34]. The role of various functional groups present in fungal biomass filtrate in the reduction and stabilizing Ag-NPs was analyzed by Fourier-transformed infrared (FT-IR) spectroscopy (Cary 630 FTIR model, Tokyo, Japan). The synthesized Ag-NPs (0.3 g) were mixed with potassium bromide (KBr) and formed a disk under high pressure followed by scanning in a range of 400 to 4000 cm^{-1} [35]. The sizes and shapes of biosynthesized Ag-NPs were detected using Transmission Electron microscopy (TEM) (TEM-JEOL 1010, Tokyo, Japan). A drop of Ag-NPs solution was added to the TEM grid (carbon-copper gride) till complete adsorption. The excess of the Ag-NPs solution on the TEM grid was removed before being analyzed [36]. The crystalline nature of synthesized Ag-NPs was assessed by X-ray Diffraction Analysis (X′ Pert Pro, Philips, Eindhoven, The Netherlands). The sample was scanned at 2θ values of 4 to 80. The XRD operates at 30 mA and 40 KV with a radiation source of Cu Ka (λ = 1.54 Å). Based on XRD analysis, the average Ag-NPs size was calculated by Debye–Scherrer equation as follows [37]:

$$D = \frac{0.9 \times 1.54}{\beta \cos \theta} \quad (1)$$

where D is the mean particle size, 0.9 is the Scherrer's constant, 1.54 is the X-ray wavelength, β is half of the maximum intensity, and θ is Bragg's angle.

2.4. Mosquito Culture

The eggs of *Aedes aegypti* were purchased from the Medical Entomology Institute, Giza, Egypt, and transferred immediately to the Animal House Institute, Mosquito Laboratory, Zoology Department, Faculty of Science, Al-Azhar University, Cairo, Egypt. The collected eggs were put into a plastic cup containing 0.5 L of tap water to hatch at optimum conditions (27 ± 2 °C, 75–85% humidity, and the light–dark photoperiod was 14:10). The hatched larvae were fed on yeast hydrolyzed–dog biscuits (1:3 *w/w*). The released pupae were put into a plastic cup filled with 0.5 L dechlorinated water followed by transfer to a chiffon cage (90 × 90 × 90 cm^3) until adults were released which fed on a 10% (*v/v*) sucrose solution [38]. The released larvae and pupae were collected for the next step which studied the toxicity of synthesized Ag-NPs.

2.5. Laboratory Larvicidal/Pupicidal Toxicity of Ag-NPs

The toxicity of biosynthesized Ag-NPs against various larval instars (I, II, III, and IV instars) and pupae of *A. aegypti* were assessed through the preparation of suspension solution in distilled H_2O at different concentrations of 5, 10, 15, 20, 25, and 30 ppm. The differences between insect instars were defined by alterations in the body proportions, growth patterns, colors, head width, and changes in the number of body segments [39]. Briefly, 25 larvae or pupae were incubated for 24 h in a glass cup containing 0.5 L dechlorinated water supplemented with the above-mentioned Ag-NPs concentration and 0.5 mg of larvae food. The experiment was conducted for each instar larvae and pupae with Ag-NPs concentration separately and then repeated three times. The control where larvae and pupa were in dechlorinated water without Ag-NPs was running with the experiment under the same conditions. After 24 h, the mortality percentages were calculated according to the following equation [7]:

$$\text{Mortality percentages } (\%) = \frac{\text{Number of dead individuals}}{\text{Number of treated individuals}} \times 100 \quad (2)$$

2.6. Field Larvicidal Bioassay

The toxicity of Ag-NPs against 3rd and 4th instar larvae of *A. aegypti* under field conditions was assessed using a knapsack sprayer in six water reservoirs at the Animal House Institute, Mosquito Laboratory, Faculty of Science, Al-Azhar University, Cairo, Egypt. The density of larvae before treatment was checked, whereas the toxicity after Ag-NPs treatment was investigated at 24, 48, and 72 h by a larval dipper. The experiment was repeated six times under the same field conditions of 80 ± 5% humidity at 28 ± 2 °C. The required Ag-NPs as a mosquitocidal agent were calculated according to the volume and surface area of reservoirs, which were prepared as 10× LC50 values as mentioned above [40]. The reduction percentages in the larvae density were calculated using the following formula [41]:

$$\text{Reduction percentages } (\%) = \frac{\text{C} - \text{T}}{\text{C}} \times 100 \quad (3)$$

where C is the total number of individuals in control and T is the total number of individuals in treatment.

2.7. Smoke Toxicity Assay

The smoke toxicity of Ag-NPs against adults of *A. aegypti* was achieved in a glass chamber (60 cm × 40 cm × 35 cm). In this method, the burning coils were composed of 1 g of biosynthesized Ag-NPs mixed with 0.5 g of binding materials (sawdust) and 0.5 g of burning materials (coconut shell powder). The components were mixed well with distilled H_2O to form a semisolid paste which was used to form a mosquito coil that remained in the shade to dry. Negative control (sawdust–coconut shell powder (0.5:0.5 *w/w*), mixed well with distilled H_2O to form a semisolid paste which remains in shade to dry) and positive control (the same previous components and mixed with pyrethrin instead of Ag-NPs) were running with the experiment under the same conditions.

In the experiment, approximately 100 adult female mosquitoes (blood-starved for an average age of three to four days) were released in a glass chamber containing a sucrose solution (10%). Additionally, a pigeon with a shaved belly was tied to the side of the chamber. This study was approved by the Ethical Committee of the Animal House Institute, Cairo, Egypt. The released adult mosquitos were exposed to the vapor of the burning coil for one hour followed by counting the number of fed and unfed (alive and dead) mosquitos. The protection of the pigeon against the bites of *A. aegypti* due to smoke from Ag-NPs was calculated as follows [42]:

$$\text{Protection }\% = \frac{\text{number of unfed mosquitos in treatment} - \text{number of unfed mosquitos in negativecontrol}}{\text{number of treated mosquitos}} \times 100 \quad (4)$$

2.8. Ovicidal Activity

The mosquito ovicidal assay with Ag-NP treatment was achieved according to the method of Su and Mulla [43]. In this assay, the eggs of *A. aegypti* were collected and placed in ovitraps (such as Petri dishes with a diameter of 60 mm containing filter papers and 100 mL distilled H_2O) and placed into the mosquito cages for 48 h. The photomicroscope was used to check the eggs loaded on the filter paper. After that, out of seven glass cups were put in the mosquito cage, six cups were filled with water supplemented with Ag-NPs concentrations (5, 10, 15, 20, 25, and 30 ppm), and the seventh cup was filled with water only (as a control). Approximately 100 *A. aegypti* eggs were put in each glass cup. The experiment was repeated three times for each treatment. After 48 h, the eggs from each treatment were transferred to a second cup containing distilled H_2O to investigate the % eggs hatched under a microscope. According to non-hatched eggs (detected by unopen opercula), the percentages of eggs hatched were calculated according to the following formula:

$$\%\text{ eggs hatched} = \frac{\text{number of hatched larvae}}{\text{Totalnumber of tested eggs}} \times 100 \quad (5)$$

2.9. Data Analysis

The data collected in the current study were analyzed using the SPSS (version 16.0). Data of acute toxicity obtained from Laboratory assays were transformed into arcsine/proportion values followed by two-way ANOVA analyses with two factors (i.e., dosage and mosquito instar). Furthermore, insect pest mortality data from laboratory assays were analyzed using probit analysis, with LC50 and LC90 calculated using Finney's method [44]. A two-way ANOVA with two factors was used to analyze the larval density data of *A. aegypti* obtained from field assays. In the smoke coil toxicity experiment, the number of fed and unfed mosquitos was analyzed using a one-way ANOVA with the treatment as the factor (i.e., the coil). Moreover, one-way ANOVA was used to analyze ovicidal data that had been transformed into arcsine proportion values. A posteriori multiple comparisons were done using Tukey's range tests at $p \leq 0.05$. All results are the means of three to five independent replicates, as specified above.

3. Results and Discussion

3.1. Mycosynthesis of Ag-NPs

In the current study, the Ag-NPs were synthesized extracellularly, this mode facilitates the purification method as compared with intracellular synthesis. The fungal strain F2 was selected in the current study due to their uncommon habitat (deteriorated archaeological manuscript) and hence predict their high activity via secretion of various enzymes and other secondary metabolites. Therefore, the biomass filtrate containing various secondary metabolites of *Aspergillus niger* F2 was used as a reducing agent for the silver ions to form silver nanoparticles (Ag-NPs) as well as being used as a capping/stabilizing agent for a new nanostructure. The pH of the reaction solution is considered one of the critical parameters during NP synthesis [45]. Herein, the alkaline condition (pH 8) of the synthesis solution was preferred over neutral and acidic conditions. This phenomenon can be attributed to the alkaline conditions enhancing the reducing activity of different functional groups that exist in the fungal biomass filtrate as well as preventing the aggregation or agglomeration of NPs [46]. Moreover, the alkaline medium can help in capping and stabilizing of NPs via reacting with amino acids and amino groups that exist on the surface of proteins [47]. The efficacy of biomass filtrate to form Ag-NPs was checked first by the change of its color to yellowish-brown because of exciting the surface plasmon vibrations [48]. Compatible with our study, *Aspergillus niger* was utilized as a biocatalyst for the biological synthesis of Ag-NPs as mentioned in the various published studies [36,45,49]. The biosynthesis of NPs using different biological sources (i.e., bacteria, fungi, actinomycetes, algae, and plants) have advantages over chemical and physical methods because of the eco-friendly, rapid, low cost, and biocompatible methods [7,50]. Recently, NPs have been fabricated by different fungal strains to itegrate into a wide range of applications [28,51,52]. The reduction process of silver ions to the nanoscale can be achieved due to the liberated electrons from the reduction of NO_3 to NO_2 by the action of metabolites present in fungal biomass filtrate, and hence, the color intensity is directly related to the number of liberated electrons and surface plasmon resonance (SPR) which varied according to the size of NPs [53,54].

3.2. Ag-NPs Characterizations

3.2.1. UV-Vis Spectroscopy

The color intensity of fungal-mediated biosynthesis of Ag-NPs was checked by measuring the absorbance at various wavelengths in the range of 300 to 600 nm to detect the maximum SPR. As shown, the maximum peak of Ag-NPs was observed at 420 nm (Figure 1), which confirms the formation of Ag-NPs [55]. The spherical shape of biosynthesized Ag-NPs is usually related to the SPR peak especially those observed at a wavelength of 410–420 nm [56]. Previously published studies reported that the optimum SPR absorption peak for biologically synthesized Ag-NPs was in the range of 400–460 nm [33,57]. Compatible with our study, Wang and co-author reported that the optimum SPR peak for Ag-NPs synthesized by harnessing metabolites of *Aspergillus sydowii* was at 420 nm [34].

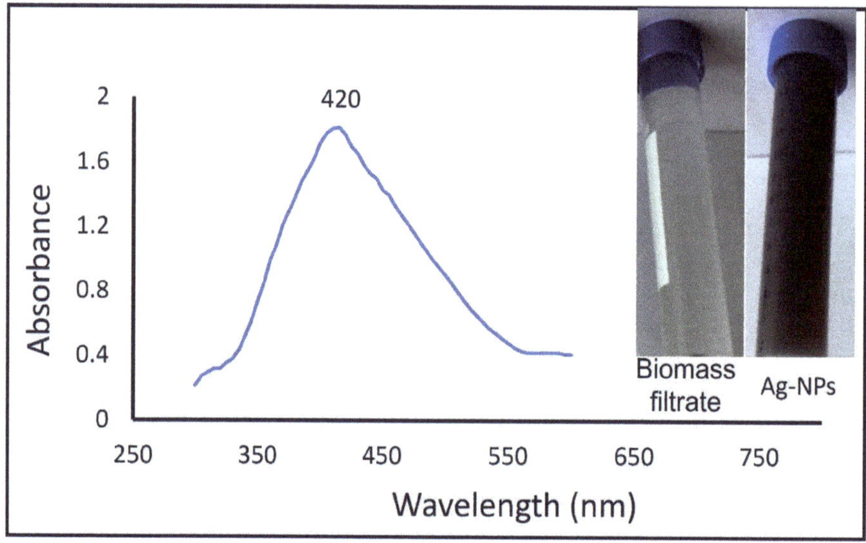

Figure 1. Color change and UV-Vis spectroscopy analysis of mycosynthesized Ag-NPs at wavelengths of 300–600 nm.

3.2.2. Fourier Transform Infrared (FT-IR) Analysis

The various functional groups present in biomass filtrate and their roles in the reduction and capping/stabilizing of mycosynthesized Ag-NPs were investigated by FT-IR. As shown, there are six peaks of fungal biomass filtrate at a wavenumber of 638, 1030, 1630, 2060, 2360, and 3430 cm^{-1} (Figure 2A). The broad peak at 538 cm^{-1} corresponds to C–Br of halo compounds, whereas the medium peak at 1030 cm^{-1} signifies the stretching C O, or CN, or bending C–H of primary amines [58,59]. The strong peak at 1630 cm^{-1} is related to C=O of polysaccharide moieties [29]. The appearance of a broad peak at 2060 cm^{-1} can be attributed to N=C=S of isothiocyanate. The weak peak at 2360 cm^{-1} signifies CO_2, whereas the broad strong peak at 3430 cm^{-1} could be related to either the O–H or N–H group of amino acids that exist in fungal biomass filtrate [59,60].

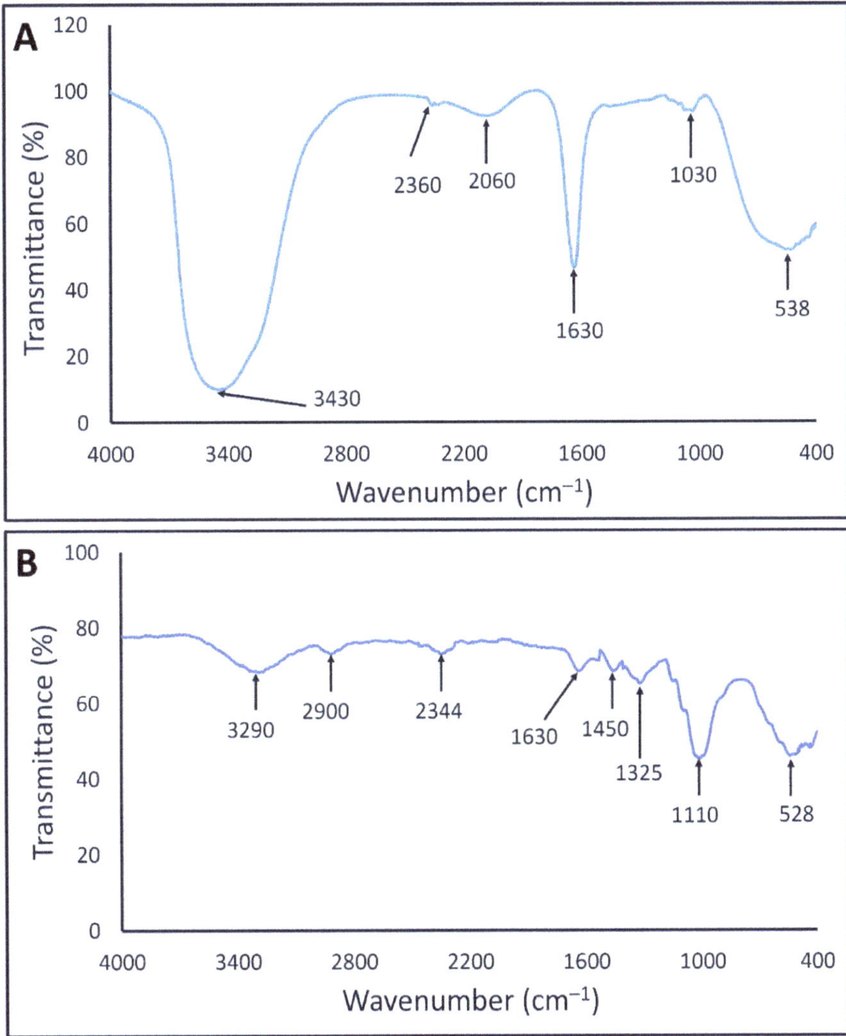

Figure 2. FT-IR spectra of fungal biomass filtrate (**A**) and Ag-NPs fabricated by *Aspergillus niger* strain F2 (**B**).

On the other hand, the FT-IR spectra of mycosynthesized Ag-NPs showed eight peaks at a wavenumber of 528, 1110, 1325, 1450, 1630, 2344, 2900, and 3290 cm^{-1} (Figure 2B). The strong, broad peak at 3290 signified the O–H stretching of carboxylic acid [61], whereas the peak at 2900 cm^{-1} can be attributed to the C–H stretching of aliphatic hydrocarbons [62,63]. The appearance of the peak at 2344 cm^{-1} is related to the O=C=O stretching of CO_2 that is adsorbed onto the surface of proteins [59,64]. On the other hand, the medium peak at 1630 cm^{-1} signified the C=O of polysaccharide moieties, whereas the peak at 1450 cm^{-1} is related to the C–H bending of the methyl group. The peak at 1325 cm^{-1} is related to the C–N stretching of aromatic amines, whereas the strong peak at 1110 cm^{-1} corresponds to the C–O stretching vibration of the carbohydrate ring [59,63]. Finally, the peak at 528 cm^{-1} represents the vibration of C–Br stretching alkyl halides [65]. The obtained FT-IR results confirm the presence of various bioactive molecules such as carbohydrates, alkenes, carboxylate, and amino acids that have been reported previously as a potential reducing agent for the biosynthesis of metal and metal oxide NPs [10,66].

3.2.3. Transmission Electron Microscopy (TEM)

The size and shape of Ag-NPs fabricated by *A. niger* F2 were investigated by TEM analysis. As shown, the as-formed NPs shape was spherical and well-dispersed with sizes in the range of 2–13 nm and an average size of 8.72 ± 2.21 nm (Figure 3A,B). According to size distribution, it can be concluded that the size of the majority of fabricated Ag-NPs was less than 10 nm. Similarly, the biomass filtrate of *A. niger* strain NRC1731 was used to fabricate spherical Ag-NPs with sizes ranging between 3 nm to 20 nm [45]. Additionally, Li et al. successfully formed well-dispersed and spherical Ag-NPs with an average size of 4.3 nm through harnessing metabolites of *Aspergillus terreus* [57]. The various applications of biosynthesized Ag-NPs were highly dependent on several parameters such as chemical compositions, sizes, shapes, and crystallographic structure [67]. The activity of as-formed metal NPs was closely related to their size, meaning the smaller size predicts high activity as previously reported [68]. For example, the activity of NPs formed by an aqueous extract of garlic with the size of 21–40 nm was higher than those with sizes of 41–50 nm [69]. Additionally, the antibacterial activity of small size Ag-NP was better than those recorded for big particle sizes as reported [70]. The obtained sizes can predict high activity for the fungal-mediated synthesis of Ag-NPs.

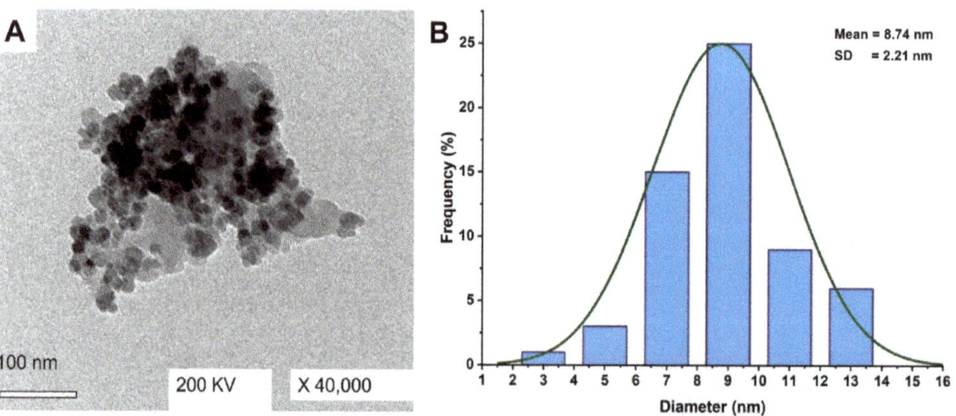

Figure 3. Characterization of Ag-NPs formed by biomass filtrate of *A. niger* strain F2. (**A**) is the TEM image, and (**B**) is the size distributions based on the TEM image.

3.2.4. X-ray Diffraction Pattern

The crystalline and amorphous nature of mycosynthesized Ag-NPs was analyzed by XRD in the 2θ range of 10–80 (Figure 4). As shown, the XRD spectra contain four intense peaks represented by (111), (200), (220), and (311) at 2θ values of 38.4°, 44.4°, 64.3°, and 77.4°, respectively, which are indexed for a face-centered-cubic (FCC) structure of biosynthesized Ag-NPs [71]. The XRD spectra are completely compatible with the JCPDS (Joint Committee on Powder Diffraction Standards) card (No. 04-0783) for the crystalline nature of Ag-NPs [57]. Thomas et al. reported the absence of other diffraction peaks in the XRD chart because of the successful crystallization of metabolites that are used to coat and stabilize Ag-NPs [72]. The obtained XRD results reflect those reported by Wang et al., who successfully fabricated the crystalline nature of Ag-NPs at 2θ values of 38.2°, 44.4°, 64.6°, and 77.8° by using a biomass filtrate of *Aspergillus sydowii* [34]. The average crystal size of synthesized Ag-NPs can be calculated using XRD analysis by the Debye–Scherrer equation which was 20 nm. TEM and XRD analysis revealed that the mycosynthesized Ag-NPs have a uniform morphological shape which was spherical and small average size.

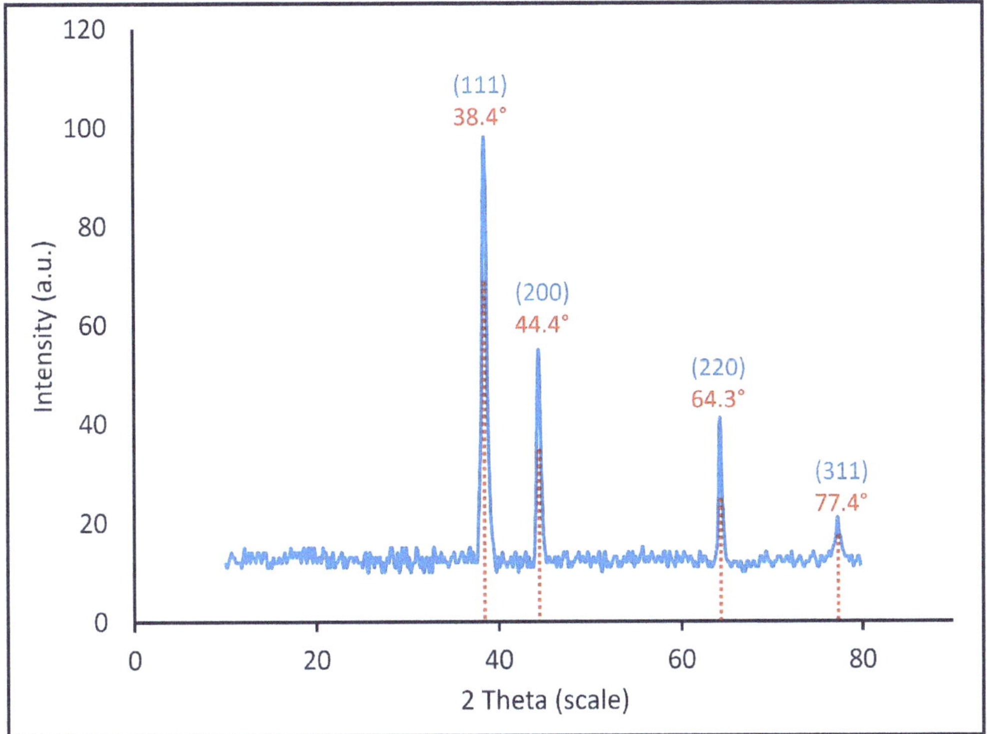

Figure 4. The XRD pattern of mycosynthesized Ag-NPs shows the crystalline nature.

3.3. *Larvicidal/Pupicidal Toxicity of Ag-NPs under Laboratory and Field Conditions*

Herein, the fungal mediated synthesis of Ag-NPs showed high efficacy as larvicide and pupicide for *A. aegypti* at various concentrations of 5, 10, 15, 20, 25, and 30 ppm under laboratory conditions. Data analysis showed that the toxicity of Ag-NPs against various instar larvae and pupae of *A. aegypti* was dose-dependent, and this phenomenon was in agreement with various published studies [41,73,74]. Recently, Ag-NPs fabricated by biological route have been showing high toxicity against a variety of mosquitos [65,75]. The LC50 values (that kill 50% of individuals) and LC90 values (that kill 90% of individuals) were 12.4, 13.6, 15.04, 20.9, and 22.9 ppm and 22.4, 24.6, 27.1, 37.6, and 41.4 ppm for I, II, III, and IV instar larvae and pupae, respectively (Table 1). Compatible with the obtained results, the Ag-NPs synthesized by an aqueous extract of *Suaeda maritima* exhibit high toxicity against first instar larvae and pupa of *A. aegypti* with LC50 values of 8.7 and 17.9 ppm, respectively [73]. Additionally, the Ag-NPs fabricated by *Turbinaria ornata* showed high activity against *Culex quinquefasciatus*, *Anopheles stephensi*, and *A. aegypti* with LC50 values of 1.5, 1.13, and 0.74 μg mL^{-1} [75]. Data reported by Jinu et al. showed that the Ag-NPs synthesized by an aqueous extract of *Strychnos nuxvomica* or *Cleistanthus collinus* were highly toxic for larvae of *Anopheles stephensi* and *A. aegypti* with IC50 values of 8.8 and 7.8 ppm and 11.1 and 11.4 ppm, respectively [11].

Table 1. The toxicity of Ag-NPs fabricated by *A. niger* strain F2 against various instar larvae and pupae of *A. aegypti*.

Target	LC50 (LC90) (ppm)	95 % Confidence Limit LC50 (LC90)	
		LCL	UCL
I instar	12.4 (22.4)	10.01 (20.3)	16.03 (25.5)
II instar	13.6 (24.6)	11.9 (25.2)	14.8 (27.1)
III instar	15.04 (27.1)	12.3 (21.1)	16.03 (29.8)
IV instar	20.9 (37.6)	18.4 (30.6)	24.4 (39.1)
Pupa	22.9 (41.4)	19.1 (37.7)	25.3 (46.3)

LC50 and LC90 are the concentration of Ag-NPs killed 50% and 90% of individuals, respectively. LCL and UCL are lower confidence and upper confidence limits, respectively.

Under field conditions, the reduction in larval density was time-dependent. Data analysis showed that the treatment of *A. aegypti* larvae in a water storage reservoir with Ag-NPs (10× LC50) fabricated by fungal strain *A. niger* F2 caused a complete reduction (100%) after 72 h (Table 2). The obtained data are in agreement with published studies. For instance, the treatment of larvae of *A. aegypti* with Ag-NPs synthesized by leaf extract of *Phyllanthus niruri* led to a complete reduction in larval density under field assay after 72 h [42]. Additionally, the larval density of *Anopheles stephensi* was reduced by a percentage of 97.7% due to treatment with Ag-NPs fabricated by *Aloe vera* in a water storage reservoir [41]. In the current study, the reduction in larval density of *A. aegypti* due to a single treatment with Ag-NPs was achieved with percentages of 59.6% and 74.7% after 24 and 48 h, respectively (Table 2). The difference between Ag-NPs activity in the laboratory and filed assay can be attributed to various reasons such as no proper distribution of the NP across the containers, aggregation of NP at a large scale, and uncontrollable environmental field conditions.

Table 2. Field assay of *A. aegypti* larval density with reduction percentages due to treatment with Ag-NPs (10× LC50) fabricated by *A. niger* strain F2.

Treatment	Larval Density with Reduction Percentages (%)							
	Before Treatment	Reduction %	After Treatment					
			24 h	Reduction %	48 h	Reduction %	72 h	Reduction %
Ag-NPs (10× LC50)	127.1 ± 4.3 [a]	0.0	51.4 ± 6.3 [b]	59.6	32.2 ± 4.1 [c]	74.7	0.0 ± 0.0 [d]	100

Data is represented by mean ± SD, different letters in the row are significantly different ($p < 0.05$).

The high toxicity of Ag-NPs synthesized by a biomass filtrate of *A. niger* F2 can be attributed to their small size (2–13 nm), which is easily adsorbed through the cuticle or digestive tract of an insect and hence interferes with the most physiological processes, ultimately leading to cell death [73]. Moreover, once Ag-NPs enter the insect cells, they can be react with various cellular macromolecules such as amino acids, proteins, and DNA, and eventually lead to cell death [76]. Baskar et al. reported that the activity of Ag-NPs toward various insects was related to their efficacy in blocking various metabolic activities through binding with hormones related to the synthesis of proteins [77]. The toxicity of Ag-NPs against pupae and adults can be attributed to the deformed abnormal morphology or swelling of the pupal integument with the exoskeleton shrinking during death as previously reported [2].

3.4. Smoke Toxicity Assay

Table 3 shows the smoke toxicity assay of Ag-NPs-based coils against the adults of *A. aegypti*. As shown, after exposure to the vapor of burning Ag-NPs-based coils, the mean percent of unfed *A. aegypti* was 81.6 ± 0.5% as compared to the percentages of negative control and pyrethrin-based coil (positive control) which were 23.9 ± 1.2% and 86.1 ± 1.1%, respectively. Analysis of variance revealed a slight change between the mortality percentages of a pyrethrin-based coil and Ag-NPs-based coils (Table 3). Based on these results, it can be concluded that the Ag-NPs-based coils can be used instead of pyrethrin-based coils as an eco-friendly approach to control the adults of *A. aegypti*. To the best of our knowledge, this is the first report to investigate the smoke toxicity of fungal mediated biosynthesis of Ag-NPs-based coils against the adults of *A. aegypti*. According to the obtained data, synthesized Ag-NPs-based coils were more efficient compared to botanical-based coils. The mortality percentages caused by exposure to the vapor of coils formed from the leaves, stems, and roots of *Suaeda maritima* were 86.6%, 79.8%, and 72.7%, respectively [73]. Moreover, the mean percentages of unfed *A. aegypti* mosquitos due to treatment with the stems, leaves, and roots of *Phyllanthus niruri*-based coils were 40%, 58%, and 61%, respectively [42]. The smoke toxicity of Ag-NPs-based coils can be attributed to their toxic effects on the central nervous system of mosquitos upon exposure to the vapor released from burning coils [78]. Moreover, the vapor released from coil burning can be causing acute irritation to the upper respiratory tracts [79]. The coil burning produces unhealthy air conditions that direct entry to internal organs, ultimately severe organs paralysis, producing smaller larvae, and reducing the fecundity [80].

Table 3. Smoke toxicity of mycosynthesized Ag-NPs-based coils against the adults of *A. aegypti*.

Treatment	Fed Mosquitoes (%)	Unfed Mosquitoes (%)		Total (%)
		Alive	Dead	
Ag-NPs based coil	14.4 ± 1.7 ^{b}	22.6 ± 1.9 ^{a}	59.0 ± 2.1 ^{a}	81.6 ± 0.5 ^{b}
Negative control	73.9 ± 0.9 ^{a}	23.9 ± 1.6 ^{a}	0.0 ± 0.0 ^{c}	23.9 ± 1.2 ^{c}
Positive control	9.03 ± 1.7 ^{c}	37.9 ± 1.7 ^{b}	48.2 ± 1.5 ^{b}	86.1 ± 1.1 ^{a}

The negative control is the coils without any active compounds; the positive control is the Pyrethrin-based coil. Data are represented by means ± SD, and different letters in the same column are significantly different ($p < 0.05$).

3.5. Ovicidal Activity

The efficacy of various concentrations of *A. niger* mediated biosynthesis of Ag-NPs (5, 10, 15, 20, 25, and 30 ppm) on the egg hatchability of *A. aegypti* was investigated. Data analysis showed that the egg hatchability of *A. aegypti* was Ag-NPs dose-dependent as reported previously [30,73]. As shown in Figure 5, the hatchability of the eggs of *A. aegypti* was completely eliminated with percentages of 100% after treatment with high concentrations (25 and 30 ppm) of Ag-NPs as compared with the control that showed hatchability percentages of 89.01 ± 0.42%. Moreover, the low concentration of biosynthesized Ag-NPs exhibits toxicity toward egg hatchability. Therefore, the egg hatchability after treatment with 5, 10, 15, and 20 ppm of Ag-NPs were 50.1 ± 0.9, 33.5 ± 1.1, 22.9 ± 1.1, and 13.7 ± 1.2%, respectively. The obtained data agree with those reported by Suresh et al., who reported that the egg hatchability percentages of *A. aegypti* after treatment with 5, 10, and 15 ppm of Ag-NPs were 53.0 ± 1.6, 36.6 ± 1.1, and 25.2 ± 1.3%, respectively, whereas the concentration of 20 and 25 ppm caused the complete reduction (100%) of egg hatchability [73]. The literature survey showed that the nanomaterials were shown ovicidal activity at low concentrations as compared to botanical extract. For instance, 100% egg mortality of *A. aegypti* was attained after treatment with an aqueous leaf extract of *Dicranopteris linearis* and Ag-NPs fabricated by the same plant at a concentration of 300 ppm and 25 ppm, respectively [81].

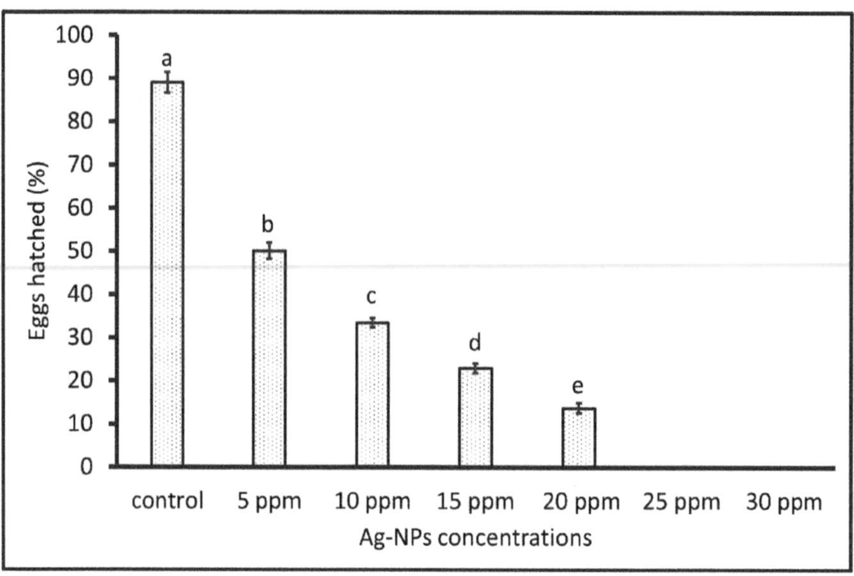

Figure 5. The eggs hatched percentages (%) of *A. aegypti* after treatment by mycosynthesized Ag-NPs at various concentrations. Data are represented by the mean ± SD (n = 5). Different letters between the column are significantly different ($p < 0.05$).

4. Conclusions

In the current study, Ag-NPs were synthesized by an environmentally friendly approach using a biomass filtrate containing metabolites of *A. niger* F2. Based on the characterization study, the FT-IR analysis revealed the role of various metabolites present in fungal biomass filtrate in the reduction and stabilization of Ag-NPs. Moreover, the crystalline nature and spherical shape of Ag-NPs with sizes ranging between 3–13 nm were confirmed by XRD and TEM analysis. Under laboratory conditions, the LC50 and LC90 for Ag-NPs against *A. aegypti* were in the range of 12.4–22.9 ppm and 22.4–41.4 ppm, respectively. In a field assay, Ag-NPs showed high reduction percentages of larval intensity with values of 100% after 72 h. Moreover, the vapor released from the burning of Ag-NPs-based coils causes a reduction percentage of 81.6 ± 0.5% in *A. aegypti* adults compared to the pyrethrin-based coils (86.1 ± 1.1%). Additionally, the Ag-NPs concentration of 25 and 30 ppm caused the complete egg mortality of *A. aegypti*. The obtained data confirmed the efficacy of Ag-NPs synthesized by the eco-friendly approach to be used as an insecticide instead of a chemical formulation that has negative impacts on the environmental eco-system and human public health.

Author Contributions: Conceptualization, M.A.A. and A.F.; methodology, M.A.A., A.M.E., T.M.Y.E., N.S., M.H.S. and A.F.; software, M.A.A., A.M.E., N.S. and A.F.; validation, M.A.A., A.M.E., T.M.Y.E., N.S., M.H.S., S.S., A.A.A.-K. and A.F.; formal analysis, M.A.A., A.M.E., N.S., M.H.S. and A.F.; investigation, M.A.A., A.M.E., T.M.Y.E. and A.F.; resources, M.A.A., A.M.E., T.M.Y.E., Z.E.A.-F., N.S., M.H.S., S.S., A.A.A.-K. and A.F.; data curation, M.A.A., T.M.Y.E., Z.E.A.-F. and A.F.; writing—original draft preparation, M.A.A., A.M.E., T.M.Y.E., Z.E.A.-F. and A.F.; writing—review and editing, M.A.A., A.M.E., S.S., A.A.A.-K. and A.F.; visualization, M.A.A., A.M.E., T.M.Y.E., S.S., A.A.A.-K. and A.F.; supervision, M.A.A., T.M.Y.E. and A.F.; project administration, M.A.A., A.M.E., T.M.Y.E., S.S., A.A.A.-K. and A.F.; funding acquisition, Z.E.A.-F., S.S. and A.A.A.-K. All authors have read and agreed to the published version of the manuscript.

Funding: This research received no external funding.

Institutional Review Board Statement: The animal study protocol was approved by the Ethical Committee of the Animal House Institute, Cairo, Egypt (protocol code, ESAZU AHI/12/2021).

Informed Consent Statement: Not applicable.

Data Availability Statement: The data presented in this study are available on request from the corresponding author.

Acknowledgments: The authors would like to thank Princess Nourah bint Abdulrahman University Researchers supporting project number (PNURSP2022R37), Princess Nourah bint Abdulrahman University, Riyadh, Saudi Arabia. Authors extend their appreciation to the Botany and Microbiology Department, Faculty of Science, Al-Azhar University, Nasr City, Cairo, Egypt, Zoology and Entomology Department, Faculty of Science, Al-Azhar University, Nasr City, Cairo, Egypt, and the Center of Environment Research and Studies, Jazan University, Jazan, Saudi Arabia, for the great cooperation in the current study.

Conflicts of Interest: The authors declare no conflict of interest.

References

1. Soliman, A.M.; Abdel-Latif, W.; Shehata, I.H.; Fouda, A.; Abdo, A.M.; Ahmed, Y.M. Green Approach to Overcome the Resistance Pattern of Candida spp. Using Biosynthesized Silver Nanoparticles Fabricated by *Penicillium chrysogenum* F9. *Biol. Trace Elem. Res.* **2021**, *199*, 800–811. [CrossRef] [PubMed]
2. Manimegalai, T.; Raguvaran, K.; Kalpana, M.; Maheswaran, R. Green synthesis of silver nanoparticle using *Leonotis nepetifolia* and their toxicity against vector mosquitoes of *Aedes aegypti* and *Culex quinquefasciatus* and agricultural pests of *Spodoptera litura* and *Helicoverpa armigera*. *Environ. Sci. Pollut. Res. Int.* **2020**, *27*, 43103–43116. [CrossRef] [PubMed]
3. Shaheen, T.I.; Fouda, A.; Salem, S.S. Integration of Cotton Fabrics with Biosynthesized CuO Nanoparticles for Bactericidal Activity in the Terms of Their Cytotoxicity Assessment. *Ind. Eng. Chem. Res.* **2021**, *60*, 1553–1563. [CrossRef]
4. Fouda, A.; Abdel-Maksoud, G.; Abdel-Rahman, M.A.; Eid, A.M.; Barghoth, M.G.; El-Sadany, M.A.-H. Monitoring the effect of biosynthesized nanoparticles against biodeterioration of cellulose-based materials by *Aspergillus niger*. *Cellulose* **2019**, *26*, 6583–6597. [CrossRef]
5. Hamza, M.F.; Hamad, N.A.; Hamad, D.M.; Khalafalla, M.S., Abdel-Rahman, A.A.H.; Zeid, I.F.; Wei, Y.; Hessien, M.M.; Fouda, A.; Salem, W.M. Synthesis of Eco-Friendly Biopolymer, Alginate-Chitosan Composite to Adsorb the Heavy Metals, Cd(II) and Pb(II) from Contaminated Effluents. *Materials* **2021**, *14*, 2189. [CrossRef]
6. Logeswari, P.; Silambarasan, S.; Abraham, J. Synthesis of silver nanoparticles using plants extract and analysis of their antimicrobial property. *J. Saudi Chem. Soc.* **2015**, *19*, 311–317. [CrossRef]
7. Salem, S.S.; Fouda, A. Green Synthesis of Metallic Nanoparticles and Their Prospective Biotechnological Applications: An Overview. *Biol. Trace Elem. Res.* **2021**, *199*, 344–370. [CrossRef]
8. Alsharif, S.M.; Salem, S.S.; Abdel-Rahman, M.A.; Fouda, A.; Eid, A.M.; El-Din Hassan, S.; Awad, M.A.; Mohamed, A.A. Multifunctional properties of spherical silver nanoparticles fabricated by different microbial taxa. *Heliyon* **2020**, *6*, e03943. [CrossRef]
9. Jeevanandam, J.; Chan, Y.S.; Danquah, M.K. Biosynthesis of Metal and Metal Oxide Nanoparticles. *ChemBioEng Rev.* **2016**, *3*, 55–67. [CrossRef]
10. Guilger-Casagrande, M.; Lima, R.D. Synthesis of Silver Nanoparticles Mediated by Fungi: A Review. *Front. Bioeng. Biotechnol.* **2019**, *7*, 287. [CrossRef]
11. Jinu, U.; Rajakumaran, S.; Senthil-Nathan, S.; Geetha, N.; Venkatachalam, P. Potential larvicidal activity of silver nanohybrids synthesized using leaf extracts of *Cleistanthus collinus* (Roxb.) Benth. ex Hook.f. and *Strychnos nuxvomica* L. *nuxvomica* against dengue, Chikungunya and Zika vectors. *Physiol. Mol. Plant Pathol.* **2017**, *101*, 163–171. [CrossRef]
12. Abdo, A.M.; Fouda, A.; Eid, A.M.; Fahmy, N.M.; Elsayed, A.M.; Khalil, A.M.A.; Alzahrani, O.M.; Ahmed, A.F.; Soliman, A.M. Green Synthesis of Zinc Oxide Nanoparticles (ZnO-NPs) by *Pseudomonas aeruginosa* and Their Activity against Pathogenic Microbes and Common House Mosquito, *Culex Pipiens*. *Materials* **2021**, *14*, 6983. [CrossRef] [PubMed]
13. Ammar, H.A.; El Aty, A.A.A.; El Awdan, S.A. Extracellular myco-synthesis of nano-silver using the fermentable yeasts *Pichia kudriavzevii* HA-NY2 and *Saccharomyces uvarum* HA-NY3, and their effective biomedical applications. *Bioprocess Biosyst. Eng.* **2021**, *44*, 841–854. [CrossRef] [PubMed]
14. Alghuthaymi, M.A.; Abd-Elsalam, K.A.; AboDalam, H.M.; Ahmed, F.K.; Ravichandran, M.; Kalia, A.; Rai, M. *Trichoderma*: An Eco-Friendly Source of Nanomaterials for Sustainable Agroecosystems. *J. Fungi* **2022**, *8*, 367. [CrossRef]
15. Salem, S.S.; Mohamed, A.; El-Gamal, M.; Talat, M.; Fouda, A. Biological Decolorization and Degradation of Azo Dyes from Textile Wastewater Effluent by *Aspergillus niger*. *Egypt. J. Chem.* **2019**, *62*, 1799–1813.
16. Khalil, A.M.A.; Hassan, S.E.; Alsharif, S.M.; Eid, A.M.; Ewais, E.E.; Azab, E.; Gobouri, A.A.; Elkelish, A.; Fouda, A. Isolation and Characterization of Fungal Endophytes Isolated from Medicinal Plant *Ephedra pachyclada* as Plant Growth-Promoting. *Biomolecules* **2021**, *11*, 140. [CrossRef]

17. Clarance, P.; Luvankar, B.; Sales, J.; Khusro, A.; Agastian, P.; Tack, J.C.; Al Khulaifi, M.M.; Al-Shwaiman, H.A.; Elgorban, A.M.; Syed, A.; et al. Green synthesis and characterization of gold nanoparticles using endophytic fungi *Fusarium solani* and its in-vitro anticancer and biomedical applications. *Saudi J. Biol. Sci.* **2020**, *27*, 706–712. [CrossRef]
18. Zaki, S.A.; Ouf, S.A.; Albarakaty, F.M.; Habeb, M.M.; Aly, A.A.; Abd-Elsalam, K.A. *Trichoderma harzianum*-Mediated ZnO Nanoparticles: A Green Tool for Controlling Soil-Borne Pathogens in Cotton. *J. Fungi* **2021**, *7*, 952. [CrossRef]
19. Benelli, G.; Romano, D. Mosquito vectors of Zika virus. *Entomol. Gen.* **2017**, *36*, 309–318. [CrossRef]
20. Lee, H.; Halverson, S.; Ezinwa, N. Mosquito-Borne Diseases. *Prim. Care* **2018**, *45*, 393–407. [CrossRef]
21. World Health Organization. Dengue and Severe Dengue. 2016. Available online: http://www.who.int/mediacentre/factsheets/fs117/en/ (accessed on 12 January 2017).
22. Benelli, G.; Pavela, R.; Maggi, F.; Petrelli, R.; Nicoletti, M. Commentary: Making green pesticides greener? The potential of plant products for nanosynthesis and pest control. *J. Clust. Sci.* **2017**, *28*, 3–10. [CrossRef]
23. Frierson, J.G. The yellow fever vaccine: A history. *Yale J. Biol. Med.* **2010**, *83*, 77–85. [PubMed]
24. Saied, E.; Fouda, A.; Alemam, A.M.; Sultan, M.H.; Barghoth, M.G.; Radwan, A.A.; Desouky, S.G.; Azab, I.H.E.; Nahhas, N.E.; Hassan, S.E. Evaluate the Toxicity of Pyrethroid Insecticide Cypermethrin before and after Biodegradation by *Lysinibacillus cresolivuorans* Strain HIS7. *Plants* **2021**, *10*, 1903. [CrossRef] [PubMed]
25. Benelli, G. Research in mosquito control: Current challenges for a brighter future. *Parasitol. Res.* **2015**, *114*, 2801–2805. [CrossRef] [PubMed]
26. Hamza, M.F.; Hamad, D.M.; Hamad, N.A.; Abdel-Rahman, A.A.H.; Fouda, A.; Wei, Y.; Guibal, E.; El-Etrawy, A.-A.S. Functionalization of magnetic chitosan microparticles for high-performance removal of chromate from aqueous solutions and tannery effluent. *Chem. Eng. J.* **2022**, *428*, 131775. [CrossRef]
27. Athanassiou, C.G.; Kavallieratos, N.G.; Benelli, G.; Losic, D.; Usha Rani, P.; Desneux, N. Nanoparticles for pest control: Current status and future perspectives. *J. Pest Sci.* **2018**, *91*, 1–15. [CrossRef]
28. Morad, M.Y.; El-Sayed, H.; Elhenawy, A.A.; Korany, S.M.; Aloufi, A.S.; Ibrahim, A.M. Myco-Synthesized Molluscicidal and Larvicidal Selenium Nanoparticles: A New Strategy to Control *Biomphalaria alexandrina* Snails and Larvae of *Schistosoma mansoni* with an In Silico Study on Induced Oxidative Stress. *J. Fungi* **2022**, *8*, 262. [CrossRef]
29. Fouda, A.; Eid, A.M.; Abdel-Rahman, M.A.; EL-Belely, E.F.; Awad, M.A.; Hassan, S.E.-D.; AL-Faifi, Z.E.; Hamza, M.F. Enhanced Antimicrobial, Cytotoxicity, Larvicidal, and Repellence Activities of Brown Algae, *Cystoseira crinita*-Mediated Green Synthesis of Magnesium Oxide Nanoparticles. *Front. Bioeng. Biotechnol.* **2022**, *10*, 849921. [CrossRef]
30. Thelma, J.; Balasubramanian, C. Ovicidal, larvicidal and pupicidal efficacy of silver nanoparticles synthesized by *Bacillus marisflavi* against the chosen mosquito species. *PLoS ONE* **2021**, *16*, e0260253.
31. Fouda, A.; Abdel-Maksoud, G.; Abdel-Rahman, M.A.; Salem, S.S.; Hassan, S.E.-D.; El-Sadany, M.A.-H. Eco-friendly approach utilizing green synthesized nanoparticles for paper conservation against microbes involved in biodeterioration of archaeological manuscript. *Int. Biodeterior. Biodegrad.* **2019**, *142*, 160–169. [CrossRef]
32. Ismail, M.A.; Amin, M.A.; Eid, A.M.; Hassan, S.E.; Mahgoub, H.A.M.; Lashin, I.; Abdelwahab, A.T.; Azab, E.; Gobouri, A.A.; Elkelish, A.; et al. Comparative Study between Exogenously Applied Plant Growth Hormones versus Metabolites of Microbial Endophytes as Plant Growth-Promoting for *Phaseolus vulgaris* L. *Cells* **2021**, *10*, 1059. [CrossRef] [PubMed]
33. Eid, A.M.; Fouda, A.; Niedbała, G.; Hassan, S.E.; Salem, S.S.; Abdo, A.M.; Hetta, H.F.; Shaheen, T.I. Endophytic *Streptomyces laurentii* Mediated Green Synthesis of Ag-NPs with Antibacterial and Anticancer Properties for Developing Functional Textile Fabric Properties. *Antibiotics* **2020**, *9*, 641. [CrossRef]
34. Wang, D.; Xue, B.; Wang, L.; Zhang, Y.; Liu, L.; Zhou, Y. Fungus-mediated green synthesis of nano-silver using *Aspergillus sydowii* and its antifungal/antiproliferative activities. *Sci. Rep.* **2021**, *11*, 10356. [CrossRef]
35. Mohamed, A.E.; Elgammal, W.E.; Eid, A.M.; Dawaba, A.M.; Ibrahim, A.G.; Fouda, A.; Hassan, S.M. Synthesis and characterization of new functionalized chitosan and its antimicrobial and in-vitro release behavior from topical gel. *Int. J. Biol. Macromol.* **2022**, *207*, 242–253. [CrossRef] [PubMed]
36. Fouda, A.; Hassan, S.E.-D.; Saied, E.; Hamza, M.F. Photocatalytic degradation of real textile and tannery effluent using biosynthesized magnesium oxide nanoparticles (MgO-NPs), heavy metal adsorption, phytotoxicity, and antimicrobial activity. *J. Environ. Chem. Eng.* **2021**, *9*, 105346. [CrossRef]
37. Borchert, H.; Shevchenko, E.V.; Robert, A.; Mekis, I.; Kornowski, A.; Grübel, G.; Weller, H. Determination of nanocrystal sizes: A comparison of TEM, SAXS, and XRD studies of highly monodisperse $CoPt_3$ particles. *Langmuir* **2005**, *21*, 1931–1936. [CrossRef]
38. Sujitha, V.; Murugan, K.; Paulpandi, M.; Panneerselvam, C.; Suresh, U.; Roni, M.; Nicoletti, M.; Higuchi, A.; Madhiyazhagan, P.; Subramaniam, J.; et al. Green-synthesized silver nanoparticles as a novel control tool against dengue virus (DEN-2) and its primary vector *Aedes Aegypti*. *Parasitol. Res.* **2015**, *114*, 3315–3325. [CrossRef]
39. Kaleka, A.; Kaur, N.; Bali, G. Larval Development and Molting. In *Edible Insects*; 2020 Edition; Mikkola, H., Ed.; Intech Open: London, UK, 2019.
40. Murugan, K.; Vahitha, R.; Baruah, I.; Das, S. Integration of botanical and microbial pesticides for the control of filarial vector, *Culex quinquefasciatus* Say (Diptera: Culicidae). *Ann. Med. Entomol.* **2003**, *12*, 12–23.
41. Dinesh, D.; Murugan, K.; Madhiyazhagan, P.; Panneerselvam, C.; Mahesh Kumar, P.; Nicoletti, M.; Jiang, W.; Benelli, G.; Chandramohan, B.; Suresh, U. Mosquitocidal and antibacterial activity of green-synthesized silver nanoparticles from *Aloe vera* extracts: Towards an effective tool against the malaria vector Anopheles stephensi? *Parasitol. Res.* **2015**, *114*, 1519–1529. [CrossRef]

42. Suresh, U.; Murugan, K.; Benelli, G.; Nicoletti, M.; Barnard, D.R.; Panneerselvam, C.; Kumar, P.M.; Subramaniam, J.; Dinesh, D.; Chandramohan, B. Tackling the growing threat of dengue: *Phyllanthus niruri*-mediated synthesis of silver nanoparticles and their mosquitocidal properties against the dengue vector *Aedes aegypti* (Diptera: Culicidae). *Parasitol. Res.* **2015**, *114*, 1551–1562. [CrossRef]
43. Su, T.; Mulla, M.S. Ovicidal activity of neem products (azadirachtin) against Culex tarsalis and Culex quinquefasciatus (Diptera: Culicidae). *J. Am. Mosq. Control Assoc.* **1998**, *14*, 204–209. [PubMed]
44. Finney, D.L. *Probit Analysis*, 2nd ed.; Cambridge University Press: New York, NY, USA, 1952; Volume 41, p. 627.
45. Elsayed, M.A.; Othman, A.M.; Hassan, M.M.; Elshafei, A.M. Optimization of silver nanoparticles biosynthesis mediated by *Aspergillus niger* NRC1731 through application of statistical methods: Enhancement and characterization. *3 Biotech* **2018**, *8*, 132. [CrossRef] [PubMed]
46. Parial, D.; Patra, H.K.; Roychoudhury, P.; Dasgupta, A.K.; Pal, R. Gold nanorod production by cyanobacteria—A green chemistry approach. *J. Appl. Phycol.* **2012**, *24*, 55–60. [CrossRef]
47. Namvar, F.; Azizi, S.; Ahmad, M.B.; Shameli, K.; Mohamad, R.; Mahdavi, M.; Tahir, P.M. Green synthesis and characterization of gold nanoparticles using the marine macroalgae *Sargassum muticum*. *Res. Chem. Intermed.* **2015**, *41*, 5723–5730. [CrossRef]
48. Sastry, M.; Mayya, K.S.; Bandyopadhyay, K. pH Dependent changes in the optical properties of carboxylic acid derivatized silver colloidal particles. *Colloids Surf. A Physicochem. Eng. Asp.* **1997**, *127*, 221–228. [CrossRef]
49. Zomorodian, K.; Pourshahid, S.; Sadatsharifi, A.; Mehryar, P.; Pakshir, K.; Rahimi, M.J.; Arabi Monfared, A. Biosynthesis and Characterization of Silver Nanoparticles by *Aspergillus* Species. *BioMed Res. Int.* **2016**, *2016*, 5435397. [CrossRef]
50. Saada, N.S.; Abdel-Maksoud, G.; Abd El-Aziz, M.S.; Youssef, A.M. Green synthesis of silver nanoparticles, characterization, and use for sustainable preservation of historical parchment against microbial biodegradation. *Biocatal. Agric. Biotechnol.* **2021**, *32*, 101948. [CrossRef]
51. Cruz-Luna, A.R.; Cruz-Martínez, H.; Vásquez-López, A.; Medina, D.I. Metal Nanoparticles as Novel Antifungal Agents for Sustainable Agriculture: Current Advances and Future Directions. *J. Fungi* **2021**, *7*, 1033. [CrossRef]
52. Salem, S.S.; Ali, O.M.; Reyad, A.M.; Abd-Elsalam, K.A.; Hashem, A.H. *Pseudomonas indica*-Mediated Silver Nanoparticles: Antifungal and Antioxidant Biogenic Tool for Suppressing Mucormycosis Fungi. *J. Fungi* **2022**, *8*, 126. [CrossRef]
53. Lee, S.H.; Jun, B.H. Silver Nanoparticles: Synthesis and Application for Nanomedicine. *Int. J. Mol. Sci.* **2019**, *20*, 865. [CrossRef]
54. Salem, S.S.; El-Belely, E.F.; Niedbała, G.; Alnoman, M.M.; Hassan, S.E.; Eid, A.M.; Shaheen, T.I.; Elkelish, A.; Fouda, A. Bactericidal and In-Vitro Cytotoxic Efficacy of Silver Nanoparticles (Ag-NPs) Fabricated by Endophytic Actinomycetes and Their Use as Coating for the Textile Fabrics. *Nanomaterials* **2020**, *10*, 2082. [CrossRef] [PubMed]
55. Iqtedar, M.; Aslam, M.; Akhyar, M.; Shehzaad, A.; Abdullah, R.; Kaleem, A. Extracellular biosynthesis, characterization, optimization of silver nanoparticles (AgNPs) using *Bacillus mojavensis* BTCB15 and its antimicrobial activity against multidrug resistant pathogens. *Prep. Biochem. Biotechnol.* **2019**, *49*, 136–142. [CrossRef] [PubMed]
56. Taha, Z.K.; Hawar, S.N.; Sulaiman, G.M. Extracellular biosynthesis of silver nanoparticles from *Penicillium italicum* and its antioxidant, antimicrobial and cytotoxicity activities. *Biotechnol. Lett.* **2019**, *41*, 899–914. [CrossRef]
57. Li, G.; He, D.; Qian, Y.; Guan, B.; Gao, S.; Cui, Y.; Yokoyama, K.; Wang, L. Fungus-mediated green synthesis of silver nanoparticles using *Aspergillus terreus*. *Int. J. Mol. Sci.* **2012**, *13*, 466–476. [CrossRef] [PubMed]
58. Hamza, M.F.; Abdel-Rahman, A.A.; Negm, A.S.; Hamad, D.M.; Khalafalla, M.S.; Fouda, A.; Wei, Y.; Amer, H.H.; Alotaibi, S.H.; Goda, A.E. Grafting of Thiazole Derivative on Chitosan Magnetite Nanoparticles for Cadmium Removal-Application for Groundwater Treatment. *Polymers* **2022**, *14*, 1240. [CrossRef] [PubMed]
59. Coates, J. Interpretation of Infrared Spectra, A Practical Approach. In *Encyclopedia of Analytical Chemistry: Applications, Theory and Instrumentation. Infrared Spectrosc*; Meyers, R.A., Ed.; Wiley Online Library: Hoboken, NJ, USA, 2006.
60. Saied, E.; Eid, A.M.; Hassan, S.E.; Salem, S.S.; Radwan, A.A.; Halawa, M.; Saleh, F.M.; Saad, H.A.; Saied, E.M.; Fouda, A. The Catalytic Activity of Biosynthesized Magnesium Oxide Nanoparticles (MgO-NPs) for Inhibiting the Growth of Pathogenic Microbes, Tanning Effluent Treatment, and Chromium Ion Removal. *Catalysts* **2021**, *11*, 821. [CrossRef]
61. Hamza, M.F.; Lu, S.; Salih, K.A.M.; Mira, H.; Dhmees, A.S.; Fujita, T.; Wei, Y.; Vincent, T.; Guibal, E. As(V) sorption from aqueous solutions using quaternized algal/polyethyleneimine composite beads. *Sci. Total Environ.* **2020**, *719*, 137396. [CrossRef]
62. Mohammadi, N.; Ganesan, A.; Chantler, C.T.; Wang, F. Differentiation of ferrocene D5d and D5h conformers using IR spectroscopy. *J. Organomet. Chem.* **2012**, *713*, 51–59. [CrossRef]
63. Hamza, M.F.; Fouda, A.; Wei, Y.; El Aassy, I.E.; Alotaibi, S.H.; Guibal, E.; Mashaal, N.M. Functionalized biobased composite for metal decontamination—Insight on uranium and application to water samples collected from wells in mining areas (Sinai, Egypt). *Chem. Eng. J.* **2022**, *431*, 133967. [CrossRef]
64. Wei, Y.; Rakhatkyzy, M.; Salih, K.A.M.; Wang, K.; Hamza, M.F.; Guibal, E. Controlled bi-functionalization of silica microbeads through grafting of amidoxime/methacrylic acid for Sr(II) enhanced sorption. *Chem. Eng. J.* **2020**, *402*, 125220. [CrossRef]
65. Sutthanont, N.; Attrapadung, S.; Nuchprayoon, S. Larvicidal Activity of Synthesized Silver Nanoparticles from *Curcuma zedoaria* Essential Oil against Culex quinquefasciatus. *Insects* **2019**, *10*, 27. [CrossRef] [PubMed]
66. Fouda, A.; Hassan, S.E.-D.; Abdel-Rahman, M.A.; Farag, M.M.S.; Shehal-deen, A.; Mohamed, A.A.; Alsharif, S.M.; Saied, E.; Moghanim, S.A.; Azab, M.S. Catalytic degradation of wastewater from the textile and tannery industries by green synthesized hematite (α-Fe_2O_3) and magnesium oxide (MgO) nanoparticles. *Curr. Res. Biotechnol.* **2021**, *3*, 29–41. [CrossRef]

67. Bansal, V.; Li, V.; O'Mullane, A.P.; Bhargava, S.K. Shape dependent electrocatalytic behaviour of silver nanoparticles. *CrystEngComm* **2010**, *12*, 4280–4286. [CrossRef]
68. Amin, M.A.; Ismail, M.A.; Badawy, A.A.; Awad, M.A.; Hamza, M.F.; Awad, M.F.; Fouda, A. The Potency of Fungal-Fabricated Selenium Nanoparticles to Improve the Growth Performance of *Helianthus annuus* L. and Control of Cutworm *Agrotis ipsilon*. *Catalysts* **2021**, *11*, 1551. [CrossRef]
69. Sribenjarat, P.; Jirakanjanakit, N.; Jirasripongpun, K. Selenium nanoparticles biosynthesized by garlic extract as antimicrobial agent. *Sci. Eng. Health Stud.* **2020**, *14*, 22–31.
70. Dong, Y.; Zhu, H.; Shen, Y.; Zhang, W.; Zhang, L. Antibacterial activity of silver nanoparticles of different particle size against *Vibrio Natriegens*. *PLoS ONE* **2019**, *14*, e0222322. [CrossRef]
71. Lashin, I.; Fouda, A.; Gobouri, A.A.; Azab, E.; Mohammedsaleh, Z.M.; Makharita, R.R. Antimicrobial and In Vitro Cytotoxic Efficacy of Biogenic Silver Nanoparticles (Ag-NPs) Fabricated by Callus Extract of *Solanum incanum* L. *Biomolecules* **2021**, *11*, 341. [CrossRef]
72. Thomas, B.; Vithiya, B.S.M.; Prasad, T.A.A.; Mohamed, S.B.; Magdalane, C.M.; Kaviyarasu, K.; Maaza, M. Antioxidant and Photocatalytic Activity of Aqueous Leaf Extract Mediated Green Synthesis of Silver Nanoparticles Using *Passiflora edulis* f. flavicarpa. *J. Nanosci. Nanotechnol.* **2019**, *19*, 2640–2648. [CrossRef]
73. Suresh, U.; Murugan, K.; Panneerselvam, C.; Rajaganesh, R.; Roni, M.; Aziz, A.T.; Naji Al-Aoh, H.A.; Trivedi, S.; Rehman, H.; Kumar, S.; et al. *Suaeda maritima*-based herbal coils and green nanoparticles as potential biopesticides against the dengue vector Aedes aegypti and the tobacco cutworm *Spodoptera litura*. *Physiol. Mol. Plant Pathol.* **2018**, *101*, 225–235. [CrossRef]
74. Hassan, S.E.; Fouda, A.; Saied, E.; Farag, M.M.S.; Eid, A.M.; Barghoth, M.G.; Awad, M.A.; Hamza, M.F.; Awad, M.F. *Rhizopus oryzae*-Mediated Green Synthesis of Magnesium Oxide Nanoparticles (MgO-NPs): A Promising Tool for Antimicrobial, Mosquitocidal Action, and Tanning Effluent Treatment. *J. Fungi* **2021**, *7*, 372. [CrossRef]
75. Deepak, P.; Sowmiya, R.; Ramkumar, R.; Balasubramani, G.; Aiswarya, D.; Perumal, P. Structural characterization and evaluation of mosquito-larvicidal property of silver nanoparticles synthesized from the seaweed, *Turbinaria ornata* (Turner) J. Agardh 1848. *Artif. Cells Nanomed. Biotechnol.* **2017**, *45*, 990–998. [CrossRef] [PubMed]
76. Badawy, A.A.; Abdelfattah, N.A.H.; Salem, S.S.; Awad, M.F.; Fouda, A. Efficacy Assessment of Biosynthesized Copper Oxide Nanoparticles (CuO-NPs) on Stored Grain Insects and Their Impacts on Morphological and Physiological Traits of Wheat (*Triticum aestivum* L.) Plant. *Biology* **2021**, *10*, 233. [CrossRef] [PubMed]
77. Baskar, K.; Muthu, C.; Ignacimuthu, S. Effect of pectolinaringenin, a flavonoid from *Clerodendrum phlomidis*, on total protein, glutathione S-transferase and esterase activities of *Earias vittella* and *Helicoverpa armigera*. *Phytoparasitica* **2014**, *42*, 323–331. [CrossRef]
78. Haldar, K.M.; Ghosh, P.; Chandra, G. Larvicidal, adulticidal, repellency and smoke toxic efficacy of *Ficus krishnae* against Anopheles stephensi Liston and *Culex vishnui* group mosquitoes. *Asian Pac. J. Trop. Dis.* **2014**, *4*, S214–S220. [CrossRef]
79. Pauluhn, J. Mosquito coil smoke inhalation toxicity. Part I: Validation of test approach and acute inhalation toxicity. *J. Appl. Toxicol. Int. J.* **2006**, *26*, 269–278. [CrossRef] [PubMed]
80. Tan, Y.Q.; Dion, E.; Monteiro, A. Haze smoke impacts survival and development of butterflies. *Sci. Rep.* **2018**, *8*, 15667. [CrossRef] [PubMed]
81. Rajaganesh, R.; Murugan, K.; Panneerselvam, C.; Jayashanthini, S.; Aziz, A.T.; Roni, M.; Suresh, U.; Trivedi, S.; Rehman, H.; Higuchi, A.; et al. Fern-synthesized silver nanocrystals: Towards a new class of mosquito oviposition deterrents? *Res. Vet. Sci.* **2016**, *109*, 40–51. [CrossRef]

Journal of
Fungi

Review

Trichoderma: An Eco-Friendly Source of Nanomaterials for Sustainable Agroecosystems

Mousa A. Alghuthaymi [1,*], Kamel A. Abd-Elsalam [2], Hussien M. AboDalam [3], Farah K. Ahmed [4], Mythili Ravichandran [5], Anu Kalia [6] and Mahendra Rai [7]

1. Biology Department, Science and Humanities College, Shaqra University, Alquwayiyah 11726, Saudi Arabia
2. Plant Pathology Research Institute, Agricultural Research Center (ARC), 9-Gamaa St., Giza 12619, Egypt; kamel.abdelsalam@arc.sci.eg
3. Plant Pathology Department, Faculty of Agriculture, Cairo University, Giza 12613, Egypt; hussien.abodlam@gmail.com
4. Biotechnology English Program, Faculty of Agriculture, Cairo University, Giza 12613, Egypt; farahkamel777@gmail.com
5. Department of Microbiology, Vivekanandha Arts and Science College for Women, Sankari 637303, Tamil Nadu, India; ms.microhoney@gmail.com
6. Electron Microscopy and Nanoscience Laboratory, Punjab Agricultural University, Ludhiana 141004, Punjab, India; kaliaanu@pau.edu
7. Department of Microbiology, Nicolaus Copernicus University, Lwowska 1, 87100 Torun, Poland; mahendra.rai@v.umk.pl

* Correspondence: malghuthaymi@su.edu.sa

Abstract: Traditional nanoparticle (NP) synthesis methods are expensive and generate hazardous products. It is essential to limit the risk of toxicity in the environment from the chemicals as high temperature and pressure is employed in chemical and physical procedures. One of the green strategies used for sustainable manufacturing is microbial nanoparticle synthesis, which connects microbiology with nanotechnology. Employing biocontrol agents *Trichoderma* and *Hypocrea* (Teleomorphs), an ecofriendly and rapid technique of nanoparticle biosynthesis has been reported in several studies which may potentially overcome the constraints of the chemical and physical methods of nanoparticle biosynthesis. The emphasis of this review is on the mycosynthesis of several metal nanoparticles from *Trichoderma* species for use in agri-food applications. The fungal-cell or cell-extract-derived NPs (mycogenic NPs) can be applied as nanofertilizers, nanofungicides, plant growth stimulators, nanocoatings, and so on. Further, *Trichoderma*-mediated NPs have also been utilized in environmental remediation approaches such as pollutant removal and the detection of pollutants, including heavy metals contaminants. The plausible benefits and pitfalls associated with the development of useful products and approaches to trichogenic NPs are also discussed.

Keywords: beneficial microbes; biocontrol agents; *Trichoderma*; *Hypocrea*; nanostructures

Citation: Alghuthaymi, M.A.; Abd-Elsalam, K.A.; AboDalam, H.M.; Ahmed, F.K.; Ravichandran, M.; Kalia, A.; Rai, M. *Trichoderma*: An Eco-Friendly Source of Nanomaterials for Sustainable Agroecosystems. *J. Fungi* **2022**, *8*, 367. https://doi.org/10.3390/jof8040367

Academic Editor: Laurent Dufossé

Received: 1 March 2022
Accepted: 31 March 2022
Published: 2 April 2022

1. Introduction

The use of myco-nanoparticles in agriculture is still in its early phases of research, especially in terms of their interactions with agriculturally beneficial microorganisms. A few in vitro experiments have been published, while in vivo investigations are currently taking place in greenhouse conditions [1–3]. Microorganisms, often known as nanofactories or nanoparticle producers, have potential because their cellular machinery may be altered to make the synthesis of NPs easier [4,5]. The biogenic synthesis based on fungi has several benefits in terms of efficiency and the generation of diverse metabolites under optimal circumstances. Furthermore, because fungi are natural producers of a wide range of antimicrobial compounds, using them as a capping agent of nanoparticles might result in a synergistic antimicrobial impact with metal NPs against pathogenic microorganisms [6].

Among the microbial agents, significant diversity of intracellular and extracellular proteins and enzymes that function as reducing agents makes fungal bioagents more suited than bacteria [7]. *Trichoderma* strains have a long history of effectiveness as biocontrol agents against a variety of pathogenic microorganisms. Furthermore, recent research has demonstrated that these fungi improve plant resilience, growth, and development, resulting in increased yield output [8,9]. One of the strategies for decreasing the harmful effects of heavy metals on plants is to use *Trichoderma harzianum*. Several *Trichoderma* species have been identified for the synthesis of a variety of important secondary metabolites such as plant growth regulators, antibiotics, and enzymes, which are mostly utilized to defend plants against pathogens [9]. Additionally, the metabolites produced and secreted by *Trichoderma* species in the culture filtrates are known to possess antimicrobial, anticancer, and antioxidant properties [10].

Recently, the use of agrochemicals incorporating nanostructured materials has emerged as a viable agricultural option [11]. Several *Trichoderma* species have been employed in nanotechnology, primarily for the synthesis of metal nanoparticles (Figure 1). Their resistance to many nano compounds has recently been discovered, but little is known about their contribution to the production of metallic NPs via tolerance to these chemicals and how these aspects affect *Trichoderma* relationships. Enzymes such as reductases, which can operate as bioreductive agents in the biofabrication of NPs, can be considered the key biocomponents for the *Trichoderma*-mediated mycosynthesis of NPs [12]. *Trichoderma* is an easy-to-manage fungus with several physiological and technological advantages [13]. The biosynthesis of metal nanoparticles using the advantageous *Trichoderma* hyphal extracts is a straightforward, environmentally friendly, and cost-effective method. Secondary metabolites released by *T. harzianum* operate as capping and reducing agents, which contribute to the biological activity and green synthesis of AgNPs using *Trichoderma* [14]. It is a quick, economically feasible, environmentally safe, nontoxic metal nanoparticle synthesis technique well-suited for large-scale production [15]. As a result, utilization of this green bio-based, environmentally friendly, and economically feasible approach can be a plausible alternative solution for the successful manufacturing of sustainable nanomaterials. In this review, the information on various uses of *Trichoderma* genus to develop techniques for the mycosynthesis of metal NPs and their uses in agroecosystems is presented. Further, the probable role of novel *Trichoderma*-NP bioconjugates as a viable alternative for sustainable agriculture will be identified.

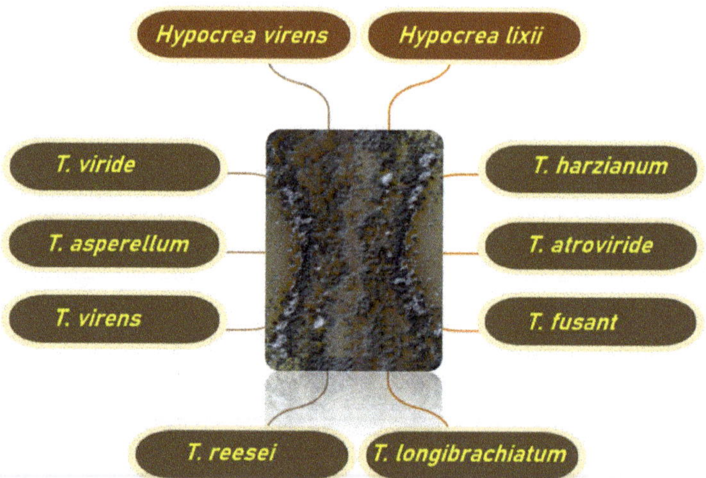

Figure 1. The top ten *Trichoderma* species used to produce safe metal nanoparticles through mycogenic synthesis.

2. Beneficial Effects of *Trichoderma* in Agroecosystems

The *Trichoderma* genus includes fungal species that are economically significant because of their plant growth and performance-promoting actions, such as enhanced nutrient availability, the mycoparasitism of plant pathogens, and the priming of plant defense. The biodiversity of the genus *Trichoderma* provides fresh insight into their uses, particularly in agriculture and industry, as well as their potential use as biofungicides and biofertilizers in the field. *Trichoderma* spp. boost plant development, initiate plant defense, and aid plant growth in response to a variety of biotic and abiotic challenges, such as soil dryness, excessive salt content, and the presence of poisonous metal ions [16]. *Trichoderma* spp. are well-known plant growth-promoting fungi (PGPFs) with the potential to compete with pathogenic microorganisms while also promoting plant health [17]. *Trichoderma* species are extremely significant as biocontrol agents against a variety of plant pathogenic fungi and, hence, serve as viable alternatives to synthetic fungicides. *Trichoderma* species have been used as biological control agents for the management of plant pathogenic microbes, including fungi and bacteria [18]. *Trichoderma* and their secondary metabolites released into the rhizosphere may have an impact on plant development and nutrition, as well as the induction of systemic resistance and the biocontrol of pathogenic bacteria [18]. Plants and *Trichoderma* communicate bidirectionally via numerous signal molecules, resulting in a beneficial symbiotic relationship [16]. It can colonize plant roots by sensing nutrients secreted from the roots in the rhizosphere; interact with plants by producing various MAMP (microbe-associated molecular pattern) and damage-associated molecular pattern (DAMP) molecules; stimulate an immune response, suppressing many plant pathogens through the use of systemic acquired resistance (SAR) and induced systemic resistance (ISR) mechanisms; and act as a systemic resistance inducer in the plant [16,17]. Colonization of roots and leaves by *Trichoderma* can prime plant defense, allowing for powerful plant responses to following pathogen threats [19]. However, in addition to biopesticide action, certain *Trichoderma* strains have been shown to exhibit biostimulant activity, plant growth promotion, enhanced yield and nutritional quality, and the ability to mitigate the negative effects of abiotic stressors [20–22]. Therefore, *Trichoderma* has evolved from a mycoparasitic biocontrol agent (BCA) to one with numerous features such as pathogen antagonism, pathogen competition for resources, induction of systemic resistance in the host, overall plant growth promotion, and reduction of abiotic stressors. Furthermore, while being previously described primarily as soil and root colonizers, it is now clear that numerous *Trichoderma* species are endophytic [23]. As a result, it is not surprising that *Trichoderma* is found to be an effective beneficial biological agent, with active ingredients in over 200 agricultural products such as biopesticides, biofertilizers, bio-growth enhancers, and bio-stimulants that are marketed across the globe [15,24,25].

Beneficial fungi, including the *Trichoderma* species, are integral components of the decomposer microflora population, which plays a vital role in bioconcentration, decontamination, and even degradation/removal of the xenobiotics added to the ecosystem due to intentional and/or unintentional incorporation of a variety of contaminants. This removal or decontamination is referred to as 'bioremediation' and involves the use of microbial enzymes to convert toxic metal compounds into nonhazardous chemicals [26]. In a research report, *T. viride* exhibited the ability to break down and use nitrogenous (trinitrotoluene, TNT) explosives at doses of 50 and 100 ppm as the N-source to meet the nitrogen requirements for normal development [27]. Besides nitrogenous explosives, the potential hydrocarbon-degrading abilities of *Trichoderma* species can be useful to bioremediate diesel oil spills in an aquatic ecosystem, ensuring the protection of the environment [28]. Another report reiterated the use of *Trichoderma* species, particularly *T. harzianum* strain T22, for ensuring the biodegradation of diesel fuel, allowing it to be used as a carbon source [29]. Further, the physiological and metabolic versatility of various *Trichoderma* fungal species can be utilized for the remediation of heavy metals from different eco-niches. *T. lixii* CR700 demonstrated excellent Cu removal capability across a wide pH range. In the removal of Cu, *T. lixii* CR700 employs simultaneous surface sorption and accumulation processes [30].

Different *Trichoderma* species are, thus, effective natural decomposition agents to speed up the degradation of organic materials. The benefit will be two-fold as the nutrients released from the xenobiotic will be available in the plant rhizosphere and can be utilized and taken up by the plant [9]. The presence of *Trichoderma* with alfalfa seedlings boosted the soil's N, P, and K content and alfalfa biomass [31]. The conjugate use of the *Trichoderma* and chemical fertilizer improved the plant's nutritional quality, productivity, and vegetative and reproductive development. This can cut agricultural expenses while also reducing pollutants [32]. Overall, *Trichoderma* acts as a nutrient mobilizer, improving the quality and yield components of the crops.

Many studies have indicated that *Trichoderma* spp. are particularly efficient at degrading pesticides and can be useful in an integrated pest control strategy as well as for reducing pesticide residual effects. *Trichoderma* spp. have been reported to degrade benzimidazole fungicide (Carbendazim) [33], chlorpyrifos [34], penthiopyrad [35], and 2,2-dichlorovinyl dimethyl phosphate (DDVP) [36]. Wu et al. [37] proposed the tolerance mechanism of *T. asperellum* TJ01 to dichlorvos. *Trichoderma* spp. produce a variety of volatile organic compounds (VOCs) possessing various chemical properties. These metabolites are critical in agricultural, food, and pharmaceutical applications [38]. *T. reesei* has not been proven to be hazardous to humans [39]. The biosynthesis of silver nanoparticles by the fungus *T. reesei* is preferable in terms of safety, economics, and large-scale production capability [39]. As a result, the use of *Trichoderma* should be encouraged since it offers sustainable agriculture by minimizing the use of hazardous chemicals in agriculture (Figure 2).

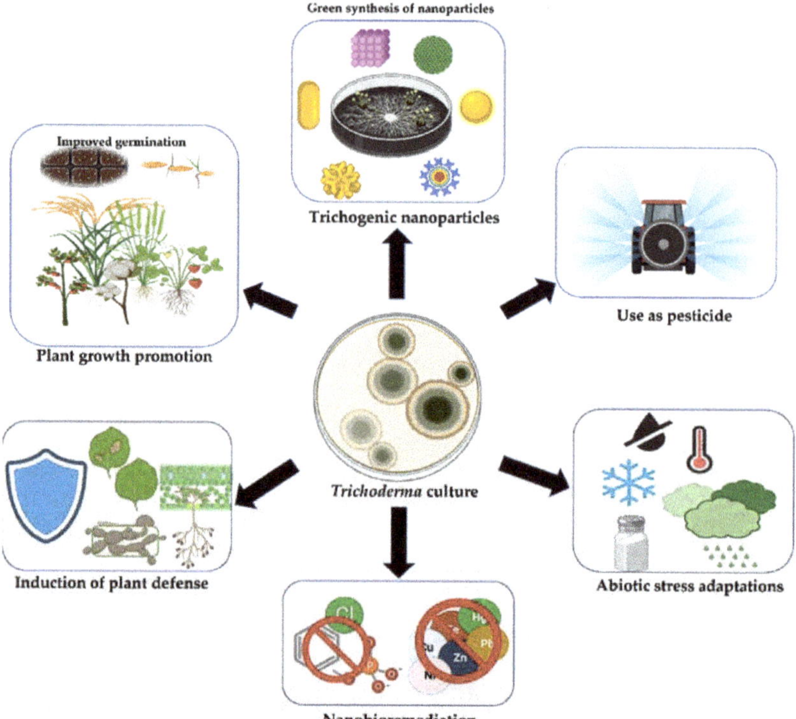

Figure 2. *Trichoderma* applications in the agricultural ecosystem. Red circle indicates stopping of the heavy metal contamination through active removal by nanomaterials. This figure was created with BioRender software.

3. Biosynthesis of Nanoparticles by *Trichoderma* Genus

Fungal nanotechnology is one of the most popular options because of the vast range of benefits it has over bacteria, actinomycetes, plants, and other organisms in terms of physicochemical qualities [40]. When it comes to the biological generation of NPs, fungi outperform the majority of microorganisms in terms of efficiency. This is due to the ability of the fungi to produce a large variety of bioactive metabolites and metal accumulation properties and have improved processes, all of which are beneficial [41]. Because of their capacity to tolerate metals and accumulate metals in their tissues, fungi have emerged as an important branch for the biosynthesis of nanoparticles [42]. Green synthesis of nanoparticles is nontoxic as it involves the utilization of safe reagents, which makes it more cost-effective and eco-friendlier compared to the traditional methods [14]. Fungi offer several other benefits, such as simplicity of management and cultivation, no requirement of complicated components, production of a large amount of biomass and metabolites, high cell wall-binding capability, and ability to absorb large amounts of metal [43]. The next section describes the production of myconanoparticles from several *Trichoderma* species.

3.1. Silver Nanoparticles

The first report on the biosynthesis of silver nanoparticles with regulated properties from a nonpathogenic and economically viable biocontrol agent involved the incubation of the cell-free filtrate of *T. asperellum* for 5 days at 25 °C containing $AgNO_3$ (1 mM) [44]. The use of UV–vis spectroscopy revealed the kinetics of the reaction with a strong surface plasmon resonance band observed at 410 nm, indicating the synthesis of silver nanoparticles. Based on the results of TEM and XRD studies, the size of the silver nanoparticles ranged from 13 to 18 nm [44]. *T. reesei* has also been used for the extracellular biosynthesis of AgNPs for the first time [39], allowing for the synthesis of AgNPs on an industrial scale. Fluorescence emission spectroscopy was used to produce detailed information on the progress of the decrease of silver nitrate (formation of silver nanoparticles) on the nanosecond timescale. The quantitative analysis of the reaction products was carried out using Fourier transform infrared spectroscopy (FTIR). The TEM images revealed the morphologically varied forms and crystalline nature of the AgNPs, with diameters ranging from 5 to 50 nm. Likewise, another study showcased *T. virens* to be the most efficient producer of AgNPs when compared to a total of 75 isolates from five distinct *Trichoderma* species. Every 24 h, the highest plasmon band was recorded at 420 nm, reaching maximum intensity at 120 h. The morphology of NPs was validated by high-resolution transmission electron microscopy (HRTEM), which revealed that nanoparticles were single or aggregated, spherical and homogeneous in shape, and ranged in size from 8 to 60 nm [45].

T. harzianum cell filtrate was used to synthesize AgNPs in a simple green and eco-friendly manner, without the need for any toxic reducing agents, capping agents, or dispersion agents. Temperature and $AgNO_3$ concentration both had a considerable impact on AgNP production, as evidenced by UV–vis spectra. AgNPs were shown to be stable for three months, as confirmed through the DLS study. The synthesized AgNPs possessed face-centered cubic symmetry with a size range of 10–51 nm [46]. *T. harzianum* extracellular filtrate is an entomopathogenic fungus used for AgNP synthesis with distinct characteristics. The reduction rate of silver ions was assessed using a UV-visible spectrophotometer, which revealed a maximum absorption at 433.5 nm. Fourier transform infrared spectroscopy identified specific fungal metabolite functional groups responsible for the synthesis of AgNPs. The linkage between amino acid residues was shown by the amide group through the vibration of the N–H bend. X-ray diffraction analysis confirmed that the mycosynthesized AgNPs were oval in shape, crystalline in nature, and monodispersed in colloidal form, as well as exhibiting high purity. The surface properties of the generated AgNPs, as depicted by scanning electron microscopy, involved no direct contact even inside aggregates, indicating that AgNPs have been stabilized by a capping agent, and the size was around 10–20 nm [47]. Elgorban et al. [48] studied the synthesis of biogenic silver nanoparticles at room temperature in darkness using the fungus *T. viride*. Centrifugation was used to increase the concentration of AgNPs. The bioreduction of silver nanoparticles (AgNPs) was seen spectrophotometrically, and the AgNPs under investigation were characterized using UV–vis, TEM, and SEM. *T. viride* AgNPs were reported to be stable and polydispersed globular particles, with diameters ranging from 1 to 50 nm. They can be used as a reducing agent in the biogenic production of silver nanoparticles. *T. harzianum*-derived AgNPs possessed a spherical shape, acceptable polydispersity, and a size distribution between 20 and 30 nm, with no aggregates [4]. Biological silver nanoparticles were generated extracellularly utilizing the fungus *T. longibrachiatum*, with the fungal cell filtrate employed as a reducing and stabilizing agent in the nanoparticle manufacturing process. Incubation at 28 °C for 72 h with fungal biomass without agitation resulted in AgNP biosynthesis. The production of spherical nanoparticles with sizes ranging from 5 to 25 nm was observed by TEM [49]. Saravanakumar and Wang [50] reported, for the first time, the synthesis of anisotropic structured AgNPs using *T. atroviride* and investigated their biomedical properties. The observation of plasmon resonance at 390–400 nm in the UV–vis spectrum verified the formation of AgNPs. FTIR, transmission electron microscopy, and EDX examination revealed a significant percentage signal of anisotropic structural AgNPs, with sizes ranging from 15 to 25 nm. The *Trichoderma* filtrate contained biochemicals capable of bio-reducing silver nitrate ($AgNO_3$), which are spherical NPs and nontoxic at low concentrations [51].

Extracellular biosynthesis of AgNPs from $AgNO_3$ solution using *T. reesei* PF biomass cell-free water extract (CFE) has been described [52]. The optimal generation of AgNPs was achieved when *T. reesei* fungi were cultured in conditions containing 0.1 percent corn steep liquor, 10% biomass extracted to produce CFE, and 10 mM $AgNO_3$. HRTEM micrographs verified the presence of Ag metallic nanoparticles in the crystal phase. The TEM pictures revealed the development of AgNPs clusters with sizes ranging from 1–4 to 15–25 nm. The NPs were stabilized by the capping action of the biomolecules, as indicated by FTIR spectroscopy. The zeta-potential of the prepared AgNPs was negative [52]. AgNPs were effectively synthesized using *T. harzianum* filtrates cultured in the presence and absence of enzymatic stimulation of the *S. sclerotiorum* cell wall, resulting in nanoparticles with distinct physicochemical properties. The DLS approach was utilized to determine hydrodynamic diameters, yielding mean hydrodynamic diameters of 57.02 ± 1.75 nm for AgNP-TS and 81.84 ± 0.67 nm for AgNP-T, respectively. The NTA approach yielded nanoparticle sizes and concentrations that differed in AgNP-TS and AgNP-T particle sizes of 88.0 ± 7.3 and 182.5 ± 6.9 nm, respectively [43]. *T. longibrachiatum* DSMZ 16,517 mycelial cell-free filtrate (MCFF) bioreduced silver ions (Ag+) to their metallic nanoparticle state (Ag^0), as shown by AgNPs. The DLS examination revealed average AgNP size and zeta potential values of 17.75 nm and 26.8 mV, suggesting the stability of the synthesized AgNPs. The crystallinity of the mycosynthesized AgNPs, with an average size of 61 nm, was confirmed by the XRD pattern. The FESEM and HRTEM images revealed non-agglomerated spherical, triangular, and cuboid AgNPs, with sizes ranging from 5 to 11 nm. The FTIR analysis of the mycosynthesized AgNPs confirmed MCFF's activity as a reducing and capping agent [53]. *T. atroviride*-hosted *Chiliadenus montanus* was shown to be the best candidate for the synthesis of mycogenic AgNPs among all investigated species. HRTEM was used to describe these AgNPs, which revealed a dispersion of spherical AgNPs ranging in size from 10 to 15 nm. By using the agar well diffusion technique, mycosynthesized AgNPs were compared to chemically synthesized AgNPs for their antibacterial, anticandidal, and antifungal activities over six pathogenic bacteria and four pathogenic fungi in vitro [54]. Chitin-induced exo-metabolites extracted from the varied and stress-tolerant *T. fusant* Fu21 were used to synthesize green AgNPs with sizes ranging from 59.66 to 4.18 nm (SEM), the spherical shape of nanoparticles with proven purity, stability (51.2 mV zeta potential), and nanoparticle polydispersity [55].

Amazon fungus *Trichoderma* sp. strain (TCH 01) was isolated from *Bertholletia excelsa* (Brazil-nut) seeds, and the soil was identified to biosynthesize the AgNP, as demonstrated by the RPSL band and UV–vis spectroscopy analysis. According to the TEM picture, the particles were polydispersed and spherical in form. AgNPs varied in size from 14 to 25 nm.

All AgNPs exhibited incipient instability for the zeta potential. When grown at pH 5 for 9 days, smaller nanoparticles and higher polydispersity indexes were obtained [56]. Among 15 isolates of the *T.* species tested for the manufacture of AgNPs, a cell-free aqueous filtrate of *T. virens* HZA14 generating gliotoxin produced the best yield for the production of AgNPs. Electron microscopy tests revealed that AgNPs were 5–50 nm in size and had spherical and oval forms with smooth surfaces [57]. UV–visible spectrophotometry, FTIR, EDS, DLS, XRD, and SEM were used to analyze the AgNPs produced from *T. harzianum* culture filtrate. The surface plasmon resonance of synthesized particles created a peak with a center wavelength of 438 nm. According to the DLS research, the average size of AgNPs is 21.49 nm. SEM was used to determine the average size of AgNPs, which was 72 nm. The cubic crystal structure determined by XRD analysis validated the particles' identification as silver nanoparticles [15]. As an alternative to traditional chemical and physical techniques, a green method for the synthesis of AgNPs has been provided. *T. reesei* fungus biomass was exploited as a green and renewable source of reductase enzymes and metabolites capable of converting Ag^+ ions into AgNPs. *T. reesei* is an appropriate reagent for the production of monodisperse AgNPs that are stabilized by the capping effect of biomolecules. Trichogenic synthesis of AgNPs mediated by the *Trichoderma* genus is shown in Figure 3.

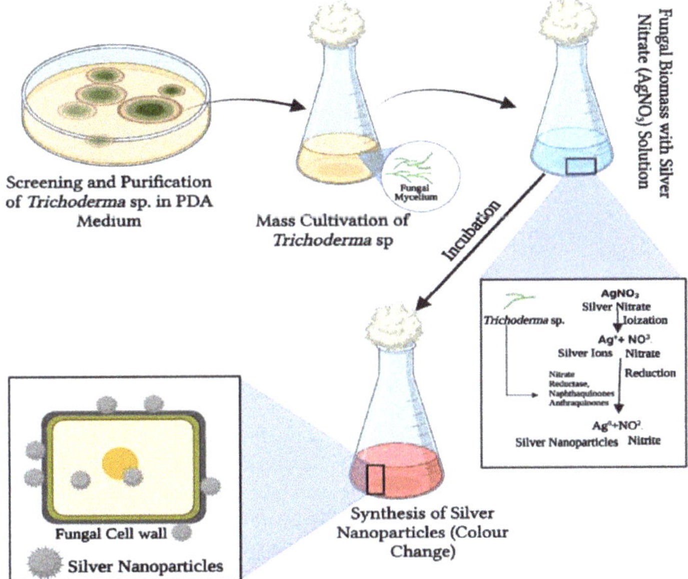

Figure 3. Schematic illustration of biosynthesis mechanism of silver nanoparticles (AgNPs) using *Trichoderma* species. The present figure was created by BioRender.com.

3.2. Zinc Oxide Nanoparticles

T. harzianum (PGT4), *T. reesei* (PGT5), and *T. reesei* (PGT13) monocultures and co-cultures (PGT4 + PGT5 + PGT13) generated secondary metabolites useful in the mycosynthesis of zinc oxide nanoparticles (ZnO NPs). *Trichoderma*—D-glucanzinc oxide nanoparticles (T—D-glu-ZnO NPs) were synthesized under optimal conditions using the fungal mycelial water extract (FWME) obtained from *T. harzianum*. The effective conjugation of D-glucan from barley with T-ZnONPs was confirmed by PACE and FTIR. The optimized T—D-glu-ZnO NPs had a spherical form with a mean size of 30.34 nm. T—D-glu-ZnONPs greatly suppressed the development of *Staphylococcus aureus* within roundworms while also enhancing roundworm growth [58]. Co-cultivation can induce the production of new secondary metabolites more effectively than monocultures. Another study involving the biosynthesis of ZnONPs showed the formation of NPs with crystalline structures free of impurities, according to PXRD analysis. The size of the crystalline particles ranged between 12 and 35 nm. These biosynthesized ZnONPs showed antibacterial efficacy against the rice cause of Bacterial Leaf Blight, *Xanthomonas oryzae* pv. *oryzae* [59]. *T. harzianum* is a possible fungal antagonist that is employed in the extracellular manufacture of ZnONPs. TEM imaging revealed that ZnONPs' crystalline structure comprises hexagonal, spherical, and rod-shaped particles in a mixture of exceptionally small particles. The size range of the generated ZnONPs was 8–23 nm [3].

3.3. Copper Nanoparticles

T. asperellum cell-free extract was used to create copper oxide nanoparticles (TA-CuONPs). TA-CuO NPs were found to be crystalline with spherical particles. The CuONPs ranged in size from 10 to 190 nm, with an average diameter of 110 nm [60]. Consolo et al. [61] revealed, for the first time, the numerous extracellular biosynthesis of NPs from *T. harzianum*, as well as the synthesis of CuO and ZnONPs from this fungus. Biogenically synthesized Ag, CuO, and ZnO nanoparticles were made by employing a cell filtrate of a strain of *T. harzianum* as a reducer and stabilizer agent. When the ZnO NPs were evaluated against different target microorganisms, the potential of Ag and CuO for phytopathogen control was emphasized. Biocontrol agents *Pseudomonas fluorescens*, *T. atroviride*, and *Streptomyces griseus* were used in the production of copper and silica nanoparticles. *T. harzianum* was effectively used in the synthesis of copper nanoparticles. CuNPs were tested for antibacterial properties against two bacteria, *Staphylococcus aureus* and *Escherichia coli*, using a simple green and eco-friendly method [62]. UV–vis spectrometry, TEM, and EDAX analysis were used to analyze silica and copper nanoparticles. Nanoparticles were discovered to be aggregated and irregularly spherical. The size of silica nanoparticles varied from 12 to 22 nm, whereas the size of copper nanoparticles ranged from 5 to 25 nm [63].

3.4. Selenium Nanoparticles

The activity of biosynthesized selenium nanoparticles (SeNPs) using *T. asperellum* culture filtrate against *Sclerospora graminicola*, the cause of mildew disease in pearl millet, was higher. SeNPs were found to be hexagonal, near-spherical, and irregular in form, with sizes ranging from 49.5 to 312.5 nm. The size of SeNPs was inversely related to their biological activity [64]. Selenium nanoparticles (TSNPs) biosynthesized from *T. harzianum* JF309 were compared to conventional SNPs. The liquid metabolites of eight *Trichoderma* strains were collected, and modified biogenic synthesis was achieved. SNPs were found to be spherical or pseudo-spherical in shape, but TSNPs were more irregular. TSNPs were discovered to be somewhat larger than standard SNPs. The antifungal impact of TSNPs was far superior to that of standard SNPs [65]. The *Trichoderma* sp. WL-Go culture broth was used to create simple and cost-effective selenium nanoparticles. SeNPs had a partial size range of 20–220 nm, with an average diameter of 147.1 nm [66].

3.5. Other Nanoparticles

Gold nanoparticles were synthesized through a very fast and environment-friendly method within 10 min; this was the first time for such rapid biosynthesis of growth-promoting and plant pathogen control AuNPs. *T. viride* cell-free extract was treated with $HAuCl_4$ at 30 °C for 10 min. TEM analysis confirmed that the NPs were widely dispersed and scattered in nature, with the bulk of them being spherical in form. The crystalline nature was verified by the SAED pattern. The size of the AuNPs ranged from 20 to 30 nm [67]. Chitosan nanoparticles were biogenically synthesized from *T. viride* and described using UV–vis spectroscopy, with FTIR verifying the functional groups of chitosan nanoparticles as OH, N–H, C–H, C=O, C–O, C–N, and P=O and electron microscopy demonstrating the roughly spherical form. DLS analysis determined the average size of chitosan nanoparticles to be 89.03 nm [68]. The *T. harzianum* (MF780864) isolate demonstrated the extracellular production of SiO2NPs from rice husks. UV, FTIR, DLS, and TEM were used to describe SiO_2 NPs, which had a size of around 89 nm and shapes of oval, rod, and cubical particles [69].

The possibility of synthesizing FeNPs from *Trichoderma* species using simple methods has also been investigated. The presence of alkene, carboxyl, and phenol groups showed that the NPs were capped by the organism following the redox event. The generation of FeNPs from fungi such as *Trichoderma* species, which has the potential to generate more FeNPs than other bioresources, was proven [70]. *T. harzianum* is a biocontrol agent that is employed in the green manufacture of biogenic iron oxide nanoparticles for stabilization. Dynamic light scattering, nanoparticle tracking analysis, scanning electron microscopy, X-ray diffraction, and Fourier transform infrared spectroscopy techniques were used to assess the physicochemical properties of nanoparticles, and the results showed that the average size diameter of hematite (Fe_2O_3) nanoparticles are 207.2 nm [71]. Table 1 presents many metal NPs produced by different *Trichoderma* species, including AgNPs, ZnONPs, CuNPs and CuONPs, SeNPs, and other NPs, each with unique features and antimicrobial potential.

Table 1. Trichoderma species employed for green synthesis metal nanoparticles.

Trichoderma Species	NPs	Size	Shape	Application	References
T. asperellum	AgNPs	13–18 nm	crystalline nature	Biomolecular detection	[44]
T. reesei	AgNPs	5–50 nm	variable morphology	Preparing many nanostructured materials and devices	[39]
T. virens	AgNPs	8–60 nm	round and uniform in shape	Crop protection	[45]
T. harzianum	AgNPs	10–51 nm	face centered cubic symmetry particles	Antioxidant properties and antibacterial activity	[46]
T. harzianum	AgNPs	10–20 nm	oval shaped, crystalline in nature	Mosquito control	[47]
T. viride	AgNPs	1–50 nm	globular particles	Antibacterial effect against human pathogenic bacteria	[48]
T. harzianum	AgNPs	20–30 nm	spherical	Control of *S. sclerotiorum*	[4]
T. longibrachiatum	AgNPs	5–25 nm	spherical	Control of many phytopathogenic fungi	[49]
T. atroviride	AgNPs	15–25 nm	anisotropic structural	Antioxidant and antibacterial against clinical pathogens	[50]
T. harzianum	AgNP-TS	57.02 ± 1.75 nm	different characteristics	Control of *S. sclerotiorum*	[43]
	AgNP-T	81.84 ± 0.67 nm			
T. reesei	AgNPs	1–4 nm 15–25 nm	crystal phase	carriers of biologically active molecules	[52]
T. longibrachiatum DSMZ 16517	AgNPs	5–11 ± 0·5 nm	spherical, triangular, and cuboid	Control of industrial microbes	[53]
Trichoderma sp.	AgNPs	14–25 nm	round	Antibacterial	[56]
T. virens HZA14	AgNPs	5–50 nm	spherical and oval with smooth surfaces	Control of *S. sclerotiorum*	[57]
T. atroviride	AgNPs	10–15 nm	spherical	Control of pathogenic bacteria and fungi	[54]
T. fusant Fu21	AgNPs	59.66 ± 4.18 nm	spherical	Control of *S. sclerotiorum*	[55]
T. harzianum	AgNPs	72 nm	cubic crystal structure	Antioxidant properties and antibacterial activity	[15]
Trichoderma spp. co-culture	ZnONPs	12–35 nm	crystal structure	Control of Bacterial Leaf Blight causative in rice	[59]
T. harzianum (SKCGW009)	ZnONPs	30.34 nm	spherical	Antibacterial activity enhanced roundworm growth	[58]
T. harzianum	ZnONPs	8–23 nm	hexagonal, spherical and rod	fungicidal action against three soil–cotton pathogenic fungi	[3]

Table 1. *Cont.*

Trichoderma Species	NPs	Size	Shape	Application	References
T. asperellum	CuONPs	10–190 nm	spherical	development of anticancer nanotherapeutics	[60]
T. harzianum	CuONPs	~20 nm	spherical structure	Antibacterial activity	[62]
T. harzianum	AgNPs	5–18 nm	spherical	Control of plant pathogens	[61]
	CuONPs	38–77 nm	Dispersed and elongated fibers in shape		
	ZnONPs	27–40 nm in width 134–200 nm in length	fan and bouquet structure	Control of microorganisms	
T. atroviride and 2 other fungi	CuONPs	5–25 nm	spherical	Management of some tea plantation diseases	[63]
	SiO_2NPs	12–22 nm			
T. asperellum	SeNPs	49.5–312.5 nm	hexagonal, near-spherical, and irregular	Control of *Sclerospora graminicola*, mildew disease causative in pearl	[64]
T. harzianum JF309	SeNPs	bigger than traditional SNP	irregular	antifungal	[65]
Trichoderma sp. WL-Go	SeNPs	An average of 147.1 nm	spherical and pseudo-spherical	ND	[66]
T. viride	AuNPs	20–30 nm	spherical	bioremediation	[67]
T. viride	Chitosan NPs	89.03 nm	nearly spherical	Control of soil borne pathogens	[68]
T. harzianum MF780864	SiO_2NPs	89 nm	oval, rod, and cubical	Bioremediation	[69]
T. harzianum	iron oxide NPs	207 ± 2 nm (α-Fe_2O_3)	Spherical shape	Control of *S. sclerotiorum*	[71]

4. Production of Metal NPs by *Hypocrea*

In pure culture, cultures obtained from *Hypocrea lixii* ascospores developed the morphological species *T. harzianum*. *T. harzianum* is known as a cosmopolitan, ubiquitous species associated with a wide variety of substrates [72]. The ability of the fungus *Hypocrea lixii* to ingest and reduce copper ions to copper NPs was investigated. This approach verified the existence of proteins as stabilizing and capping agents around the copper nanoparticles. These results showed that the dead biomass of *H. lixii* is an economically and technically feasible solution for wastewater bioremediation and a possible candidate for industrial-scale synthesis of copper NPs [73]. Rapid synthesis of gold nanoparticles is possible in less than a minute using cell-free extracts of *T. viride* and recombinant *H. lixii* at various reaction temperatures. Another study was the first report of *T.* sp. involved in the rapid synthesis of gold nanoparticles that may function as an antimicrobial agent as well as an effective biocatalyst [67]. The fungus *H. lixii's* dead biomass was discovered to be an effective instrument for the extracellular and intracellular synthesis of nickel oxide NPs as well as the absorption of hazardous metal ions from an aqueous solution. The fungus's dead biomass played a significant role in the formation of metallic NPs, which are easily oxidized to nickel oxide in the media. The dead biomass also functioned as a stabilizer during the creation of nickel oxide NPs and so might be employed for nickel ion uptake during bioremediation procedures [74]. In addition, a new species, *H. virens*, has been discovered to be a teleomorph of *T. virens*, a species commonly used in biological control applications. *H. virens*, for example, is a possible biocontrol agent agonist of the *Ceratocystis paradoxa* pathogen that causes pineapple disease on sugarcane. The fungus *H. virens* was tested for its ability to synthesize AgNPs. A *H. virens* fungal biomass of extracellular and intracellular extract to aqueous silver nitrate solution for 72 h revealed that the appearance of a dark-brown color in the fungal biomass after reaction with Ag^+ ions was a clear indicator of metal ion reduction and the formation of AgNPs by the fungal biomass [45].

5. Toxicity

In a recent study, the blood kinetics and tissue distribution of 20, 80 and 110 nm AgNPs were studied in rats [75]. The silver nanoparticles disappeared quickly from the blood and were distributed to all animal organs, independent of the NP size [75]. As a result, it can be identified that the AgNPs may have different toxicity and, hence, be associated with a distinct health risk. After repeated intravenous injections of silver nanoparticles, accumulation was observed [76]. The toxicity of AgNPs was determined by the dose and particle size [50]. In terms of toxicity, comparisons of cell lines 3T3 (mouse embryo fibroblasts), HeLa (human cervical adenocarcinoma), HaCaT (human keratinocytes), V79 (Chinese hamster pulmonary fibroblasts), and A549 (human epithelial adenocarcinoma) with controls revealed that the biogenic nanoparticles exhibited cytotoxic and genotoxic effects that varied depending on the cell line used and the exposure concentration. An interesting observation was that in the majority of these assays, the effects increased in direct proportion to the doses employed and were most strong at concentrations higher than those used in *S. sclerotiorum* inhibition evaluations. Concerning the impacts of nanoparticles in soil, there was evidence of some partial effects on bacteria, with probable microbiota recovery over time. Because the AgNP-T nanoparticles had no detrimental impacts on soybean germination and development, it is possible to conclude that this approach constitutes the first step towards the potential control of white mold in soybean crops using nanotechnology [4]. *T. atroviride*-induced AgNPs triggered cell changes, inhibition, or damage, as well as cytoplasmic compression in a concentration-dependent way. AgNPs significantly reduced cell viability with an inhibitory dose of 16.5 g/mL against human breast cancer cells (MDA-MB-231). Cell viability increased significantly with drug dosage concentration (NPs) against IMR 90, U251, and A549 lung cells, and human breast cancer cells MCF-7 and MDA-MB-231 [50]. The cytotoxicity and genotoxicity of AgNP-TS and AgNP-T nanoparticles were shown to be low in V79, 3T3, and HaCat cell lines. Guilger-Casagrande and colleagues [43] used, for the first time, photothermolysis to confirm the anticancer activity of biogenic CuO NPs. The TA-CuO NPs caused the photothermolysis of A549 cancer cells through ROS production, nucleus damage, mitochondrial membrane potential (m), and regulatory protein expression [60]. When SNPs and TSNPs were tested for cytotoxic action using human viability and mortality, neither SNPs nor TSNPs demonstrated hypertoxicity or lethality for the three cells [65]. At a concentration of 200 g ml^{-1}, it possessed multi-mode antimicrobial action and has been determined to be nontoxic to humans. The lowest inhibitory concentration of green nanoformulations for the greater degrading activity of the pathogen fungal mycelium was determined to be 20 g Ag/mL green nanoformulations [55]. T—D-glu-ZnO NPs were not hazardous to NIH3T3 cells but had a dose-dependent inhibitory impact on human pulmonary carcinoma A549 cells, according to a cytotoxicity investigation. T-ZnO NPs and T—D-glu-ZnO NPs caused cancer cell death by necrosis and apoptosis, respectively [58]. The zebrafish model was used to assess the toxicity; there was no toxicity of nanocopper and nanosilica when exposed to 0.5, 3, and 30 g concentrations in respect of embryo viability, hatching rate, body mass index, and heartbeat counts [63]. The cytotoxicity of biogenic iron oxide nanoparticles was evaluated using several cell lines and *Allium cepa* assays to measure the genotoxicity. They did not impact cell viability when compared to controls and did not cause changes in the mitotic index at the quantities used. The presence of iron oxide nanoparticles did not affect seed germination [71].

6. Understanding the Mechanism of Synthesis of Trichogenic NPs

Fungal biosynthetic techniques can be grouped into intracellular and extracellular synthesis based on where nanoparticles are produced. Extracellular synthesis of nanoparticles, for example, is still being developed in terms of understanding the mechanisms of synthesis, simple downstream processing, and quick scale-up processing.

6.1. Bioactive Metabolites

The use of *T. longibrachiatum* for AgNP biosynthesis revealed the existence of proteins and their binding to AgNPs via carbonyl groups of amino acid residues and peptides, which might have contributed to their stability and prevention of agglomeration. Proteins on the surface of AgNPs function as capping agents [49]. There are 35 metabolites that have been discovered as significant variations between selenium nanoparticles biosynthesized from *T. harzianum* (TSNP) and standard SNPs, which included organic acids, sugars, amino acids, and carbohydrate metabolism intermediates. Among these, 27 are strong antifungal agents with higher levels in TSNPs than in SNPs. Many acids, sugars, and their derivatives are used in the coating of TSNPs, including heptonic acid, ferulate, fumaric acid, threonic acid, glucose, and mannitol. All of these organic compounds in aqueous formulations capped the selenium NPs and acted as stabilizers while also increasing the antagonistic capability of biosynthesized selenium NPs against pathogens [65]. While using *T. virens* HZA14 for AgNP biosynthesis revealed the interaction patterns of protein, carbohydrate, and heterocyclic compound molecules with AgNPs, the maximum yield was associated with gliotoxin [57]. Aromatic amino acids such as tyrosine and tryptophan were found in the FTIR spectra of *Trichoderma* spp. medium. The electrostatic attraction of negatively charged carboxylate groups by free amine groups, cysteine residues, or some proteins secreted by the fungus during the formation of AgNPs can bind to them. The release of extracellular protein molecules from fungus is responsible for the synthesis and stability of AgNPs [56]. The interaction of different functional groups of exometabolites with Ag was validated using FTIR for the production of green AgNPs and may be responsible for stabilization. *T. fusant* Fu21 inoculated with SM-containing chitin was used for GC–MS profiling to identify functional groups and compounds, which were discovered to be alkanes, dicarboxylic acid, aromatic ketones, amino acid, hetero-acyclic compounds, ketose sugar, sugar alcohol, aliphatic amines, polyol compounds, steroidal pheromones, and carbocyclic sugars [55]. Secondary metabolites released by *T. harzianum* operated as capping and reducing agents, providing consistency and contributing to biological activity determined by LC-MS/MS. The most common compounds included 1-benzoyl-3-[(S)-((2S,4R,8R)-8-ethylquinuclidin-2-yl] thiourea (6-methoxyquinolin-4-yl) methyl, puerarin, genistein, isotalatizidine, and ginsenoside [15]. The existence of numerous functional groups of biomolecules and capping protein, enclosing biosynthesized SiO_2 NPs corresponding to carbonyl residues, alcohol, nitrile, acid chloride, alkene bands, and peptide bonds of the proteins involved, was verified by FTIR analysis [69]. The presence of functional groups of alkene, alkane, and alcohol in the FTIR test suggested that these may have participated in the SeNP production processes [66]. The presence of functional groups was verified by FTIR analysis of biosynthesized silica and copper nanoparticles using biocontrol agents, resulting in effective synthesis [63]. According to the FTIR analysis results, phenolic, proteins, amino acids, aldehydes, ketone, and other functional groups were involved in the reduction, capping, and stability of zinc oxide NPs [3].

6.2. Enzymes

While many microbial species can produce metal NPs, the mechanism of nanoparticle biosynthesis has not been identified as yet [77,78]. For its survival, *T. harzianum* produces enzymes and metabolites that are involved in the breakdown of silver nitrate into Ag^+ ions and NO_3, owing to the possible action of hydrolytic/nitrate reductase enzymes. Through the catalytic activity of extracellular fungal secondary metabolites, the toxic Ag^+ ions are further converted to nontoxic (Ag^0 = biosilver) metallic nanoparticles in this procedure. The bioreduction of silver is based on the presence of relevant functional groups in the extracellular filtrate of *T. harzianum*, which is more effective than other fungi and is nontoxic to humans [47,79]. FTIR and surface-enhanced resonance were used. Raman spectroscopy revealed a viable mechanism for the synthesis of silver nanoparticles in *T. asperellum* [44]. The procedure consisted of two critical steps: bioreduction of $AgNO_3$ to produce AgNPs, followed by stabilization and/or encapsulation with a suitable capping agent. The cell's defense mechanism for silver detoxification has been proposed as a biological process for AgNP production [80]. Extracellular enzyme secretion provides the advantage of obtaining large quantities in a relatively pure state, free of other cellular proteins associated with the organism, and can be easily processed by filtering the cells and isolating the enzyme for nanoparticle synthesis from the cell-free filtrate. *T. reesei* is thought to be the most effective extracellular enzyme producer compared to other filamentous fungi [39], and it has a long history of producing industrial enzymes [81]. The NADH co-enzyme, as well as NADH-dependent enzymes such as nitrate reductase, are present in *Trichoderma* genus strains and are essential in the formation of nanoparticles and capping that gives greater stability [39,82]. The determination of hydrolytic enzyme-specific activity revealed that AgNP-TS had a greater specific activity of NAGase and chitinase than AgNP-T. Concerning the filtrates, *T. harzianum* exposure to the cell wall of *S. sclerotiorum* enhanced the specific activity of the enzyme NAGase [43]. *T. harzianum* produces enzymes and metabolites involved in breaking strong ionic bonds between silver and nitrate ions for its own survival, perhaps by the action of hydrolytic/nitrate reductase enzymes. Extracellular fungal secondary metabolite enzymatic activity transforms dangerous Ag+ ions into nontoxic biosilver NPs [14].

7. Applications of *Trichoderma*-Mediated NPs in the Agri-Food Sector

Crop production invariably involves rampant use of chemical pesticides, insecticides, and herbicides that have been exhibiting bioaccumulation of toxic chemical residues potentially hazardous to plant ecosystems [40]. This problem has shifted the research foci to the identification of more effective alternatives such as nano-based agrochemicals. The intrusion of nano-interventions in the agri-food sector is being extended to extract the benefits showcased by nanotechnology applications in electronics, pharmaceutical, biomedical, paint, and cosmetics industries. However, the chemically synthesized nanoparticles have a high cost and substantial eco-toxicity (https://www.reportlinker.com/p06193716/Nanotechnology-Services-Global-Market-Report.html?utm_source=GNW, accessed on 20 February 2022). Their rampant use in open field conditions is anticipated to enhance the negative ecological and health hazards, besides the huge input costs. Therefore, green nanoparticles have been speculated to address these issues related to synthetic NPs. Various fungal species may create mycogenic nanoparticles that might stimulate growth and protect crops against diseases in some prospective agricultural uses through antioxidant, antimicrobial, and plant-stimulating properties. The primary benefits of the mycogenic NP synthesis protocol(s) will be the one-pot, cost-effective, and less environmentally corrosive features. Still, the widespread usage of myco-nanoparticles may lead to a few complications. For example, on multiple applications of these NPs, the nano-toxicity aspects will render the plant beneficial microbes vulnerable to the applied nanoparticles, which might compromise their viability as well as the biocontrol and plant growth promotion benefits. Farms and agricultural consumers will soon be able to utilize myconano-functional agrochemicals, pre-harvest and post-harvest crop protection agents, sensors used in genetic equipment, and crop protection components. Additionally, the fungus-derived nanoparticles can function as useful myco-nano-sorbents and would be an attractive way to perform heavy metal biosorption from contaminated wastewater. The employment of *Trichoderma* genus for the synthesis of metallic nanoparticles is ecologically advantageous, time-saving, and cost-effective. Furthermore, potential agrochemicals can be designed by amalgamating NPs and *Trichoderma* strains to produce more sustainable products. The agri-food applications of this technology outlined below will help you understand how the *Trichoderma* genus is used to make nanodevices and how these trichogenic NPs are being used in various applications.

7.1. Antifungal Activity

The effect of metal chitosan nanocomposites at 100 g mL^{-1} in combination with Cu-tolerant *Trichoderma longibrachiatum* strains on cotton seedling damping-off under greenhouse conditions was also investigated. In vitro, the bimetallic blends (BBs) and Cu-chitosan nanocomposite had the best antifungal effectiveness against both *R. solani* anastomosis groups. These findings suggested that BBs, the Cu chitosan nanocomposite, and BBs mixed with *Trichoderma* may inhibit *R. solani*-caused cotton seedling disease in vivo. *R. solani* was evaluated in a greenhouse with a *Trichoderma* strain and shown to have a synergistic inhibitory effect with BBs [83]. The biogenic synthesis of AgNPs was carried out by use of mycelial extracts of *T. harzianum* supplemented in aqueous silver nitrate (1×10^{-3} mol L^{-1}). In a potato dextrose agar-based poison food assay, the use of concentrations ranging from 0.15×10^{12} and 0.31×10^{12} NPs/mL reduced the mycelial development and the generation of new sclerotia in *S. sclerotiorum* [4]. Antifungal application of AgNPs resulted in a considerable decrease in the number of forming colonies for several plant pathogenic fungi, with an efficiency of up to 90% against *Fusarium verticillioides, Fusarium moniliforme, Penicillium brevicompactum, Helminthosporium oryzae,* and *Pyricularia grisea* [49]. Because mycelial development was inhibited and no new sclerotia were formed, both AgNP-TS and AgNP-T nanoparticles showed promise for controlling *S. sclerotiorum*. The most effective inhibition of mycelial development was accomplished with AgNP-TS, which might be attributable to the nanoparticles' reduced hydrodynamic diameter as well as a potential biomolecule effect from the nanoparticles' capping [43]. With 20 g Ag/mL green nanoformulations for increased pathogen fungal mycelium breakdown activity, green AgNPs boost antifungal action to reduce the phytopathogen *Sclerotium rolfsii* producing stem rot in groundnut [55]. In vitro testing of AgNPs produced with *T. virens* HZA14 antifungal efficacy against *S. sclerotiorum* revealed that hyphal development, sclerotial formation, and myceliogenic germination of sclerotia were all inhibited by 100%, 93.8%, and 100%, respectively. The direct interaction between nanoparticles and fungal cells, including AgNPs' contact, accumulation, lamellar fragment creation, and micropore or fissure development on fungal cell walls, was demonstrated using SEM/EDS technologies [57]. When compared to *T. asperellum* alone and carbendazim @0.1%, chitosan nanoparticles in combination with *T. asperellum* were found to be superior in suppressing the mycelial development of soil-borne pathogens such as *F. oxysporum, R. solani,* and *S. rolfsii* [68]. Copper and silica nanoparticles biosynthesized with biocontrol agents suppressed *P. hypolateritia* and *P. theae* growth, suggesting that this might be a unique way of managing diseases that affect tea plantations while also improving tea quality parameters. Various nanoformulations were prepared using suitably inert and eco-friendly carrier materials employing nano copper and nanosilica [63]. In the laboratory and greenhouse, *T. harizinum*-mediated ZnONPs were demonstrated, for the first time, to exhibit fungicidal activity against three soil–cotton pathogenic fungi (*Fusarium* sp., *R. solani,* and *Macrophomina phaseolina*) [3].

Pseudomonas fluorescens and *T. viride,* two distinct biocontrol agents, were used to make microbial CuONPs, which were then evaluated using various analytical methods. In terms of in vitro antifungal efficacies, CuONPs synthesized from *T. viride* showed the highest percent growth inhibition compared to CuONPs generated from *P. fluorescens.* CuONPs produced from *T. viride,* on the other hand, demonstrated considerably stronger antifungal activity in vivo than the commonly used Bordeaux mixture [84]. Antifungal activity against three soil-borne pathogens was proven in vitro and in the greenhouse. In the three fungal infections, AgNPs greatly reduced hyphal growth. The ability of *T. harzianum* isolates to synthesize a wide spectrum of proteins and enzymes without the need for chemical reducers and stabilizers has been proven. Biosynthesized AgNPs have shown high potential in protecting cotton plants from the fungal invasion induced by damping-off [14]. *Trichoderma* species have a wide range of biotechnological uses, including acting as biofungicides to manage various plant diseases, biofertilizers to promote plant development, and synthesizing and bioremediating metal nanoparticles. Possible applications of mycogenic NPs generated by the *Trichoderma* genus are demonstrated in Figure 4.

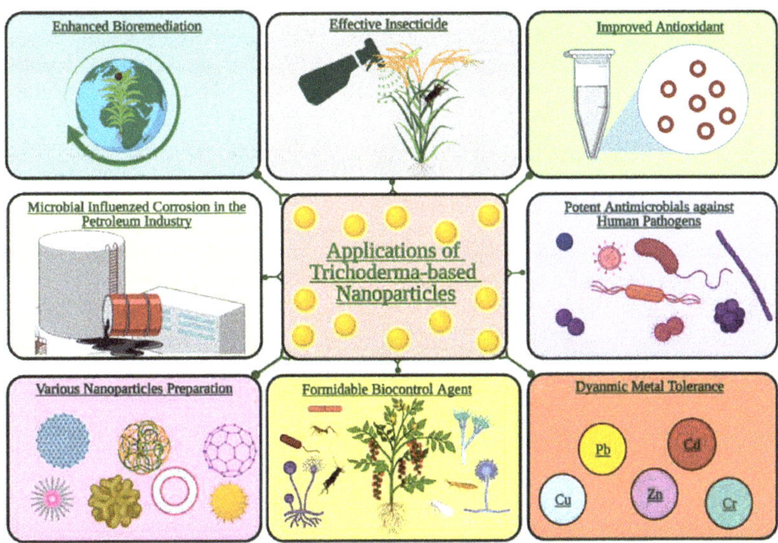

Figure 4. Various potential applications of *Trichoderma*-mediated nanoparticles in agroecosystems. The present figure was created by BioRender.com.

7.2. Antibacterial Activity

The antibacterial activity of green nano-biosilver *T. harzianum* AgNPs was demonstrated against *S. aureus* and *K. aeruginosa*, as well as Gram-positive and Gram-negative bacteria, with the Gram-negative bacterium (*K. pneumoniae*) displaying better sensitivity. *T. atroviride*'s AgNPs were found to have antibacterial action against Gram-positive and Gram-negative clinical pathogens such as *E. coli*, *P. aeruginosa*, and *S. aureus* [46]. AgNPs had stronger antibacterial activity than $AgNO_3$ and were similar to the positive control kanamycin. They also demonstrated DPPH scavenging action in a dose-dependent manner, with an IC50 of 45.6 g/mL. The cell-free filtrate, on the other hand, showed no inhibition against the pathogens, probably due to metabolites with the antibiotic property. As a result, the biosynthesized AgNPs can be employed as a natural antioxidant to prevent human cell damage and degenerative diseases by regulating antioxidants, pro-oxidants, and ROS levels [50]. The antibacterial activity of AgNP was evaluated against a variety of bacteria types. Gram-negative (*E. coli* and *P. aeruginosa*) bacteria had lower MBC and MIC values than Gram-positive (*S. aureus* and *E. faecalis*) bacteria, and this difference was ascribed to the bacterial cell wall structure. This suggests that Gram-negative bacteria have better antibacterial activity due to bacterial surface adsorption and oxidative stress induction; nevertheless, more research is required [56]. Gram-positive (*S. aureus* and *B. subtilis*) and Gram-negative (*E. coli* and *R. solanacearum*) pathogenic bacteria were tested, and the AgNPs synthesized using *T. harzianum* filtrate were reported to have significant antioxidant properties and antibacterial activity against both Gram-positive (*S. aureus* and *B. subtilis*) and Gram-negative (*E. coli* and *R. solanacearum*) bacteria, with higher activity against the Gram-negative bacteria [15]. CuNPs were effectively produced using *T. harzianum*, an agriculturally beneficial fungus, in a simple green and environmentally favorable way. CuNPs and *T. harzianum* fungus were shown to have antibacterial action against both Gram-positive and Gram-negative bacteria [62]. The antibacterial activity of the generated AgNPs against *Escherichia coli* was investigated [85].

7.3. Plant Growth Promotion

In greenhouse settings, *Trichoderma*-mediated AgNPs were evaluated against tomato wilt caused by *Fusarium* species. The treated plants with various doses of AgNPs showed a promoting effect on all the tested parameters in comparison with the control and *Trichoderma* formulation [86]. With the prolonged soaking time of silver nanoparticles solution, *T. harzianum*-synthesized AgNPs demonstrated an increase in the percentage of seed germination. AgNPs produced by *T. harzianum* had a positive effect on oilseed germination. As a result, mycogenic AgNPs have a biological assay for increasing seed viability in agriculture [87]. The biocontrol agent *T. harzianum* was used as a stabilizing agent in the green manufacture of biogenic iron oxide nanoparticles. The nanoparticles' antifungal effectiveness against *S. sclerotiorum* (white mold) was tested in vitro. The impact of the NPs on seed germination was also examined. They were also able to promote the proliferation of *Trichoderma*, which inhibited the establishment of the pathogen *S. sclerotiorum* while having no effect on seed germination [71].

7.4. Trichoderma-Based Nanobioremediation

Fungi-based nanotechnology is quickly evolving as an effective technology to treat industrial wastewaters. *T. harzianum*-derived CdS-NPs were used to photocatalyze the breakdown of methylene blue in a photocatalytic cell [88]. The production of electron (e) and hole (h+) pairs, which function as powerful oxidants and reductants, caused the degradation. Adsorbed water on the CdS-NPs traps the hole (h^+), resulting in the formation of a hydroxyl radical, while oxygen takes electrons from the conduction band, resulting in the formation of an anion radical. These radicals target the dye's azo bond, causing it to degrade and produce CO_2, H_2O, and NH_4^+. Due to an effective biocatalyst that converted 4-nitrophenol to 4-aminophenol in the presence of $NaBH_4$ and the capacity to inhibit pathogenic bacteria, biosynthesized gold nanoparticles present new hope for green bioremediation [67]. The test human pathogenic bacteria were strongly suppressed by biogenic AgNPs made from the fungus *T. viride* [48]. Because of their biocidal action against halotolerant planktonic sulfate-reducing bacteria, mycosynthesized AgNPs have become an appealing alternative for controlling microbially driven corrosion in the petroleum industry [53]. *T. harzianum*-biosynthesized SiO_2 NPs were effective as a lead adsorbent from water during the bioremediation process. The content of lead in water and the muscles of Nile tilapia (*Oreochromis niloticus*) decreased. *O. niloticus'* immune system, liver, and renal functions improved [69]. The *Trichoderma* genus can be employed as a possible metal biosorbent and as a powerful bioremediation agent. Some *Trichoderma* strains have high metal tolerance and bioaccumulation ability, making them a viable mycoremediation agent of heavy metal contamination in environmental safety. Figure 5 illustrates the pathways of tolerance to metals and nanoparticles (NPs) in *Trichoderma*.

7.5. Miscellaneous Advantages and Disadvantages of Trichogenic Nanoparticles

The biogenic AgNPs synthesized from the extracellular filtrate of *T. harzianum* have an additional benefit over physical and chemical approaches that need high pressure, energy, chemical precursors, and a high cost. This filamentous fungus has a larger capacity for binding Ag^+ Ag^0, making this procedure easier and less expensive for the large-scale manufacturing of NPs at the industrial level, making it cost-effective [47]. *T. viride's* capacity to generate AgNPs is extremely promising for green, sustainable nanomaterial synthesis [48]. The nanoparticles demonstrated little size fluctuation as well as high physicochemical stability. As the beans were exposed to biogenic silver nanoparticles, there were no significant changes in germination and seedling development compared to the negative control [4]. *T. longibrachiatum* showed a possibility for the extracellular and reliable biosynthesis of silver nanoparticles. The AgNPs synthesized using this biosystem technique were generally stable up to two months after synthesis [49], and, even after six months of storage, nanocrystalline silver particles created by *T. asperellum* did not display significant aggregation [15,44]. The use of fungi in the biogenic synthesis of silver nanoparticles has several benefits, including the development of a capping agent from fungal biomolecules, which provides stability and can assist in biological activity. The cytotoxicity and genotoxicity of AgNP-TS and AgNP-T nanoparticles to V79, 3T3, and

HaCat cell lines were both low [43]. When compared to normal pesticides, the use of nano-based formulations will be a preferable option for avoiding excess chemicals in soil [45]. *T. reesei* possesses simpler and less expensive culture conditions, as well as greater growth rates on both industrial and laboratory sizes, resulting in lower costs in large-scale manufacturing [39]. *T. atroviride*-based biosynthesized AgNPs are an economically effective and ecologically sustainable way of producing AgNPs [50]. The small-sized nanoparticles generated, their narrow size distribution, the stability attained without the use of chemicals as capping agents, and the low number of chemical reagents used are the key benefits of the described extracellular biosynthesis of AgNPs. Because of the negative zeta potential that might aid the attachment of physiologically active compounds, the generated AgNPs were appropriate carriers of such compounds [52,54]. The use of green AgNPs as an antifungal agent is seen as an environmentally friendly resource, an alternative to fungicides, and a cost-effective method [55]. The surface characteristics of the biogenic chitosan nanoparticles generated from *T. viride* were discovered to be positively charged and stable [68].

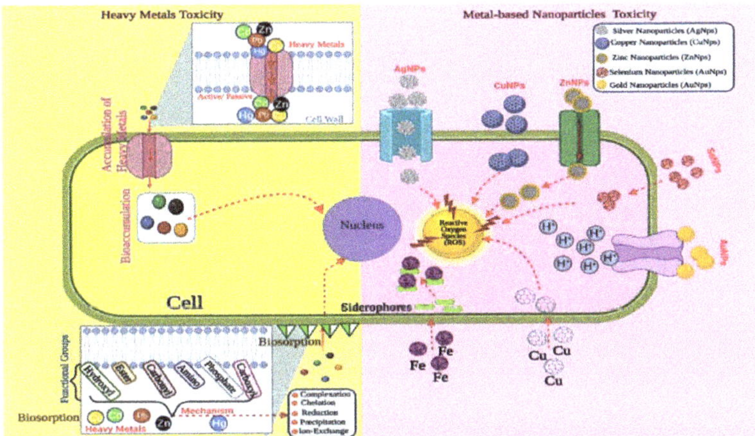

Figure 5. The adsorption and absorption mechanisms in the cell walls caused by the presence of functional groups, proteins, or compounds that serve as chelating agents, as well as the accumulation of these in the vacuoles, are illustrated schematically in the *Trichoderma* tolerance mechanisms to micro- and nano-metals. Mechanisms involving antioxidant enzyme activity that reduces the damage produced by reactive oxygen species may also be present. The arrows indicate movement of the metal ions or nanoparticles affecting specific organelle in the cell. The present figure was created by BioRender.com.

However, there are several drawbacks that must be resolved in order to properly employ fungus for biogenic synthesis. The engagement of the host organism for protein expression, immature synthesis techniques, and limited know-how on large-scale production technique(s) may possibly limit the benefits associated with fungus-based nanoparticle synthesis. The most critical factors that may affect the final product include knowing which fungus to employ, its growth parameters, the necessity for sterile settings, and the time it takes for the fungus to develop and finish the synthesis. Scaling-up can also create challenges, such as the need for research focusing on the mechanism associated with the development of capping layers and effective nucleation [68]. All these challenges associated with the green synthesis of various metal NPs through the use of different types of fungal biomass should be solved, and the feasibility of green synthesized nanoparticles for obtaining useful commercial applications in the agri-food industry should be identified.

8. Challenges

Mycogenic nanoparticles can be synthesized with a wide range of particle size distributions. There are issues with reproducibility as well as a lack of control over particle size, shape, and dispersion. As a result, in future efforts, the parameters of synthesis should be thoroughly investigated [89]. The development of innovative nanofungicides, nanofertilizers, and nanosorbents generated by fungi and the evaluation of these nanoformulations for the remediation of heavy metal-contaminated industrial wastewater are still in their early phases. Several problems must be overcome to scale up mycogenic nanoparticles and fulfill the demands of the real world. For example, a new nanosorbent created by diverse microorganisms must be economically advantageous and socially acceptable while also being able to meet various water quality regulations to ensure human health and environmental safety [90]. Scaling up the method is difficult since the commercial synthesis of nanoparticles still needs more study. No NP residue must be discharged into the environment as a result of the treatment method. For example, the environmental tracking of silver in the field is critical for evaluating its influence on the environment and human health. Furthermore, further study should be conducted to assess the hazardous effect of AgNPs before widespread manufacture and usage in agricultural applications [49]. The toxicity of NPs remains a key hurdle in transferring these materials from lab to industry for the time being, and this has to be addressed further.

9. Future Trends

The biogenic silver nanoparticles exhibited no deleterious effects on *T. harzianum* culture compared to the negative control. This shows that combining *T. harzianum* with nanoparticles would not affect the fungus' development [4]. It is hoped that techniques involving the biogenic synthesis of metallic nanoparticles will be developed, taking into account not only the potential biological activity of the metal nanoparticles but also the biomolecules and organic compounds obtained from the organisms used in the synthesis, which compose the capping on the nanoparticles [43]. In order to have better control over the size and polydispersity of these nanoparticles, more research is needed to understand the specific chemical pathway leading to their production using biological means. More studies should be carried out on the creation of AgNPs-based compounds and their combination with fungicides [49]. Extracellular biosynthesized AgNPs might be used as carriers for different agrochemicals and natural bioactive molecules. There is a greater need for field trials of *Trichoderma*-based nanoparticles to understand nanoparticle–plant–microbe interactions and toxicity issues to develop safer and user-friendly agrochemicals.

10. Conclusions

Based on the literature review, we conclude on whether developing agrochemicals that employ *Trichoderma* strains to produce green NPs could be more sustainable agrochemicals. Given that the *Trichoderma* genus is one of the most often used fungi for the mycosynthesis of NPs, it is possible that it possesses tolerance mechanisms to these structures that might be used in combination to generate products that boost agricultural productivity while simultaneously treating plant infections in the field. The synthesis of NPs by beneficial fungi such as *Trichoderma* is novel, cost-effective, and eco-friendly compared to the synthesis by chemical and physical methods. However, more study is required to assess the *Trichoderma* and heavy metals interaction in a heterogeneous system under field circumstances. Additionally, to comprehend the potential advantages of nanoparticles produced by beneficial microbes in the agricultural sector, it is necessary to comprehend nanoparticle penetration and transport routes in plants. The shape and size of the nanoparticles have a significant impact on their reactivity, stability, and behavior. More effective agrochemical applications using composite materials are desirable because these can decrease potential economic losses and, in particular, environmental damage. For example, combining the antimicrobial impact of mycogenic nanoparticles with biofungicides for plant disease control may boost antifungal efficacy through synergistic interaction, allowing for a reduced fungicide dose and, as a result, avoiding the development of fungal pathogen resistance.

Author Contributions: K.A.A.-E., M.A.A. and H.M.A. conceptualized, reviewed, edited, and acquired funding. A.K. and M.A.A. performed the investigations. A.K. and M.R. (Mahendra Rai) also acquired the resources. M.A.A. curated the data, H.M.A., M.R. (Mythili Ravichandran) and F.K.A. wrote the first draft and edited it. M.A.A. acquired the funding. All authors have read and agreed to the published version of the manuscript.

Funding: The authors would like to thank the Deanship of Scientific Research at Shaqra University for supporting this work.

Institutional Review Board Statement: Not applicable.

Informed Consent Statement: Not applicable.

Data Availability Statement: Not applicable.

Acknowledgments: The authors would like to thank the Deanship of Scientific Research at Shaqra University for supporting this work.

Conflicts of Interest: The authors declare no conflict of interest.

References

1. Alghuthaymi, M.A.; Rajkuberan, C.; Rajiv, P.; Kalia, A.; Bhardwaj, K.; Bhardwaj, P.; Abd-Elsalam, K.A.; Valis, M.; Kuca, K. Nanohybrid antifungals for control of plant diseases: Current status and future perspectives. *J. Fungi* **2021**, *7*, 48. [CrossRef]
2. Hashem, A.H.; Abdelaziz, A.M.; Askar, A.A.; Fouda, H.M.; Khalil, A.M.A.; Abd-Elsalam, K.A.; Khaleil, M.M. *Bacillus megaterium*-Mediated Synthesis of Selenium Nanoparticles and Their Antifungal Activity against *Rhizoctonia solani* in Faba Bean Plants. *J. Fungi* **2021**, *7*, 195. [CrossRef] [PubMed]
3. Zaki, S.A.; Ouf, S.A.; Albarakaty, F.M.; Habeb, M.M.; Aly, A.A.; Abd-Elsalam, K.A. *Trichoderma harzianum*-Mediated ZnO Nanoparticles: A Green Tool for Controlling Soil-Borne Pathogens in Cotton. *J. Fungi* **2021**, *7*, 952. [CrossRef]
4. Guilger, M.; Pasquoto-Stigliani, T.; Bilesky-Jose, N.; Grillo, R.; Abhilash, P.C.; Fraceto, L.F.; De Lima, R. Biogenic silver nanoparticles based on *Trichoderma harzianum*: Synthesis, characterization, toxicity evaluation and biological activity. *Sci. Rep.* **2017**, *7*, 44421. [CrossRef] [PubMed]
5. Mahawar, H.; Prasanna, R. Prospecting the interactions of nanoparticles with beneficial microorganisms for developing green technologies for agriculture. *Environ. Nanotechnol. Monit. Manag.* **2018**, *10*, 477–485. [CrossRef]
6. Chauhan, A.; Anand, J.; Parkash, V.; Rai, N. Biogenic synthesis: A sustainable approach for nanoparticles synthesis mediated by fungi. *Inorg. Nano-Metal Chem.* **2022**, 1–14. [CrossRef]
7. Misra, M.; Chattopadhyay, S.; Sachan, A.; Sachan, S.G. Microbially synthesized nanoparticles and their applications in environmental clean-up. *Environ. Technol. Rev.* **2022**, *11*, 18–32. [CrossRef]
8. Khan, R.A.A.; Najeeb, S.; Mao, Z.; Ling, J.; Yang, Y.; Li, Y.; Xie, B. Bioactive Secondary Metabolites from *Trichoderma* spp. against Phytopathogenic Bacteria and Root-Knot Nematode. *Microorganisms* **2020**, *8*, 401. [CrossRef] [PubMed]
9. Zin, N.A.; Badaluddin, N.A. Biological functions of *Trichoderma* spp. for agriculture applications. *Ann. Agric. Sci.* **2020**, *65*, 168–178. [CrossRef]
10. El-Rahman, A.A.E.-M.A.; El-Shafei, S.M.A.E.-A.; Ivanova, E.V.; Fattakhova, A.N.; Pankova, A.V.; El-Shafei, M.A.E.-A.; El-Morsi, E.-M.A.E.-F.; Alimova, F.K. Cytotoxicity of *Trichoderma* spp. Cultural Filtrate Against Human Cervical and Breast Cancer Cell Lines. *Asian Pac. J. Cancer Prev.* **2014**, *15*, 7229–7234. [CrossRef] [PubMed]
11. Singh, R.P.; Handa, R.; Manchanda, G. Nanoparticles in sustainable agriculture: An emerging opportunity. *J. Control. Release* **2021**, *329*, 1234–1248. [CrossRef] [PubMed]
12. Elegbede, J.A.; Lateef, A.; Azeez, M.A.; Asafa, T.B.; Yekeen, T.A.; Oladipo, I.C.; Aina, D.A.; Beukes, L.S.; Gueguim-Kana, E.B. Biofabrication of Gold Nanoparticles Using Xylanases Through Valorization of Corncob by *Aspergillus niger* and *Trichoderma longibrachiatum*: Antimicrobial, Antioxidant, Anticoagulant and Thrombolytic Activities. *Waste Biomass Valoriz.* **2020**, *11*, 781–791. [CrossRef]
13. Ramírez-Valdespino, C.A.; Orrantia-Borunda, E. *Trichoderma* and Nanotechnology in Sustainable Agriculture: A Review. *Front. Fungal Biol.* **2021**, *2*. [CrossRef]
14. Zaki, S.A.; Ouf, S.A.; Abd-Elsalam, K.A.; Asran, A.A.; Hassan, M.M.; Kalia, A.; Albarakaty, F.M. Trichogenic Silver-Based Nanoparticles for Suppression of Fungi Involved in Damping-Off of Cotton Seedlings. *Microorganisms* **2022**, *10*, 344. [CrossRef] [PubMed]
15. Konappa, N.; Udayashankar, A.C.; Dhamodaran, N.; Krishnamurthy, S.; Jagannath, S.; Uzma, F.; Pradeep, C.K.; De Britto, S.; Chowdappa, S.; Jogaiah, S. Ameliorated antibacterial and antioxidant properties by *Trichoderma harzianum* mediated green synthesis of silver nanoparticles. *Biomolecules* **2021**, *11*, 535. [CrossRef]

16. Madbouly, A.K. Biodiversity of Genus *Trichoderma* and Their Potential Applications. In *Industrially Important Fungi for Sustainable Development. Fungal Biology*; Abdel-Azeem, A.M., Yadav, A.N., Yadav, N., Usmani, Z., Eds.; Springer: Cham, Switzerland, 2021; pp. 429–460.
17. Bitas, V.; Kim, H.-S.; Bennett, J.W.; Kang, S. Sniffing on Microbes: Diverse Roles of Microbial Volatile Organic Compounds in Plant Health. *Mol. Plant-Microbe Interact.* **2013**, *26*, 835–843. [CrossRef] [PubMed]
18. Contreras-Cornejo, H.A.; Macías-Rodríguez, L.; Del-Val, E.; Larsen, J. Ecological functions of *Trichoderma* spp. and their secondary metabolites in the rhizosphere: Interactions with plants. *FEMS Microbiol. Ecol.* **2016**, *92*, fiw036. [CrossRef] [PubMed]
19. Gupta, R.; Bar, M. Plant Immunity, Priming, and Systemic Resistance as Mechanisms for *Trichoderma* spp. Biocontrol. In *Trichoderma*; Sharma, A., Sharma, P., Eds.; Springer: Singapore, 2020; pp. 81–110.
20. Hermosa, R.; Viterbo, A.; Chet, I.; Monte, E. Plant-beneficial effects of *Trichoderma* and of its genes. *Microbiology* **2012**, *158*, 17–25. [CrossRef] [PubMed]
21. López-Bucio, J.; Pelagio-Flores, R.; Herrera-Estrella, A. *Trichoderma* as biostimulant: Exploiting the multilevel properties of a plant beneficial fungus. *Sci. Hortic.* **2015**, *196*, 109–123. [CrossRef]
22. Nakkeeran, S.; Rajamanickam, S.; Vanthana, M.; Renukadevi, P.; Muthamilan, M. Harnessing the Perception of *Trichoderma* Signal Molecules in Rhizosphere to Improve Soil Health and Plant Health. In *Trichoderma*; Springer: Singapore, 2020; pp. 61–79.
23. Chakraborty, B.N.; Chakraborty, U.; Sunar, K. Induced Immunity Developed by *Trichoderma* species in Plants. In *Trichoderma*; Springer: Singapore, 2020; pp. 125–147.
24. Woo, S.L.; Ruocco, M.; Vinale, F.; Nigro, M.; Marra, R.; Lombardi, N.; Pascale, A.; Lanzuise, S.; Manganiello, G.; Lorito, M. *Trichoderma*-based Products and their Widespread Use in Agriculture. *Open Mycol. J.* **2014**, *8*, 71–126. [CrossRef]
25. Meher, J.; Rajput, R.S.; Bajpai, R.; Teli, B.; Sarma, B.K. *Trichoderma*: A Globally Dominant Commercial Biofungicide. In *Trichoderma: Agricultural Applications and Beyond*; Springer: Cham, Switzerland, 2020; pp. 195–208.
26. Ram, R.M.; Vaishnav, A.; Singh, H.B. *Trichoderma* spp.: Expanding Potential beyond Agriculture. In *Trichoderma: Agricultural Applications and Beyond*; Springer: Cham, Switzerland, 2020; pp. 351–367.
27. Weber, R.W.S.; Ridderbusch, D.C.; Anke, H. 2,4,6-Trinitrotoluene (TNT) tolerance and biotransformation potential of microfungi isolated from TNT-contaminated soil. *Mycol. Res.* **2002**, *106*, 336–344. [CrossRef]
28. Nazifa, T.H.; Bin Ahmad, M.A.; Hadibarata, T.; Salmiati; Aris, A. Bioremediation of diesel oil spill by filamentous fungus *Trichoderma reesei* H002 in aquatic environment. *Int. J. Integr. Eng.* **2018**, *10*, 14–19. [CrossRef]
29. Elshafie, H.S.; Camele, I.; Sofo, A.; Mazzone, G.; Caivano, M.; Masi, S.; Caniani, D. Mycoremediation effect of *Trichoderma harzianum* strain T22 combined with ozonation in diesel-contaminated sand. *Chemosphere* **2020**, *252*, 126597. [CrossRef] [PubMed]
30. Kumar, V.; Dwivedi, S.K. Bioremediation mechanism and potential of copper by actively growing fungus *Trichoderma lixii* CR700 isolated from electroplating wastewater. *J. Environ. Manag.* **2021**, *277*, 111370. [CrossRef] [PubMed]
31. Zhang, F.; Xu, X.; Huo, Y.; Xiao, Y. *Trichoderma*-Inoculation and Mowing Synergistically Altered Soil Available Nutrients, Rhizosphere Chemical Compounds and Soil Microbial Community, Potentially Driving Alfalfa Growth. *Front. Microbiol.* **2019**, *9*, 3241. [CrossRef] [PubMed]
32. Molla, A.H.; Manjurul Haque, M.; Amdadul Haque, M.; Ilias, G.N.M. *Trichoderma*-Enriched Biofertilizer Enhances Production and Nutritional Quality of Tomato (*Lycopersicon esculentum* Mill.) and Minimizes NPK Fertilizer Use. *Agric. Res.* **2012**, *1*, 265–272. [CrossRef]
33. Singh, U.B.; Malviya, D.; Wasiullah; Singh, S.; Pradhan, J.K.; Singh, B.P.; Roy, M.; Imram, M.; Pathak, N.; Baisyal, B.M.; et al. Bio-protective microbial agents from rhizosphere eco-systems trigger plant defense responses provide protection against sheath blight disease in rice (*Oryza sativa* L.). *Microbiol. Res.* **2016**, *192*, 300–312. [CrossRef] [PubMed]
34. Willian, G.A. Biodegradation of Chlorpyrifos by Whole Cells of Marine-Derived Fungi *Aspergillus sydowii* and *Trichoderma* sp. *J. Microb. Biochem. Technol.* **2015**, *7*, 133–139. [CrossRef]
35. Podbielska, M.; Kus-Liśkiewicz, M.; Jagusztyn, B.; Piechowicz, B.; Sadło, S.; Słowik-Borowiec, M.; Twarużek, M.; Szpyrka, E. Influence of Bacillus subtilis and *Trichoderma harzianum* on Penthiopyrad Degradation under Laboratory and Field Studies. *Molecules* **2020**, *25*, 1421. [CrossRef]
36. Sun, J.; Yuan, X.; Li, Y.; Wang, X.; Chen, J. The pathway of 2,2-dichlorovinyl dimethyl phosphate (DDVP) degradation by *Trichoderma atroviride* strain T23 and characterization of a paraoxonase-like enzyme. *Appl. Microbiol. Biotechnol.* **2019**, *103*, 8947–8962. [CrossRef]
37. Wu, Q.; Ni, M.; Wang, G.; Liu, Q.; Yu, M.; Tang, J. Omics for understanding the tolerant mechanism of *Trichoderma asperellum* TJ01 to organophosphorus pesticide dichlorvos. *BMC Genom.* **2018**, *19*, 596. [CrossRef]
38. Salwan, R.; Rialch, N.; Sharma, V. Bioactive Volatile Metabolites of *Trichoderma*: An overview. In *Secondary Metabolites of Plant Growth Promoting Rhizomicroorganisms*; Springer: Singapore, 2019; pp. 87–111.
39. Vahabi, K.; Mansoori, G.A.; Karimi, S. Biosynthesis of Silver Nanoparticles by Fungus *Trichoderma reesei* (A Route for Large-Scale Production of AgNPs). *Insciences J.* **2011**, *1*, 65–79. [CrossRef]
40. Abd-Elsalam, K.A. Special Issue: Fungal Nanotechnology. *J. Fungi* **2021**, *7*, 583. [CrossRef] [PubMed]
41. Zhao, X.; Zhou, L.; Riaz Rajoka, M.S.; Yan, L.; Jiang, C.; Shao, D.; Zhu, J.; Shi, J.; Huang, Q.; Yang, H.; et al. Fungal silver nanoparticles: Synthesis, application and challenges. *Crit. Rev. Biotechnol.* **2018**, *38*, 817–835. [CrossRef]
42. Ayad, F.; Matallah-Boutiba, A.; Rouane–Hacene, O.; Bouderbala, M.; Boutiba, Z. Tolerance of *Trichoderma* sp. to Heavy Metals and its Antifungal Activity in Algerian Marine Environment. *J. Pure Appl. Microbiol.* **2018**, *12*, 855–870. [CrossRef]

43. Guilger-Casagrande, M.; Lima, R. de Synthesis of Silver Nanoparticles Mediated by Fungi: A Review. *Front. Bioeng. Biotechnol.* **2019**, *7*, 287. [CrossRef] [PubMed]
44. Mukherjee, P.; Roy, M.; Mandal, B.P.; Dey, G.K.; Mukherjee, P.K.; Ghatak, J.; Tyagi, A.K.; Kale, S.P. Green synthesis of highly stabilized nanocrystalline silver particles by a non-pathogenic and agriculturally important fungus *T. asperellum*. *Nanotechnology* **2008**, *19*, 075103. [CrossRef] [PubMed]
45. Prameela Devi, T.; Kulanthaivel, S.; Kamil, D.; Borah, J.L.; Prabhakaran, N.; Srinivasa, N. Biosynthesis of silver nanoparticles from *Trichoderma* species. *Indian J. Exp. Biol.* **2013**, *51*, 543–547.
46. Ahluwalia, V.; Kumar, J.; Sisodia, R.; Shakil, N.A.; Walia, S. Green synthesis of silver nanoparticles by *Trichoderma harzianum* and their bio-efficacy evaluation against *Staphylococcus aureus* and *Klebsiella pneumonia*. *Ind. Crops Prod.* **2014**, *55*, 202–206. [CrossRef]
47. Sundaravadivelan, C.; Padmanabhan, M.N. Effect of mycosynthesized silver nanoparticles from filtrate of *Trichoderma harzianum* against larvae and pupa of dengue vector *Aedes aegypti* L. *Environ. Sci. Pollut. Res.* **2014**, *21*, 4624–4633. [CrossRef] [PubMed]
48. Elgorban, A.M.; Al-Rahmah, A.N.; Sayed, S.R.; Hirad, A.; Mostafa, A.A.F.; Bahkali, A.H. Antimicrobial activity and green synthesis of silver nanoparticles using *Trichoderma viride*. *Biotechnol. Biotechnol. Equip.* **2016**, *30*, 299–304. [CrossRef]
49. Elamawi, R.M.; Al-Harbi, R.E.; Hendi, A.A. Biosynthesis and characterization of silver nanoparticles using *Trichoderma longibrachiatum* and their effect on phytopathogenic fungi. *Egypt. J. Biol. Pest Control* **2018**, *28*, 28. [CrossRef]
50. Saravanakumar, K.; Wang, M.-H. *Trichoderma* based synthesis of anti-pathogenic silver nanoparticles and their characterization, antioxidant and cytotoxicity properties. *Microb. Pathog.* **2018**, *114*, 269–273. [CrossRef] [PubMed]
51. Adebayo-Tayo, B.; Ogunleye, G.; Ogbole, O. Biomedical application of greenly synthesized silver nanoparticles using the filtrate of *Trichoderma viride*: Anticancer and immunomodulatory potentials. *Polym. Med.* **2020**, *49*, 57–62. [CrossRef] [PubMed]
52. Gemishev, O.T.; Panayotova, M.I.; Mintcheva, N.N.; Djerahov, L.P.; Tyuliev, G.T.; Gicheva, G.D. A green approach for silver nanoparticles preparation by cell-free extract from *Trichoderma reesei* fungi and their characterization. *Mater. Res. Express* **2019**, *6*, 095040. [CrossRef]
53. Omran, B.A.; Nassar, H.N.; Younis, S.A.; Fatthallah, N.A.; Hamdy, A.; El-Shatoury, E.H.; El-Gendy, N.S. Physiochemical properties of *Trichoderma longibrachiatum* DSMZ 16517-synthesized silver nanoparticles for the mitigation of halotolerant sulphate-reducing bacteria. *J. Appl. Microbiol.* **2019**, *126*, 138–154. [CrossRef] [PubMed]
54. Abdel-Azeem, A.; Nada, A.A.; O'Donovan, A.; Kumar Thakur, V.; Elkelish, A. Mycogenic Silver Nanoparticles from Endophytic *Trichoderma atroviride* with Antimicrobial Activity. *J. Renew. Mater.* **2020**, *8*, 171–185. [CrossRef]
55. Hirpara, D.G.; Gajera, H.P. Green synthesis and antifungal mechanism of silver nanoparticles derived from chitin- induced exometabolites of *Trichoderma interfusant*. *Appl. Organomet. Chem.* **2020**, *34*, e5407. [CrossRef]
56. Ramos, M.M.; dos Morais, E.S.; da Sena, I.S.; Lima, A.L.; de Oliveira, F.R.; de Freitas, C.M.; Fernandes, C.P.; de Carvalho, J.C.T.; Ferreira, I.M. Silver nanoparticle from whole cells of the fungi *Trichoderma* spp. isolated from Brazilian Amazon. *Biotechnol. Lett.* **2020**, *42*, 833–843. [CrossRef] [PubMed]
57. Tomah, A.A.; Alamer, I.S.A.; Li, B.; Zhang, J.-Z. Mycosynthesis of Silver Nanoparticles Using Screened *Trichoderma* Isolates and Their Antifungal Activity against *Sclerotinia sclerotiorum*. *Nanomaterials* **2020**, *10*, 1955. [CrossRef] [PubMed]
58. Saravanakumar, K.; Jeevithan, E.; Hu, X.; Chelliah, R.; Oh, D.-H.; Wang, M.-H. Enhanced anti-lung carcinoma and anti-biofilm activity of fungal molecules mediated biogenic zinc oxide nanoparticles conjugated with β-D-glucan from barley. *J. Photochem. Photobiol. B Biol.* **2020**, *203*, 111728. [CrossRef] [PubMed]
59. Shobha, B.; Lakshmeesha, T.R.; Ansari, M.A.; Almatroudi, A.; Alzohairy, M.A.; Basavaraju, S.; Alurappa, R.; Niranjana, S.R.; Chowdappa, S. Mycosynthesis of ZnO Nanoparticles Using *Trichoderma* spp. Isolated from Rhizosphere Soils and Its Synergistic Antibacterial Effect against *Xanthomonas oryzae* pv. *oryzae*. *J. Fungi* **2020**, *6*, 181. [CrossRef] [PubMed]
60. Saravanakumar, K.; Shanmugam, S.; Varukattu, N.B.; MubarakAli, D.; Kathiresan, K.; Wang, M.-H. Biosynthesis and characterization of copper oxide nanoparticles from indigenous fungi and its effect of photothermolysis on human lung carcinoma. *J. Photochem. Photobiol. B Biol.* **2019**, *190*, 103–109. [CrossRef] [PubMed]
61. Consolo, V.F.; Torres-Nicolini, A.; Alvarez, V.A. Mycosinthetized Ag, CuO and ZnO nanoparticles from a promising *Trichoderma harzianum* strain and their antifungal potential against important phytopathogens. *Sci. Rep.* **2020**, *10*, 20499. [CrossRef]
62. Ahmadi-Nouraldinvand, F.; Afrouz, M.; Elias, S.G.; Eslamian, S. Green synthesis of copper nanoparticles extracted from guar seedling under Cu heavy-metal stress by *Trichoderma harzianum* and their bio-efficacy evaluation against Staphylococcus aureus and Escherichia coli. *Environ. Earth Sci.* **2022**, *81*, 54. [CrossRef]
63. Natesan, K.; Ponmurugan, P.; Gnanamangai, B.M.; Manigandan, V.; Joy, S.P.J.; Jayakumar, C.; Amsaveni, G. Biosynthesis of silica and copper nanoparticles from *Trichoderma, Streptomyces* and *Pseudomonas* spp. evaluated against collar canker and red root-rot disease of tea plants. *Arch. Phytopathol. Plant Prot.* **2021**, *54*, 56–85. [CrossRef]
64. Nandini, B.; Hariprasad, P.; Prakash, H.S.; Shetty, H.S.; Geetha, N. Trichogenic-selenium nanoparticles enhance disease suppressive ability of *Trichoderma* against downy mildew disease caused by *Sclerospora graminicola* in pearl millet. *Sci. Rep.* **2017**, *7*, 2612. [CrossRef]
65. Hu, D.; Yu, S.; Yu, D.; Liu, N.; Tang, Y.; Fan, Y.; Wang, C.; Wu, A. Biogenic *Trichoderma harzianum*-derived selenium nanoparticles with control functionalities originating from diverse recognition metabolites against phytopathogens and mycotoxins. *Food Control* **2019**, *106*, 106748. [CrossRef]
66. Diko, C.S.; Zhang, H.; Lian, S.; Fan, S.; Li, Z.; Qu, Y. Optimal synthesis conditions and characterization of selenium nanoparticles in *Trichoderma* sp. WL-Go culture broth. *Mater. Chem. Phys.* **2020**, *246*, 122583. [CrossRef]

67. Mishra, A.; Kumari, M.; Pandey, S.; Chaudhry, V.; Gupta, K.C.; Nautiyal, C.S. Biocatalytic and antimicrobial activities of gold nanoparticles synthesized by *Trichoderma* sp. *Bioresour. Technol.* **2014**, *166*, 235–242. [CrossRef]
68. Boruah, S.; Dutta, P. Fungus mediated biogenic synthesis and characterization of chitosan nanoparticles and its combine effect with *Trichoderma asperellum against Fusarium oxysporum, Sclerotium rolfsii* and *Rhizoctonia solani. Indian Phytopathol.* **2021**, *74*, 81–93. [CrossRef]
69. El-Gazzar, N.; Almanaa, T.N.; Reda, R.M.; El Gaafary, M.N.; Rashwan, A.A.; Mahsoub, F. Assessment the using of silica nanoparticles (SiO_2NPs) biosynthesized from rice husks by *Trichoderma harzianum* MF780864 as water lead adsorbent for immune status of Nile tilapia (*Oreochromis niloticus*). *Saudi J. Biol. Sci.* **2021**, *28*, 5119–5130. [CrossRef] [PubMed]
70. Kareem, S.; Adeleye, T.; Ojo, R. Effects of pH, temperature and agitation on the biosynthesis of iron nanoparticles produced by *Trichoderma* species. *IOP Conf. Ser. Mater. Sci. Eng.* **2020**, *805*, 012036. [CrossRef]
71. Bilesky-José, N.; Maruyama, C.; Germano-Costa, T.; Campos, E.; Carvalho, L.; Grillo, R.; Fraceto, L.F.; de Lima, R. Biogenic α-Fe_2O_3 Nanoparticles Enhance the Biological Activity of *Trichoderma* against the Plant Pathogen *Sclerotinia sclerotiorum. ACS Sustain. Chem. Eng.* **2021**, *9*, 1669–1683. [CrossRef]
72. Chaverri, P.; Samuels, G.J. *Hypocrea lixii,* the teleomorph of *Trichoderma harzianum. Mycol. Prog.* **2002**, *1*, 283–286. [CrossRef]
73. Salvadori, M.R.; Lepre, L.F.; Ando, R.A.; Oller do Nascimento, C.A.; Corrêa, B. Biosynthesis and Uptake of Copper Nanoparticles by Dead Biomass of *Hypocrea lixii* Isolated from the Metal Mine in the Brazilian Amazon Region. *PLoS ONE* **2013**, *8*, e80519. [CrossRef] [PubMed]
74. Salvadori, M.R.; Ando, R.A.; Oller Nascimento, C.A.; Corrêa, B. Extra and Intracellular Synthesis of Nickel Oxide Nanoparticles Mediated by Dead Fungal Biomass. *PLoS ONE* **2015**, *10*, e0129799. [CrossRef]
75. Kakakhel, M.A.; Sajjad, W.; Wu, F.; Bibi, N.; Shah, K.; Yali, Z.; Wang, W. Green synthesis of silver nanoparticles and their shortcomings, animal blood a potential source for silver nanoparticles: A review. *J. Hazard. Mater. Adv.* **2021**, *1*, 100005. [CrossRef]
76. Lankveld, D.P.K.; Oomen, A.G.; Krystek, P.; Neigh, A.; de Jong, A.T.; Noorlander, C.W.; Van Eijkeren, J.C.H.; Geertsma, R.E.; De Jong, W.H. The kinetics of the tissue distribution of silver nanoparticles of different sizes. *Biomaterials* **2010**, *31*, 8350–8361. [CrossRef]
77. Durán, N.; Marcato, P.D.; Alves, O.L.; De Souza, G.I.; Esposito, E. Mechanistic aspects of biosynthesis of silver nanoparticles by several *Fusarium oxysporum* strains. *J. Nanobiotechnol.* **2005**, *3*, 8. [CrossRef]
78. Das, S.K.; Marsili, E. A green chemical approach for the synthesis of gold nanoparticles: Characterization and mechanistic aspect. *Rev. Environ. Sci. Bio/Technol.* **2010**, *9*, 199–204. [CrossRef]
79. Benítez, T.; Rincón, A.M.; Limón, M.C.; Codón, A.C. Biocontrol mechanisms of *Trichoderma* strains. *Int. Microbiol.* **2004**, *7*, 249–260. [CrossRef]
80. Thakkar, K.N.; Mhatre, S.S.; Parikh, R.Y. Biological synthesis of metallic nanoparticles. *Nanomed. Nanotechnol. Biol. Med.* **2010**, *6*, 257–262. [CrossRef]
81. Oksanen, T.; Pere, J.; Paavilainen, L.; Buchert, J.; Viikari, L. Treatment of recycled kraft pulps with *Trichoderma reesei* hemicellulases and cellulases. *J. Biotechnol.* **2000**, *78*, 39–48. [CrossRef]
82. Anil Kumar, S.; Abyaneh, M.K.; Gosavi, S.W.; Kulkarni, S.K.; Pasricha, R.; Ahmad, A.; Khan, M.I. Nitrate reductase-mediated synthesis of silver nanoparticles from $AgNO_3$. *Biotechnol. Lett.* **2007**, *29*, 439–445. [CrossRef]
83. Abd-Elsalam, K.A.; Vasil'kov, A.Y.; Said-Galiev, E.E.; Rubina, M.S.; Khokhlov, A.R.; Naumkin, A.V.; Shtykova, E.V.; Alghuthaymi, M.A. Bimetallic blends and chitosan nanocomposites: Novel antifungal agents against cotton seedling damping-off. *Eur. J. Plant Pathol.* **2018**, *151*, 57–72. [CrossRef]
84. Sawake, M.M.; Moharil, M.P.; Ingle, Y.V.; Jadhav, P.V.; Ingle, A.P.; Khelurkar, V.C.; Paithankar, D.H.; Bathe, G.A.; Gade, A.K. Management of *Phytophthora parasitica* causing gummosis in citrus using biogenic copper oxide nanoparticles. *J. Appl. Microbiol.* **2022**. [CrossRef] [PubMed]
85. Gemishev, O.; Panayotova, M.; Gicheva, G.; Mintcheva, N. Green Synthesis of Stable Spherical Monodisperse Silver Nanoparticles Using a Cell-Free Extract of *Trichoderma reesei. Materials* **2022**, *15*, 481. [CrossRef]
86. Pandey, S.; Shahid, M.; Srivastava, M.; Singh, A.; Kumar, V.; Trivedi, S.; Srivastava, Y.K. Biosynthesis of SILVER nanoparticles by *Trichoderma harzianum* and its effect on the germination, growth and yield of tomato plant. *J. Pure Appl. Microbiol.* **2015**, *9*, 3335–3342.
87. Shelar, G.B.; Chavan, A.M. Myco-synthesis of silver nanoparticles from *Trichoderma harzianum* and its impact on germination status of oil seed. *Biolife* **2015**, *3*, 109–113. [CrossRef]
88. Bhadwal, A.S.; Tripathi, R.M.; Gupta, R.K.; Kumar, N.; Singh, R.P.; Shrivastav, A. Biogenic synthesis and photocatalytic activity of CdS nanoparticles. *RSC Adv.* **2014**, *4*, 9484. [CrossRef]
89. Pereira, L.; Dias, N.; Carvalho, J.; Fernandes, S.; Santos, C.; Lima, N. Synthesis, characterization and antifungal activity of chemically and fungal-produced silver nanoparticles against *Trichophyton rubrum. J. Appl. Microbiol.* **2014**, *117*, 1601–1613. [CrossRef] [PubMed]
90. Shakya, M.; Rene, E.R.; Nancharaiah, Y.V.; Lens, P.N.L. Fungal-Based Nanotechnology for Heavy Metal Removal. In *Nanotechnology,* Food Security and Water Treatment; Springer: Berlin/Heidelberg, Germany, 2018; pp. 229–253.

Journal of *Fungi*

Article

Antifungal Effect of Copper Nanoparticles against *Fusarium kuroshium*, an Obligate Symbiont of *Euwallacea kuroshio* Ambrosia Beetle

Enrique Ibarra-Laclette [1,*], Jazmín Blaz [1,2], Claudia-Anahí Pérez-Torres [1,3], Emanuel Villafán [1], Araceli Lamelas [1], Greta Rosas-Saito [1], Luis Arturo Ibarra-Juárez [1,3], Clemente de Jesús García-Ávila [4], Arturo Isaías Martínez-Enriquez [5] and Nicolaza Pariona [1,*]

1 Instituto de Ecología, A.C. (INECOL), Red de Estudios Moleculares Avanzados (REMAV), Xalapa 91073, Veracruz, Mexico; jazmin-itzel.blaz-sanchez@u-psud.fr (J.B.); claudia.perez@inecol.mx (C.-A.P.-T.); emanuel.villafan@inecol.mx (E.V.); araceli.lamelas2@gmail.com (A.L.); greta.rosas@inecol.mx (G.R.-S.); luis.ibarra@inecol.mx (L.A.I.-J.)
2 Centre National de la Recherche Scientifique (CNRS), Unité d'Ecologie Systématique et Evolution, Diversity, Ecology and Evolution of Microbes Team (DEEM), Université Paris-Saclay, AgroParisTech, 91405 Orsay, France
3 Investigador por México-CONACYT en el Instituto de Ecología, A.C. (INECOL), Xalapa 91073, Veracruz, Mexico
4 Dirección General de Sanidad Vegetal-Centro Nacional de Referencia Fitosanitaria (DGSV-CNRF), Tecamac 55740, Mexico; clemente.garcia@senasica.gob.mx
5 Centro de Investigación y de Estudios Avanzados del IPN Unidad Saltillo, Coahuila 25900, Mexico; arturo.martinez@cinvestav.edu.mx
* Correspondence: enrique.ibarra@inecol.mx (E.I.-L.); nicolaza.pariona@inecol.mx (N.P.)

Citation: Ibarra-Laclette, E.; Blaz, J.; Pérez-Torres, C.-A.; Villafán, E.; Lamelas, A.; Rosas-Saito, G.; Ibarra-Juárez, L.A.; García-Ávila, C.d.J.; Martínez-Enriquez, A.I.; Pariona, N. Antifungal Effect of Copper Nanoparticles against *Fusarium kuroshium*, an Obligate Symbiont of *Euwallacea kuroshio* Ambrosia Beetle. *J. Fungi* **2022**, *8*, 347. https://doi.org/10.3390/jof8040347

Academic Editor: Kamel A. Abd-Elsalam

Received: 21 February 2022
Accepted: 18 March 2022
Published: 27 March 2022

Publisher's Note: MDPI stays neutral with regard to jurisdictional claims in published maps and institutional affiliations.

Abstract: Copper nanoparticles (Cu-NPs) have shown great antifungal activity against phytopathogenic fungi, making them a promising and affordable alternative to conventional fungicides. In this study, we evaluated the antifungal activity of Cu-NPs against *Fusarium kuroshium*, the causal agent of *Fusarium* dieback, and this might be the first study to do so. The Cu-NPs (at different concentrations) inhibited more than 80% of *F. kuroshium* growth and were even more efficient than a commercial fungicide used as a positive control (cupric hydroxide). Electron microscopy studies revealed dramatic damage caused by Cu-NPs, mainly in the hyphae surface and in the characteristic form of macroconidia. This damage was visible only 3 days post inoculation with used treatments. At a molecular level, the RNA-seq study suggested that this growth inhibition and colony morphology changes are a result of a reduced ergosterol biosynthesis caused by free cytosolic copper ions. Furthermore, transcriptional responses also revealed that the low- and high-affinity copper transporter modulation and the endosomal sorting complex required for transport (ESCRT) are only a few of the distinct detoxification mechanisms that, in its conjunction, *F. kuroshium* uses to counteract the toxicity caused by the reduced copper ion.

Keywords: nanofungicide; antifungal activity; ambrosial complex

1. Introduction

The applications of nanotechnology have significantly increased over the last few years. Currently, different nanomaterials are being used in agriculture, creating a new field known as nanoagriculture. Various nanomaterials with antimicrobial activity have been tested for the control of infectious diseases, such as Ag nanoparticles (NPs) [1], Au-NPs [2], TiO_2-NPs [3], and ZnO-NPs [4]. Copper-based nanoparticles have drawn particular interest due to their low cost, excellent antimicrobial properties, and minimal environmental impact when used correctly (low concentrations with highly efficient modifications/formulations) [5]. For centuries, copper salts have been used for disease control [6].

One of their main advantages is that pathogens do not develop resistance to them, as occurs with most antibiotics [7]. However, due to their high dissolution in water, the cumulative dosages may be toxic to fish and other organisms [8].

Due to their unique physicochemical features, copper-based NPs have shown high antifungal properties against a broad spectrum of fungi species, including *Phoma destructive, Curvularia lunata, Alternaria alternate, Fusarium oxysporum, Saccharomyces cerevisiae,* among others [9,10]. Previous studies [10–14] have demonstrated that copper NPs antifungal activity depends on their shape, size, and concentration, which could vary depending on the fungal species. Previously, we evaluated the antifungal activity of five Cu/Cu_xO-NPs with different phase compositions and sizes, using a *Fusarium oxysporum* strain as a study case. The results showed that with a low concentration (0.25 mg/mL) of Cu/Cu_xO-NPs, with a high proportion of Cu_2O phase and relatively small size particles, more than 90% of fungal growth was inhibited. Meanwhile, copper salts reached only 5% growth inhibition [11]. Differences were also observed in antifungal activity of Cu-NPs even against species belonging to the same genus (e.g., *Fusarium* sp. AF-6, AF-8, *F. oxysporum,* and *F. solani*).

Fusarium kuroshium [15] is a member of the Ambrosia *Fusarium* Clade (AFC) [16,17] and is recognized as one of the symbionts of the Asian Kuroshio shot hole borer (*Euwallacea kuroshio* Gomez and Huler. Since its introduction into the United States of America, this pest has spread from Southern California's west coast to Northeastern Mexico [18]. The fungus–beetle complex is responsible for causing significant damage to several tree species distributed in urban, natural, agricultural, and riparian areas [17,19–23]. As a control strategy, fungicides from the azole family are commonly used even when they are inefficient. These chemicals can negatively impact ecological interactions and the environment [24]. Hence, it is necessary to find alternatives.

For the first time and based on the framework mentioned above, in this study, we describe the antifungal activity of Cu-NPs exerted against *Fusarium kuroshium* [15]. We analyzed the fungal morphological (growth and development) and molecular response in the presence of Cu-NPs, combining RNA-seq methodology and field emission scanning electron microscopy (FE-SEM).

2. Materials and Methods

2.1. Source of Fungal Symbionts of Ambrosia Beetles, Media, and Culture Conditions

Under strict biosecurity conditions, all in-vitro assays were carried out in the mycology laboratory at 'Centro Nacional de Referencia Fitosanitaria (CNRF)'. CNRF is a Mexican institution belonging to 'Dirección General de Sanidad Vegetal (DGSV)' and 'Servicio Nacional de Sanidad, Inocuidad y Calidad Agroalimentaria (SENASICA)', both dependencies of 'Secretaría de Agricultura y Desarrollo Rural (SADER)'. The strain HFEW-16-IV-019 of *Fusarium kuroshium* species was used in the present study [19,25,26]. This strain was isolated from the Kuroshio shot hole borer (KSHB), collected in Tijuana, B.C., Mexico, and stored in 25% glycerol at −80 °C [18]. Conidia from *F. kuroshium* were propagated on potato dextrose agar (PDA) (Sigma-Aldrich, St. Louis, MO, USA). Plates were incubated for 5–7 days at 28 °C in darkness, and fungal spores were collected by gently shaking the plate with 3–5 mL of sterile water at room temperature. After the conidia were washed twice with sterile water, they were collected and stored in an aqueous solution (at 5×10^6 colony forming unit (CFU)/mL) and used on the antifungal activity assay.

2.2. In Vitro Antifungal Activity Assay

As recently reported, the Cu-NPs used for the in vitro assays were synthesized [27]. These Cu-NPs are faceted particles of 200 nm in size, coated with citrate groups, water dispersible, and stable in the open atmosphere. The commercial fungicide product (Cupravit® Hidro, Bayer de México, CDMX, México) containing the active ingredient cupric hydroxide was used as the positive control and reference of antifungal activity. Sterile distilled water was used as a solvent to prepare both the Cu-NPs suspension and the cupric hydroxide solution. The Cu-NPs suspension was sonicated for 30 min to ensure good dispersion of NPs in the PDA culture medium.

The antifungal activity of Cu-NPs against *F. kuroshium* was evaluated using the poisoned food method [28]. Briefly, PDA was mixed with different amounts of Cu-NPs to obtain the following final concentrations: 0.1, 0.25, 0.5, 0.75, and 1.0 mg/mL. Cupric hydroxide was used at the same concentrations as Cu-NPs, and non-amended media were used as control. Spore suspensions (1×10^6 CFU/mL) were inoculated at the center of each PDA plate and incubated in darkness at 28 °C for six days. All treatments were carried out in triplicate. Colony diameters were measured three and six days after inoculation (dai). The percentage of growth inhibition was calculated by measuring the average area of the fungal colonies in the treatments and compared to the negative control.

2.3. Analysis of Fungal Morphology through FE-SEM

Six-day-old fungal from treatment and control cultures were used to determine mycelial radial growth and morphology. Mycelial discs of 10 mm diameter were cut, fixed, and processed as previously described [27,29]. The images were collected using an FE-SEM FEI Quanta 250-FEG (Brno, Czech Republic).

2.4. RNA Extraction

Three and six dai mycelium were collected from the Cu-NPs treatments (0.5, 0.75, and 1.0 mg/mL) and control. Samples were immediately frozen in liquid nitrogen and stored at −80 °C for posterior extraction. Total RNA was isolated from 200 mg of pulverized mycelia using Norgen RNA Purification Kit (Nor-gen Biotek Corporation, Thorold, Canada). RNA was quantified using a NanoDrop 2000 c spectrophotometer (Thermo Scientific, Thermo Fisher Scientific, Waltham, MA, USA) and assessed for purity by UV absorbance measurements at 260 and 280 nm. Total RNA integrity was confirmed by capillary electrophoresis using an Agilent 2100 Bioanalyzer (Agilent Technologies, Santa Clara, CA, USA).

2.5. RNA-seq Analysis: cDNA Library Preparation and Sequencing

cDNA libraries were prepared by the Massive Sequencing Unit of the Ecology Institute (INECOL, Xalapa, Ver., Mexico) using the TruSeq RNA Sample Preparation Kit (Illumina, San Diego, CA, USA) following the manufacturer's instructions. A total of 24 samples consisting of three biological replicates of Cu-NPs treatments 0.5, 0.75, and 1 mg/mL and negative control collected at 3 and 6 dai were sequenced. All samples were sequenced together on a single flow cell (High Output Kit v2.5; 300 Cycles) using the NextSeq500 platform (Illumina, San Diego, CA, USA). Paired-end reads (2×150 bp) were generated, and index codes were used to identify each sample independently. The RNA-seq data were deposited in the Short Read Archive (SRA) database of the National Center for Biotechnology Information (NCBI). Accession numbers were placed at the end of the manuscript in the data availability statement section.

2.6. Data Processing

The resulting raw paired-end reads from the sequencing process were cleaned using Trimmomatic v0.38 [30] to use only high-quality sequences. Reads alignment to the reference genome (*Fusarium kuroshium*; [25,26]) and transcript abundance estimation were performed using Bowtie2 v2.3.5.1 [31] and RNA-Seq by Expectation-Maximization (RSEM) v1.3.1 [32] software packages, respectively. The transcript abundance matrix created contains each of *F. kuroshium* genes (rows) and the expected count (EC) values calculated for each sampling point (3 and 6 dai) at the different concentrations of Cu-NPs employed (0, 0.5, 0.75, and 1 mg/mL; all represented in the corresponding columns). The EC values represent the expression levels and are calculated by the maximum likelihood estimation approach and posterior mean estimates with 95% credibility intervals. RSEM uses these EC values to calculate transcripts per million (TPM) and fragments per kilobase per million mapped reads (FPKM) values. It has been reported that TPM values are highly consistent among samples [33]. These values were used to perform principal component analysis to detect the significant sources of variance underlying the selected sampling points and the Cu-NPs treatments. The DESeq2 v1.2.4.0 R/Bioconductor package performed a differential expression analysis, using a negative binomial model to perform pairwise Wald tests, and the Benjamini–Hochberg method to perform multiple testing [34]. A $\log_2$ fold change (FC) value $\pm$ 1.0 and an adjusted p value of ≤ 0.05 were the criteria for identifying differentially expressed genes (DEGs) across treatments.

Considering that gene models predicted in the *F. kuroshium* genome lack annotation [29], its homologs were identified by BLAST searches. Only the best hit in unidirectional pairwise comparisons was considered (*F. kuroshium* versus some other available *Fusarium* species: *F. vanettenii* 77-13-4, *F. graminearum* PH-1, *F. pseudograminearum* CS3096, *F. verticillioides* 7600, *F. fujikuroi* IMI 58289, and *F. oxysporum* NRRL 32931). *Neurospora crassa* OR74A and *Saccharomyces cerevisiae* S288C were also included as outgroups. The names of species mentioned and those used as references are accompanied by the strain identifier (e.g., 77-13-4). The latest versions of these reference genomes, all available in the GenBank database (https://www.ncbi.nlm.nih.gov/; accessed on 17 February 2022), were those used in this study. Gene Ontology (GO) terms [35], eukaryotic orthologous group (KOG), the Enzyme Commission (EC) numbers [36], and Kyoto Encyclopedia of Genes and Genomes (KEGG) pathways [37] were inherited to each *F. kuroshium* gene. InterProScan [38,39] was used for this purpose. The g: Profiler web tool (http://biit.cs.ut.ee/gprofiler/; accessed on 15 February 2022; [40]) was used to identify the enriched functional categories (GO terms) and deep-represented metabolic pathways (KEGG) by genes that respond to the Cu-NPs treatments, significantly changing their transcription level (differentially expressed genes). Finally, GO and KEGG enrichment analysis of the identified DEGs was performed by g: Profiler web tool (http://biit.cs.ut.ee/gprofiler/; accessed on 15 February 2022) using the hypergeometric distribution adjusted by set count sizes (SCS) for multiple hypothesis correction [40]. Based on the method mentioned above (g: SCS), p-adjusted values ≤ 0.05 were used as a threshold after performing multiple correction tests.

3. Results

3.1. Antifungal Activity of Cu-NPs on Mycelial Growth

Both treatments, Cu-NPs and cupric hydroxide, were found to inhibit mycelial growth in a dose-dependent manner. As seen in Figure 1, Cu-NPs had more antifungal activity than cupric hydroxide. Figure 1 shows the radial mycelial growth of *F. kuroshium* exposed to different Cu-NPs and cupric hydroxide concentrations in both sampling points (3 and 6 dai).

At three dai, changes in colony pigmentation and mycelial growth inhibition started to be visible in both treatments (Cu-NPs and cupric hydroxide, respectively). However, colony morphology and percentage of mycelial radial growth inhibition were more evident at 6 dai (Figure 1). At this late sampling time, the *F. kuroshium* colony showed a cotton-like texture and pale orange pigmentation in the negative control (plates with PDA culture medium). In the presence of 0.1 and 0.25 mg/mL Cu-NPs, the color of the colony became white and dark cherry color, and the pigment disappeared when the concentration of Cu-NPs increased from 0.5 to 1 mg/mL. Changes in the colony morphology were also observed (irregular growth), being significant at 0.5 mg/mL.

Regarding the cupric hydroxide treatments (positive control), colony pigmentation changes were also observed from the lowest concentrations (0.1 and 0.25 mg/mL). Still, it turned dark purple g at 0.5 mg/mL (Figure 1).

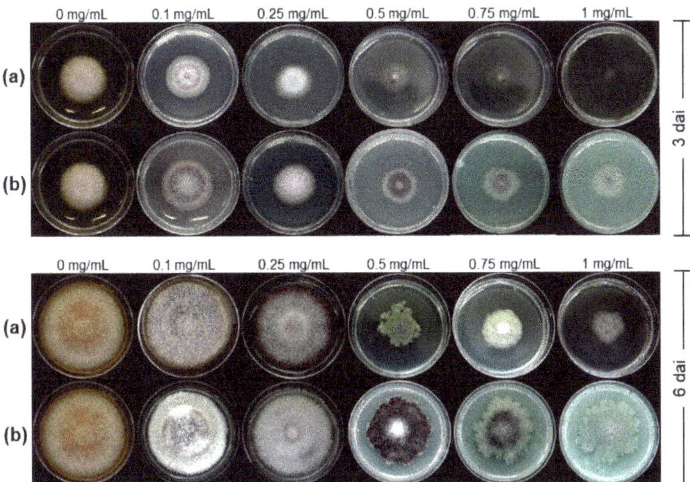

Figure 1. *F. kuroshium* mycelial growth inhibition assays. The colony morphology of *F. kuroshium* wild-type strains grown on plates with PDA culture medium supplemented with different concentrations (0 mg/mL (control), 0.1, 0.25, 0.5, 0.75, and 1 mg/mL) of (**a**) Cu-NPs and (**b**) cupric hydroxide at 3 and 6 days after inoculation (dai).

Additionally, the mycelial radial growth inhibition percentage was quantified (Figure 2). For both treatments (Cu-NPs and cupric hydroxide), growth inhibition became evident at six dai for 0.5, 0.75, and 1 mg/mL concentrations. The mycelial radial growth percentages resulted as higher for Cu-NPs than for cupric hydroxide. As seen in Figure 2, at 0.5 and 0.75 mg/mL of Cu-NPs, ~80% of the fungal growth was inhibited, while at the highest concentration (1 mg/mL), more than 90% inhibition was reached. These growth inhibition percentages were even higher than those observed for the cupric hydroxide treatments, with only 46% inhibition at 0.5 mg/mL and no increase at higher concentrations. As mentioned above, these results revealed that the Cu-NPs treatments at concentrations as low as 0.5 mg/mL might inhibit the growth of *F. kuroshium*, and this treatment seems to perform better than the commercial products available, such as cupric hydroxide, here used as a positive control.

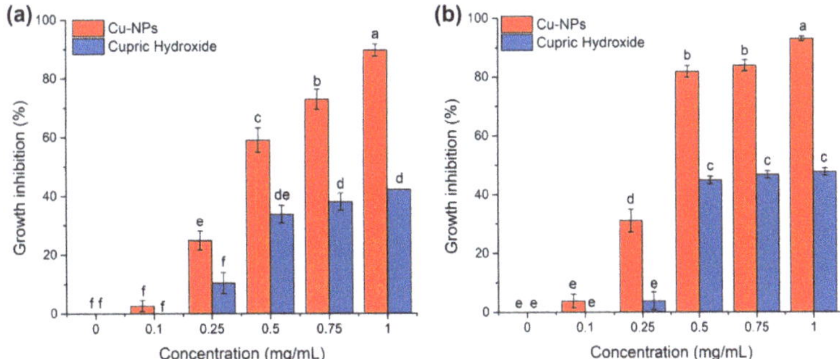

Figure 2. The mycelial radial growth inhibition percentage from F. kuroshium was quantified at (**a**) 3 and (**b**) 6 dai in both treatments, Cu-NPs, and cupric hydroxide, respectively. A one-way ANOVA with a Tukey's test was used to determine significance across all the treatments. Different letters on top of the bars indicate significant differences ($p \leq 0.01$). Error bars represent the standard error ($n = 3$).

3.2. Analysis of Fungal Morphology through FE-SEM

FE-SEM micrographs were used to study the structural changes of the fungal hyphae after the treatment by Cu-NPs. In the supplemented control, healthy hyphae exhibited a tubular morphology with a smooth surface and the characteristic formation of fusiform-clavate macroconidia (Figure 3a). In contrast, *F. kuroshium* growing on Cu-NPs treatments showed multiple alterations in the hyphae and macroconidia morphology (Figure 3b–f). At 0.1 mg/mL (Figure 3b), both the hyphae and the macroconidia showed morphological distortion. A reduction in hypha thickness, irregular shrinkages, and peanut shape were observed (yellow arrow). For the 0.25 mg/mL Cu-NPs treatment, there was no production of macroconidia; in addition, hyphae lost their smoothness and exhibited peeling (see red arrows in Figure 3c). At 1 mg/mL of Cu-NPs, *F. kuroshium* hyphae were swollen, deformed, fractured, and broken (pink arrows), leading to the outflow of intracellular components (Figure 3d–f).

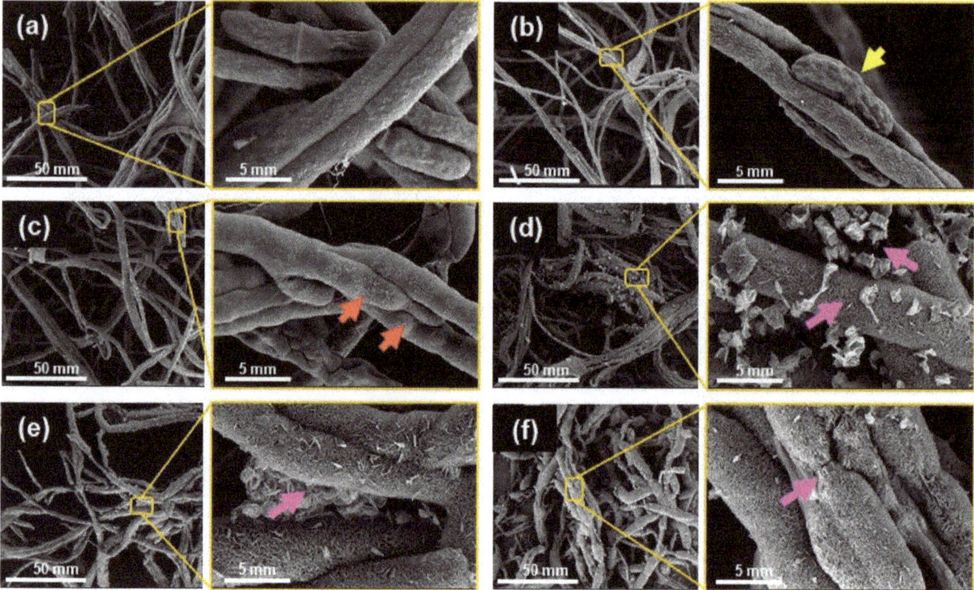

Figure 3. SEM micrographs of *F. kuroshium* hyphae after growing 6 days in PDA culture medium supplemented with different concentrations of Cu-NPs: (**a**) 0 (control), (**b**) 0.1, (**c**) 0.25, (**d**) 0.5, (**e**) 0.75, and (**f**) 1 mg/mL. The Cu-NPs treatments at concentrations as low as 0.1 mg/mL provoked changes in the hyphae morphology, ranging from an apparent loss of turgor to a loss of cell wall integrity. At 0.25 mg/mL, peeling hyphae (red arrow) indicated the loss of cell wall integrity. At concentrations of 0.5 mg/mL or greater, the hyphae cell wall showed higher porosity and leakage of the cytoplasmic contents (pink arrows). The yellow arrow indicates the morphological changes observed in the macroconidia, only found in the control and the 0.1 mg/mL treatment.

3.3. Differential Gene Expression of F. kuroshium in Response to Cu-NPs Treatments

A total of 481,775,061 high-quality (HQ) paired-end reads were obtained from the 24 RNA-seq sequenced libraries (around 20 million reads per library on average; Table S1). These HQ reads were mapped against the published *F. kuroshium* genome [25]. From the total of *F. kuroshium* predicted protein-coding genes (13,777), 97.39% were annotated based on *Fusarium vanettenii* (equivalent: *F. solani* f. sp. *pisi*) homologs proteins (Table S2). Homologs proteins were also detected for *Fusarium graminearum* (92.04%), *Fusarium pseudograminearum* (92.80%), *Fusarium verticillioides* (94.27%), *Fusarium fujikuroi* (94.56%), *Fusarium oxysporum* (94.90%), *Saccharomyces cerevisiae* (50.07%), *Neurospora crassa* (80.81%)

(Table S2; see Methods for details). The results mentioned above show that, as expected, the amount of homologs proteins identified during the annotation process (homology-based inference) increased as the species they were compared against were phylogenetically more closely related (details in [16]). Tables S3–S5.

The principal component analysis (PCA) using the estimated TPM values (Table S6; see Methods for more detail) was conducted to determine the differential expression and to detect the major sources of variance underlying the sampling points (3 and 6 dai) and the Cu-NPs concentrations (0, 0.5, 0.75, and 1 mg/mL). The two-dimensional PC plot in which the first two principal components (PC1 and PC2) were included was the one that best illustrated the variance, with explanatory values of 49% (PC1) and 22% (PC2), respectively (Figure 4a). Since all libraries were independently included in the analysis, the PCA plot indicates that not only the employed biological replicates have high reproducibility values but also, regarding the Cu-NPs treatments, they can be grouped in at least two major distinguishable discriminating groups: Group 1, which represents the control treatments (that is, without Cu-NPs), and Group 2, representing those treatments in which Cu-NPs were added to the culture media (PC2, at 0.5, 0.75 and 1 mg/mL, respectively). Regarding the sampling points (3 and 6 dai), despite the visible differences, they only explain a low percentage of the variance (PC1; Figure 4a). Based on these results, pairwise comparisons were performed to identify differentially expressed genes involved in Cu-NPs responses. Comparisons performed were 0.5, 0.75, and 1 mg/mL versus 0 mg/mL (control) at 3 and 6 dai, respectively. DEseq2 R package was used to calculate differential expression between these pairs of compared samples. In total, there were 5476 *F. kuroshium* genes with differential expression of two-fold or greater ($Log_2FC = \pm 1$) and an adjusted significant p value of ≤ 0.05 at three dai (Table S7).

Conversely, the DEGs were slightly more abundant (6787) once six days after inoculation elapsed (Table S8). Venn diagram comparison of DEGs showed that a high percentage of DEG was shared at both sampling points analyzed (3 and 6 dai, respectively). There is a similar percentage of up- and downregulated genes (53.9% and 65.3%; Figure 4c). The DEGs resulted as higher as the concentrations of Cu-NPs increased (Figure 4 and Tables S7 and S8). These data suggest that even when colony morphology and mycelial growth inhibition are more significant at six dai, fungal molecular responses to overcome toxic stress and maintain cell viability are triggered at earlier stages and probably kept over time, while the stress is present and the fungal cells lose their viability.

Pairwise Pearson's correlation coefficients (r) were estimated using the lists of DEGs to compare transcriptional responses (at global level) between the distinct Cu-NPs concentrations. That is, for each sampling point (3 and 6 dai), coefficients (r) were estimated between 0.5–0.75 mg/mL, 0.5–1 mg/mL, and 0.75–1 mg/mL. Student's t test was used to assess whether correlations were significant (t test, $p \leq 0.05$). The transcriptional responses seem to be similar based on these analyses. According to r values (ranging from 0.845 to 0.985), no significant differences exist between the distinct Cu-NPs concentrations or the sampling time points (Table S9). Similar to the transcriptional responses, colony morphology and mycelial radial growth inhibition percentages showed that the Cu-NPs at concentrations as low as 0.5 mg/mL have a comparable effect to those with higher concentrations (0.75 and 1 mg/mL). Transcriptional responses that may be involved are similar regardless of the time point analyzed, 3 or 6 dai.

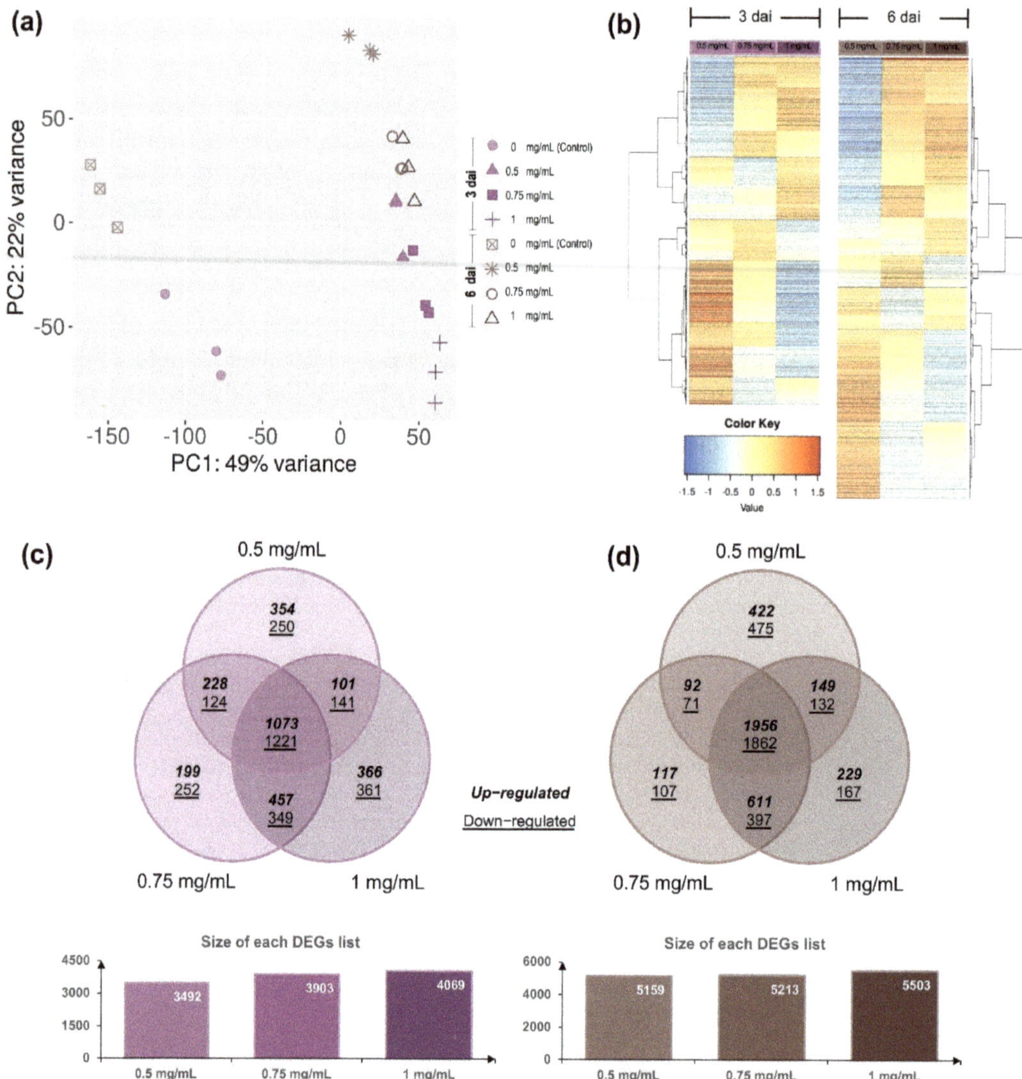

Figure 4. Expression profiles of *Fusarium kuroshium* Differentially Expressed Genes (DEGs) in response to Cu-NPs. (**a**) Principal component analysis (PCA) plot displaying all 24 RNA-seq sequenced libraries used in the presented study, the three independent replicates of the distinct concentrations of Cu-NPs used (0 (control), 0.5, 0.75 and 1 mg/mL) and evaluated at 3 and 6 days after inoculation (dai). PCA was performed using the transcripts per million (TPM) values. (**b**) Heatmaps of the average linkage hierarchical clustering based on the correlation distance measurements. Log_2FC values (±1) that resulted in significance (adjusted p value of ≤ 0.05) were used to represent the lists of DEGs obtained from both the 3 and 6 dai. DEGs lists were generated from pairwise comparisons in which each of the Cu-NPs treatments (0.5, 0.75, and 1 mg/mL) were compared against the control sample (0 mg/mL). The Venn diagram represents the shared amount of up- and downregulated genes in each Cu-NPs treatment at the two sampling points evaluated, 3 dai (**c**) and six dai (**d**).

3.4. Gene Ontology Enrichment Analysis of Cu-NPs Responsive Genes

To further examine the functions of the DEGs, an enrichment analysis of GO functional categories and KEGG metabolic pathways was performed using g: Profiler web server (see Methods for details). Nineteen molecular function (MF) terms, 32 biological processes (BP) terms, and 23 cellular components (CC) terms were significantly enriched by 4028 of the DEGs (66.4% of total), which were identified at both sampling points (Table S10). The top three GO terms enriched on each of these three major categories (Figure 5) included for MF were: oxidoreductase activity (GO:0016491), active (ion) transmembrane transporter activity (GO:0022804), and catalytic activity (GO:0003824); for BP: oxidation-reduction process (GO:0055114), organic acid metabolic process (GO:0006082), and transmembrane transport (GO:0055085); and for CC: cytoplasm (GO:0005737), organelle (GO:0043226), and intracellular membrane-bounded organelle (GO:0043231). Three KEGG metabolic pathways which were significantly enriched (p value ≥ 0.05) by DEGs were biosynthesis of secondary metabolites (KEGG:01110), tryptophan metabolism (KEGG:00380), and propanoate metabolism (KEGG:00640) (Figure 5 and Table S10).

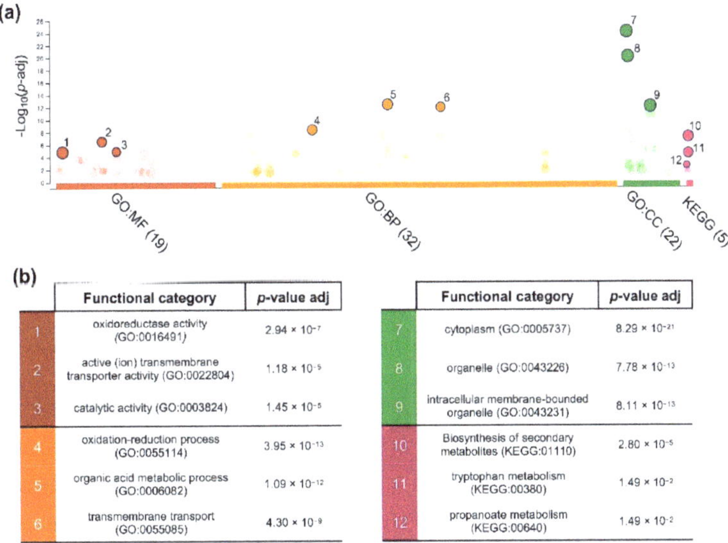

	Functional category	p-value adj
1	oxidoreductase activity (GO:0016491)	2.94×10^{-7}
2	active (ion) transmembrane transporter activity (GO:0022804)	1.18×10^{-5}
3	catalytic activity (GO:0003824)	1.45×10^{-5}
4	oxidation-reduction process (GO:0055114)	3.95×10^{-13}
5	organic acid metabolic process (GO:0006082)	1.09×10^{-12}
6	transmembrane transport (GO:0055085)	4.30×10^{-9}

	Functional category	p-value adj
7	cytoplasm (GO:0005737)	8.29×10^{-21}
8	organelle (GO:0043226)	7.78×10^{-13}
9	intracellular membrane-bounded organelle (GO:0043231)	8.11×10^{-13}
10	Biosynthesis of secondary metabolites (KEGG:01110)	2.80×10^{-5}
11	tryptophan metabolism (KEGG:00380)	1.49×10^{-2}
12	propanoate metabolism (KEGG:00640)	1.49×10^{-2}

Figure 5. Enrichment of the GO terms and KEGG metabolic pathways by DEGs responsive to Cu-NPs. (**a**) Manhattan plot illustrating the significantly enriched (g: SCS threshold, p value ≤ 0.05) terms. The top three terms (solid colored dots) were numbered to distinguish them from the rest (dimmed colored dots). The name of each of these categories and its statistical significance are also shown (**b**).

3.5. *Fusarium kuroshium* Genes Involved in Transport, Homeostasis, and Cooper Toxicity and Resistance

There is still a limited understanding of the resistance mechanisms deployed by fungi to cope with the toxicity caused by Cu-NPs. Some of these molecular mechanisms have been studied mainly in yeast (Saccharomyces cerevisiae), but filamentous fungi reports are scarce. Downregulation of metal ion importers, utilization of metallothionein, metallothionein-like structures, and ion sequestration to the vacuole have been implicated in yeast's resistance to metals (zinc, copper, iron, and silver, among others). In filamentous fungi, however, metal resistance relies heavily upon the export of these ions [41]. Therefore, we extensively searched genes involved in copper resistance using previous reports and recent reviews as a starting point [41–46]. *F. kuroshium* homologs of the genes from either yeast or filamentous fungi were identified on the lists of DEGs (Table S11). We found several homologs of enzymes involved in copper transport and homeostasis previously reported in yeasts, for example, some *F. kuroshium* genes homologs to FRE1 (FuKu07004) and FRE7 (Fu-Ku03123, FuKu04041, FuKu10175), both ferric/cupric-chelate reductases that, except for FuKu03123, were strongly downregulated (Log_2FC values ranged from −3.32 to −10.17) in all Cu-NPs analyzed treatments (0.5, 0.75, and 1 mg/mL). FRE1 [47] (and other members of this gene family [46]) are metallo-reductases that reduce both cupric (Cu^{2+}) and ferric (Fe^{3+}) ions by bounding to two distinct transcription factors, MAC1, and ATF1, respectively [48–50]. No homologs to these transcription factors were identified in the *F. kuroshium* proteins coding genes set, suggesting that, perhaps in Fusarium species, distinct transcription factors are involved in a similar response. In addition to FRE proteins, homologs to low- (CRT2; FuKu05634) and high- (CRT3; FuKu05575, Fu-Ku07307) affinity copper transporters were also repressed or downregulated in all tested Cu-NPs treatments. Similar expression patterns (significant downregulated) were found for other homologs to copper transporters such as PIC2 (FuKu08121) and CCC2 (FuKu08773), proteins which shuttle Cu^{+} from the cytoplasm to the mitochondrial matrix and Golgi bodies, respectively [51–53].

Other enzymes such as ferroxidases 3 (FET3; FuKu00497, FuKu00629, FuKu01416, FuKu05480) and 5 (FET5; FuKu12927, FuKu08718) were significantly upregulated even when they were required for uptake and oxidation of ferrous iron. It is known that they require copper as a cofactor for properly functioning [54]. As expected, an ortholog to the *CrpA* gene from *Aspergillus fumigatus* (FuKu02881) was also significantly induced (Log_2FC values > 8). This gene participates as a copper export and is an intermediate of copper's reactive oxygen species responses [55].

Other groups of upregulated genes were those involved in the biosynthesis of cell wall components such as chitin (BioCyc ID: PWY-6981; enzymes: NTH1; FuKu06343, HXK2; FuKu07788 and FuKu11848, PCM1; FuKu08072, and QRI1; FuKu07424) and b-glucans (GO-term: fungal-type cell wall beta-glucan biosynthetic process (GO:0070880); genes: *Rot2; FuKu01065, FuKu04120, FuKu08774* and *FuKu09863, Cwh41; FuKu09879, KAR2; FuKu03662*, and *Kre5; FuKu01731*), besides those which participate in copper detoxification by Golgi-to-vacuole transport by the AP-3 adapter complex in the alkaline phosphatase pathway and in the carboxipeptidase Y pathway, which transport cargo to the vacuole through endosomal intermediates ([43] proteins: GDA1; FuKu09812, GYP1; FuKu05874 and FuKu06284, RUD3; FuKu02081, HOC1; FuKu04569, HOC1; FuKu11979, IMH1; FuKu00391, VPS25; FuKu03105, SNF7; FuKu06560, PEP1; FuKu07887, NHX1; FuKu09337, APS3; FuKu05521, CCC1; FuKu05129).

Consistent with previous studies show that exposure of yeasts to trace amounts of metals such as copper, lead, iron, or zinc produce toxicity or death by interfering with several biological processes, including the ergosterol biosynthesis [41,56]. We found that *F. kuroshium* downregulated most of the genes involved in this biosynthetic pathway, even some of those represented in multi copies (paralogs) in *Fusarium* genomes in response to the majority of Cu-NPs concentrations (Figure 6 and Table S12). Similarly, Candida albican's nine sterol-response elements (ERG1, ERG2, ERG5, ERG6, ERG10, ERG11, ERG24, ERG26, and ERG27) are regulated by UPC2 transcription factor [57,58]. In *F. kuroshium*, most of these enzymes (Figure 6) show downregulated patterns in response to the Cu-NPs treatments. While it is true, it has been proven that the efficiency of ergosterol biosynthesis is determined by some limiting enzymes, and more crucially by the optimal coordination of the regulation of encoding genes involved in this biosynthetic pathway [59]. In ascomycetes and basidiomycetes, there is a positive correlation between the synthesized metabolites (ergosterol and its precursors) and expression profiles of genes codifying for enzymes involved in its biosynthesis, mainly in those genes related to the post-squalene pathway [60,61].

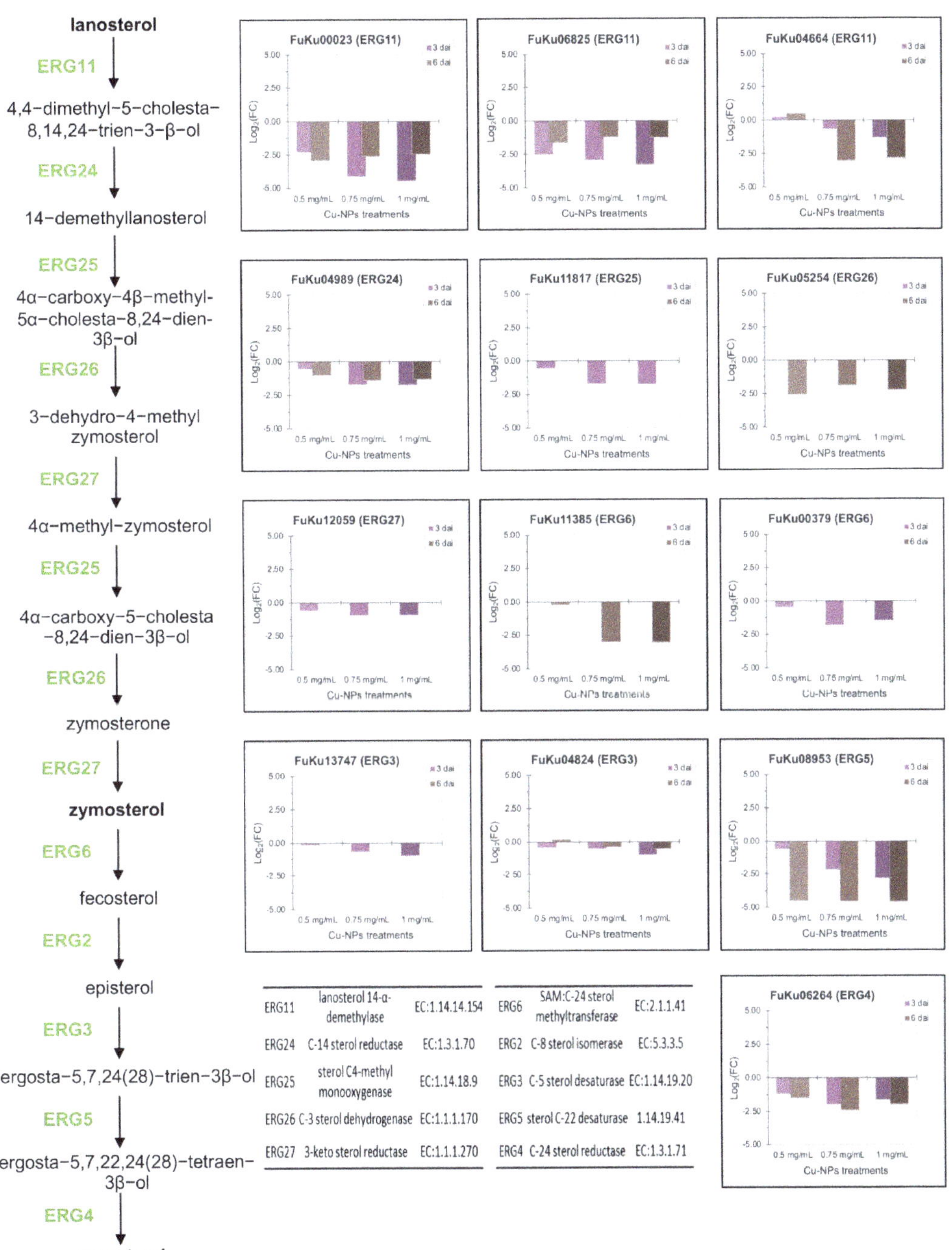

Figure 6. Ergosterol biosynthesis pathway (left) shows the expression profiles of *F. kuroshium* genes (ERG enzymes) involved. These expression profiles were significant in RNAseq differential expression analysis and are represented as Log_2FC values. Enzyme names and corresponding Enzymatic Commission (EC) numbers are also shown.

4. Discussion

This study showed that Cu-NPs exhibit a better antifungal activity against *F. kuroshium* than cupric hydroxide. Some effects that were observed at concentrations ranging from 0.1 to 0.5 mg/mL were color changes of the fungal colony (Figure 1). This effect has been observed in other plant pathogenic fungi, such as *F. solani*, *Neofusicoccum sp.*, and *F. oxysporum* [27]. In fungi, pigment production is related to melanin and carotenoid synthesis and is considered a defense mechanism against external stress [62]. In addition, it has been proven that the roles of fungal melanin include, among others, the scavenging of free radicals [63]. This is consistent with our results since the synthesis of pigments might be a mechanism by which *F. kuroshium* seeks to counteract the oxidative stress produced by Cu-NPs.

DEGs' enrichment analyses of GO terms and KEGG metabolic pathways show that the top three enriched terms in the BP category defend against copper toxicity. Those processes are related to each other and correspond to oxidation–reduction processes, organic acid biosynthesis, and the active transport of ions through the plasma membrane and the membranes that bound the organelles.

Organic acid production has been suggested to give a competitive advantage to filamentous fungi over other organisms by decreasing the pH and impacting metal detoxification [64,65]. The decrease in pH upon their secretion may give a competitive advantage to the acid-tolerant filamentous fungi, depending on the environment in which they grow [66]. For saprophytic and wood-decaying fungi, pH acidification, caused by oxalic acid production (another significantly enriched GO term; GO:0043436), leads to acid-catalyzed hydrolysis of holocellulose [67–69]. Depending on their concentration, type of metal, and pH, organic acids can also be complex with di- and tri-valent metals (Fe, Cu, Al, among others), explaining their essential role in metal detoxification [65]. The degree of complexation is also dependent on the organic acid involved (number and proximity of carboxyl groups). This result suggests that *F. kuroshium*, at least in part, seeks to counteract the toxicity caused by Cu-NPs by synthesizing some organic acids.

Fungal–copper interactions are necessary for the activation of metalloproteins involved in biochemical processes. This includes the activation of superoxide dismutase, which is responsible for cellular detoxification of reactive oxygen species (ROS) and activation of cytochrome c oxidate, a catalyst within the electron transport chain [41]. Copper [56], zinc [70], and silver [71,72] NPs interfere with ergosterol biosynthesis, increasing leakage of the cytoplasmic contents, depolarization, occurrence of ROS, and reducing cell wall integrity in yeasts. This explains the significant enrichment of the oxidation–reduction processes (GO:0055114) and enzymes with oxidoreductase activity (GO:0016491. In addition, metallothioneins (proteins that use metal ions as cofactors that possess a cysteine-rich domain) bind free cytosolic ions as a mechanism of ion storage or detoxification. In metal-deficient conditions, ions may be released back into the cellular environment [73]. Specific protein intracellular transporters are involved in this movement of ions to organelles either for storage or as cofactors for protein functioning [41,53] It is known that interference with these systems causes a homeostatic imbalance, resulting in toxicity [41].

Regarding KEGG terms, we consider that the secondary metabolites pathway (KEGG:01110) could be significantly enriched due to the pigments produced by *F. kuroshium* (Figure 1). Meanwhile, the enrichment of the tryptophan biosynthetic pathway (KEGG:00380) is consistent with Jo et al., who in 2017 [43], used microarrays and deletion mutants to identify genes in *Saccharomyces cerevisiae* involved in the toxic response against iron and copper. In that study, the changes in the expression of genes in the tryptophan biosynthesis pathway were specific to the copper response, suggesting that at least in yeasts, the mechanisms to deal with high concentrations of these two metals are specific for each of them. The role of the tryptophan biosynthetic pathway in the overload of copper in yeasts and some fungi such as *F. kuroshium* is still unknown. However, it has been suggested that its involvement is associated with the metabolites produced during degradation in the kynurenine pathway, which have antioxidant properties [74], or its radical-scavenging activity, as superoxide radicals are used as a cofactor to cleave the pyrrole ring in tryptophan [75]. Alternatively, it is also possible that tryptophan may be required as a critical residue in specific proteins involved in the defense against copper toxicity [43].

Based on the expression profile of some DEGs, our data suggest that *F. kuroshium* counteracts the toxicity caused by Cu-NPs through several mechanisms as shown in Figure 7, including a significant decrease in the transcription of genes codifying both the reductase that reduces extracellular copper (Cu^{2+}), and the low- and high-affinity membrane transporters that shuttle the reduced copper (Cu^{+}) to the cytoplasm. In addition, several transporters in intracellular membrane-bounded organelles are also downregulated. These results suggest that *F. kuroshium* tries to considerably reduce the shuttle of Cu^{+} to some organelles as Golgi bodies and the mitochondria. In contrast, in toxic copper concentrations, the overexpression of the CrpA transporter may occur as a defense mechanism to prolong its life by exporting Cu^{+} from the cytoplasm to the extracellular space. The overexpression of some metalloproteins and other proteins that use copper ions as cofactors (e.g., ferroxidases) can also be considered as copper storage or a detoxification mechanism because these proteins bind free cytosolic ions, releasing them back into the cellular environment in metal-deficient conditions [41,73].

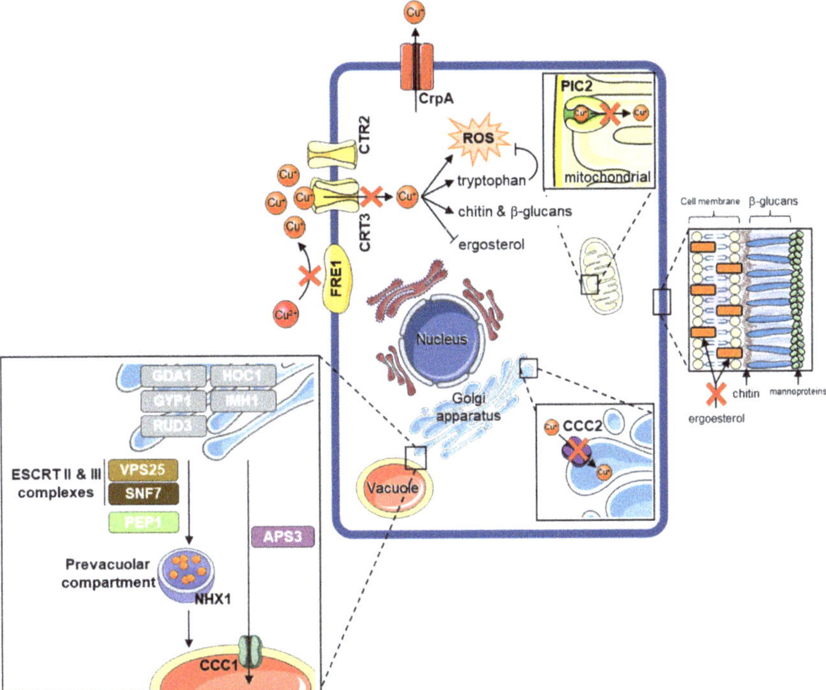

Figure 7. Schematic representation of mechanism involved in detoxification and the resistance to Cu-NPs treatments in *F. kuroshium* species. The membrane transporters and other proteins represented in the cell are named based on its yeast *(Saccharomyces cerevisiae)* homologs.

Considering the expression profile (upregulated) of several genes whose coding proteins form the endosomal sorting complex are required for transport (ESCRT), we suggest that both *F. kuroshium* such as *S. cerevisiae* (and probably another eucaryotic organism), employ this detoxification pathway in response to the copper overload [43]. No DEGs were found for the retromer complex; this suggests that intracellular traffic of copper ions (or proteins that bind it) may occur preferably in one way (from Golgi to vacuole). In addition, high levels of Aps3 suggest that the AP-3 complex (which, similar to ESCRT, also converges toward the vacuole) is also involved in copper detoxification. It has been reported that even when yeast molecular responses to iron and copper share some mechanisms, the AP-3 adapter complex in the alkaline phosphatase pathway is mainly involved in iron overload resistance [41,43].

Particular concentrations of copper cytosolic ions also interfere with the redox balance and increase the generation of reactive oxygen species [76]. High amounts of reactive oxygen species (ROS) can induce autophagy, apoptosis, and cell death [77]. Other consequences of free cytosolic Cu^+ ions reduce ergosterol biosynthesis and increase tryptophan synthesis. As mentioned above, it has been discussed that the participation of tryptophan in the response to copper-induced toxicity could be through antioxidant properties of the metabolites produced during degradation in the kynurenine pathway, which has radical-scavenging activity as a superoxide radical (a radical that contributes to oxidative stress) [43,74,75]. The reduction in ergosterol biosynthesis decreases cell wall integrity, increases cellular leakage and depolarization, and increases the occurrence of ROS [70]. We found that the genes involved in chitin and b-glucans biosynthesis are upregulated. This suggests that maybe *F. kuroshium*, faced with a constant block in the synthesis of ergosterol, seeks to maintain the cell wall integrity by increasing the production of its other primary components (e.g., chitin and β-glucans).

SEM micrographs (Figure 3) show a loss in cell wall integrity. Our analyses discussed before can explain this phenomenon by observed changes in the transcript levels of the genes involved in ergosterol biosynthesis. However, SEM micrographs also revealed that macroconidia, such as hyphae, were severely damaged and can only be found at concentrations as low as 0.1 mg/mL of Cu-NPs. This suggests that concentrations slightly higher ($\geq$0.25 mg/mL) not only inhibit *F. kuroshium* growth but also interfere in the formation of asexual spores such as macroconidia. We cannot explain this observation in light of the generated results; however, this effect of Cu-NPs treatments will be addressed in future works. Together all these results suggest that the toxicity of Cu-NPs affects several biological processes that compromise cell viability.

5. Conclusions

The presented work proves that using Cu-NPs could be considered as a highly efficient alternative with better antifungal properties than other formulations commonly proposed and commercially available fungicides such as cupric hydroxide. Molecular responses to Cu-NPs treatments analyzed by RNA-seq suggest that *F. kurhosium* counteracts the toxicity caused by free cytosolic copper ions through different mechanisms. These mechanisms include avoiding copper reduction, internalization, and intracellular movement. For this purpose, the amount of high- and low-affinity transporters and other specific transporters decreases considerably. In addition, free copper cytosolic ions also decrease by binding to copper-dependent proteins, which are strongly induced, including metallothionein. The overexpression of other transporters exporting Cu^+ from the cytoplasm to the extracellular space is also essential in the detoxification process. These detoxification mechanisms seek to maintain cell viability, which is ultimately compromised due to the loss of cell wall integrity resulting from reduced ergosterol synthesis. Cytosolic leakage and depolarization increase the occurrence of ROS, which induces autophagy, apoptosis, and cell death.

Supplementary Materials: The following are available online at https://www.mdpi.com/article/10.3390/jof8040347/s1, Table S1: Summary of Illumina sequencing, Table S2: Annotation of the Fusarium kuroshium genes by homology-based inference, Table S3: Gene Ontology-based functional characterization of the Fusarium kuroshium genes, Table S4: KOG terms inherited to Fusarium kuroshium genes, Table S5: KEGG pathways inherited to Fusarium kuroshium genes, Table S6: Expression profile matrix of Fusarium kuroshium genes, Table S7: Fusarium kuroshium differentially expressed genes (DEGs) at 3 dai, Table S8: Fusarium kuroshium differentially expressed genes (DEGs) at 6 dai, Table S9: Pearson's correlation matrix, Table S10: Over-represented GO terms on the list of Fusarium kuroshium DEGs (Gene Ontology enrichment analysis), Table S11: Common copper-responsive genes shared between yeast and Fusarium kuroshium, Table S12: Fusarium kuroshium differentially expressed genes (DEGs) involved in ergosterol biosynthesis.

Author Contributions: Conceptualization, N.P. and A.L.; Methodology, A.L., G.R.-S. and L.A.I.-J.; Formal Analysis, E.I.-L., J.B., C.-A.P.-T., E.V., A.L., G.R.-S., L.A.I.-J., C.d.J.G.-Á., A.I.M.-E. and N.P.; Investigation, E.I.-L. and N.P.; Data Curation, E.I.-L.; Writing—Original Draft Preparation, E.I.-L., A.I.M.-E. and N.P.; Writing—Review and Editing, E.I.-L. and N.P.; Funding Acquisition, N.P. All authors have read and agreed to the published version of the manuscript.

Funding: This work was supported by Consejo Nacional de Ciencia y Tecnología (CONACYT) through the Fondo Institucional de Fomento Regional para el Desarrollo Científico, Tecnológico y de Innovación (FORDECYT) grant 292399 (currently named Programa Presupuestario F003).

Data Availability Statement: The RNA-seq data were deposited in the Short Read Archive (SRA) of the National Center for Biotechnology Information (NCBI) with accession number PRJNA805244.

Acknowledgments: We would like to thank Alexandro Alonso Sánchez for his help in providing us with the sequencing services. The authors would also like to thank the Institute of Ecology A.C. (INECOL) for all the facilities provided, including access to the high-performance computing system (HUTZILIN). We also thank the staff from the Mycology Laboratory at Centro Nacional de Referencia Fitosanitaria (CNRF) for access to the fungus (*F. kuroshium*) and for allowing us to perform our experiments in their laboratories under biosafety conditions. A special thanks goes to Diana Sánchez Rangel and Martín Aluja Schuneman Hofer for management and all relations to coordination and tracking of the project that funded this research (FORDECyT-CONACYT, grant 292399). We thank Luisa Cruz from the Tropical Research and Education Center, University of Florida, for editing the paper.

Conflicts of Interest: The authors declare no conflict of interest. The funders had no role in the design of the study; in the collection, analyses, or interpretation of data; in the writing of the manuscript, or in the decision to publish the results.

References

1. Kim, S.W.; Jung, J.H.; Lamsal, K.; Kim, Y.S.; Min, J.S.; Lee, Y.S. Antifungal Effects of Silver Nanoparticles (AgNPs) against Various Plant Pathogenic Fungi. *Mycobiology* **2012**, *40*, 53–58. [CrossRef] [PubMed]
2. López-Lorente, Á.I.; Cárdenas, S.; González-Sánchez, Z.I. Effect of synthesis, purification and growth determination methods on the antibacterial and antifungal activity of gold nanoparticles. *Mater. Sci. Eng. C Mater. Biol. Appl.* **2019**, *103*, 109805. [CrossRef] [PubMed]
3. Rekha, R.; Divya, M.; Govindarajan, M.; Alharbi, N.S.; Kadaikunnan, S.; Khaled, J.M.; Al-Anbr, M.N.; Pavela, R.; Vaseeharan, B. Synthesis and characterization of crustin capped titanium dioxide nanoparticles: Photocatalytic, antibacterial, antifungal and insecticidal activities. *J. Photochem. Photobiol. B Biol.* **2019**, *199*, 111620. [CrossRef] [PubMed]
4. Pariona, N.; Paraguay-Delgado, F.; Basurto-Cereceda, S.; Morales-Mendoza, J.E.; Hermida-Montero, L.A.; Mtz-Enriquez, A.I. Shape-dependent antifungal activity of ZnO particles against phytopathogenic fungi. *Appl. Nanosci.* **2020**, *10*, 435–443. [CrossRef]
5. Abd-Elsalam, K.A. *Copper Nanostructures: Next-Generation of Agrochemicals for Sustainable Agroecosystems*; Elsevier Science: Amsterdam, The Netherlands, 2022.
6. Gadi, B.; Jeffrey, G. Copper, An Ancient Remedy Returning to Fight Microbial, Fungal and Viral Infections. *Curr. Chem. Biol.* **2009**, *3*, 272–278. [CrossRef]
7. Keller, A.A.; Adeleye, A.S.; Conway, J.R.; Garner, K.L.; Zhao, L.; Cherr, G.N.; Hong, J.; Gardea-Torresdey, J.L.; Godwin, H.A.; Hanna, S.; et al. Comparative environmental fate and toxicity of copper nanomaterials. *NanoImpact* **2017**, *7*, 28–40. [CrossRef]
8. Adam, N.; Vakurov, A.; Knapen, D.; Blust, R. The chronic toxicity of CuO nanoparticles and copper salt to Daphnia magna. *J. Hazard. Mater.* **2015**, *283*, 416–422. [CrossRef] [PubMed]
9. Giannousi, K.; Sarafidis, G.; Mourdikoudis, S.; Pantazaki, A.; Dendrinou-Samara, C. Selective synthesis of Cu^2O and Cu/Cu^2O NPs: Antifungal activity to yeast Saccharomyces cerevisiae and DNA interaction. *Inorg. Chem.* **2014**, *53*, 9657–9666. [CrossRef]
10. Kanhed, P.; Birla, S.; Gaikwad, S.; Gade, A.; Seabra, A.B.; Rubilar, O.; Duran, N.; Rai, M. In vitro antifungal efficacy of copper nanoparticles against selected crop pathogenic fungi. *Mater. Lett.* **2014**, *115*, 13–17. [CrossRef]
11. Hermida-Montero, L.A.; Pariona, N.; Mtz-Enriquez, A.I.; Carrión, G.; Paraguay-Delgado, F.; Rosas-Saito, G. Aqueous-phase synthesis of nanoparticles of copper/copper oxides and their antifungal effect against Fusarium oxysporum. *J. Hazard. Mater.* **2019**, *380*, 120850. [CrossRef] [PubMed]
12. Shen, Y.; Borgatta, J.; Ma, C.; Elmer, W.; Hamers, R.J.; White, J.C. Copper Nanomaterial Morphology and Composition Control Foliar Transfer through the Cuticle and Mediate Resistance to Root Fungal Disease in Tomato (Solanum lycopersicum). *J. Agric. Food Chem.* **2020**, *68*, 11327–11338. [CrossRef]

13. Borgatta, J.; Ma, C.; Hudson-Smith, N.; Elmer, W.; Plaza Pérez, C.D.; De La Torre-Roche, R.; Zuverza-Mena, N.; Haynes, C.L.; White, J.C.; Hamers, R.J. Copper Based Nanomaterials Suppress Root Fungal Disease in Watermelon (Citrullus lanatus): Role of Particle Morphology, Composition and Dissolution Behavior. *ACS Sustain. Chem. Eng.* **2018**, *6*, 14847–14856. [CrossRef]
14. Xiong, L.; Tong, Z.H.; Chen, J.J.; Li, L.L.; Yu, H.Q. Morphology-dependent antimicrobial activity of Cu/CuxO nanoparticles. *Ecotoxicology* **2015**, *24*, 2067–2072. [CrossRef] [PubMed]
15. O'Donnell, K.; Al-Hatmi, A.M.S.; Aoki, T.; Brankovics, B.; Cano-Lira, J.F.; Coleman, J.J.; de Hoog, G.S.; Di Pietro, A.; Frandsen, R.J.N.; Geiser, D.M.; et al. No to Neocosmospora: Phylogenomic and Practical Reasons for Continued Inclusion of the Fusarium solani Species Complex in the Genus Fusarium. *Msphere* **2020**, *5*, e00810-20. [CrossRef]
16. Geiser, D.M.; Al-Hatmi, A.M.S.; Aoki, T.; Arie, T.; Balmas, V.; Barnes, I.; Bergstrom, G.C.; Bhattacharyya, M.K.; Blomquist, C.L.; Bowden, R.L.; et al. Phylogenomic Analysis of a 55.1-kb 19-Gene Dataset Resolves a Monophyletic Fusarium that Includes the Fusarium solani Species Complex. *Phytopathology* **2021**, *111*, 1064–1079. [CrossRef] [PubMed]
17. Na, F.; Carrillo, J.D.; Mayorquin, J.S.; Ndinga-Muniania, C.; Stajich, J.E.; Stouthamer, R.; Huang, Y.-T.; Lin, Y.-T.; Chen, C.Y.; Eskalen, A. Two novel fungal symbionts Fusarium kuroshium sp. nov. and Graphium kuroshium sp. nov. of Kuroshio shot hole borer (Euwallacea sp. nr. fornicatus) cause Fusarium dieback on woody host species in California. *Plant Dis.* **2018**, *102*, 1154–1164. [CrossRef] [PubMed]
18. García-Avila, C.D.J.; Trujillo-Arriaga, F.J.; López-Buenfil, J.A.; González-Gómez, R.; Carrillo, D.; Cruz, L.F.; Ruiz-Galván, I.; Quezada-Salinas, A.; Acevedo-Reyes, N. First Report of Euwallacea nr. fornicatus (Coleoptera: Curculionidae) in Mexico. *Fla. Entomol.* **2016**, *99*, 555–556. [CrossRef]
19. Pérez-Torres, C.A.; Ibarra-Laclette, E.; Hernández-Domínguez, E.E.; Rodríguez-Haas, B.; Pérez-Lira, A.J.; Villafán, E.; Alonso-Sánchez, A.; García-Ávila, C.J.; Ramírez-Pool, J.A.; Sánchez-Rangel, D. Molecular evidence of the avocado defense response to Fusarium kuroshium infection: A deep transcriptome analysis using RNA-Seq. *PeerJ* **2021**, *9*, e11215. [CrossRef] [PubMed]
20. Boland, J.M. The impact of an invasive ambrosia beetle on the riparian habitats of the Tijuana River Valley, California. *PeerJ* **2016**, *4*, e2141. [CrossRef]
21. Eskalen, A.; Stouthamer, R.; Lynch, S.C.; Rugman-Jones, P.F.; Twizeyimana, M.; Gonzalez, A.; Thibault, T. Host Range of Fusarium Dieback and Its Ambrosia Beetle (Coleoptera: Scolytinae) Vector in Southern California. *Plant Dis.* **2013**, *97*, 938–951. [CrossRef] [PubMed]
22. Freeman, S.; Sharon, M.; Maymon, M.; Mendel, Z.; Protasov, A.; Aoki, T.; Eskalen, A.; O'Donnell, K. Fusarium euwallaceae sp. nov.–a symbiotic fungus of Euwallacea sp., an invasive ambrosia beetle in Israel and California. *Mycologia* **2013**, *105*, 1595–1606. [CrossRef]
23. Eskalen, A. Invasive Shot Hole Borers. Available online: https://sdmmp.com/upload/SDMMP_Repository/0/rzxynghs6qc2vtw57mdbk9p018j4f3.pdf (accessed on 10 February 2022).
24. Mayorquin, J.S.; Carrillo, J.D.; Twizeyimana, M.; Peacock, B.B.; Sugino, K.Y.; Na, F.; Wang, D.H.; Kabashima, J.N.; Eskalen, A. Chemical Management of Invasive shot hole borer and Fusarium Dieback in California sycamore (Platanus racemosa) in Southern California. *Plant Dis.* **2018**, *102*, 1307–1315. [CrossRef] [PubMed]
25. Ibarra-Laclette, E.; Sánchez-Rangel, D.; Hernández-Domínguez, E.; Pérez-Torres, C.A.; Ortiz-Castro, R.; Villafán, E.; Alonso-Sánchez, A.; Rodríguez-Hass, B.; López-Buenfil, A.; García-Avila, C.; et al. Draft Genome Sequence of the Phytopathogenic Fungus Fusarium euwallaceae, the Causal Agent of Fusarium Dieback. *Genome Announc.* **2017**, *5*, e00881-17. [CrossRef] [PubMed]
26. Sánchez-Rangel, D.; Hernandez-Dominguez, E.E.; Perez-Torres, C.A.; Ortiz-Castro, R.; Villafan, E.; Rodriguez-Haas, B.; Alonso-Sanchez, A.; Lopez-Buenfil, A.; Carrillo-Ortiz, N.; Hernandez-Ramos, L.; et al. Environmental pH modulates transcriptomic responses in the fungus Fusarium sp. associated with KSHB Euwallacea sp. near fornicatus. *BMC Genom.* **2018**, *19*, 721. [CrossRef] [PubMed]
27. Pariona, N.; Mtz-Enriquez, A.I.; Sánchez-Rangel, D.; Carrión, G.; Paraguay-Delgado, F.; Rosas-Saito, G. Green-synthesized copper nanoparticles as a potential antifungal against plant pathogens. *RSC Adv.* **2019**, *9*, 18835–18843. [CrossRef]
28. Balouiri, M.; Sadiki, M.; Ibnsouda, S.K. Methods for in vitro evaluating antimicrobial activity: A review. *J. Pharm. analysis* **2016**, *6*, 71–79. [CrossRef]
29. Bozzola, J.J.; Russell, L.D. *Electron Microscopy: Principles and Techniques for Biologists*; Jones and Bartlett: Boston, MA, USA, 1999.
30. Bolger, A.M.; Lohse, M.; Usadel, B. Trimmomatic: A flexible trimmer for Illumina sequence data. *Bioinformatics* **2014**, *30*. [CrossRef] [PubMed]
31. Langmead, B.; Salzberg, S.L. Fast gapped-read alignment with Bowtie 2. *Nat. Methods* **2012**, *9*, 357–359. [CrossRef] [PubMed]
32. Li, B.; Dewey, C.N. RSEM: Accurate transcript quantification from RNA-Seq data with or without a reference genome. *BMC Bioinform.* **2011**, *12*, 323. [CrossRef]
33. Wagner, G.P.; Kin, K.; Lynch, V.J. Measurement of mRNA abundance using RNA-seq data: RPKM measure is inconsistent among samples. *Theory Biosci.* **2012**, *131*, 281–285. [CrossRef]
34. Love, M.I.; Huber, W.; Anders, S. Moderated estimation of fold change and dispersion for RNA-seq data with DESeq2. *Genome Biol.* **2014**, *15*, 550. [CrossRef] [PubMed]
35. Ashburner, M.; Ball, C.A.; Blake, J.A.; Botstein, D.; Butler, H.; Cherry, J.M.; Davis, A.P.; Dolinski, K.; Dwight, S.S.; Eppig, J.T.; et al. Gene ontology: Tool for the unification of biology. The Gene Ontology Consortium. *Nat. Genet.* **2000**, *25*, 25–29. [CrossRef] [PubMed]

36. Barrett, A.J. Nomenclature Committee of the International Union of Biochemistry and Molecular Biology (NC-IUBMB). Enzyme Nomenclature. Recommendations 1992. Supplement 4: Corrections and additions (1997). *Eur. J. Biochem.* **1997**, *250*, 1–6. [CrossRef] [PubMed]
37. Kanehisa, M.; Sato, Y.; Kawashima, M.; Furumichi, M.; Tanabe, M. KEGG as a reference resource for gene and protein annotation. *Nucleic Acids Res.* **2016**, *44*, D457–D462. [CrossRef] [PubMed]
38. Jones, P.; Binns, D.; Chang, H.Y.; Fraser, M.; Li, W.; McAnulla, C.; McWilliam, H.; Maslen, J.; Mitchell, A.; Nuka, G.; et al. InterProScan 5: Genome-scale protein function classification. *Bioinformatics* **2014**, *30*, 1236–1240. [CrossRef]
39. Finn, R.D.; Attwood, T.K.; Babbitt, P.C.; Bateman, A.; Bork, P.; Bridge, A.J.; Chang, H.-Y.; Dosztányi, Z.; El-Gebali, S.; Fraser, M.; et al. InterPro in 2017—beyond protein family and domain annotations. *Nucleic Acids Research* **2017**, *45*, D190–D199. [CrossRef]
40. Raudvere, U.; Kolberg, L.; Kuzmin, I.; Arak, T.; Adler, P.; Peterson, H.; Vilo, J. g:Profiler: A web server for functional enrichment analysis and conversions of gene lists (2019 update). *Nucleic Acids Res.* **2019**, *47*, W191–W198. [CrossRef] [PubMed]
41. Robinson, J.R.; Isikhuemhen, O.S.; Anike, F.N. Fungal-Metal Interactions: A Review of Toxicity and Homeostasis. *J. Fungi* **2021**, *7*, 225. [CrossRef] [PubMed]
42. Ren, M.; Li, R.; Han, B.; You, Y.; Huang, W.; Du, G.; Zhan, J. Involvement of the High-Osmolarity Glycerol Pathway of Saccharomyces Cerevisiae in Protection against Copper Toxicity. *Antioxidants* **2022**, *11*, 200. [CrossRef] [PubMed]
43. Jo, W.J.; Loguinov, A.; Chang, M.; Wintz, H.; Nislow, C.; Arkin, A.P.; Giaever, G.; Vulpe, C.D. Identification of genes involved in the toxic response of Saccharomyces cerevisiae against iron and copper overload by parallel analysis of deletion mutants. *Toxicol. Sci. Off. J. Soc. Toxicol.* **2008**, *101*, 140–151. [CrossRef] [PubMed]
44. Dong, K.; Addinall, S.G.; Lydall, D.; Rutherford, J.C. The yeast copper response is regulated by DNA damage. *Mol. Cell Biol.* **2013**, *33*, 4041–4050. [CrossRef] [PubMed]
45. Oc, S.; Eraslan, S.; Kirdar, B. Dynamic transcriptional response of Saccharomyces cerevisiae cells to copper. *Sci. Rep.* **2020**, *10*, 18487. [CrossRef]
46. Yasokawa, D.; Murata, S.; Kitagawa, E.; Iwahashi, Y.; Nakagawa, R.; Hashido, T.; Iwahashi, H. Mechanisms of copper toxicity in Saccharomyces cerevisiae determined by microarray analysis. *Environ. Toxicol.* **2008**, *23*, 599–606. [CrossRef] [PubMed]
47. Georgatsou, E.; Alexandraki, D. Two distinctly regulated genes are required for ferric reduction, the first step of iron uptake in Saccharomyces cerevisiae. *Mol. Cell Biol.* **1994**, *14*, 3065–3073. [CrossRef]
48. Hassett, R.; Kosman, D.J. Evidence for Cu(II) reduction as a component of copper uptake by Saccharomyces cerevisiae. *J. Biol. Chem.* **1995**, *270*, 128–134. [CrossRef] [PubMed]
49. Yamaguchi-Iwai, Y.; Dancis, A.; Klausner, R.D. AFT1: A mediator of iron regulated transcriptional control in Saccharomyces cerevisiae. *Embo J.* **1995**, *14*, 1231–1239. [CrossRef] [PubMed]
50. Georgatsou, E.; Mavrogiannis, L.A.; Fragiadakis, G.S.; Alexandraki, D. The yeast Fre1p/Fre2p cupric reductases facilitate copper uptake and are regulated by the copper-modulated Mac1p activator. *J. Biol. Chem.* **1997**, *272*, 13786–13792. [CrossRef] [PubMed]
51. Beaudoin, J.; Ekici, S.; Daldal, F.; Ait-Mohand, S.; Guérin, B.; Labbé, S. Copper transport and regulation in Schizosaccharomyces pombe. *Biochem. Soc. Trans.* **2013**, *41*, 1679–1686. [CrossRef] [PubMed]
52. Yuan, D.S.; Dancis, A.; Klausner, R.D. Restriction of copper export in Saccharomyces cerevisiae to a late Golgi or post-Golgi compartment in the secretory pathway. *J. Biol. Chem.* **1997**, *272*, 25787–25793. [CrossRef]
53. Vest, K.E.; Leary, S.C.; Winge, D.R.; Cobine, P.A. Copper import into the mitochondrial matrix in Saccharomyces cerevisiae is mediated by Pic2, a mitochondrial carrier family protein. *J. Biol. Chem.* **2013**, *288*, 23884–23892. [CrossRef]
54. De Silva, D.M.; Askwith, C.C.; Eide, D.; Kaplan, J. The FET3 gene product required for high affinity iron transport in yeast is a cell surface ferroxidase. *J. Biol. Chem.* **1995**, *270*, 1098–1101. [CrossRef]
55. Wiemann, P.; Perevitsky, A.; Lim, F.Y.; Shadkchan, Y.; Knox, B.P.; Landero Figueora, J.A.; Choera, T.; Niu, M.; Steinberger, A.J.; Wüthrich, M.; et al. Aspergillus fumigatus Copper Export Machinery and Reactive Oxygen Intermediate Defense Counter Host Copper-Mediated Oxidative Antimicrobial Offense. *Cell Rep.* **2017**, *19*, 1008–1021. [CrossRef]
56. Cheong, Y.K.; Arce, M.P.; Benito, A.; Chen, D.; Luengo Crisóstomo, N.; Kerai, L.V.; Rodríguez, G.; Valverde, J.L.; Vadalia, M.; Cerpa-Naranjo, A.; et al. Synergistic Antifungal Study of PEGylated Graphene Oxides and Copper Nanoparticles against Candida albicans. *Nanomaterials* **2020**, *10*, 819. [CrossRef] [PubMed]
57. Dunkel, N.; Liu, T.T.; Barker, K.S.; Homayouni, R.; Morschhäuser, J.; Rogers, P.D. A gain-of-function mutation in the transcription factor Upc2p causes upregulation of ergosterol biosynthesis genes and increased fluconazole resistance in a clinical Candida albicans isolate. *Eukaryot. Cell* **2008**, *7*, 1180–1190. [CrossRef] [PubMed]
58. MacPherson, S.; Akache, B.; Weber, S.; De Deken, X.; Raymond, M.; Turcotte, B. Candida albicans zinc cluster protein Upc2p confers resistance to antifungal drugs and is an activator of ergosterol biosynthetic genes. *Antimicrob. Agents Chemother.* **2005**, *49*, 1745–1752. [CrossRef] [PubMed]
59. Jyothi Lekshmi, O.B.; Amrutha, P.R.; Jeeva, M.L.; Asha Devi, A.; Veena, S.S.; Sreelatha, G.L.; Sujina, M.G.; Syriac, T. Development of an Efficient Real-time PCR Assay to Accurately Quantify Resistant Gene Analogue Expression in Taro (Colocasia esculenta). *J. Root Crop.* **2020**, *44*, 3–11.
60. Wang, R.; Ma, P.; Li, C.; Xiao, L.; Liang, Z.; Dong, J. Combining transcriptomics and metabolomics to reveal the underlying molecular mechanism of ergosterol biosynthesis during the fruiting process of Flammulina velutipes. *BMC Genom.* **2019**, *20*, 999. [CrossRef] [PubMed]

61. Veen, M.; Stahl, U.; Lang, C. Combined overexpression of genes of the ergosterol biosynthetic pathway leads to accumulation of sterols in Saccharomyces cerevisiae. *FEMS Yeast Res.* **2003**, *4*, 87–95. [CrossRef]
62. Thabet, S.; Simonet, F.; Lemaire, M.; Guillard, C.; Cotton, P. Impact of photocatalysis on fungal cells: Depiction of cellular and molecular effects on Saccharomyces cerevisiae. *Appl. Environ. Microbiol.* **2014**, *80*, 7527–7535. [CrossRef]
63. Cordero, R.J.; Casadevall, A. Functions of fungal melanin beyond virulence. *Fungal Biol. Rev.* **2017**, *31*, 99–112. [CrossRef] [PubMed]
64. Dutton, M.V.; Evans, C.S. Oxalate production by fungi: Its role in pathogenicity and ecology in the soil environment. *Can. J. Microbiol.* **1996**, *42*, 881–895. [CrossRef]
65. Jones, D.L. Organic acids in the rhizosphere—A critical review. *Plant Soil* **1998**, *205*, 25–44. [CrossRef]
66. Liaud, N.; Giniés, C.; Navarro, D.; Fabre, N.; Crapart, S.; Gimbert, I.H.; Levasseur, A.; Raouche, S.; Sigoillot, J.-C. Exploring fungal biodiversity: Organic acid production by 66 strains of filamentous fungi. *Fungal Biol. Biotechnol.* **2014**, *1*, 1. [CrossRef]
67. Tanaka, N.; Akamatsu, Y.; Hattori, T.; Shimada, M.J.W.R. Effect of Oxalic Acid on the Oxidative Breakdown of Cellulose by the Fenton Reaction. *Wood Res. Bull. Wood Res. Inst. Kyoto Univ.* **1994**, *81*, 8–10.
68. Shimada, M.; Akamtsu, Y.; Tokimatsu, T.; Mii, K.; Hattori, T. Possible biochemical roles of oxalic acid as a low molecular weight compound involved in brown-rot and white-rot wood decays. *J. Biotechnol.* **1997**, *53*, 103–113. [CrossRef]
69. Green, F.; Highley, T.L. Mechanism of brown-rot decay: Paradigm or paradox. *Int. Biodeterior. Biodegrad.* **1997**, *39*, 113–124. [CrossRef]
70. Galván Márquez, I.; Ghiyasvand, M.; Massarsky, A.; Babu, M.; Samanfar, B.; Omidi, K.; Moon, T.W.; Smith, M.L.; Golshani, A. Zinc oxide and silver nanoparticles toxicity in the baker's yeast, Saccharomyces cerevisiae. *PLoS ONE* **2018**, *13*, e0193111. [CrossRef]
71. Horstmann, C.; Campbell, C.; Kim, D.S.; Kim, K. Transcriptome profile with 20 nm silver nanoparticles in yeast. *FEMS Yeast Res.* **2019**, *19*. [CrossRef]
72. Das, D.; Ahmed, G. Silver nanoparticles damage yeast cell wall. *J. Biotechnol.* **2012**, *3*, 36–39.
73. Butt, T.R.; Sternberg, E.; Herd, J.; Crooke, S.T. Cloning and expression of a yeast copper metallothionein gene. *Gene* **1984**, *27*, 23–33. [CrossRef]
74. Christen, S.; Peterhans, E.; Stocker, R. Antioxidant activities of some tryptophan metabolites: Possible implication for inflammatory diseases. *Proc. Natl. Acad. Sci. USA* **1990**, *87*, 2506–2510. [CrossRef] [PubMed]
75. Hayaishi, O.; Hirata, F.; Ohnishi, T.; Henry, J.P.; Rosenthal, I.; Katoh, A. INDOLEAMINE 2,3-DIOXYGENASE—INCORPORATION OF (02)-0-18- AND (02)-0-18 INTO REACTION-PRODUCTS. *J. Biol. Chem.* **1977**, *252*, 3548–3550. [CrossRef]
76. Antsotegi-Uskola, M.; Markina-Iñarrairaegui, A.; Ugalde, U. Copper Resistance in Aspergillus nidulans Relies on the P(I)-Type ATPase CrpA, Regulated by the Transcription Factor AceA. *Front. Microbiol.* **2017**, *8*, 912. [CrossRef] [PubMed]
77. Sharon, A.; Finkelstein, A.; Shlezinger, N.; Hatam, I. Fungal apoptosis: Function, genes and gene function. *FEMS Microbiol. Rev.* **2009**, *33*, 833–854. [CrossRef] [PubMed]

Journal of
Fungi

Article

Mechanism of Wheat Leaf Rust Control Using Chitosan Nanoparticles and Salicylic Acid

Mohsen Mohamed Elsharkawy [1,*], Reda Ibrahim Omara [2], Yasser Sabry Mostafa [3], Saad Abdulrahman Alamri [3], Mohamed Hashem [3,4], Sulaiman A. Alrumman [3] and Abdelmonim Ali Ahmad [5]

1 Agricultural Botany Department, Faculty of Agriculture, Kafrelsheikh University, Kafr Elsheikh 33516, Egypt
2 Wheat Diseases Research Department, Plant Pathology Research Institute, Agricultural Research Center, Giza 12619, Egypt; redaomara43@gmail.com
3 Department of Biology, College of Science, King Khalid University, Abha 62529, Saudi Arabia; ysolhasa1969@hotmail.com (Y.S.M.); amri555@yahoo.com (S.A.A.); drmhashem69@yahoo.com (M.H.); salrumman@kku.edu.sa (S.A.A.)
4 Department of Botany and Microbiology, Faculty of Science, Assiut University, Assiut 71515, Egypt
5 Department of Plant Pathology, Faculty of Agriculture, Minia University, El Minia 61519, Egypt; abdelmonim.ali@mu.edu.eg
* Correspondence: mohsen.abdelrahman@agr.kfs.edu.eg; Tel.: +20-106-577-2170

Abstract: Wheat leaf rust is one of the world's most widespread rusts. The progress of the disease was monitored using two treatments: chitosan nanoparticles and salicylic acid (SA), as well as three application methods; spraying before or after the inoculation by 24 h, and spraying both before and after the inoculation by 24 h. Urediniospore germination was significantly different between the two treatments. Wheat plants tested for latent and incubation periods, pustule size and receptivity and infection type showed significantly reduced leaf rust when compared to untreated plants. *Puccinia triticina* urediniospores showed abnormalities, collapse, lysis, and shrinkage as a result of chitosan nanoparticles treatment. The enzymes, peroxidase and catalase, were increased in the activities. In both treatments, superoxide (O_2^-) and hydrogen peroxide (H_2O_2), were apparent as purple and brown discolorations. Chitosan nanoparticles and SA treatments resulted in much more discoloration and quantitative measurements than untreated plants. In anatomical examinations, chitosan nanoparticles enhanced thickness of blade (μ), thickness of mesophyll tissue, thickness of the lower and upper epidermis and bundle length and width in the midrib compared to the control. In the control treatment's top epidermis, several sori and a large number of urediniospores were found. Most anatomical characters of flag leaves in control plants were reduced by biotic stress with *P. triticina*. Transcription levels of *PR1-PR5* and *PR10* genes were activated in chitosan nanoparticles treated plants at 0, 1 and 2 days after inoculation. In light of the data, we suggest that the prospective use of chitosan nanoparticles might be an eco-friendly strategy to improve growth and control of leaf rust disease.

Keywords: *Puccinia triticina*; wheat; salicylic acid; chitosan nanoparticles; enzymes; ROS; anatomical characters

Citation: Elsharkawy, M.M.; Omara, R.I.; Mostafa, Y.S.; Alamri, S.A.; Hashem, M.; Alrumman, S.A.; Ahmad, A.A. Mechanism of Wheat Leaf Rust Control Using Chitosan Nanoparticles and Salicylic Acid. *J. Fungi* **2022**, *8*, 304. https://doi.org/10.3390/jof8030304

Academic Editor: Kamel A. Abd-Elsalam

Received: 12 February 2022
Accepted: 14 March 2022
Published: 16 March 2022

1. Introduction

Wheat is one of Egypt's and the world's most vital nutritional winter crops. Stripe, stem and leaf rust diseases induced by *Puccinia striiformis* f.sp. *tritici*, *P. graminis* f.sp. *tritici*, and *P. triticina* f.sp. *tritici*, respectively, may attack wheat throughout the season [1]. Leaf rust is Egypt's most frequent disease of wheat. It appears annually on wheat varieties and causes annual losses in grain yield. Several epidemics of leaf rust on wheat crop have been reported in the past, and this disease is a major hazard to future wheat production [2]. Moreover, it results in a significant reduction in grain production, which may be as high as 23% on susceptible wheat cultivars under ideal climatic conditions [2,3]. Since the fungus is an obligate parasite, it can continuously produce infectious urediniospores. Urediniospores are spread by the wind across great distances, infecting new wheat crops in the spring. For infection and disease development, temperatures between 10 and 18 °C with six h of dew are ideal. A new generation of pustules and spores may emerge every two weeks under these conditions [2]. The treatments with some chemical compounds and pre–inoculation with beneficial microbes may be used to systematically generate resistance in certain susceptible plants [4]. Tilt and Crown 25% fungicides were utilized in the management of yellow rust disease and showed excellent results in reducing disease severity [4]. However, the way the farmers deal with the fungicide may have a negative impact on their lives, and there can be residual effects of fungicides on the environment.

Wheat leaf rust disease has yet to be cured; therefore, the first line of defense in the resistance is the cultivation of resistant varieties. In recent years, the cultivation of more than one resistant variety was utilized [1]. The emergence of new disease races and the breaking of resistance in the varieties increase disease losses [5]. Therefore, the second line of disease control is to rely on chemical control. However, chemical control causes pollution to the environment, humans and animals. Furthermore, using the same fungicides regularly may raise the possibility of establishing aggressive fungicide-resistant strains [6]. Additionally, increased fungicide use has negative impacts on human health, food safety, and environmental hazards, as well as the potential for toxicity to non-target beneficial bacteria. As a result, fungicide-based management methods are not long-term effective. This encourages us to search for some safe alternative methods to combat this disease. Compounds such as, benzothiadiazole (BTH), other chemical inducers, *Artemisia cina* extract, salicylic acid and chitosan were used to control several diseases [7–10]. However, the effect of chitosan nanoparticles on the severity of wheat rust diseases and the involved mechanisms in disease resistance remain unclear. Sustainable techniques are needed to develop innovative alternative control approaches that combine safe and environmentally acceptable methods such as chitosan nanoparticles and salicylic acid to decrease the use of fungicides completely or partly.

Bacteria, fungi, and viruses are often associated with the buildup of reactive oxygen species (ROS), which causes oxidative stress in plants. Up-regulation of antioxidant defense mechanisms in plants appears to be a general reaction to oxidative stress under natural circumstances [6]. However, plant cells produce different antioxidant enzymes to reduce the harmful effects of oxidation. Superoxide dismutase, which converts superoxide (O_2^-) into hydrogen peroxide (H_2O_2), and catalase, which converts H_2O_2 into water and oxygen gas, are two key players [11]. O_2^- and H_2O_2 have the potential to degrade DNA, proteins, and lipids, making them toxic to the pathogen. ROS metabolism during pathogen attack includes several antioxidant enzymes such as ascorbate peroxidase (APX), peroxidase (POX), superoxide dismutase (SOD), catalase (CAT), polyphenol oxidase (PPO), and glutathione reductase (GR). Plants may be protected from oxidative stress by developing an antioxidant defense mechanism that detoxifies ROS [9,11].

Nanotechnology is regarded as a vital method having economic, social, and environmental implications [12–14]. The area of nanotechnology is one of the most active fields of recent research [15]. Nanoparticles are defined by certain features such as size, shape and distribution with new or improved characteristics [16]. Nanoparticles and nanomaterials are swiftly used for novel applications. Nanotechnology is now being utilized in today's antimicrobial industry [17]. Chitosan is a naturally present cationic biopolymer consisting of N-acetyl-D-glucosamine and D-glucosamine units linked together by β-1,4-glycosidic bonds [18,19]. A previous study has evaluated the antibacterial properties of chitosan, and more recently, several kinds of chitosan derivates have been made to boost its natural antimicrobial properties [20,21]. Moreover, chitosan treatment affects a number of genes in plants, including the genes involved in defense pathway activation, resulting in the accumulation of defense proteins [22].

The research aims to test the potential of chitosan nanoparticles and salicylic acid against wheat leaf rust disease. A variety of immune-related responses to chitosan nanoparticles and salicylic acid treatments (before inoculation, after inoculation and before and after inoculation) was investigated. Therefore, the effects of chitosan nanoparticles and salicylic acid on stimulating systemic resistance (activation of CAT, POX and ROS) and transcription levels of defense-related genes, as well as the direct effect on urediniospores were evaluated to determine how chitosan nanoparticles and salicylic acid affect rust disease.

2. Materials and Methods

This research was performed in the leaf rust greenhouse, ARC, Giza, Egypt (20–22 °C, 14/10 light/dark cycle, 50–55% relative humidity). Chitosan nanoparticles (purity 99%, Nanoshel, Congleton, UK) were mixed with acetic acid (1%) and held overnight under magnetic stirring for the full dissolving of its particles before diluting with distilled water to get the appropriate volume. The concentration of nano chitosan and salicylic acid is 150 μg/ml. The same concentration of acetic acid (1%) was mixed with water for the control treatment. Under greenhouse conditions, the effects of chitosan nanoparticles and salicylic acid treatments on the incidence of leaf rust disease were studied.

2.1. Cultivation of Wheat Plants

Morocco, a susceptible variety, was cultivated in plastic pots (10 cm in diameter, filled with clay soil) using 10 grains per pot in the greenhouse. The inoculation and incubation procedures were performed after 7 days of planting [23]. To induce spore germination and development of infection, seedlings were rubbed carefully between wet fingers with water, and infected samples were scraped using sterile spatulas and applied to these seedlings and carefully sprayed again with water. Finally, the infected seedlings were incubated for 24 h in moist chambers at 18–20 °C and 100% RH before being transferred to benches in a greenhouse for 14 days at a temperature 20 ± 2 °C with 50–55% relative humidity and 7500 Lux light intensity (14 h light and 10 h dark). Twelve days after pustules appeared, rust data were collected. After pustules rupture, rust data were recorded after 12 days. Rust symptoms were graded as infection type, with resistance (=0, 0; 1 and 2) and susceptible (=3 and 4) indicating low infection type and high infection type, respectively (Supplementary Table S1) [24]. The application methods were A= spray 24 h before inoculation, B = spray 24 h after inoculation C = spray 24 h before and after inoculation. Chitosan nanoparticles and salicylic acid treatments were sprayed at a rate of 10 mL per plant.

2.2. Morphogenesis of the Disease on the Susceptible Variety

2.2.1. Effect of Chitosan Nanoparticles and Salicylic Acid Treatments on *Puccinia triticina* Spores

Chitosan nanoparticles and salicylic acid treatments were sprayed directly on the susceptible variety, then (after 6 h) leaf samples (6 cm long) were chopped and observed under a light microscope to observe the morphology of the spores.

On glass slides, urediniospores were placed according to the general method [25]. Slides were washed with ethyl alcohol and air-dried before being covered with a thin smear of 2% water agar that had been supplemented with chitosan nanoparticles, salicylic acid and non-treated control. In sterilized Petri dishes containing many layers of water-saturated filter papers, slides were put on V-shaped glass rods. Slides holding spores were incubated at 25 °C for 12 h under continuous illumination before being examined microscopically at ×100 magnification to evaluate spore germination [26]. A germ tube longer than the spore's width was considered valid for spores' germination [27]. For 100 spores on a slide, germination percentages were determined. For each treatment, three slides were analyzed. Water agar slides without treatments were used as a control.

2.2.2. Latent and Incubation Periods and Number of Pustules

Incubation period was determined by counting the number of visible pustules on marked leaves per day until no more pustules developed [28]. The latent period was determined by the time between inoculation and 50% of pustules that were evident or emerged. On the top surface of the leaves, the number of pustules per unit leaf area (2.0×0.5 cm^2) was counted [29].

2.2.3. Measurement of Pustule Size

Due to the evident variable forms of *P. triticina* pustules, length and width measurements were used to compare pustule sizes. The dimensions of 20 pustules on the first leaf of susceptible plants were assessed in length and width.

2.3. Biochemical Assays of Antioxidant Enzymes

Fresh wheat leaf material (0.5 g) was homogenized in 50 mM Tris buffer (3 mL, pH 7.8, with 1 mM EDTA-Na2) and 7.5% polyvinylpyrrolidone for enzyme analyses. The homogenates were centrifuged at 12,000 rpm for 20 min [8]. The UV-160A spectrophotometer (Shimadzu, Japan) was used for all measurements, which were performed at 25 °C. Catalase activity was measured [30]. Sodium phosphate buffer (2 mL, 0.1 M, pH 6.5), 100 μL of H_2O_2 (0.02M) and enzyme extract (50 μL) were added to the reaction mixture. The activity was estimated using the extinction coefficient (0.04 mM^{-1} cm^{-1} at 240 nm). For 3 min, changes at 240 nm absorbance were measured every 30 sec. Molecular hydrogen peroxide g FW^{-1} min^{-1} was used to measure enzyme activity.

Peroxidase was assessed in the crude enzyme extract [31]. Sodium phosphate buffer (2 mL, pH 6.0) with 0.25% (V/V) guaiacol (2-methoxyphenol, 100 μL) and 100 mM H_2O_2 (100 μL) were used in the process. A crude enzyme extract (100 μL) was added to start the procedure. For 3 min, the changes in the absorbance (470 nm) were determined at 30 s intervals. The tetra-guaiacol extinction coefficient (26.6 mM^{-1} cm^{-1} at 470 nm) was employed to measure the activity. The activity of enzymes was measured in μmol tetraguaiacol g FW^{-1} min^{-1}.

2.4. Histochemical Analysis of Reactive Oxygen Species (ROS)

Nitro blue tetrazolium (NBT) and 3,3-diaminobenzidine (DAB) were used to detect superoxide (O_2^-) and hydrogen peroxide (H_2O_2), respectively. Infiltration of the leaves was done with potassium salicylate buffer (10 mM, pH 7.8, containing 0.1% NBT or DAB, Sigma–Aldrich, Steinheim, Germany). To remove the trichloroacetic acid, the samples were purified in ethanol:chloroform (4:1, v/v) for a day using trichloroacetic acid in NBT-and DAB-treated samples (0.15 w/v%) [32]. Before being evaluated, cleared samples were rinsed with water and transferred into 50% glycerol. Using Chemilmager 4000 digital system, discoloration of leaves was measured.

2.5. Anatomical Studies

At the age of 15 days, flag leaves measuring 0.5 cm in length were collected. The materials were cleaned in 50% ethyl alcohol and dehydrated in a standard butyl alcohol series after treatment with the killing and fixation solution (FAA). The specimens were then coated in paraffin wax (56–58 °C). The rotary microtome type 820 was used to cut transverse sections that were 12 microns thick. Albumin was used to fix the pieces, which were then dyed with safranin and mounted in canada balsam [33]. The sections were inspected microscopically and photographed.

2.6. Defense-Related Genes Transcriptional Levels

Wheat samples (the first leaf) were used to extract RNA. For all treatments, 100 mg of wheat leaves were collected at 0, 1, and 2 dpi (days post inoculation) for total RNA extraction. RNA was purified using the kit Thermo Scientific, Fermentas, #K0731 [34]. After verifying the concentration, integrity and purity of RNA were assessed using agarose gel electrophoresis and Nano SPECTROstar. The reverse transcription process was done using the reverse transcription kits (Thermo Scientific, Fermentas, #EP0451). The generated cDNA was employed for qRT-PCR amplification using specific primers to identify the expression patterns of the six wheat genes (*PR1-PR5* and *PR10*) (Table 1) [35]. All genes transcripts were amplified using a real-time cycler [35]. To normalize gene expressions, the *β-tubulin* reference gene (Table 1) was used. Three technical and biological replicates were used for each treatment. The relative expression levels were calculated using Livak and Schmittgen's method [36].

Table 1. The nucleotide sequences of primers utilized in this investigation.

Primer Name	Forward Primer (5′ . . . . . 3′)	Reverse Primer (5′ . . . . . 3′)
PR1 (basic)	CTGGAGCACGAAGCTGCAG	CGAGTGCTGGAGCTTGCAGT
PR2	CTCGACATCGGTAACGACCAG	GCGGCGATGTACTTGATGTTC
PR3	AGAGATAAGCAAGGCCACGTC	GGTTGCTCACCAGGTCCTTC
PR4	CGAGGATCGTGGACCAGTG	GTCGACGAACTGGTAGTTGACG
PR5	ACAGCTACGCCAAGGACGAC	CGCGTCCTAATCTAAGGGCAG
PR10	TTAAACCAGCACGAGAAACATCAG	ATCCTCCCTCGATTATTCTCACG
β-tubulin	GCCATGTTCAGGAGGAAGG	CTCGGTGAACTCCATCTCGT

2.7. Statistical Analysis

Three repetitions of a randomized complete block design (RCBD) were utilized. Analysis of variance (ANOVA) was used to statistically examine the data using the SPSS software V22.0 22 (SPSS Inc., Chicago, IL, USA).

3. Results

3.1. Effect of Chitosan Nanoparticles and Salicylic Acid on Urediniospores Germination

The effect of salicylic acid and chitosan nanoparticles on *P. triticina* urediniospores germination on water-agar medium was studied (Figure 1). All treatments gave significant differences in urediniospores germination. The best treatment was chitosan nanoparticles in the inhibition of germination (Figure 1A,D), followed by salicylic acid treatment (Figure 1B,D) compared to the control (Figure 1C,D).

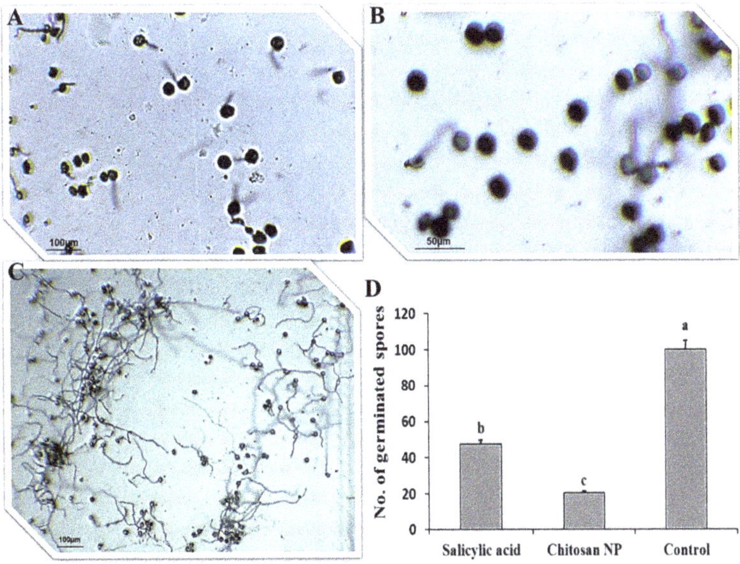

Figure 1. *Puccinia triticina* spore germination on water agar media with two treatments; salicylic acid (**A**), chitosan nanoparticles (**B**) and control non-treated (**C**) and the number of germinated spores (**D**). The letters (a, b and c) denote significant difference.

3.2. The Effect of Chitosan Nanoparticles and Salicylic Acid on the Development of Wheat Leaf Rust

Data illustrated in (Figure 2) show the effect of salicylic acid and chitosan nanoparticles and three application methods; spray before inoculation by 24 h, spray after inoculation by 24 h and spray before and after inoculation by 24 h on incubation period, latent period and infection type to assess disease development on treated and untreated wheat plants. The two treatments were significantly effective in controlling leaf rust compared with the check control (treated with water). Chitosan nanoparticles treatment revealed a substantial rise in latent and incubation periods compared to the control treatment, which displayed the highest latent and incubation periods with three application methods (Figure 2A,B). Salicylic acid significantly increased incubation and latent periods compared to the control treatment but with less order. On the other hand, chitosan nanoparticles treatment decreased infection type (IT) with three applications compared to control (Figures 2C and 3). It was also noted that the best application method was spray before and after inoculation by 24 h on increased incubation and latent periods and decreased infection type, followed by the application method of spray after inoculation by 24 h.

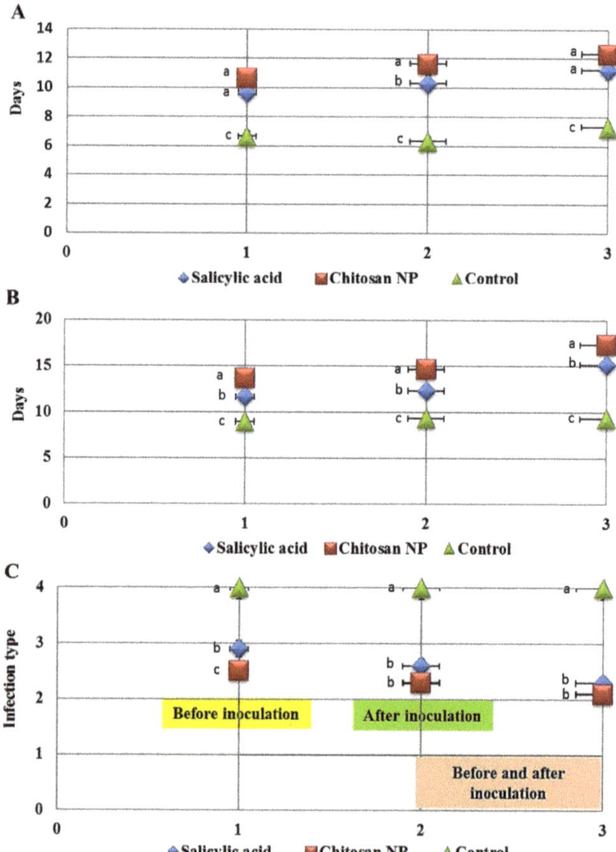

Figure 2. Effect of salicylic acid and chitosan nanoparticles application methods (1 = spray before inoculation by 24 h, 2 = spray after inoculation by 24 h and 3 = spray before and after inoculation by 24 h) on incubation period (**A**), latent periods (**B**) and infection type (**C**) of wheat leaf rust disease. The letters (a, b and c) denote significant difference.

Figure 3. Effect of salicylic acid (1) and chitosan nanoparticles (2) applications (spray before inoculation by 24 h, spray after inoculation by 24 h and spray before and after inoculation by 24 h) on infection type of wheat leaf rust disease.

3.3. Effect of Chitosan Nanoparticles and Salicylic acid on Pustules Size and Receptivity

It was clear from the study that chitosan nanoparticles and salicylic acid achieved significant results in influencing the pustule size (pustules length and width) and the number of pustules compared to the control with the three application methods (Figure 4). Chitosan nanoparticles treatment was the best treatment in reducing pustules length and width compared to the control, followed by salicylic acid treatment (Figure 4A,B). Similarly, chitosan nanoparticles treatment was the best treatment in decreasing the number of pustules (1 cm^2) compared to the control (Figure 4C). However, spray before and after inoculation by 24 h was the best application method.

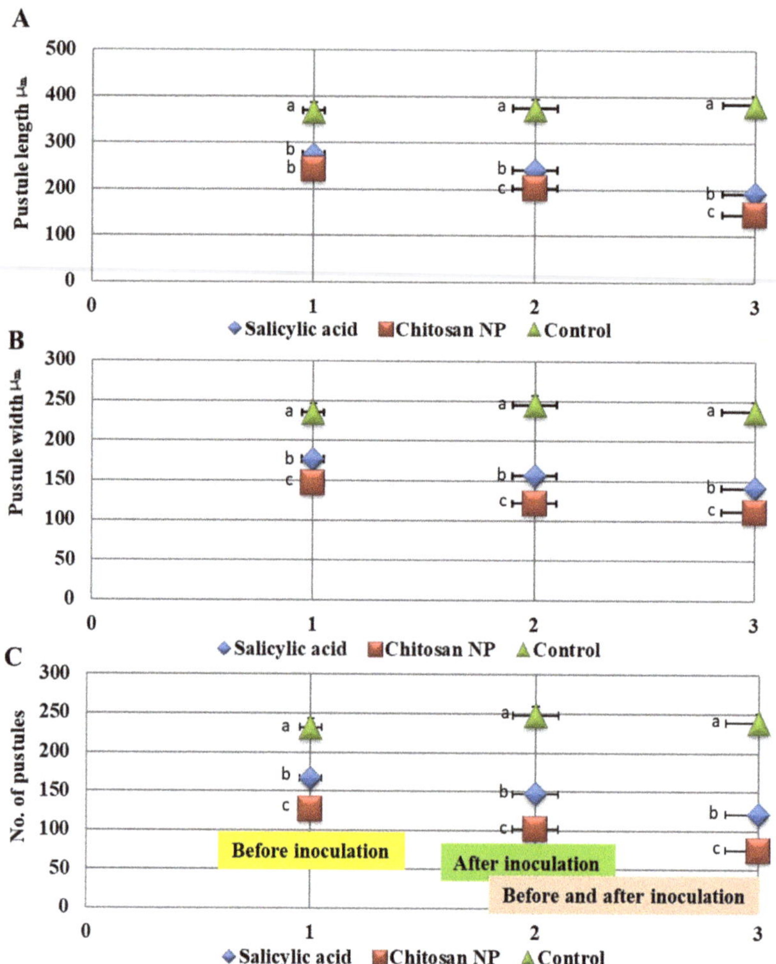

Figure 4. Effect of salicylic acid and chitosan nanoparticles applications (1 = spray before inoculation by 24 h, 2 = spray after inoculation by 24 h and 3 = spray before and after inoculation by 24 h) on pustules length (**A**), pustules width (**B**) and receptivity (**C**). The letters (a, b and c) denote significant difference.

3.4. Effect of Chitosan Nanoparticles and Salicylic Acid on Disease Symptoms

The direct effect of the tested treatments on symptoms was studied (Figure 5). The effect was evident through abnormalities, collapse, lysis and shrinking of urediniospores in *P. triticina*. Chitosan nanoparticles treatment resulted in lysis, abnormalities and shrinking of urediniospores (Figure 5A), while salicylic acid treatment resulted in abnormalities, collapse and shrinking of urediniospores (Figure 5B) compared to the control (Figure 5C). Moreover, the number of abnormal urediniospores was high with chitosan nanoparticles followed by salicylic acid compared to the control (Figure 5D).

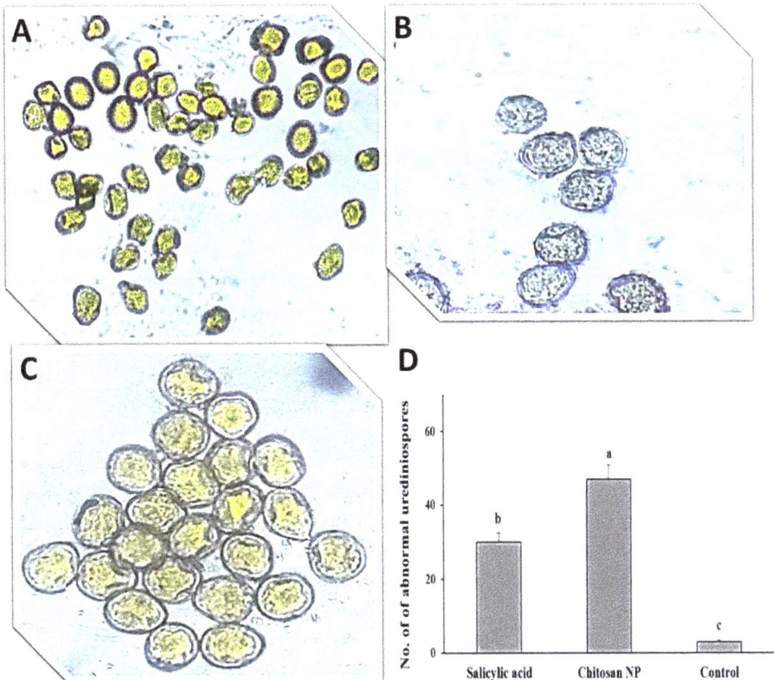

Figure 5. Light microscopy of *P. triticina* spores on wheat leaves treated with salicylic acid (**A**), chitosan nanoparticles (**B**) and control (untreated) (**C**) and the number of collapses, lysis, abnormalities and shrinking of urediniospores (**D**). The letters (a, b and c) denote significant difference.

3.5. Effect of Chitosan Nanoparticles and Salicylic Acid Treatments on Enzyme Activities

The activity of the enzymes represented by peroxidase and catalase was demonstrated with different application methods (Figure 6). The highest values were achieved in before and after inoculation by 24 h followed by after inoculation by 24 h in increasing the activity of enzymes. The study also showed that the maximum increase in enzyme activities was achieved using chitosan nanoparticles, followed by salicylic acid, and all this compared to control (Figure 6).

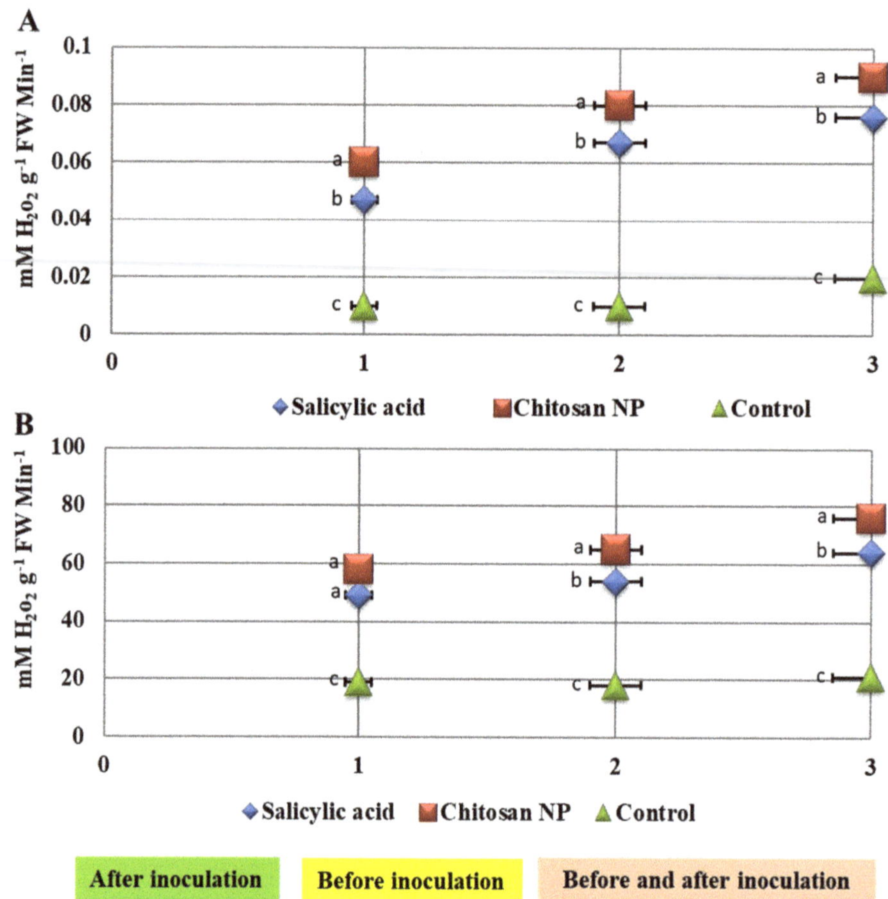

Figure 6. Effect of salicylic acid and chitosan nanoparticles application methods (1 = spray before inoculation by 24 h, 2 = spray after inoculation by 24 h and 3 = spray before and after inoculation by 24 h) on peroxidase (**A**) and catalase (**B**) activities in wheat leaves infected with *Puccinia triticina*. The letters (a, b and c) denote significant difference.

3.6. Histochemical Analysis of Reactive Oxygen Species (ROS)

Hydrogen peroxide (H_2O_2) and superoxide (O_2^-) were visualized as purple and brown discoloration in the salicylic acid and chitosan nanoparticles treatments and they were also quantified (Figure 7). The discoloration was significantly increased in chitosan nanoparticles treatment compared to the control treatment (Figure 7A). For quantitative measurements, the chitosan nanoparticles treatment was the highest in hydrogen peroxide (H_2O_2), and superoxide (O_2^-) compared to the control treatment (Figure 7B). Moreover, the highest values of all treatments were achieved in the application method of before and after inoculation by 24 h.

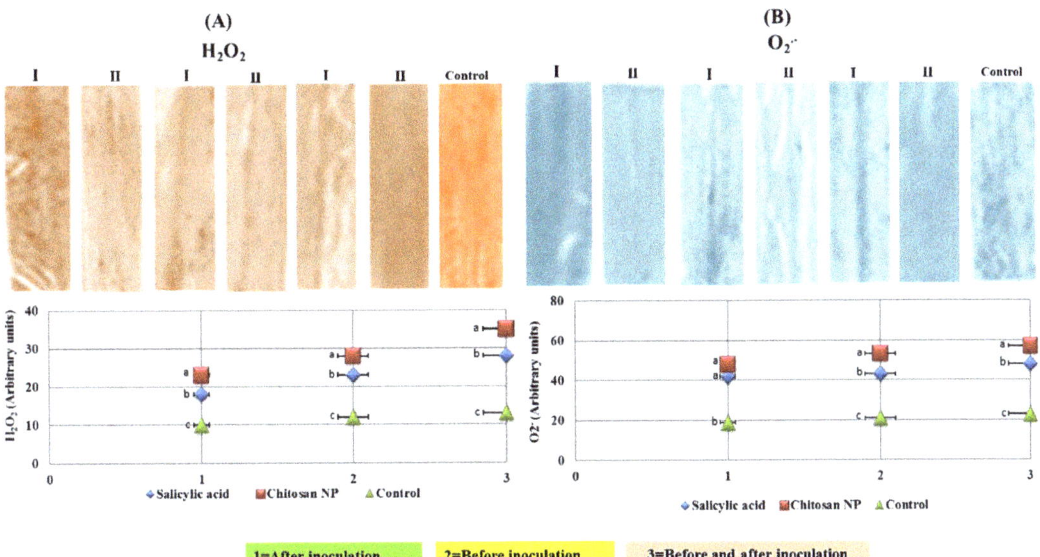

Figure 7. Effect of salicylic acid (I) and chitosan nanoparticles (II) on hydrogen peroxide (**A**) and superoxide anion (**B**) in infected plants with *Puccinia triticina* with three application methods (1 = spray before inoculation by 24 h, 2 = spray after inoculation by 24 h and 3 = spray before and after inoculation by 24 h). The letters (a, b and c) denote significant difference.

3.7. Effect of Chitosan Nanoparticles and Salicylic Acid on Anatomical Traits

The best application method was chosen through the previous results to evaluate the effect of the treatments on anatomical traits of infected wheat leaves. Data illustrated that flag leaf anatomical traits were reduced in control plants exposed to *P. triticina* stress (Figure 8). Salicylic acid and chitosan nanoparticles treatments increased the thickness of blade (μ), the thickness of the lower and upper epidermis, the thickness of mesophyll tissue and bundle length and width in the midrib compared to the control. Additionally, urediniospores and sori are abundant in the upper epidermis of the control leaves (Figure 8). The impact of chitosan nanoparticles treatment was higher than salicylic acid treatment in decreasing the number of urediniospores and increasing the thickness of blade (μ), the thickness of the lower and upper epidermis, the thickness of mesophyll tissue and bundle length and width in the midrib (Figure 8).

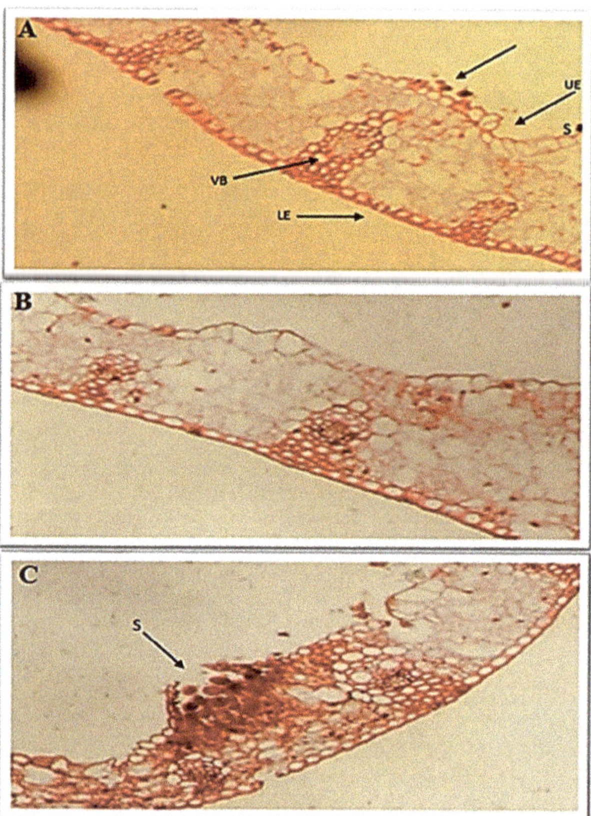

Figure 8. Transverse sections of salicylic acid (**A**) and chitosan nanoparticles (**B**) compared to the control (**C**). (Magnification × 100). Details:—UE: upper epidermis, VB: vascular bundle, MT: mesophyll tissue, S: urediniospores.

4. Defense-Related Genes Transcriptional Levels

Pathogenesis-related genes (*PR1-PR5* and *PR10*) transcription levels were examined at time intervals of 0, 1 and 2 days after inoculation. One day after treatments (0 day after inoculation), induction of the genes *PR1, PR3* and *PR4* was significantly greater in the chitosan nanoparticles treated plants compared to salicylic acid-treated plants, and no significant differences was found in the case of *PR2* and *PR10* gene expressions (Figure 9). One day after inoculation, the genes *PR1, PR3, PR4* and *PR10* were significantly stimulated to higher levels in chitosan nanoparticles treated plants, and no significant difference was found between both treatments in the case of *PR5* (Figure 10). This is principally striking for *PR3* and *PR4,* which had transcription levels around 5 times higher in the chitosan nanoparticles treatment than in salicylic acid treatment. *PR1, PR3, PR4* and *PR10* had the greatest expression levels in chitosan nanoparticles at 2 dpi when compared to salicylic acid and a mock-inoculated control (Figure 11). Both chitosan nanoparticles and salicylic acid dramatically increased transcriptions of *PR2* and *PR5,* with relative expression levels almost 8-fold higher than the mock-inoculated control.

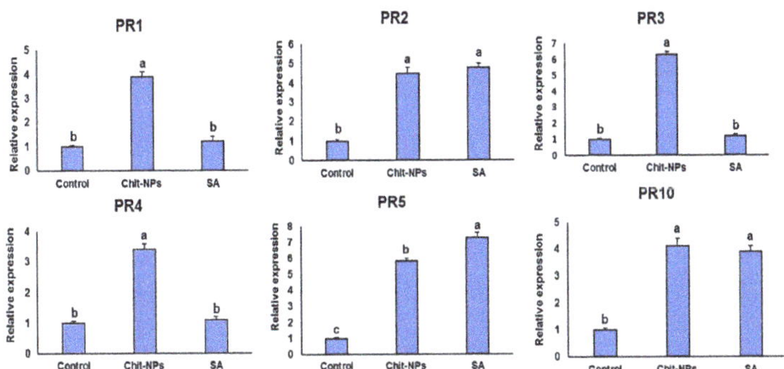

Figure 9. Effect of chitosan nanoparticles (Chit-NPs) and salicylic acid (SA) on the relative transcription levels of *PR1-PR5* and *PR10* genes in infected wheat plants at 1 day after treatment (0-day post inoculation). The letters (a, b and c) denote significant difference.

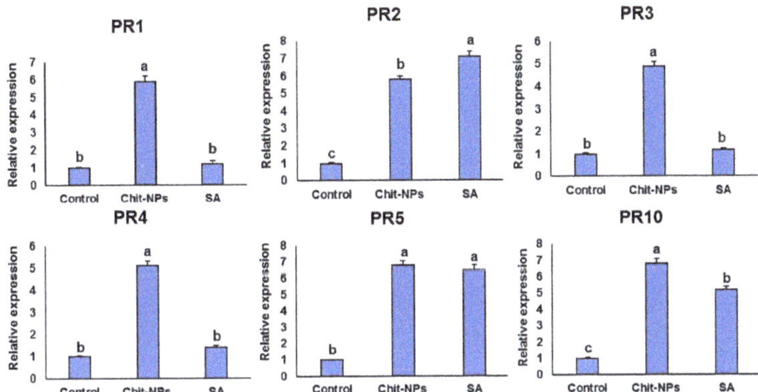

Figure 10. Effect of chitosan nanoparticles (Chit-NPs) and salicylic acid (SA) on the relative transcription levels of *PR1-PR5* and *PR10* genes in infected wheat plants at 1 day post inoculation. The letters (a, b and c) denote significant difference.

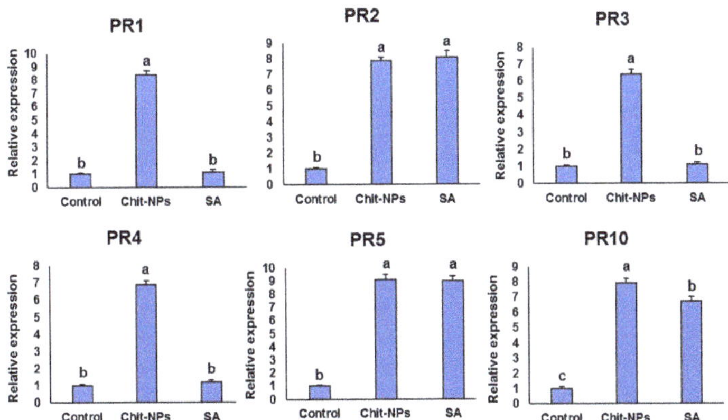

Figure 11. Effect of chitosan nanoparticles (Chit-NPs) and salicylic acid (SA) on the relative transcription levels of *PR1-PR5* and *PR10* genes in infected wheat plants at 2 days post inoculation. The letters (a, b and c) denote significant difference.

5. Discussion

Leaf rust is a common disease in wheat in all growing areas of the world. Field observations showed that it appears annually at varying magnitudes in the different areas [3]. The severity of infection varies according to the sensitivity of each variety to this disease. Therefore, in the case of planting a highly susceptible variety, chemical resistance must be used to reduce the resulting losses [37]. This can also be achieved by developing rust resistant genotypes or by successful varietal manipulation of the available genotypes throughout the country to avoid heavy infection to the susceptible genotypes [1]. The emergence of new races capable of breaking the resistance in the new varieties, such as the emergence of the TTTST race on Shandweel-1, Sakha-94 and Sakha-95 varieties was reported [5]. Therefore, the second line of control methods should be used, which is chemical fungicides. The use of these chemical fungicides results in pollution to the environment and is dangerous to humans, especially people who are not familiar with the precautionary measures to deal with pesticides [38]. Therefore, the study turned to the use of some safe materials to combat this disease, such as salicylic acid and chitosan nanoparticles [10]. The effect of these chemicals on urediniospores germination and disease development was evaluated. All treatments gave significant differences in urediniospores germination. The best treatment in increasing the incubation and latent periods was chitosan nanoparticles. Plant pathogens have been reported to exhibit chitinolytic and chitosanolytic activities [39]. Variations in chitosan's impact on hyphal development were discovered in a previous study across nine plant pathogens, although *V. dahliae* was the most tolerant among them [40].

Chitosan nanoparticles treatment increased latent and incubation periods as well as decreased infection type, pustule size and the number of pustules compared to the control treatment. Salicylic acid was also effective in increasing latent and incubation periods and decreasing infection type, pustule size and number of pustules compared to the control treatment, but it had lower effectiveness compared to chitosan nanoparticles. The high effectiveness of chitosan and salicylic acid in inducing resistance against *Botrytis cinerea* under the plastic house was elucidated in a previous study [10]. The active role of chemical inducers such as BTH and salicylic acid in reducing the impact of sugar beet rust was reported [41]. Natural polymer chitosan is widely believed to be the most prevalent natural polymer with dual functions. It inhibits pathogen growth, viability, sporulation, germination and cell alterations, as well as stimulating and/or suppressing diverse defensive responses in host plants [42]. Previous studies confirmed that the application of chitosan led to a reduction in the pathological severity of root diseases [43,44]. Chitosan's antifungal effect is associated with its capacity to interfere with the plasma membrane of fungal cells and fungal DNA and/or RNA [45,46]. The treatments also clearly increased the anatomical characters of flag leaves in wheat plants. Clear increases in the thickness of blade (μ), the thickness of the lower and upper epidermis, the thickness of mesophyll tissue and bundle length and width in the midrib compared to untreated plants (control) were observed. Lower concentrations of chitosan nanoparticles were more effective than higher concentrations. Chitosan nanoparticles improved seed germination, water uptake and transport, activation of water channels proteins, and increased the absorption of nutrients in the plant. All of these alterations disappeared when the concentration of chitosan nanoparticles was increased [47].

It was necessary to explain how chitosan nanoparticles and salicylic acid work in the development of the disease, by studying the activity of some enzymes such as peroxidase and catalase as well as by studying reactive oxygen species such as hydrogen peroxide (H_2O_2) and superoxide (O_2^-). The maximum increase in enzyme activities was achieved by chitosan nanoparticles, specially the application method of before and after inoculation by 24 h compared to the control. Salicylic acid treatment resulted in increases in chitinase and peroxidase activity on both the local and systemic levels [48]. Superoxide and hydrogen peroxide discolorations were significantly increased using chitosan nanoparticles and salicylic acid treatments compared to the control treatment. For quantitative measurements, chitosan nanoparticles treatment was the highest in superoxide and hydrogen peroxide compared to the control. Antioxidants, both non-enzymatic and enzymatic, effectively remove ROS in non-stressful situations however, under stress, the synthesis of ROS and antioxidant enzymes may be affected. ROS have recently been shown to be beneficial for biosystems as signaling molecules and immune defense stimulants [49]. ROS, on the other hand, have the potential to damage organs and tissues. Studies focusing on diseases linked to ROS are controversial. There are increasing numbers of studies showing that chitosan and its derivatives have several mechanisms of action. It suppresses pathogen development and alter the defensive response of host plants [50]. Salicylic acid plays an important role in signal transduction of resistance in various plant pathogen interactions. Salicylic acid activated various defense reactions in plants against pathogen [51]. Salicylic acid induces rapid transient-generation of reactive O_2 through oxidative burst in incompatible interaction [10]. Significant increases in the activity of peroxidase and polyphenol oxidase were found after spraying wheat and sugar beet plants with salicylic acid [41,52].

Pathogenesis related proteins, β-1, 4-glucanase, peroxidase, and chitinase were activated in resistant plants [53]. Lignification in wheat seems to be of special value in induced resistance. Lignin biosynthesis in wheat is related to defense enzymes [54]. The increased lignification's rate was reported through accomplished hypersensitive reaction due to foliar application of BABA [55]. The microbial defense-enhancing activities of chitinase and beta 1,3-glucanase were useful in developing resistance to fungi [56]. The transcript levels of *PR1-PR5* and *PR10* genes were assessed using RT-qPCR relative to the reference gene *β-tubulin*. Defense genes were activated after chitosan nanoparticles and salicylic acid treatments compared to mock-inoculated controls at all three time periods analyzed. After being treated with chitosan nanoparticles for 9 h, downy mildew-infected pearl millet plants showed increased levels of defense enzymes [57]. Many genes in plants are regulated by chitosan treatment, including the stimulation of phytoprotective pathways [22]. Chitosan has been shown to increase the defense enzyme activities in *Pinus koraiensis* seedlings to their maximum levels at 2 days post-inoculation (dpi) [58]. Peroxidase, phenylalanine ammonium lyase and pathogen-related protein-1 transcriptional levels were all shown to be associated with systemic resistance induction by chitosan nanoparticles.

Through this study, it was found that salicylic acid treatment was less effective in reducing the development of the wheat leaf rust disease than chitosan nanoparticles, and this could be due to that salicylic acid required more time and high dosage before induction of resistance. The increased activities of peroxidase and catalase were associated with the induction of resistance. *PR1-PR5* and *PR10* gene transcriptions were considerably greater in plants treated with chitosan nanoparticles than in controls. Multiple mechanisms were established to mediate the resistance elicited by chitosan nanoparticles, resulting in a full decrease in the disease. The nano product of chitosan increased its effectiveness in the process of combating this disease. To manage plant pathogens in agricultural crops, nanotechnology seems to have great promise. For novel formulations of plant disease-control fungicides, concentration, molecular weight, particle size, and dose are all essential considerations that should be regarded. More research into field applications is required to evaluate the potential of chitosan nanoparticles on yield traits.

Supplementary Materials: The following supporting information can be downloaded at: https://www.mdpi.com/article/10.3390/jof8030304/s1, Supplementary Table S1: Wheat leaf rust infection types used in disease assessment for seedling stage according to Johnston and Browder.

Author Contributions: Conceptualization, M.M.E. and R.I.O.; methodology M.M.E. and R.I.O.; software, M.M.E. and R.I.O.; validation, M.M.E. and R.I.O.; formal analysis, M.M.E. and R.I.O.; investigation, M.M.E., R.I.O. and A.A.A.; resources, M.M.E. and A.A.A.; data curation, M.M.E. and A.A.A.; writing—original draft preparation, M.M.E., R.I.O. and A.A.A.; writing—review and editing, M.M.E.; visualization, Y.S.M., S.A.A. (Saad Abdulrahman Alamri), S.A.A. (Sulaiman A. Alrumman) and M.M.E.; supervision, Y.S.M., S.A.A. (Saad Abdulrahman Alamri) and M.M.E.; funding acquisition, Y.S.M., S.A.A. (Saad Abdulrahman Alamri), M.H., S.A.A. (Sulaiman A. Alrumman). All authors have read and agreed to the published version of the manuscript.

Funding: The work was funded by the Deputyship for Research & Innovation, Ministry of Education in Saudi Arabia through the project number IFP-KKU-2020/2.

Institutional Review Board Statement: Not applicable.

Informed Consent Statement: Not applicable.

Data Availability Statement: Available upon request from the corresponding author.

Acknowledgments: The authors extend their appreciation to the Deputyship for Research & Innovation, Ministry of Education in Saudi Arabia for funding this research work through the project number IFP-KKU-2020/2.

Conflicts of Interest: The authors declare no conflict of interest.

References

1. Omara, R.I.; Shahin, A.A.; Ahmed, S.M.; Mostafa, Y.S.; Alamri, S.A.; Hashem, M.; Elsharkawy, M.M. Wheat resistance to stripe and leaf rusts conferred by introgression of slow rusting resistance genes. *J. Fungi* **2021**, *7*, 622. [CrossRef] [PubMed]
2. Nazim, M.; El-Shehidi, A.A.; Abdou, Y.A.; El-Daoudi, Y.H. Yield loss caused by leaf rust on four wheat cultivars under epiphytotic levels. In Proceedings of the 4th Conference of Microbiology, Cairo, Egypt, 24–28 December 1980; pp. 17–27.
3. Ali, R.G.; Omara, R.I.; Ali, Z.A. Effect of leaf rust infection on yield and technical properties in grains of some Egyptian wheat cultivars. *Menoufia J. Plant Prot.* **2016**, *1*, 19–35. [CrossRef]
4. Gad, A.M.; Abdel-Halim, K.Y.; Seddik, F.A.; Soliman, H.M.A. Comparative of fungicidal efficacy against yellow rust disease in wheat plants in compatibility with some biochemical alterations. *Menoufia J. Plant Prot.* **2020**, *5*, 29–38. [CrossRef]
5. Omara, R.I.; Nehela, Y.; Mabrouk, O.I.; Elsharkawy, M.M. The emergence of new aggressive leaf rust races with the potential to supplant the resistance of wheat cultivars. *Biology* **2021**, *10*, 925. [CrossRef]
6. Xi, K.; Kumar, K.; Holtz, M.D.; Turkington, T.K.; Chapman, B. Understanding the development and management of stripe rust in central Alberta. *Can. J. Plant Pathol.* **2015**, *37*, 21–39. [CrossRef]
7. Hafez, Y.M.; El-Baghdady, N.A. Role of reactive oxygen species in suppression of barley powdery mildew fungus, *Blumeria graminis* f. sp. *hordei* with benzothiadiazole and riboflavin. *Egypt J. Biol. Pest. Cont.* **2013**, *23*, 125–132.
8. Hafez, Y.M.; Soliman, N.K.; Saber, M.M.; Imbaby, I.A.; Abd-Elaziz, A.S. Induced resistance against *Puccinia triticina* the causal agent of wheat leaf rust by chemical inducers. *Egypt J. Biol. Pest. Cont.* **2014**, *24*, 173–181.
9. Omara, R.I.; Kamel, S.M.; Hafez, Y.M.; Morsy, S.Z. Role of non-traditional control treatments in inducing resistance against wheat leaf rust caused by *Puccinia triticina*. *Egypt J. Biol. Pest. Cont.* **2015**, *25*, 335–344.
10. Al-Juboory, H.; Al-Hadithy, H. Evaluation of chitosan and salicylic acid to induce systemic resistance in eggplant against under plastic *Botrytis cinerea* house conditions. *Indian J. Ecol.* **2021**, *48*, 213–221.
11. Omara, R.I.; Abdelaal, K.A.A. Biochemical, histopathological and genetic analysis associated with leaf rust infection in wheat plants (*Triticum aestivum* L.). *Physiol. Mol. Plant Pathol.* **2018**, *104*, 48–57. [CrossRef]
12. El-Saadony, M.T.; Alkhatib, F.M.; Alzahrani, S.O.; Shafi, M.E.; El Abdel-Hamid, S.; Taha, T.F.; Aboelenin, S.M.; Soliman, M.M.; Ahmed, N.H. Impact of mycogenic zinc nanoparticles on performance, behavior, immune response, and microbial load in Oreochromis niloticus. *Saudi J. Biol. Sci.* **2021**, *28*, 4592–4604. [CrossRef] [PubMed]
13. Bano, A. Interactive effects of Ag-nanoparticles, salicylic acid, and plant growth promoting rhizobacteria on the physiology of wheat infected with yellow rust. *J. Plant Pathol.* **2020**, *102*, 1215–1225. [CrossRef]
14. Jasrotia, P.; Kashyap, P.L.; Bhardwaj, A.K.; Kumar, S.; Singh, G.P. Scope and applications of nanotechnology for wheat production: A review of recent advances. *Wheat Barley Res.* **2018**, *10*, 1–14.
15. El-Saadony, M.T.; Saad, A.M.; Najjar, A.A.; Alzahrani, S.O.; Alkhatib, F.M.; Shafi, M.E.; Selem, E.; Desoky, E.-S.; Fouda, S.E.E.; El-Tahan, A.M.; et al. The use of biological selenium nanoparticles to suppress *Triticum aestivum* L. crown and root rot diseases induced by *Fusarium* species and improve yield under drought and heat stress. *Saudi J. Biol. Sci.* **2021**, *28*, 4461–4471. [CrossRef] [PubMed]
16. El-Saadony, M.T.; Sitohy, M.Z.; Ramadan, M.F.; Saad, A.M. Green nanotechnology for preserving and enriching yogurt with biologically available iron (II). *Innov. Food Sci. Emerg. Technol.* **2021**, *69*, 102645. [CrossRef]
17. Elizabath, A.; Babychan, M.; Mathew, A.M.; Syriac, G.M. Application of nanotechnology in agriculture. *Int. J. Pure Appl. Biosci.* **2019**, *7*, 131–139. [CrossRef]
18. Elieh-Ali-Komi, D.; Hamblin, M.R. Chitin and chitosan: Production and application of versatile biomedical nanomaterials. *Int J Adv. Res.* **2016**, *4*, 411–427.
19. Kocięcka, J.; Liberacki, D. The potential of using chitosan on cereal crops in the face of climate change. *Plants* **2021**, *10*, 1160. [CrossRef]
20. Abd El-Hack, M.E.; El-Saadony, M.T.; Shafi, M.E.; Zabermawi, N.M.; Arif, M.; Batiha, G.E.; Khafaga, A.F.; Abd El-Hakim, Y.M.; Al-Sagheer, A.A. Antimicrobial and antioxidant properties of chitosan and its derivatives and their applications: A review. *Int. J. Biol. Macromol.* **2020**, *164*, 2726–2744. [CrossRef]

21. Omar, H.S.; Al Mutery, A.; Osman, N.H.; Reyad, N.E.H.A.; Abou-Zeid, M.A. Genetic diversity, antifungal evaluation and molecular docking studies of Cu-chitosan nanoparticles as prospective stem rust inhibitor candidates among some Egyptian wheat genotypes. *PLoS ONE* **2021**, *16*, e0257959. [CrossRef]
22. Pichyangkura, R.; Chadchawan, S. Biostimulant activity of chitosan in horticulture. *Sci. Hortic.* **2015**, *196*, 49–65. [CrossRef]
23. Soleiman, N.H.; Solis, I.; Ammar, A.; Dreisigacker, S.; Soleiman, M.H.; Martinez, F. Resistance to leaf rust in a set of durum wheat cultivars and landraces in Spain. *J. Plant Pathol.* **2014**, *96*, 353–362.
24. Johnston, C.O.; Browder, L.E. Seventh revision of the international register of physiologic races of *Puccinia recondita* f.sp. *tritici*. *Plant Dis. Report* **1966**, *50*, 756–760.
25. Nair, K.R.S.; Ellingboe, A.H. A method of controlled inoculations with condiospores of *Erysiphe graminis* var. *tritici*. *Phytopathology* **1962**, *52*, 417.
26. Reifshneider, F.J.B.; Bolitexa, L.S.; Occhiena, E.M. Powdery mildew of melon (*Cucumis melo*) caused by *Sphaerotheca fuliginea* in Brazil. *Plant Dis.* **1985**, *69*, 1069–1070.
27. Parlevliet, J.E. Partial resistance of barley to leaf rust, *Puccinia hordei*. I. effect of cultivars and development stage on latent period. *Euphytica* **1975**, *24*, 21–27. [CrossRef]
28. Parlevliet, J.E.; Kuiper, H.J. Partial resistance of barley to leaf rust, *Puccinia hordei*. IV. Effect of cultivars and development stage on infection frequency. *Euphytica* **1977**, *26*, 249–255. [CrossRef]
29. Menzies, J.G.; Belanger, R.R. Recent advances in cultural management of diseases of greenhouse crops. *Can. J. Plant Pathol.* **1996**, *18*, 186–193. [CrossRef]
30. Aebi, H. Catalase in vitro. *Methods Enzymol.* **1984**, *105*, 121–126.
31. Hammerschmidt, R.; Nuckles, E.M.; Kuć, J. Association of enhanced peroxidase activity with induced systemic resistance of cucumber to *Colletotrichum lagenarium*. *Physiol. Plant Pathol.* **1982**, *20*, 73–82. [CrossRef]
32. Huckelhoven, R.; Fodor, J.; Preis, C.; Kogel, K.H. Hypersensitive cell death and papilla formation in barley attacked by the powdery mildew fungus are associated with hydrogen peroxide but not with salicylic acid accumulation. *Plant Physiol.* **1999**, *119*, 1251–1260. [CrossRef] [PubMed]
33. Nassar, M.A.; El-Sahhar, K.F. *Botanical Preparations and Microscopy (Microtechnique)*; Academic Bookshop: Dokki, Giza, Egypt, 1998; p. 219.
34. Elsharkawy, M.; Derbalah, A.; Hamza, A.; El-Shaer, A. Zinc oxide nanostructures as a control strategy of bacterial speck of tomato caused by *Pseudomonas syringae* in Egypt. *Environ. Sci. Pollut Res.* **2020**, *27*, 19049–19057. [CrossRef]
35. Desmond, O.J.; Edgar, C.I.; Manners, J.M.; Maclean, D.J.; Schenk, P.M.; Kazan, K. Methyl jasmonate induced gene expression in wheat delays symptom development by the crown rot pathogen *Fusarium pseudograminearum*. *Physiol. Mol. Plant Pathol.* **2006**, *67*, 171–179. [CrossRef]
36. Livak, K.J.; Schmittgen, T.D. Analysis of relative gene expression data using real-time quantitative PCR and the $2^{-\Delta\Delta CT}$ method. *Methods* **2001**, *25*, 402–408. [CrossRef] [PubMed]
37. Chen, X.M. Epidemiology and control of stripe rust (*Puccinia striiformis* f sp. *tritici*) on wheat. *Can. J. Plant Pathol.* **2005**, *27*, 314–337. [CrossRef]
38. Hossard, L.; Philibert, A.; Bertrand, M.; Colnenne-David, C.; Debaeke, P.; Munier-Jolain, N.; Jeuffroy, M.H.; Richard, G.; Makowski, D. Effects of halving pesticide use on wheat production. *Sci. Rep.* **2015**, *4*, 4405. [CrossRef]
39. El Hadrami, A.; Adam, L.R.; El Hadrami, I.; Daayf, F. Chitosan in plant protection. *Mar. Drugs* **2010**, *8*, 968–987. [CrossRef] [PubMed]
40. Xu, J.; Zhao, X.; Han, X.; Du, Y. Antifungal activity of oligo chitosan against *Phytophthora capsici* and other plant pathogenic fungi *in vitro*. *Pestic. Biochem. Physiol.* **2007**, *87*, 220–228. [CrossRef]
41. Ata, A.A.; El-Samman, M.G.; Moursy, M.A.; Mostafa, M.H. Inducing resistance against rust disease of sugar beet by certain chemical compounds. *Egypt J. phytopathol.* **2008**, *36*, 113–132.
42. Hassan, O.; Chang, T. Chitosan for eco-friendly control of plant disease. *Asian J. Plant Pathol.* **2017**, *11*, 53–70. [CrossRef]
43. Benhamou, N.; Kloepper, J.W.; Tuzan, B.B. Induction of resistance against Fusarium wilt of tomato by combination of chitosan with an endophytic bacterial stain: Ultrastructure and cytochemistry of host response. *Planta* **1998**, *204*, 153–168. [CrossRef]
44. Ragab, M.M.; Ragab, M.M.; El-Nagar, M.A.; Farrag, E.S. Effect of chitosan and its derivatives as an antifungal and preservative agent on storage of tomato fruits. *Egypt J. Phytopathol.* **2001**, *29*, 107–116.
45. Bautista, B.S.; Hern, N.; Indez, L.; Bosquez, M.E.; Wil, C.L. Effect of chitosan and plant extracts on growth *of Colletotrichum gloeosporioides*, anthracnose level and quality of papaya fruit. *Crop Prot.* **2003**, *22*, 1087–1092. [CrossRef]
46. Rabea, E.; Badawy, M.T.; Stevens, C.; Smagghe, G.; Steurbaut, W. Chitosan as antimicrobial agent: Applications and Mode of action. *Biomacromolecules* **2003**, *4*, 1457–1465. [CrossRef] [PubMed]
47. Verma, S.K.; Das, A.K.; Gantait, S.; Kumar, V.; Gurel, E. Applications of carbon nanomaterials in the plant system: A perspective view on the pros and cons. *Sci. Total Environ.* **2019**, *667*, 485–499. [CrossRef]
48. Catherine, B.; Marc, O.; Philippe, T.; Jacques, D. Cloning and expression analysis of cDNAs corresponding to genes activated in cucumber showing systemic acquired resistance after BTH treatment. *BMC Plant Biol.* **2004**, *4*, 15.
49. Yang, S.; Gaojian, L. ROS and diseases: Role in metabolism and energy supply. *Mol. Cell. Biochem.* **2020**, *467*, 1–12. [CrossRef] [PubMed]

50. Xing, K.; Zhu, X.; Peng, X.; Qin, S. Chitosan antimicrobial and eliciting properties for pest control in agriculture: A review. *Agron. Sustain. Dev.* **2015**, *35*, 569–588. [CrossRef]
51. Christian, N.; Jean, P.M. Salicylic acid induction defiant mutants of *Arabidiopsis* express *PR*-2 and *PR-5* and accumulate high levels of camalexin after pathogen inoculation. *Plant Cell* **1999**, *11*, 1393–1404.
52. Omara, R.I.; Essa, T.A.; Khalil, A.A.; Elsharkawy, M.M. A case study of non-traditional treatments for the control of wheat stem rust disease. *Egypt J. Biol. Pest. Cont.* **2020**, *30*, 83. [CrossRef]
53. Quiroga, M.; Guerrero, C.; Botella, M.A.; Barcelo, A.; Amaya, I.; Meding, M.; Alonso, F.J.; Forcheti, S.M.; Tigir, H.; Vulpuesta, V.A. Tomato peroxidase involved in the synthesis of lignin and suberin. *Plant Physiol.* **2000**, *122*, 1119–1127. [CrossRef] [PubMed]
54. Bruno, M.M.; Ulrik, N.; Lilliane, G.; Hans-joachim, R. Specific inhibition of lignifications breaks hypersensitive resistance of wheat to stem rust. *Plant Physiol.* **1990**, *93*, 465–470.
55. Cohen, Y. The BABA story of induced resistance. *Phytoparasitica* **2001**, *29*, 375–378. [CrossRef]
56. Chang, M.M.; Hadwiger, L.A.; Horovits, D. Molecular characterization of a pea 1,3 glucanase induced by *Fusarium solani* and chitosan challenge. *Plant Mol. Biol.* **1992**, *20*, 609–618. [CrossRef] [PubMed]
57. Siddaiah, C.N.; Prasanth, K.V.H.; Satyanarayana, N.R.; Mudili, V.; Gupta, V.K.; Kalagatur, N.K.; Satyavati, T.; Dai, X.-F.; Chen, J.-Y.; Mocan, A. Chitosan nanoparticles having higher degree of acetylation induce resistance against pearl millet downy mildew through nitric oxide generation. *Sci. Rep.* **2018**, *8*, 2485. [CrossRef]
58. Liu, R.; Wang, Z.-Y.; Li, T.-T.; Wang, F.; An, J. The role of chitosan in polyphenols accumulation and induction of defense enzymes in *Pinus koraiensis* seedlings. *Chin. J. Plant Ecol.* **2014**, *38*, 749.

Journal of
Fungi

Article

Myco-Synthesized Molluscicidal and Larvicidal Selenium Nanoparticles: A New Strategy to Control *Biomphalaria alexandrina* Snails and Larvae of *Schistosoma mansoni* with an In Silico Study on Induced Oxidative Stress

Mostafa Y. Morad [1], Heba El-Sayed [2], Ahmed A. Elhenawy [3,4], Shereen M. Korany [5], Abeer S. Aloufi [5,*] and Amina M. Ibrahim [6]

1 Zoology and Entomology Department, Faculty of Science, Helwan University, Helwan 11795, Egypt; myame_mostafa@yahoo.com
2 Botany and Microbiology Department, Faculty of Science, Helwan University, Helwan 11795, Egypt; drhebaelsayed39@gmail.com
3 Chemistry Department, Faculty of Science, Al-Azhar University, Nasr City, Cairo 11884, Egypt; elhenawy_sci@hotmail.com
4 Chemistry Department, Faculty of Science and Art, Al Baha University, Mukhwah, Al Baha 6531, Saudi Arabia
5 Department of Biology, College of Science, Princess Nourah bint Abdulrahman University, P.O. Box 84428, Riyadh 11671, Saudi Arabia; smkorany@pnu.edu.sa
6 Medical Malacology Department, Theodor Bilharz Research Institute, Giza 12411, Egypt; aminamd.ibrahim@yahoo.com
* Correspondence: asaloufi@pnu.edu.sa

Citation: Morad, M.Y.; El-Sayed, H.; Elhenawy, A.A.; Korany, S.M.; Aloufi, A.S.; Ibrahim, A.M. Myco-Synthesized Molluscicidal and Larvicidal Selenium Nanoparticles: A New Strategy to Control *Biomphalaria alexandrina* Snails and Larvae of *Schistosoma mansoni* with an In Silico Study on Induced Oxidative Stress. *J. Fungi* **2022**, *8*, 262. https://doi.org/10.3390/jof8030262

Academic Editor: Kamel A. Abd-Elsalam

Received: 11 February 2022
Accepted: 27 February 2022
Published: 4 March 2022

Abstract: Schistosomiasis is a tropical disease with socioeconomic problems. The goal of this study was to determine the influence of myco-synthesized nano-selenium (SeNPs) as a molluscicide on *Biomphlaria alexandrina* snails, with the goal of reducing disease spread via non-toxic routes. In this study, *Penicillium chrysogenum* culture filtrate metabolites were used as a reductant for selenium ions to form nano-selenium. The SeNPs were characterized via UV-Vis spectrophotometer, Fourier transform infrared (FT-IR) spectroscopy, transmission electron microscopy (TEM), dynamic light scattering (DLS), and X-ray diffraction (XRD). Myco-synthesized SeNPs had a significant molluscicidal effect on *B. alexandrina* snails after 96 h of exposure at a concentration of 5.96 mg/L. SeNPs also had miracidicidal and cercaricidal properties against *S. mansoni*. Some alterations were observed in the hemocytes of snails exposed to SeNPs, including the formation of pseudopodia and an increasing number of granules. Furthermore, lipid peroxide, nitric oxide (NO), malondialdehyde (MDA), and glutathione s-transferase (GST) increased significantly in a dose-dependent manner, while superoxide dismutase (SOD) decreased. The comet assay revealed that myco-synthesized SeNPs could cause breaks in the DNA levels. In silico study revealed that SeNPs had promising antioxidant properties. In conclusion, myco-synthesized SeNPs have the potential to be used as molluscicides and larvicides.

Keywords: *Penicillium chrysogenum*; *Biomphalaria alexandrina*; *Schistosoma mansoni*; selenium nanoparticles; molluscicide; larvicide; docking study

1. Introduction

Schistosomiasis is a serious illness that has impacted the lives of people and animals around the world [1]. *Schistosoma mansoni* is a widely distributed parasitic species in many regions in Africa, the Middle East, and South America, where the intermediate host, a freshwater snail named *Biomphalaria* (phylum Mollusca, class Gastropoda), is located [2]. Snails have great medical, veterinary, and economic importance as they are the causative agent in transmitting diseases that affect many animals [3]. Until now, praziquantel has been widely used to treat adult trematode as well as cestode worms, but it is less effective against juvenile stages [4]. Thus, the urgent need for control strategies has evolved to control the snail population [5,6]. The chemical control of snail populations has many disadvantages, especially that it is very expensive, toxic to the non-target organism, and could accumulate in the environment [7]. Biological control of snail populations is a low-cost and effective alternative to chemical molluscicides [8].

Nanotechnology is an emerging technology that has been rapidly developed over the last two decades to enhance and manage a wide range of issues and problems in fields, such as health, food, the environment, agriculture, and numerous industries [9,10]. The emergence of nano-sized components (1–100 nm) is accomplished through three main methods: physical, chemical, and biological reactions [11]. Chemical and physical synthesis methods generally involve unique processing components and difficult conditions, such as hazardous chemicals, pressure, controlled pH and temperature, and large equipment. Moreover, these production techniques are expensive and produce undesired by-products that cause difficulties in the environment [12]. In contrast to chemical and physical procedures, the biological strategy is distinguished by its simplicity, speed, safety for the environment, and relatively inexpensive [13]. As a result, researchers are focusing on the biological approach, or green technique, of producing nanomaterials, which employs fungi, bacteria, yeast, algae, actinomycetes, and plants [14–16].

Fungi are one of the living organisms involved in the green synthesis of nanoparticles [17]. Furthermore, because of fungi's versatility, high metal tolerance, ease of handling, high biomass output, and commercial feasibility, fungi are well-suited for the production of a wide range of nanoparticles [18]. *Penicillium chrysogenum* is one of the most common fungi that produces a large variety of metabolites, such as various enzymes, roquefortines, siderophores, fungisporin, penitric acid, indole-3-acetic acid, chrysogine, hydroxyemodin, and chrysogenin [19–21]. As a consequence, it could be used to fabricate a variety of metal and metal oxide nanoparticles.

Selenium is an essential nutrient that is necessary for good health and regulates a variety of cellular processes via selenium proteins [22]. Selenium is essential in the prevention of a range of diseases, including infectious diseases, hypercholesterolemia, cardiovascular disease, and some malignancies. SeNPs have significant antibacterial activity in naked [23,24] and conjugated forms, such as the selenium nanoparticles-lysozyme nanohybrid system [25]. Despite these numerous benefits, large doses of selenium can have negative side effects. As a result, reports are now focusing on the use of nanomaterials to avoid high doses of Se metal while retaining biological effects [26]. SeNPs have received a lot of recent interest due to their unique properties and biological activities. Selenium nanoparticles have shown biomedical and larvicidal effects, and they can be used for infection control [27]. The strategy of NPs' toxicity at the cellular level has not been completely identified, but it may result in membrane disruption, protein oxidation, interruption of energy transduction, genotoxicity, the release of toxic constituents, and the formation of reactive oxygen species (ROS) [28]. The current work was conducted to test the use of myco-synthesized SeNPs by exposing *Penicillium chrysogenum* culture filtrate metabolites to sodium selenite as a tool for biological control against the intermediate host, *Biomphalaria alexandrina* snails, as well as larval stages of *Schistosoma mansoni*.

2. Materials and Methods

2.1. Isolation Conditions of the Fungal Isolate

The studied fungus was isolated from stones collected on the Mediterranean's southern coast, at Alexandria, Egypt. Isolation was done on a medium comprising 10.0 g saja peptone, 3.0 g of yeast extract, 3.0 g of malt extract, 10.0 g of glucose, 30.0 g of NaCl, 25.0 g of agar, 1000 mL of distilled water, pH 7.5. One thousand microliters per litre of streptomycin was provided after the media had cooled. The culture was incubated for five days at 25 °C.

2.2. Molecular Identification

The fungal isolate was cultured in Czapek's yeast extract agar medium (CYA) for 7 days at 28 °C [29]. The fungal mat was filtered through Whatman's filter paper No. 1, and mycelial DNA was isolated using the Patho-gene-spin DNA/RNA extraction kit as per the manufacturer's instructions (Intron Biotechnology Company, Korea). SolGent Company in Daejeon, South Korea, performed the polymerase chain reaction (PCR) and rRNA gene sequencing. ITS1 (forward) and ITS4 (reverse) primers were incorporated into the reaction mixture for PCR. ITS1 (5′-TCC GTA GGT GAA CCT GCG G-3′) and ITS4 (5′-TCC TCC GCT TAT TGA TAT GC-3′) are the two primers used. With the addition of ddNTPs to the reaction mixture, the PCR product was sequenced using the same primers [30]. The obtained sequences were analyzed using the National Center of Biotechnology Information's (NCBI) website's Basic Local Alignment Search Tool (BLAST).

2.3. SeNPs Myco-Synthesis by *P. chrysogenum*

The mycosynthesis of SeNPs was carried out according to Amin et al. [31] with some modifications. *P. chrysogenum* was cultivated on potato dextrose broth and incubated in a static incubator for 7 days at 25 °C. Mycelia were then separated from the culture supernatant by centrifugation (Sigma, 3-16PK, Osterode am Harz, Germany) at 10,000 rpm for 10 min [32]. After that, sodium selenite (Na_2SeO_3) was mixed with 15 mL of culture supernatant to obtain a final concentration of 3 mM, which was consequently incubated at 40 °C for 30 min until the formation of SeNPs. A change in the color of the culture supernatant from yellow to red confirmed the synthesis of SeNPs [33]. Finally, the biosynthesized SeNPs were centrifuged three times at 10,000 rpm for 10 min with double distilled water to purify them prior to being oven-dried at 60 °C for 48 h. A culture supernatant was used as a control under the same experimental conditions. The nanoparticles were kept in the refrigerator and re-dispersed with distilled water during use.

2.4. Characterization of the Myco-Synthesized SeNPs

A UV–Visible spectrophotometer, Zetasizer analyzer, X-ray diffraction instrument, transmission electron microscope (TEM), and Fourier transform infrared (FTIR) spectrophotometer were used to characterize the optical, morphological, structural, elemental, and functional characteristics of the synthesized SeNPs. The SeNPs absorbance was examined using a UV–visible spectrophotometer at wavelengths in the range of 400 to 800 nm (PerkinElmer Life and Analytical Sciences, CT, Ohio, USA). The average diameter size and distribution, as well as zeta potential charges, were determined by the particle size analyzer Dynamic Light Scattering (DLS) (Zetasizer Nano ZN, Malvern Panalytical Ltd., Malvern, UK) at a fixed angle of 173° at 25 °C. XRD was performed using a Bruker D8 DISCOVER Diffractometer, USA, with Cu-K radiation (λ = 1.54060 Angstrom) to determine the particles' crystalline size. The relative intensity information was analyzed throughout a 2θ range of 5–100°. 2θ values and relative intensities (I/Io) were obtained from the chart, and core materials minerals were characterized using JCPDS carts. A high-resolution transmission electron microscope (HR-TEM; JEOL 2100, Japan) equipped with an electron diffraction pattern was also used to take transmission electron photographs. Fourier transform infrared spectroscopy was employed to investigate the elemental structure of SeNPs as well as the functional groups (FTIR; PerkinElmer, Ohio, USA). Triplicate samples were analyzed.

2.5. Investigation of Molluscicidal Activity of SeNPs

2.5.1. Snails

The snails, *B. alexandrina* (Ehrenberg, 1831) (8–10 mm in diameter), were acclimatized in the Medical Malacology Laboratory, Theodor Bilharz Research Institute (TBRI), Giza, Egypt. Snails were housed in plastic aquaria (16 × 23 × 9 cm) and fed with oven-dried lettuce leaves, blue-green algae (*Nostoc muscorum*), and tetramin. The following properties were used in the experiment: dechlorinated aerated tap water (10 snails/L), pH: 7 ± 0.2, and temperature (25 ± 2 °C) were covered with glass plates. Thirty mg/L of calcium carbonate was added to the water to achieve its optimum hardness for snail fecundity, shell length, and growth [34].

2.5.2. Molluscicidal Activity of SeNPs

A series of concentrations were prepared from the stock solution of SeNPs (95, 80, 65, 50, 35, and 25 mg/L) to calculate LC_{90} [35]. The snails (180) were subjected to a 96-h exposure period followed by a 24-h recovery period. Only dechlorinated water was used to keep the three control groups (30 snails) the same size. For each concentration, three replicates were used, each with ten snails. The mortality rate was measured and analyzed. [36].

2.5.3. Miracidicidal and Cercaricidal Activity

As a control group, 10 mL of dechlorinated tap water was added to 102 newly formed miracidia or cercariae. To assess the effect of biosynthesized SeNPs on the newly formed miracidia or cercariae, about 5 mL of LC_{10} (31.826 mg/L) and LC_{25} (44.15 mg/L) of biosynthesized SeNPs were added to 102 newly formed miracidia or cercariae found in 5 mL of water [37]. Using the dissecting microscope, the vitality of miracidiae and cercariae was scored after 15, 30, 45, 60, 120, 180, 240, 300, and 360 min [38].

2.6. Experimental Design

Ten snails in each aquarium were exposed to the sub-lethal concentrations of SeNPs at LC_{10} (31.82 mg/L) or LC_{25} (44.15 mg/L) for 96 h (exposure), followed by another 24 h of recovery, two weeks of repeating, followed by two weeks of recovery.

2.6.1. Hemolymph and Light Microscopy Preparation

The hemolymph was withdrawn from the heart using a capillary tube [39]. Part of the collected hemolymph was put on a glass slide, making a monolayer of hemocytes. Then the slides were dried in a moist chamber for 15 min at room temperature, followed by 5 min of dehydration in methanol, and finally stained for 20 min with 10% Giemsa stain (Aldrich) [40].

2.6.2. Comet Assay

After the exposure of *B. alexandrina* snails (8–10 mm) to SeNPs at concentrations of LC_{10} (31.28 mg/L) and LC_{25} (44.15 mg/L) for 96 h, the head feet of 10 snails from each group were cut and kept at −80 °C until they were needed. The single-cell gel assay was used to measure the comet assay for the detection of DNA breaks, as described by Grazeffe et al. [41] and Ibrahim and Sayed [42]. The slides were coded independently and scored independently.

2.6.3. Tissue Preparation

The soft tissues of the exposed and control groups were obtained by crushing the shells of the snails using two slides, weighing (1 g tissue/10 mL of phosphate buffer), and homogenizing with a glass Dounce homogenizer. Then, the tissue homogenates were centrifuged (Sigma, 3-16PK, Osterode am Harz, Germany) at 3000 rpm for 10 min, and the supernatants were stored at −80 °C until used.

Determination of Testosterone (T) and Estradiol (E2) Hormones Concentrations

The hormonal activity of T and E2 was estimated following the manufacturer's instructions, in which T concentrations were measured using an EIA kit (Abia Testosterone, REF, DK. 040.01.3), while for E2 concentration, an immunoassay test kit (BioCheck, Inc., South San Francisco, CA 94080, USA) was used [42].

Investigation of the Antioxidant Responses: Superoxide Dismutase (SOD); Glutathione S-Transferase (GST), Nitric Oxide (NO), Malondialdehyde (MDA), and Total Antioxidant Capacity (TAC)

For each group, biochemical changes in the tissue homogenate's supernatant were monitored. Biodiagnostic kits (Biodiagnostic, Dokki, Giza, Egypt) were used to assess SOD [43]. In addition, cell MDA (lipid peroxide) was measured using the Ohkawa et al. [44] method, and GST was detected using the Beutler [45] method. TAC was determined with the kit (Cat. No. TA 2513) [46]. According to Bellos et al. [47], NO was estimated.

2.7. The Molecular Docking Study

The inhibitory potential of Na_2SeO_3 was investigated using two enzymes from the cellular anti-oxidant mechanism, SOD and GST. Three-dimensional crystals of human SOD 1 (PDB id: 5YTU) and SOD 2 (PDB id: 13GS) complexed with isoproterenol and sulfasalazine, respectively. These catalases were obtained from (www.rcsb.org/, 10 February 2022) in.pdb format. Na_2SeO_3 and reference inhibitors (isoproterenol (for SOD) and sulfasalazine (for GST) were selected. (www.pubchem.ncbi.nm.nih.gov/, 10 February 2022) was used for generating the 3D structures for ligands in .sdf format. MOE2015 is advanced computational modelling software for evaluating ligand $\rightarrow$ active site interactions. MOE 2015 conducted a docking experiment, which was used to correct errors in active sites during the structure preparation reaction. Hydrogens were added after the correction, and partial charges (AMBER12: EHT) were calculated. Energy minimization was carried out (AMBER12: EHT, root mean square gradient: 0.100). The MOE Site Finder program, which uses a geometric approach to calculate putative binding sites in a protein starting from its tridimensional structure, was used to find the receptor's binding site. This method is based on alpha spheres, which are a generalization of convex hulls, rather than energy models. The binding sites predicted by the MOE Site Finder module in the holo-forms of the investigated proteins confirmed the binding sites defined by the co-crystallized ligands.

2.8. Statistical Analysis

The Probit facility analyzed the median lethal and lethal concentration values [48]. The mean values of the experimental and control groups were compared using the Student's *t*-test [49]. The data was analyzed using the statistical software SPSS version 20 for Windows (SPSS, Inc., Chicago, IL, USA). The results were expressed as the average value $\pm$ S.E.

3. Results

3.1. Molecular Identification of the Fungal Strain

The ITS sequences of the fungal isolate's rDNA were aligned with strains from GenBank that were genetically related. It demonstrated 99–100% identity and 100% coverage with several *Penicillium chrysogenum* strains, including the type strain CBS 306.48 (GB no.: NR 077145). The sequence was registered in the GenBank database under the accession number MZ945518 (Figure 1).

3.2. The Myco-Synthesis of SeNPs by P. chrysogenum

After 30 min of incubation, the color of the culture medium changed from yellow to brick-red after the culture filtrate was treated with 1 mM Na_2SeO_3 (Figure 2). Following incubation, the presence of a red-brick color inside the culture medium was clear evidence that the extracellular metabolites rapidly reduced selenite ions to the elemental Se (Se0) form [33]. The SeNPs' productivity was calculated to be around 38 mg/100 mL.

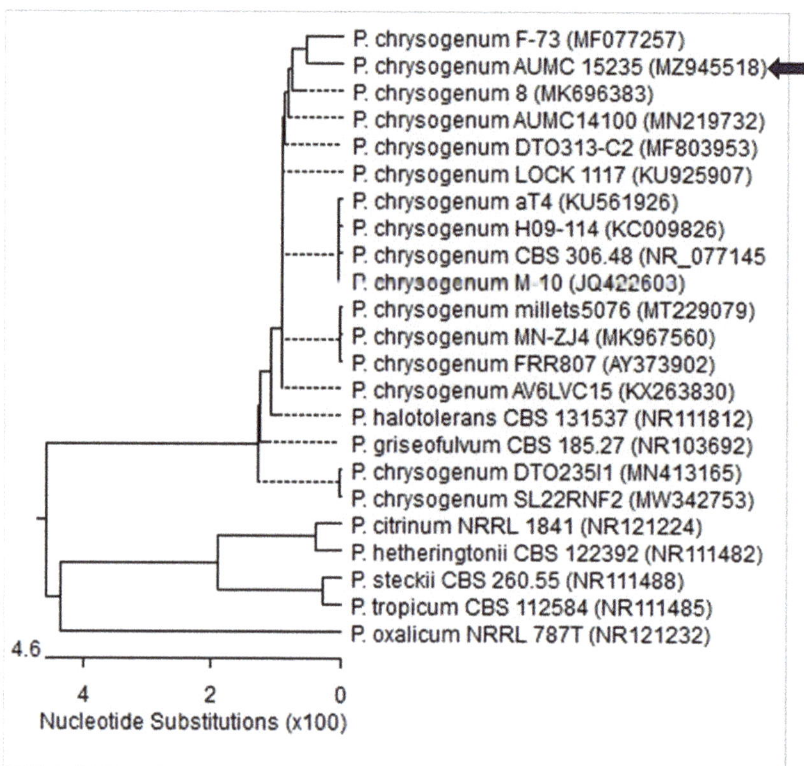

Figure 1. Phylogenetic tree of the fungal isolate depending on ITS sequence (GenBank accession no. MZ945518, arrowed) aligned with closely similar strains in the GenBank. (*P.* = *Penicillium*). The sequences were phylogenetically analyzed utilizing MegAlign (DNA Star) software version 5.05.

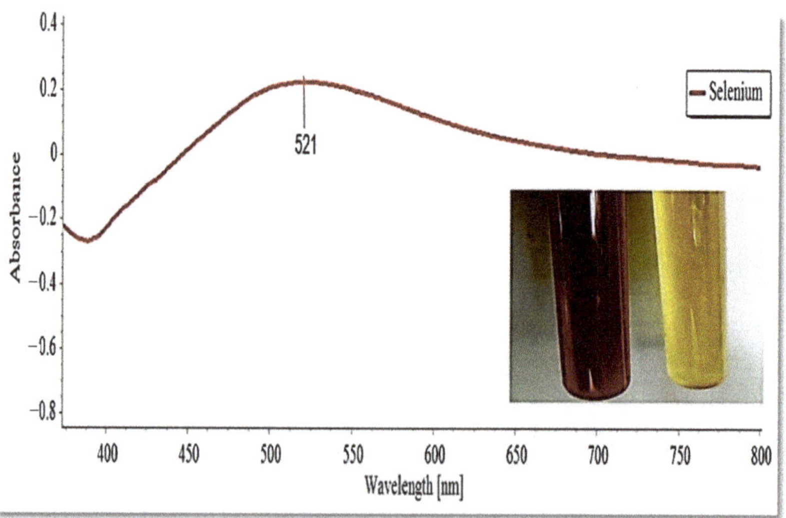

Figure 2. UV–Visible absorption spectrum and brick-red color of the myco-synthesized selenium nanoparticle.

3.3. Characterization of the Myco-Synthesized SeNPs

UV–Visible spectrophotometry was used to monitor the SeNPs production in the culture filtrate, revealing a strong and broad surface plasmon resonance (SPR) peak at 521 nm, which is an SeNPs feature (Figure 2). In the control, however, no absorption peak corresponding to the SeNPs was found.

The synthesis of polydispersed spherical SeNPs with diameter sizes ranging from 44 to 78 nm was revealed by TEM analysis of a colloidal solution of myco-synthesized SeNPs (Figure 3a). Figure 3b shows the average size distribution in the SeNP solution as determined by DLS. According to the results obtained, the myco-synthesized SeNPs were measured to have an average diameter of 207 nm. DLS determines the size of SeNPs, which is influenced by biomolecules coated on their surfaces as stabilizers as well as their metallic cores. The stability of SeNPs was assessed using the zeta potential assessment of particle surface charge, which revealed a mean zeta potential of −32.4 mV (Figure 3c).

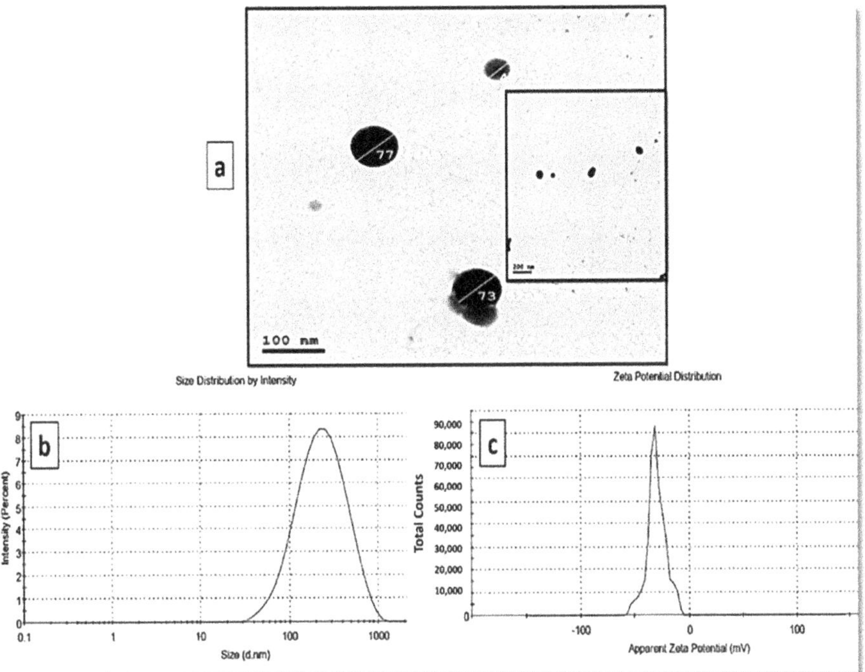

Figure 3. Morphological characterization of myco-synthesized SeNPs (**a**) TEM photographs of myco-synthesized selenium nano-selenium using *P. chrysogenum* culture filtrate at the scale of 100 nm and 200 nm, (**b**) size distribution pattern, and (**c**) zeta potential distribution.

The XRD results revealed a broad pattern with no clear Bragg peaks. While there were no significant peaks, smaller peaks at 2Θ values were found at 12.658°, 19.146°, 20.712°, 21.011°, 25.352°, 29.402°, 31.830°, 53.938°, 55.218°, 58.039°, and 61.540°. The results demonstrated that myco-synthesized SeNPs are rather more amorphous than crystalline (Figure 4).

The presence of various functional groups in metabolites that are responsible for SeNP myco-synthesis, capping, and stabilization was determined using FTIR measurements. The FTIR for the culture supernatant of *P. chrysogenum* was analyzed and showed five intense peaks observed at 3307.57, 2107.89, 1635.22, 431.08, and 407.76 cm^{-1} (Figure 5a). These peaks were shifted to eight peaks in the chart of SeNPs. The interaction of metabolites with

SeNPs is illustrated by wavenumbers at 3307.05, 2114.39, 1635.50, 451.03, 442.54, 429.87, 419.61, and 403.05 cm^{-1} (Figure 5b).

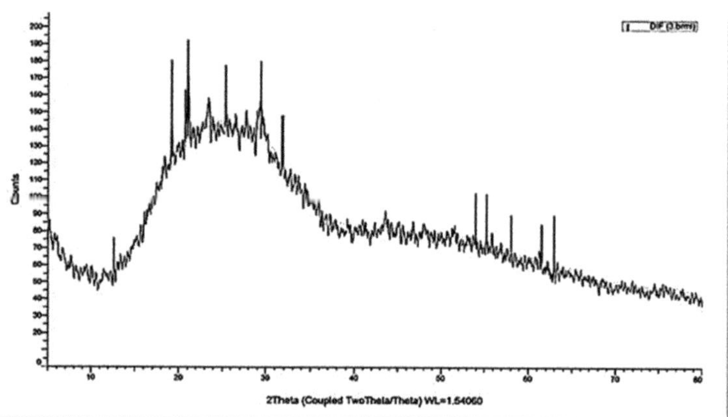

Figure 4. XRD spectrum of myco-synthesized selenium nanospheres.

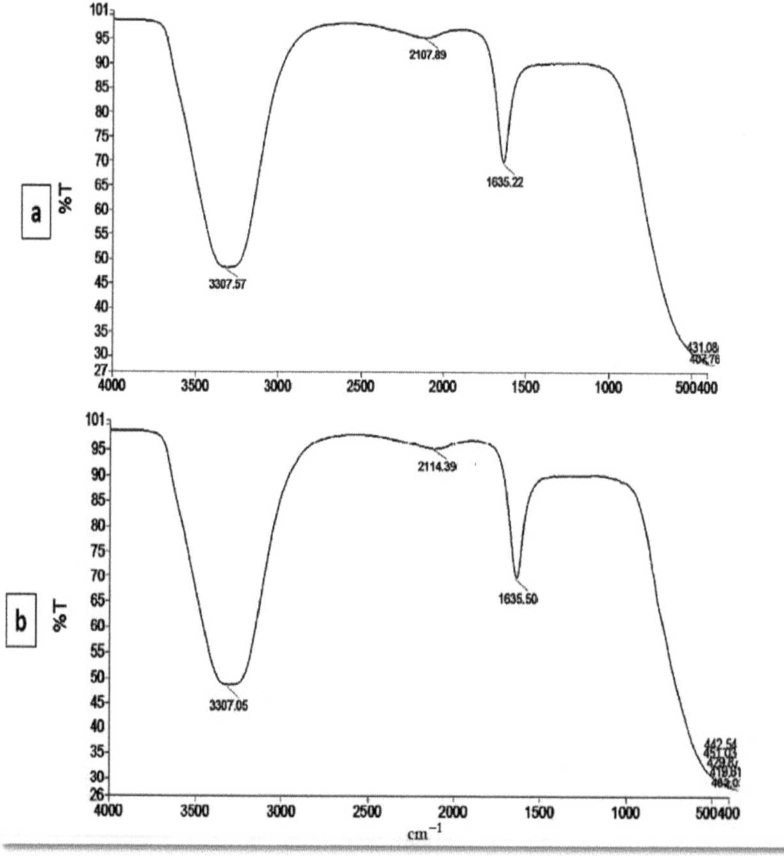

Figure 5. FTIR pattern (**a**) culture supernatant, and (**b**) myco-synthesized selenium nanoparticles, where Y-axis represented the transmission (%T) and X-axis represented the wavenumber (cm^{-1}).

3.4. Effects of Selenium Nanoparticles against *B. alexandrina* Snails

The present findings revealed that SeNPs have a molluscicidal effect against adult *B. alexandrina* snails after 96 h of exposure at LC_{50} 5.96 mg/L (Table 1 and Figure 6).

Table 1. Myco-synthesized SeNPs' molluscicidal activity against *B. alexandrina* snails after 96 h of exposure.

Lethal Concentration Doses	LC_{10}	LC_{25}	LC_{50}	LC_{90}	Slope
Concentration (mg/L)	31.826	44.15	57.85	83.87	1.4

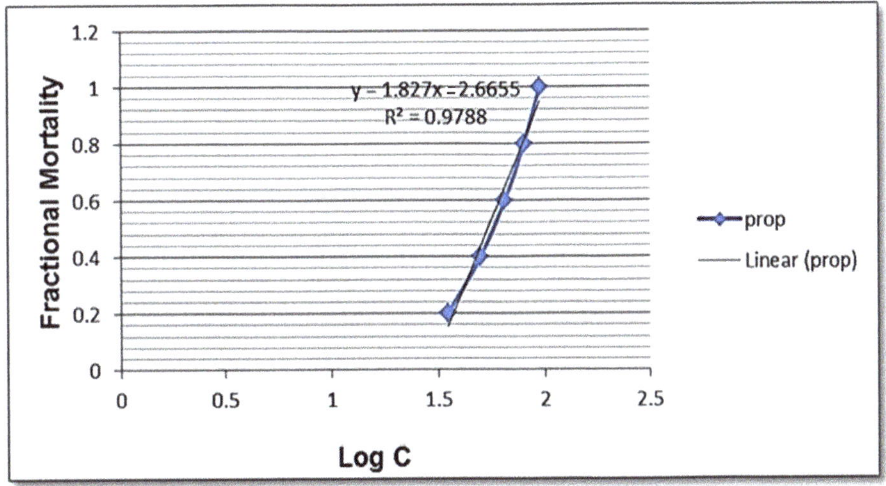

Figure 6. Molluscicidal effect of SeNPs against *B. alexandrina* snails as shown in probit analysis.

The current results showed that SeNPs have a toxic effect on *S. mansoni* stages, as shown in Figure 7. All miracidae exposed to SeNPs died after 60 min, compared to only 20% of the deaths in the control group (Figure 7A), while 45 min was enough to kill all exposed cercariae, compared to 10% of the control group.

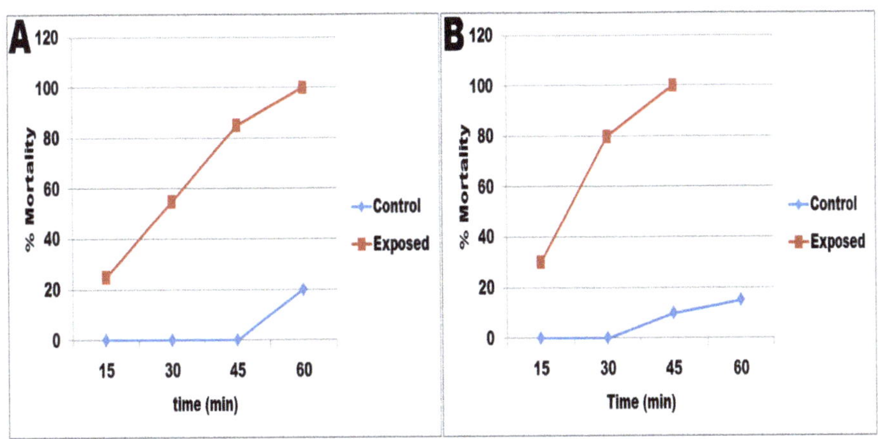

Figure 7. The miracidicidal (**A**) and cercaricidal (**B**) activities of SeNPs.

The examination of hemocyte monolayers by light microscope showed the presence of three cell types of hemocytes in the control group: small hemocytes, hyalin oocytes, and granulocytes (Figure 8A). Hemocytes of exposed *B. alexandrina* snails suffered from many changes. The exposure to LC_{10} (31.826 mg/L) of SeNPs showed numerous granules and newly formed pseudopodia in the granulocytes, while hyalinocytes suffered from incomplete cell division where the nucleus divided, forming two nuclei without cell membrane separation (Figure 8B). The exposure to LC_{25} (44.15 mg/L) of SeNPs resulted in increased observed granules with an irregular cell membrane in granulocytes, while newly small pseudopodia were seen in hyalinocytes (Figure 8C).

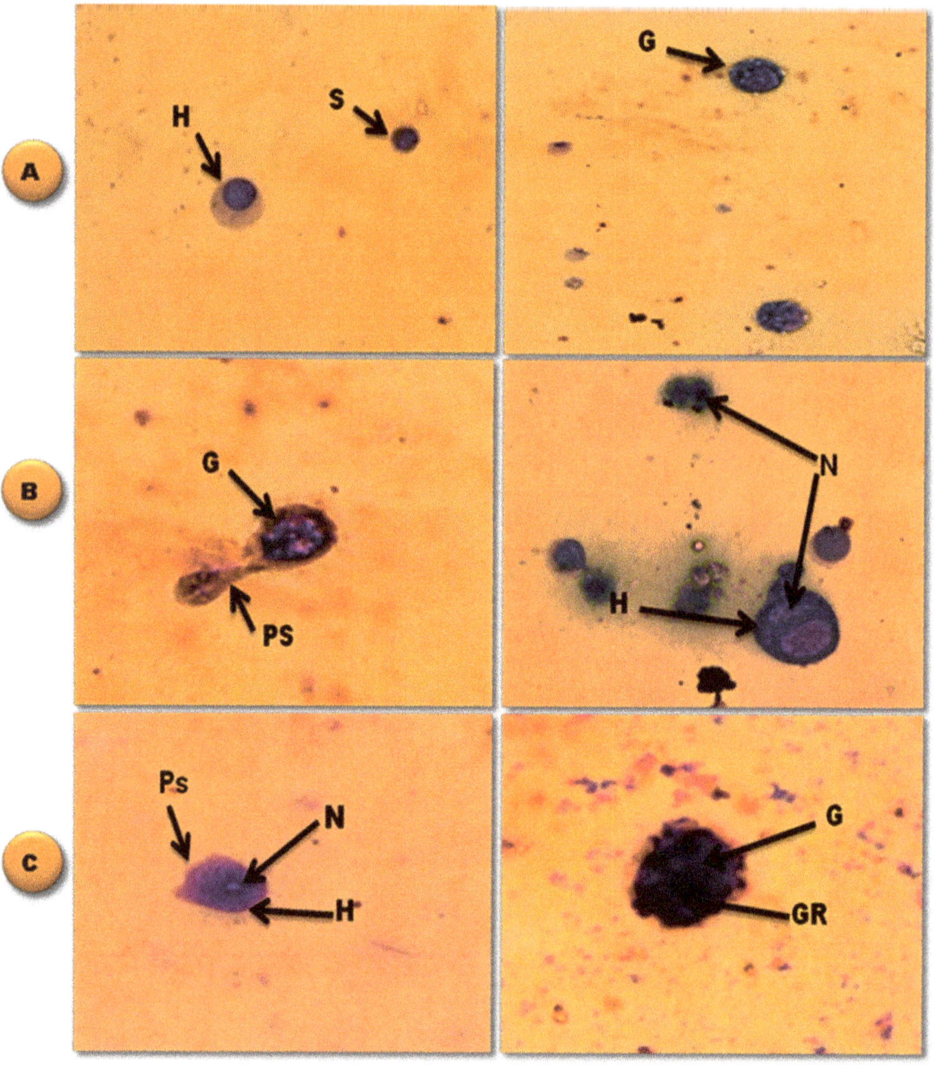

Figure 8. Photomicrographs (×40) of different types of hemocytes.(**A**) hemocytes of control group of adult *B. alexandrina* snails, (**B**) *B. alexandrina* snails hemocytes after exposure to LC_{10} of SeNPs (31.826 mg/L), (**C**) hemocytes of adult *B. alexandrina* snails after exposure to LC_{25} of SeNPs (44.15 mg/L). Abbreviations, G: granulocyte, GR: granules, H: hyalinocyte, N: nucleus, PS: pseudopodia, S: small round.

The current study found that after the exposure to sublethal concentrations of SeNPs, there were DNA breaks where the percentage of the comet, tail length, percent DNA in tail, and tail moment were increased ($p < 0.05$ and 0.01) compared to control snails (Table 2 and Figure 9).

Table 2. DNA breaks after SeNPs exposure to *B. alexandrina* snails.

	Comet %	Tail Length (px)	% DNA in Tail	Tail Moment
Control	14.95	8.15 ± 0.42	23.79 ± 3.75	1.955 ± 0.402
LC_{10}	15.85	10.507 ± 0.54 *	33.66 ± 1.43 *	3.529 ± 0.03 *
LC_{25}	19.5	11.96 ± 1.37 *	33.588 ± 5.7 *	4.094 ± 1.15 *

1 px = 0.24 µm; * = significant compared to control at $p < 0.05$.

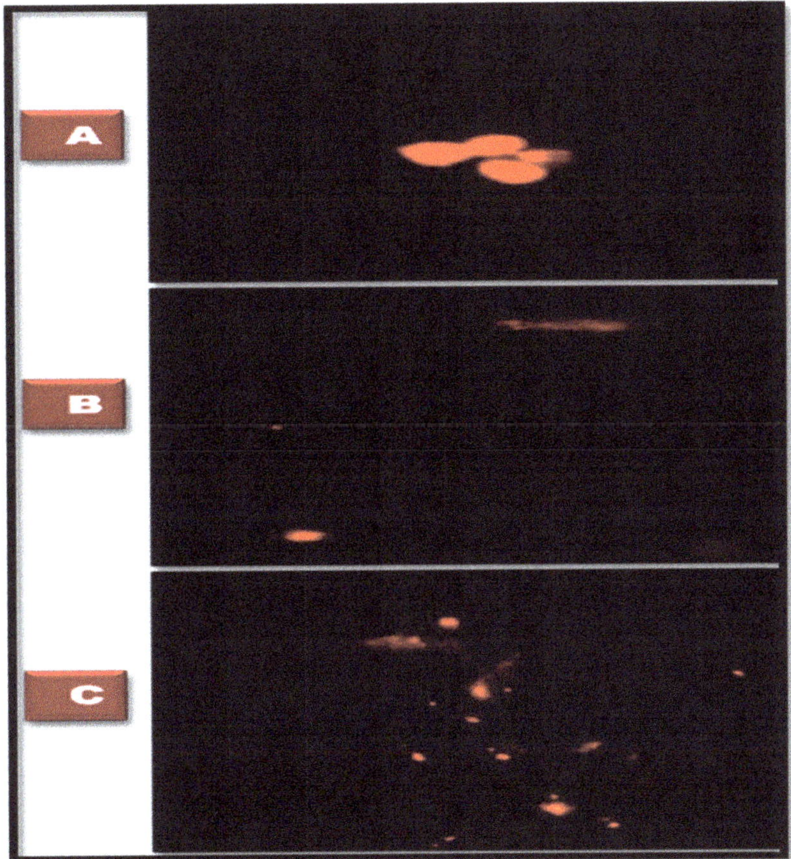

Figure 9. Ranks of comet according to the percent of DNA in the tail. (**A**) Control group, (**B**) *B. alexandrina* snails after exposure to LC_{10} of SeNPs (31.826 mg/L), (**C**) adult *B. alexandrina* snails after exposure to LC_{25} of SeNPs (44.15 mg/L).

The current findings revealed that testosterone (T) and estradiol (E2) levels were significantly higher ($p < 0.05$) after exposure to sublethal concentrations when compared to the control group (Table 3).

Significant increases in a concentration-dependent manner ($p < 0.05$) of MDA, NO, and GST were noticed after in vivo exposure of *B. alexandrina* snails to sub-lethal concentrations of SeNPs. On the other hand, SOD activity was significantly decreased ($p < 0.05$) while TAC was insignificantly decreased (Table 4).

Table 3. SeNP exposure effects on (T) and (E2) concentration of *B. alexandrina* snails.

Groups	Testosterone (nmol/L)	Estradiol (pg/mL)
Control	20 ± 0.52	100 ± 3.1
LC_{10}	30 ± 0.71 *	300 ± 5.2 *
LC_{25}	35 ± 0.56 *	1000 ± 4.6 *

* = significant compared to control at $p < 0.05$.

Table 4. Effects of SeNP exposure on some biochemical parameters of *B. alexandrina* snails.

Biochemical Parameters	MDA (nmol/g.tissue)	NO (µmol/L)	SOD (U/g.tissue)	GST (U/g.tissue)	TAC (mM/L)
Control	9.811 ± 0.012	136.75 ± 0.42	8.36 ± 0.1	0.85 ± 0.05	1.505 ± 0.4
LC_{10}	10.612 ± 0.4 *	148 ± 0.34 *	5.045 ± 0.02 *	0.927 ± 0.1	1.488 ± 0.21
LC_{25}	14.47 ± 0.3 *	322 ± 0.21 *	3.624 ± 0.2 *	1.0545 ± 0.03 *	1.375 ± 0.3

* = significant compared to control at $p < 0.05$.

3.5. Molecular Docking Study

The two enzymes, SOD and GST, were selected for the cellular antioxidant mechanism to determine the inhibitory potential of sodium selenite. Three-dimensional structures of human SOD 1 complexed with isoproterenol (PDB id: 5YTU) and human GST complexed with sulfasalazine (PDB id: 13GS) were used for molecular docking. The tested compound revealed high efficiency against the respective receptor binding sites of SOD and GST (Figure 10). The molecular docking showed that Na_2SeO_3 succeeded in binding to similar active sites in the same spot as the original inhibitor, which suggests that Na_2SeO_3 blocks the receptors in the same way.

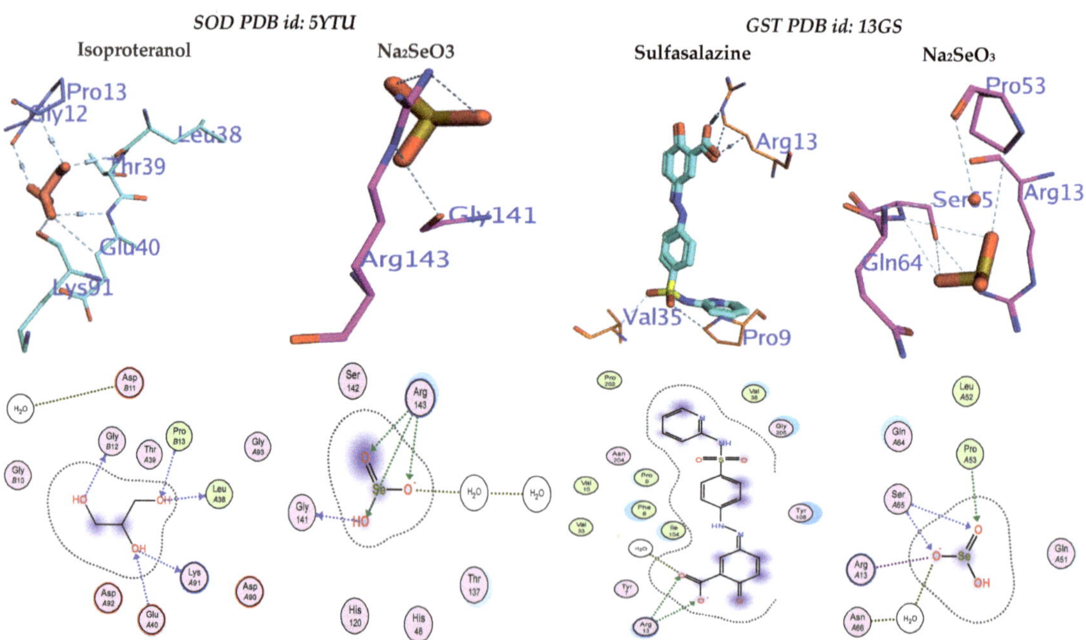

Figure 10. 2D and 3D docked interaction map for the Na_2SeO_3 compounds into the active site of SOD and GST.

The inhibition activities for the enzymes were examined using interaction-free energy, which is known as the docking score. Na_2SeO_3 showed promising docking and H-interaction scores (−5.03 and −4.70 Kcal/mol), respectively, against SOD and GST, and its compound displayed promising scores (Table 5).

Table 5. The docking energy scores (kcal/mol) for Na_2SeO_3.

	ΔG	rmsd	E.vdw	E.Int	*E.H.B.*	Eele
5YTU	−5.03	3.86	−266.94	−1.86	−11.10	−27.82
13GS	−4.70	1.53	−266.52	−2.18	−10.09	−27.31

ΔG: The ligand's free binding energy from a given conformer, E.Int.: affinity binding energy of hydrogen bond interaction with the receptor; *E.H.B.*: hydrogen bonding energy between protein and ligand; Eele: electrostatic interaction with the receptor; Evdw: Van der Waals energies between the ligand and the receptor.

4. Discussion

The application of green synthesis nanoparticles for freshwater snail control is novel. In this approach, natural metabolites extracted from plants and microorganisms are used to synthesize nanoparticles. Natural molecules are thought to produce safer nanoparticles than the more toxic chemical polymers used in the fabrication process [50]. The activity of metabolites released by various organisms, including plants, fungi, actinomycetes, bacteria, and algae, were exploited in metal ion reduction, capping, and stability for the biosynthesis of metal nanoparticles [18]. In the current study, the reducing capacity of the culture filtrate metabolites of *P. chrysogenum* was used to synthesize SeNPs. *P. chrysogenum* is capable of producing a diverse spectrum of secondary biomolecules that work as a biocatalyst to reduce and stabilize nanoparticles. Proteins, several enzymes, and carbohydrates are some of these molecules [10,51]. Joshi et al. [52] observed that the fungal strain's ability to reduce Se ion and synthesize SeNPs is due to the numerous extracellular proteins and enzymes. Amin et al. [31] reported that *Penicillium chrysogenum F9* was employed as a biocatalyst for SeNPs biosynthesis. The resultant color shifting was caused by the surface plasmon resonance of monoclinic Se particles [53]. The production of SeNPs in culture filtrate was measured using UV–Visible spectroscopy, which revealed a strong and broad peak at 521 nm. It was reported that the absorbance values of SeNPs were determined at 300 nm and another value at 540 nm [54,55]. SeNPs displayed absorbance at approximately 520 nm due to Mie scattering [56]. Ullah et al. [57] stated that the maximum absorption peak of selenium nanoparticles synthesized by *Bacillus subtilis* BSN313 was at 650 nm. The current findings were consistent with those of Ranjitha and Ravishankar [58], who indicated that selenium nanoparticles synthesized by *Streptomyces griseoruber* had a maximum peak at 575 nm. The sizes determined by TEM (44 to 78 nm) and DLS (207 nm) differed because DLS examines the hydrodynamic quantity while TEM examines the solid core [59]. The negative charge of particles indicates the electrostatic stability of the synthesized nanoparticles as reported by [32]. The electrostatic stability of the synthesized colloidal nanoparticle solution is indicated by a zeta potential greater than +30 mV or less than −30 mV [60]. The XRD results revealed that the myco-synthesized SeNPs are rather more amorphous than crystalline. This amorphous nature is consistent with previous research with lycopene [61], *Pseudomonas stutzeri* [62], and *Withania somnifera* [63]. FTIR analysis revealed the interaction between *P. chrysogenum* metabolites and SeNPs. The signal band at 3307.05 cm^{-1} corresponds to N–H, C–H, and O–H stretching vibrations, suggesting the presence of primary amine in the fungal proteins [31], alkyne, and alcohol, respectively [64]. This demonstrates the importance of N–H-containing proteins in the reduction of Se ions and the formation of SeNPs. The band at 2114.39 cm^{-1} corresponds to the presence of alkyne. Furthermore, the peak at 1635.50 cm^{-1} was associated with various peptide linkage and polysaccharide ring moieties such as N-H, C=N, C=O, and C=C [33]. The bands at 451.03, 442.54, 429.87, 419.61, and 403.05 cm^{-1} revealed the binding of SeNPs with the metabolites of *P. chrysogenum* culture filtrate. According to these results, many functional groups of organic compounds, such as proteins and polysaccharides, present in the culture filtrate of *P. chrysogenum* are involved in the capping, stability, and reduction of SeNPs.

The present study revealed that SeNPs have molluscicidal activity against *B. alexandrina* after 96 h of exposure. These findings are consistent with those of Osman et al. [65], who observed that *B. alexandrina* snails treated with *Aspergillus fumigatus* fungal extract showed a high molluscicidal effect. In addition, Abdel-Hamid and Mekawey [65] reported that the myco-biosynthesis of silver nanoparticles (AgNps) from the two fungi, *Paecilomyces variotii* and *Aspergillus niger*, has molluscicidal activity against *B. alexandrina* snails.

Moreover, the current research found that SeNPs had miracidicidal and cercaricidal properties. Many similar studies have revealed the effective role of SeNPs against many other parasitic species. Mahmoudvand et al. [66] showed that Se NPs could kill promastigote and amastigote stages of *Leishmania major*. Similarly, Alkhudhayri et al. [67] found that Se NPs have anti-coccidial, anti-apoptotic, and anti-inflammatory effects against the Eimeria parasite in the jejunum of mice. Furthermore, Se nanoparticles showed promising protective roles against mice infected with *Schistosoma mansoni* [68]. Comparing the Se nanoparticles' activity against meracidia and cercariae, the present results showed a faster mortality rate of cercariae than the meracidial rate. Unlike the present study, Osman et al. [65] showed that the fungal extract of *Aspergillus fumigatus* caused a higher mortality rate of miracidia than that of cercariae after exposure to the same experimental period. The mortality rates of both larval stages appear to differ depending on their biological nature and their internal structure [69].

One of the most sensitive tools to detect DNA defects is the comet assay, which detects DNA single-strand breaks [70]. The present results of the comet assay revealed that SeNPs induced DNA damage in *B. alexandrina*. Similarly, Ali [71] found that TiO_2NP induced DNA damage in the freshwater snail *Lymnea leuteola*. Moreover, the author reasoned that these DNA breaks might be related to the oxidative stress that might be generated after the treatment. In good accordance with the present results, Ibrahim and Ghoname [72] found that exposure of *B. alexandrina* snails to the leaves of *Anagalis arvensis* aqueous extracts caused DNA breaks revealed by the comet assay. Inline supporting these results, Wang et al. [73] named the process of metal oxide NP transport ions onto the cells as "Trojan-horse type carriers", causing serious damage of metal oxide NPs affecting the DNA molecules.

In gastropods, hemocytes, found in the hemolymph, represent the main component of the immune response [74]. According to their morphology, Cavalcanti et al. [75] reported three different types of haemocytes in *B. alexandrina*; spherical, small (undifferentiated), hyalinocytes, and granulocytes. The current study revealed that after exposure to LC_{10} (31.82 mgL^{-1}) of SeNPs, the granulocytes showed pseudopodia and hyalinocytes had incomplete cell division, At exposure to 44.15 mg/L, some granulocytes formed many granules with an irregular cell membrane, and some hyalinocytes formed pseudopodia. These results are similar to those obtained by Abdel-Hamid and Mekawey [76], who found that hemocytes of treated *B. alexandrina* with LC_{25} of both *P. variotii* and *A. niger* AgNPs showed many alterations in their morphology, such as apoptotic hemolymph cells and fragmented, vacuolated, and degenerated cytoplasm. In agreement with the current study, Ibrahim et al. [77] exposed *B. alexandrina* to butralin, glyphosate isopropyl ammonium, or pendimethalin herbicides, and they observed that many granules and pseudopodia were produced by granulocytes while the hyalinocytes revealed a shrunken nucleus. Ray et al. [78] reported that hemocytes are the main immune cells working by phagocyting foreign particles in Mollusca species. Donaghy et al. [79] suggested that the observed pseudopodia of granulocytes may be a method of phagocytosis used to eliminate these foreign particles.

In the current study, the exposure of *B. alexandrina* to SeNPs induced oxidative stress that was noticed in the high levels of MDA in a concentration-dependent manner. Ohkawa et al. [44] reported that MDA is the most catastrophic effect of ROS, which is formed by the peroxidation of lipid membranes. Like the current study, Khalil [80] reported that MDA was significantly increased in the snail *Lanistes carinatus* after exposure to chlorpyrifos. These findings are similar to those of Ibrahim and Sayed [42], who found that malondialdehyde (MDA) increased in a concentration-dependent manner after the treatment of *B. alexandrina* snails with sub-lethal doses of oxyfluorfen herbicide (LC_0, LC_{10}, or LC_{25}).

The antioxidant enzymes SOD and GST have a pivotal role in the elimination of ROS and modulate the response of living organisms to oxidative conditions. GST levels increased significantly after LC_{25} exposure, whereas SOD levels decreased significantly after LC_{10} and LC_{25} exposure. These results are consistent with the observations obtained by Khalil [80], who found that the GST activity was significantly increased in the adult freshwater snail *L. carinatus* when treated with chlorpyrifos. Similarly, the levels of SOD decreased in the snail *B. alexandrina* after treatment with atrazine and Roundup [81]. In contrast, Ibrahim and Sayed [42] found that the activity of SOD increased after the exposure of *B. alexandrina* to sub-lethal concentrations of oxyfluorfen herbicide. The unexpected suppression of SOD might result from protein degradation through oxidative damage to SOD or gene expression modifications [82]. One of the most effective innate immune defence mechanisms is the production of nitric oxide, which leads to the cytotoxicity of the invading pathogens in mollusks [78]. In the present study, the NO concentration significantly increased after the exposure to sublethal LC_{10} and LC_{25} of SeNPs. These results are in agreement with the findings of Wang et al. [83] who reported that the exposure of snail *B. straminea* to pyridyl phenyl urea derivatives led to high activity of NO. Also, Saleh et al. [84] found increasing NO activity after the treatment of *B. alexandrina* with the veterinary antibiotics oxytetracycline and trimethoprim-sulphadiazine.

The molecular docking method for examining receptor-ligand interactions is an important tool for predicting the inhibition actions of the enzymes associated with antioxidant activity. MOE 2016 [85] was used to perform the docking experiment. The enzymatic components include (SOD; PDB id: 5YTU [86]) and (GST; PDB id: 13GS [87]), the most potent non-enzymatic cellular antioxidant that is used by GST and GPx to neutralize oxidants [88]. SOD catalyzes the transformation of ($O_2{}^{\cdot -)}$ $\rightarrow$ (H_2O_2), which is responsible for reducing the RONS levels [89]. The docking study aimed to determine the potential of Na_2SeO_3 in altering the cellular antioxidant defence system using molecular modeling. The observed results suggested the inhibition potency of Na_2SeO_3 against SOD and GST by interfering with their active important amino acids for catalytic sites.

The physiological and genotoxicological properties of *Biomphalaria alexandrina* snails were negatively impacted by myco-synthesized SeNPs. However, more research should be conducted to determine their effects on freshwater zooplanktonic species, such as the water flea *Daphnia magna*, which is used as a non-target organism for toxicity assessment in aquatic ecosystems found in the same habitat as *B. alexandrina* [90].

5. Conclusions

Myco-synthesized SeNPs have molluscicidal activity against the snail *Biomphalaria alexandrina* and larvicidal activity against *Schistosoma mansoni* larval stages, which could lead to a decrease in the spread of schistosomiasis. Further studies, including the effects on other (non-target) organisms and sub-lethal effects on snail fecundity/and fertility, are needed to determine whether the myco-synthesized SeNPs can be used as a promising tool for biological control and to replace the toxic synthetic chemical molluscicides.

Author Contributions: Conceptualization, M.Y.M., H.E.-S. and A.M.I.; methodology, M.Y.M., H.E.-S. and A.M.I.; software, M.Y.M., H.E.-S., A.S.A. and A.M.I.; validation, M.Y.M., H.E.-S. and A.M.I.; formal analysis, S.M.K. and A.S.A.; investigation, M.Y.M., H.E.-S. and A.M.I.; resources, M.Y.M., H.E.-S., A.A.E. and A.M.I.; data curation, S.M.K. and A.S.A.; writing—original draft preparation, M.Y.M., H.E.-S. and A.M.I.; writing—review and editing, M.Y.M., H.E.-S. and A.M.I. All authors have read and agreed to the published version of the manuscript.

Funding: Princess Nourah bint Abdulrahman University Researchers Supporting Project number (PNURSP2022R214), Princess Nourah bint Abdulrahman University, Riyadh, Saudi Arabia.

Institutional Review Board Statement: Not applicable.

Informed Consent Statement: Not applicable.

Data Availability Statement: All data generated or analyzed during this study are included in this article.

Conflicts of Interest: The authors declare no conflict of interest.

References

1. WHO Schistosomiasis. 2020. Available online: https://www.who.int/news-room/fact-sheets/detail/schistosomiasis (accessed on 11 February 2022).
2. Ibrahim, A.M.; Sayed, S.S.M. Assessment of the molluscicidal activity of the methanolic seed extracts of Ziziphus spina-christi and Carica papaya on immunological and molecular aspects of *Biomphalaria alexandrina* snails. *Aquac. Res.* **2021**, *52*, 2014–2024. [CrossRef]
3. Loker, E.S.; Mkoji, G.M. Schistosomes and Their Snail Hosts BT—Schistosomiasis. In *Schistosomiasis*; Secor, W.E., Colley, D.G., Eds.; Springer: Boston, MA, USA, 2005; pp. 1–11. ISBN 978-0-387-23362-8.
4. Faust, C.L.; Crotti, M.; Moses, A.; Oguttu, D.; Wamboko, A.; Adriko, M.; Adekanle, E.K.; Kabatereine, N.; Tukahebwa, E.M.; Norton, A.J.; et al. Two-year longitudinal survey reveals high genetic diversity of *Schistosoma mansoni* with adult worms surviving praziquantel treatment at the start of mass drug administration in Uganda. *Parasites Vectors* **2019**, *12*, 607. [CrossRef] [PubMed]
5. Abdel-Tawab, H.; Ibrahim, A.M.; Hussein, T.; Mohamed, F. Mechanism of action and toxicological evaluation of engineered layered double hydroxide nanomaterials in *Biomphalaria alexandrina* snails. *Environ. Sci. Pollut. Res.* **2021**, *29*, 11765–11779. [CrossRef] [PubMed]
6. Ibrahim, A.M.; Bakry, F.A. Assessment of the molluscicidal impact of extracted chlorophyllin on some biochemical parameters in the nervous tissue and histological changes in *Biomphalaria alexandrina* and *Lymnaea natalensis* snails. *Invertebr. Neurosci.* **2019**, *19*, 7. [CrossRef] [PubMed]
7. Elsareh, F.; Abdalla, R.; Abdalla, E. The effect of aqueous leaves extract of *Solenostemma argel* (Del Hayne) on egg masses and neonates of *Biomphalaria pfeifferi* snails. *J. Med. Plants* **2016**, *4*, 271–274.
8. Abd El Ghaffar, M.M.; Sadek, G.S.; Harba, N.M.; Abd El Samee, M.F. Evaluation of the effect of some plant molluscicides on the infectivity of *Schistosoma mansoni* cercariae. *Menoufia Med. J.* **2018**, *31*, 1448–1455. [CrossRef]
9. Rani, P.; Kaur, G.; Rao, K.V.; Singh, J.; Rawat, M. Impact of Green Synthesized Metal Oxide Nanoparticles on Seed Germination and Seedling Growth of *Vigna radiata* (Mung Bean) and *Cajanus cajan* (Red Gram). *J. Inorg. Organomet. Polym. Mater.* **2020**, *30*, 4053–4062. [CrossRef]
10. Fouda, A.; Abdel-Maksoud, G.; Abdel-Rahman, M.A.; Salem, S.S.; Hassan, S.E.D.; El-Sadany, M.A.H. Eco-friendly approach utilizing green synthesized nanoparticles for paper conservation against microbes involved in biodeterioration of archaeological manuscript. *Int. Biodeterior. Biodegrad.* **2019**, *142*, 160–169. [CrossRef]
11. Pham, D.T.N.; Khan, F.; Phan, T.T.V.; Park, S.-k.; Manivasagan, P.; Oh, J.; Kim, Y.M. Biofilm inhibition, modulation of virulence and motility properties by FeOOH nanoparticle in *Pseudomonas aeruginosa*. *Braz. J. Microbiol.* **2019**, *50*, 791–805. [CrossRef]
12. Raliya, R.; Tarafdar, J.C.; Choudhary, K.; Mal, P.; Raturi, A.; Gautam, R.K.S.; Singh, S.K. Synthesis of MgO Nanoparticles Using *Aspergillus tubingensis* TFR-3. *J. Bionanosci.* **2014**, *8*, 34–38. [CrossRef]
13. Fatimah, I. Biosynthesis and characterization of ZnO nanoparticles using rice bran extract as low-cost templating agent. *J. Eng. Sci. Technol.* **2018**, *13*, 409–420.
14. El-Belely, E.F.; Farag, M.M.S.; Said, H.A.; Amin, A.S.; Azab, E.; Gobouri, A.A.; Fouda, A. Green synthesis of zinc oxide nanoparticles (Zno-nps) using *Arthrospira platensis* (class: Cyanophyceae) and evaluation of their biomedical activities. *Nanomaterials* **2021**, *11*, 95. [CrossRef] [PubMed]
15. Lashin, I.; Fouda, A.; Gobouri, A.A.; Azab, E.; Mohammedsaleh, Z.M.; Makharita, R.R. Antimicrobial and in vitro cytotoxic efficacy of biogenic silver nanoparticles (Ag-nps) fabricated by callus extract of *Solanum incanum* L. *Biomolecules* **2021**, *11*, 341. [CrossRef]
16. Velgosova, O.; Dolinská, S.; Mražíková, A.; Briančin, J. Effect of *P. kessleri* extracts treatment on AgNPs synthesis. *Inorg. Nano-Met. Chem.* **2020**, *50*, 842–852. [CrossRef]
17. Spagnoletti, F.N.; Spedalieri, C.; Kronberg, F.; Giacometti, R. Extracellular biosynthesis of bactericidal Ag/AgCl nanoparticles for crop protection using the fungus *Macrophomina phaseolina*. *J. Environ. Manag.* **2019**, *231*, 457–466. [CrossRef]
18. Shafiq, T.; Uzair, M.; Iqbal, M.J.; Zafar, M.; Hussain, S.J.; Shah, S.A.A. Green synthesis of metallic nanoparticles and their potential in bio-medical applications. *Nano Biomed. Eng.* **2021**, *13*, 191–206. [CrossRef]
19. Fouda, A.; Abdel-Maksoud, G.; Saad, H.A.; Gobouri, A.A.; Mohammedsaleh, Z.M.; El-Sadany, M.A.H. The efficacy of silver nitrate ($AgNO_3$) as a coating agent to protect paper against high deteriorating microbes. *Catalysts* **2021**, *11*, 310. [CrossRef]
20. Fouda, A.; Awad, M.A.; Eid, A.M.; Saied, E.; Barghoth, M.G.; Hamza, M.F.; Awad, M.F.; Abdelbary, S.; Hassan, S.E.D. An eco-friendly approach to the control of pathogenic microbes and *Anopheles stephensi* malarial vector using magnesium oxide nanoparticles (Mg-nps) fabricated by *Penicillium chrysogenum*. *Int. J. Mol. Sci.* **2021**, *22*, 5096. [CrossRef]

21. Khalil, A.M.A.; Hassan, S.E.D.; Alsharif, S.M.; Eid, A.M.; Ewais, E.E.D.; Azab, E.; Gobouri, A.A.; Elkelish, A.; Fouda, A. Isolation and characterization of fungal endophytes isolated from medicinal plant *Ephedra pachyclada* as plant growth-promoting. *Biomolecules* **2021**, *11*, 140. [CrossRef]
22. Rahman, A.U.; Wei, Y.; Ahmad, A.; Khan, A.U.; Ali, R.; Ullah, S.; Ahmad, W.; Yuan, Q. Selenium Nanorods Decorated Gold Nanostructures: Synthesis, Characterization and Biological Applications. *J. Clust. Sci.* **2020**, *31*, 727–737. [CrossRef]
23. Geoffrion, L.D.; Hesabizadeh, T.; Medina-Cruz, D.; Kusper, M.; Taylor, P.; Vernet-Crua, A.; Chen, J.; Ajo, A.; Webster, T.J.; Guisbiers, G. Naked Selenium Nanoparticles for Antibacterial and Anticancer Treatments. *ACS Omega* **2020**, *5*, 2660–2669. [CrossRef] [PubMed]
24. Filipović, N.; Ušjak, D.; Milenković, M.T.; Zheng, K.; Liverani, L.; Boccaccini, A.R.; Stevanović, M.M. Comparative Study of the Antimicrobial Activity of Selenium Nanoparticles With Different Surface Chemistry and Structure. *Front. Bioeng. Biotechnol.* **2021**, *8*, 624621. [CrossRef] [PubMed]
25. Vahdati, M.; Tohidi Moghadam, T. Synthesis and Characterization of Selenium Nanoparticles-Lysozyme Nanohybrid System with Synergistic Antibacterial Properties. *Sci. Rep.* **2020**, *10*, 510. [CrossRef] [PubMed]
26. Liao, W.; Yu, Z.; Lin, Z.; Lei, Z.; Ning, Z.; Regenstein, J.M.; Yang, J.; Ren, J. Biofunctionalization of Selenium Nanoparticle with *Dictyophora indusiata* Polysaccharide and Its Antiproliferative Activity through Death-Receptor and Mitochondria-Mediated Apoptotic Pathways. *Sci. Rep.* **2016**, *5*, 18629. [CrossRef]
27. Menon, S.; Agarwal, H.; Venkat Kumar, S.; Rajeshkumar, S. Chapter 8—Biomemetic synthesis of selenium nanoparticles and its biomedical applications. In *Green Synthesis, Characterization and Applications of Nanoparticles*; Micro and Nano Technologies; Shukla, A.K., Iravani, S., Eds.; Elsevier: Amsterdam, The Netherlands, 2019; pp. 165–197. ISBN 978-0-08-102579-6.
28. Klaine, S.J.; Alvarez, P.J.J.; Batley, G.E.; Fernandes, T.F.; Handy, R.D.; Lyon, D.Y.; Mahendra, S.; McLaughlin, M.J.; Lead, J.R. Nanomaterials in the environment: Behavior, fate, bioavailability, and effects. *Environ. Toxicol. Chem.* **2008**, *27*, 1825–1851. [CrossRef] [PubMed]
29. Pitt, J.I.; Hocking, A.D. *Fungi and Food Spoilage*; Springer: Berlin/Heidelberg, Germany, 2009; Volume 519.
30. White, T.J.; Bruns, T.; Lee, S.; Taylor, J. Amplification and direct sequencing of fungal ribosomal RNA genes for phylogenetics. *PCR Protoc. Guide Methods Appl.* **1990**, *18*, 315–322.
31. Amin, M.A.; Ismail, M.A.; Badawy, A.A.; Awad, M.A.; Hamza, M.F.; Awad, M.F.; Fouda, A. The Potency of Fungal-Fabricated Selenium Nanoparticles to Improve the Growth Performance of *Helianthus annuus* L. and Control of Cutworm *Agrotis ipsilon*. *Catalysts* **2021**, *11*, 1551. [CrossRef]
32. Vahidi, H.; Kobarfard, F.; Kosar, Z.; Mahjoub, M.A.; Saravanan, M.; Barabadi, H. Mycosynthesis and characterization of selenium nanoparticles using standard *Penicillium chrysogenum* PTCC 5031 and their antibacterial activity: A novel approach in microbial nanotechnology. *Nanomed. J.* **2020**, *7*, 315–323. [CrossRef]
33. Salem, S.S.; Fouda, M.M.G.; Fouda, A.; Awad, M.A.; Al-Olayan, E.M.; Allam, A.A.; Shaheen, T.I. Antibacterial, Cytotoxicity and Larvicidal Activity of Green Synthesized Selenium Nanoparticles Using *Penicillium corylophilum*. *J. Clust. Sci.* **2021**, *32*, 351–361. [CrossRef]
34. Eveland, L.K.; Haseeb, M.A. Laboratory Rearing of *Biomphalaria glabrata* Snails and Maintenance of Larval Schistosomes In Vivo and In Vitro. In *Biomphalaria Snails and Larval Trematodes*; Springer: New York, NY, USA, 2011; pp. 33–55.
35. WHO. *Report of the Scientific Working Group on Plant Molluscicide & Guidelines for Evaluation of Plant Molluscicides*; TDR/SCHSWE (4)/83.3; WHO: Geneva, Switzerland, 1983.
36. WHO. Molluscicide screening and evaluation. *Bull. World Health Organ.* **1965**, *33*, 567–581.
37. Hamdi, S.A.H.; Ibrahim, A.M.; Ghareeb, M.A.; Fol, M.F. Chemical characterization, biocidal and molluscicidal activities of chitosan extracted from the crawfish *Procambarus clarkii* (Crustacea: Cambaridae). *Egypt. J. Aquat. Biol. Fish.* **2021**, *25*, 355–371. [CrossRef]
38. Eissa, M.M.; El Bardicy, S.; Tadros, M. Bioactivity of miltefosine against aquatic stages of *Schistosoma mansoni*, *Schistosoma haematobium* and their snail hosts, supported by scanning electron microscopy. *Parasit. Vectors* **2011**, *4*, 73. [CrossRef] [PubMed]
39. Nduku, W.K.; Harrison, A.D. Cationic responses of organs and haemolymph of *Biomphalaria pfeifferi* (Krauss), *Biomphalaria glabrata* (Say) and *Helisoma trivolvis* (Say)(Gastropoda: Planorbirdae) to cationic alterations of the medium. *Hydrobiologia* **1980**, *68*, 119–138. [CrossRef]
40. Abdul-Salam, J.M.; Michelson, E.H. *Biomphalaria glabrata* amoebocytes: Effect of *Schistosoma mansoni* infection on in vitro phagocytosis. *J. Invertebr. Pathol.* **1980**, *35*, 241–248. [CrossRef]
41. Grazeffe, V.S.; de Freitas Tallarico, L.; de Sa Pinheiro, A.; Kawano, T.; Suzuki, M.F.; Okazaki, K.; de Bragança Pereira, C.A.; Nakano, E. Establishment of the comet assay in the freshwater snail *Biomphalaria glabrata* (Say, 1818). *Mutat. Res.—Genet. Toxicol. Environ. Mutagen.* **2008**, *654*, 58–63. [CrossRef] [PubMed]
42. Ibrahim, A.M.; Sayed, D.A. Toxicological impact of oxyfluorfen 24% herbicide on the reproductive system, antioxidant enzymes, and endocrine disruption of *Biomphalaria alexandrina* (Ehrenberg, 1831) snails. *Environ. Sci. Pollut. Res. Int.* **2019**, *26*, 7960–7968. [CrossRef]
43. Aebi, H. Catalase in vitro. *Methods Enzymol.* **1984**, *105*, 121–126.
44. Ohkawa, H.; Ohishi, N.; Yagi, K. Assay for lipid peroxides in animal tissues by thiobarbituric acid reaction. *Anal. Biochem.* **1979**, *95*, 351–358. [CrossRef]

45. Beutler, E.; Duron, O.; Kelly, B.M. Improved method for the determination of blood glutathione. *J. Lab. Clin. Med.* **1963**, *61*, 882–888.
46. Koracevic, D.; Koracevic, G.; Djordjevic, V.; Andrejevic, S.; Cosic, V. Method for the measurement of antioxidant activity in human fluids. *J. Clin. Pathol.* **2001**, *54*, 356–361. [CrossRef]
47. Bellos, J.K.; Perrea, D.N.; Theodotopoulou, E.; Vlachos, I.; Kostakis, A.I. Nitric oxide production as an indicator of recurrence of focal and segmental glomerulosclerosis following kidney transplantation. *Saudi J. Kidney Dis. Transpl.* **2011**, *22*, 1030–1032. [PubMed]
48. Finney, D.J. *Probit Analysis*, 3rd ed.; Combridge University Press: New York, NY, USA, 1971.
49. Murray, R. Speigel: Statistical estimation theory. In *Schaum's Outline Series: Theory and Problems of Statistics in SI Units*, 1st ed.; McGraw Hill Book Company: Singapore, 1981; Volume 156.
50. Oyeyemi, O.T. Application of nanotized formulation in the control of snail intermediate hosts of schistosomes. *Acta Trop.* **2021**, *220*, 105945. [CrossRef] [PubMed]
51. Pohl, C.; Polli, F.; Schütze, T.; Viggiano, A.; Mózsik, L.; Jung, S.; de Vries, M.; Bovenberg, R.A.L.; Meyer, V.; Driessen, A.J.M. A *Penicillium rubens* platform strain for secondary metabolite production. *Sci. Rep.* **2020**, *10*, 7630. [CrossRef] [PubMed]
52. Joshi, S.M.; De Britto, S.; Jogaiah, S.; Ito, S.I. Mycogenic selenium nanoparticles as potential new generation broad spectrum antifungal molecules. *Biomolecules* **2019**, *9*, 419. [CrossRef]
53. Ramamurthy, C.H.; Sampath, K.S.; Arunkumar, P.; Kumar, M.S.; Sujatha, V.; Premkumar, K.; Thirunavukkarasu, C. Green synthesis and characterization of selenium nanoparticles and its augmented cytotoxicity with doxorubicin on cancer cells. *Bioprocess Biosyst. Eng.* **2013**, *36*, 1131–1139. [CrossRef]
54. Worrall, E.A.; Hamid, A.; Mody, K.T.; Mitter, N.; Pappu, H.R. Nanotechnology for plant disease management. *Agronomy* **2018**, *8*, 285. [CrossRef]
55. Zare, B.; Babaie, S.; Setayesh, N.; Shahverdi, A.R.; Shahverdi, A. Isolation and characterization of a fungus for extracellular synthesis of small selenium nanoparticles Extracellular synthesis of selenium nanoparticles using fungi. *Nanomed. J.* **2013**, *1*, 13–19.
56. Ahmad, M.S.; Yasser, M.M.; Sholkamy, E.N.; Ali, A.M.; Mehanni, M.M. Anticancer activity of biostabilized selenium nanorods synthesized by *Streptomyces bikiniensis* strain Ess_amA-1. *Int. J. Nanomed.* **2015**, *10*, 3389–3401. [CrossRef]
57. Ullah, A.; Yin, X.; Wang, F.; Xu, B.; Mirani, Z.A.; Xu, B.; Chan, M.W.H.; Ali, A.; Usman, M.; Ali, N.; et al. Biosynthesis of selenium nanoparticles (via *Bacillus subtilis* bsn313), and their isolation, characterization, and bioactivities. *Molecules* **2021**, *26*, 5559. [CrossRef]
58. Ranjitha, V.R.; Ravishankar, V.R. Extracellular Synthesis of Selenium Nanoparticles from an Actinomycetes *Streptomyces griseoruber* and Evaluation of its Cytotoxicity on HT-29 Cell Line. *Pharm. Nanotechnol.* **2018**, *6*, 61–68. [CrossRef]
59. Salem, S.S. Bio-fabrication of Selenium Nanoparticles Using Baker's Yeast Extract and Its Antimicrobial Efficacy on Food Borne Pathogens. *Appl. Biochem. Biotechnol.* **2022**. [CrossRef] [PubMed]
60. Barabadi, H.; Kobarfard, F.; Vahidi, H. Biosynthesis and Characterization of Biogenic Tellurium Nanoparticles by Using *Penicillium chrysogenum* PTCC 5031: A Novel Approach in Gold Biotechnology. *Iran. J. Pharm. Res. IJPR* **2018**, *17*, 87–97. [PubMed]
61. Al-Brakati, A.; Alsharif, K.F.; Alzahrani, K.J.; Kabrah, S.; Al-Amer, O.; Oyouni, A.A.; Habotta, O.A.; Lokman, M.S.; Bauomy, A.A.; Kassab, R.B.; et al. Using green biosynthesized lycopene-coated selenium nanoparticles to rescue renal damage in glycerol-induced acute kidney injury in rats. *Int. J. Nanomed.* **2021**, *16*, 4335–4349. [CrossRef] [PubMed]
62. Rajkumar, K.; Sandhya, M.V.S.; Koganti, S.; Burgula, S. Selenium nanoparticles synthesized using *Pseudomonas stutzeri* (Mh191156) show antiproliferative and anti-angiogenic activity against cervical cancer cells. *Int. J. Nanomed.* **2020**, *15*, 4523–4540. [CrossRef] [PubMed]
63. Alagesan, V.; Venugopal, S. Green Synthesis of Selenium Nanoparticle Using Leaves Extract of *Withania somnifera* and Its Biological Applications and Photocatalytic Activities. *Bionanoscience* **2019**, *9*, 105–116. [CrossRef]
64. Lian, S.; Diko, C.S.; Yan, Y.; Li, Z.; Zhang, H.; Ma, Q.; Qu, Y. Characterization of biogenic selenium nanoparticles derived from cell-free extracts of a novel yeast *Magnusiomyces ingens*. *3 Biotech* **2019**, *9*, 221. [CrossRef]
65. Osman, G.Y.; Ahmed, M.M.; Abdel Kader, A.; Mohamed, A.A. Biological and biochemical impacts of the fungal extract of *Aspergillus fumigatus* on *Biomphalaria alexandrina* snails infected with *Schistosoma mansoni*. *Biosciences* **2013**, *7*, 473–484.
66. Mahmoudvand, H.; Shakibaie, M.; Tavakoli, R.; Jahanbakhsh, S.; Sharifi, I. In Vitro Study of Leishmanicidal Activity of Biogenic Selenium Nanoparticles against Iranian Isolate of Sensitive and Glucantime-Resistant *Leishmania tropica*. *Iran. J. Parasitol.* **2014**, *9*, 452–460.
67. Alkhudhayri, A.; Al-Shaebi, E.M.; Qasem, M.A.A.; Murshed, M.; Mares, M.M.; Al-Quraishy, S.; Dkhil, M.A. Antioxidant and anti-apoptotic effects of selenium nanoparticles against murine eimeriosis. *An. Acad. Bras. Cienc.* **2020**, *92*, e20191107. [CrossRef]
68. Dkhil, M.A.; Khalil, M.F.; Diab, M.S.M.; Bauomy, A.A.; Santourlidis, S.; Al-Shaebi, E.M.; Al-Quraishy, S. Evaluation of nanoselenium and nanogold activities against murine intestinal schistosomiasis. *Saudi J. Biol. Sci.* **2019**, *26*, 1468–1472. [CrossRef]
69. De Carvalho Augusto, R.; Merad, N.; Rognon, A.; Gourbal, B.; Bertrand, C.; Djabou, N.; Duval, D. Molluscicidal and parasiticidal activities of *Eryngium triquetrum* essential oil on *Schistosoma mansoni* and its intermediate snail host *Biomphalaria glabrata*, a double impact. *Parasites Vectors* **2020**, *13*, 486. [CrossRef] [PubMed]
70. Ibrahim, A.M.; Ahmed, A.K.; Bakry, F.A.; Abdel-Ghaffar, F. Hematological, physiological and genotoxicological effects of Match 5% EC insecticide on *Biomphalaria alexandrina* snails. *Ecotoxicol. Environ. Saf.* **2018**, *147*, 1017–1022. [CrossRef] [PubMed]

71. Ali, D. Evaluation of environmental stress by comet assay on freshwater snail *Lymnea luteola* L. exposed to titanium dioxide nanoparticles. *Toxicol. Environ. Chem.* **2014**, *96*, 1185–1194. [CrossRef]
72. Ibrahim, A.M.; Ghoname, S.I. Molluscicidal impacts of *Anagallis arvensis* aqueous extract on biological, hormonal, histological and molecular aspects of *Biomphalaria alexandrina* snails. *Exp. Parasitol.* **2018**, *192*, 36–41. [CrossRef] [PubMed]
73. Wang, Z.; Zhang, L.; Zhao, J.; Xing, B. Environmental processes and toxicity of metallic nanoparticles in aquatic systems as affected by natural organic matter. *Environ. Sci. Nano* **2016**, *3*, 240–255. [CrossRef]
74. Prokhorova, E.E.; Serebryakova, M.K.; Tokmakova, A.S.; Ataev, G.L. Hemocytes of mollusc *Biomphalaria glabrata* (Gastropoda, pulmonata). *Invertebr. Surviv. J.* **2018**, *15*, 346–351. [CrossRef]
75. Cavalcanti, M.G.S.; Filho, F.C.; Mendonça, A.M.B.; Duarte, G.R.; Barbosa, C.C.G.S.; De Castro, C.M.M.B.; Alves, L.C.; Brayner, F.A. Morphological characterization of hemocytes from *Biomphalaria glabrata* and *Biomphalaria straminea*. *Micron* **2012**, *43*, 285–291. [CrossRef]
76. Abdel-Hamid, H.; Mekawey, A.A.I. Biological and hematological responses of *Biomphalaria alexandrina* to mycobiosynthsis silver nanoparticles. *J. Egypt. Soc. Parasitol.* **2014**, *44*, 627–637. [CrossRef]
77. Ibrahim, A.M.; Ahmed, A.K.; Bakry, F.A.; Rabei, I.; Abdel-Ghaffar, F. Toxicological impact of butralin, glyphosate-isopropylammonium and pendimethalin herbicides on physiological parameters of *Biomphalaria alexandrina* snails. *Molluscan Res.* **2019**, *39*, 224–233. [CrossRef]
78. Ray, M.; Bhunia, N.S.; Bhunia, A.S.; Ray, S. A comparative analyses of morphological variations, phagocytosis and generation of cytotoxic agents in flow cytometrically isolated hemocytes of Indian molluscs. *Fish Shellfish Immunol.* **2013**, *34*, 244–253. [CrossRef]
79. Donaghy, L.; Hong, H.-K.; Lambert, C.; Park, H.-S.; Shim, W.J.; Choi, K.-S. First characterisation of the populations and immune-related activities of hemocytes from two edible gastropod species, the disk abalone, *Haliotis discus discus* and the spiny top shell, *Turbo cornutus*. *Fish Shellfish Immunol.* **2010**, *28*, 87–97. [CrossRef] [PubMed]
80. Khalil, A.M. Toxicological effects and oxidative stress responses in freshwater snail, *Lanistes carinatus*, following exposure to chlorpyrifos. *Ecotoxicol. Environ. Saf.* **2015**, *116*, 137–142. [CrossRef] [PubMed]
81. Barky, F.A.; Abdelsalam, H.A.; Mahmoud, M.B.; Hamdi, S.A.H. Influence of Atrazine and Roundup pesticides on biochemical and molecular aspects of *Biomphalaria alexandrina* snails. *Pestic. Biochem. Physiol.* **2012**, *104*, 9–18. [CrossRef]
82. Sayed, S.S.M.; Abdel-Wareth, M.T.A. The comparative effect of chlorine and Huwa-san as disinfecting agents on *Biomphalaria alexandrina* snails and free larval stages of *Schistosoma mansoni*. *Parasitol. Res.* **2017**, *116*, 2627–2635. [CrossRef]
83. Wang, W.; Mao, Q.; Yao, J.; Yang, W.; Zhang, Q.; Lu, W.; Deng, Z.; Duan, L. Discovery of the pyridylphenylureas as novel molluscicides against the invasive snail *Biomphalaria straminea*, intermediate host of *Schistosoma mansoni*. *Parasites Vectors* **2018**, *11*, 291. [CrossRef]
84. Saleh, H.A.; Abdel-Motleb, A.; Habib, M.R. Neurotoxicity and genotoxicity of the veterinary antibiotics oxytetracycline and trimethoprim-sulphadiazine to *Biomphalaria alexandrina* snails. *Int. J. Environ. Stud.* **2021**, *78*, 983–1002. [CrossRef]
85. Chemical Computing Group Inc. Molecular operating environment (MOE). 2016. Available online: https://www.chemcomp.com/Products.htm (accessed on 10 February 2022).
86. Manjula, R.; Wright, G.S.A.; Strange, R.W.; Padmanabhan, B. Assessment of ligand binding at a site relevant to SOD1 oxidation and aggregation. *FEBS Lett.* **2018**, *592*, 1725–1737. [CrossRef]
87. Oakley, A.J.; Lo Bello, M.; Nuccetelli, M.; Mazzetti, A.P.; Parker, M.W. The ligandin (non-substrate) binding site of human Pi class glutathione transferase is located in the electrophile binding site (H-site). *J. Mol. Biol.* **1999**, *291*, 913–926. [CrossRef]
88. Kim, G.H.; Kim, J.E.; Rhie, S.J.; Yoon, S. The Role of Oxidative Stress in Neurodegenerative Diseases. *Exp. Neurobiol.* **2015**, *24*, 325–340. [CrossRef]
89. Bastos, F.F.; Tobar, S.A.L.; Dantas, R.F.; Silva, E.S.; Nogueira, N.P.A.; Paes, M.C.; Righi, B.D.P.; Bastos, J.C.; Bastos, V.L.F.C. Melatonin affects conjugation of 4-hydroxynonenal with glutathione in liver of pacu, a hypoxia-tolerant fish. *Fish Physiol. Biochem.* **2013**, *39*, 1205–1214. [CrossRef]
90. Montassir, L.; Berrebaan, I.; Mellouki, F.; Zkhiri, F.; Boughribil, S.; Bessi, H. Acute toxicity and reprotoxicity of aqueous extract of a Moroccan plant (*Tetraclinis articulata*) on freshwater cladoceran *Daphnia magna*. *J. Mater. Environ. Sci.* **2017**, *8*, 770–776.

Journal of *Fungi*

Article

Pseudomonas indica-Mediated Silver Nanoparticles: Antifungal and Antioxidant Biogenic Tool for Suppressing Mucormycosis Fungi

Salem S. Salem [1], Omar M. Ali [2,*], Ahmed M. Reyad [3,4], Kamel A. Abd-Elsalam [5,*] and Amr H. Hashem [1,*]

1 Department of Botany and Microbiology, Faculty of Science, Al-Azhar University, Nasr City, Cairo 11884, Egypt; salemsalahsalem@azhar.edu.eg
2 Department of Chemistry, Turabah University College, Turabah Branch, Taif University, Taif 21944, Saudi Arabia
3 Biology Department, Faculty of Science, Jazan University, Jazan 82817, Saudi Arabia; areyadegy@yahoo.com
4 Botany and Microbiology Department, Faculty of Science, Beni-Suef University, Beni-Suef 62511, Egypt
5 Plant Pathology Research Institute, Agricultural Research Centre, Giza 12619, Egypt
* Correspondence: om.ali@tu.edu.sa (O.M.A.); kamelabdelsalam@gmail.com (K.A.A.-E.); amr.hosny86@azhar.edu.eg (A.H.H.)

Abstract: Mucormycosis is considered one of the most dangerous invasive fungal diseases. In this study, a facile, green and eco-friendly method was used to biosynthesize silver nanoparticles (AgNPs) using *Pseudomonas indica* S. Azhar, to combat fungi causing mucormycosis. The biosynthesis of AgNPs was validated by a progressive shift in the color of *P. indica* filtrate from colorless to brown, as well as the identification of a distinctive absorption peak at 420 nm using UV-vis spectroscopy. Fourier-transform infrared spectroscopy (FTIR) results indicated the existence of bioactive chemicals that are responsible for AgNP production. AgNPs with particle sizes ranging from 2.4 to 53.5 nm were discovered using transmission electron microscopy (TEM). Pattern peaks corresponding to the 111, 200, 220, 311, and 222 planes, which corresponded to face-centered cubic forms of metallic silver, were also discovered using X-ray diffraction (XRD). Moreover, antifungal activity measurements of biosynthesized AgNPs against *Rhizopus Microsporus*, *Mucor racemosus*, and *Syncephalastrum racemosum* were carried out. Results of antifungal activity analysis revealed that the biosynthesized AgNPs exhibited outstanding antifungal activity against all tested fungi at a concentration of 400 µg/mL, where minimum inhibitory concentrations (MIC) were 50, 50, and 100 µg/mL toward *R. microsporus*, *S. racemosum*, and *M. racemosus* respectively. In addition, the biosynthesized AgNPs revealed antioxidant activity, where IC_{50} was 31 µg/mL when compared to ascorbic acid (0.79 µg/mL). Furthermore, the biosynthesized AgNPs showed no cytotoxicity on the Vero normal cell line. In conclusion, the biosynthesized AgNPs in this study can be used as effective antifungals with safe use, particularly for fungi causing mucormycosis.

Keywords: silver nanoparticles; green biosynthesis; antifungal activity; mucormycosis; antioxidant activity

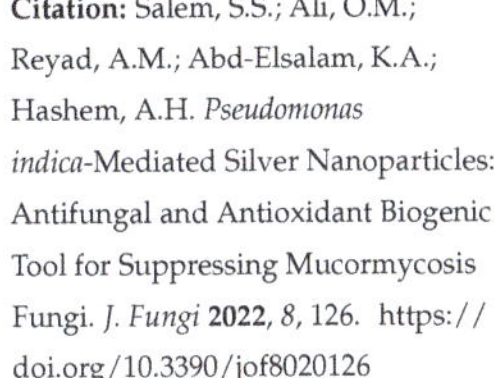

Citation: Salem, S.S.; Ali, O.M.; Reyad, A.M.; Abd-Elsalam, K.A.; Hashem, A.H. *Pseudomonas indica*-Mediated Silver Nanoparticles: Antifungal and Antioxidant Biogenic Tool for Suppressing Mucormycosis Fungi. *J. Fungi* **2022**, *8*, 126. https://doi.org/10.3390/jof8020126

Academic Editor: Vincent Bruno

Received: 14 January 2022
Accepted: 25 January 2022
Published: 27 January 2022

1. Introduction

More than 1.2 billion people are infected by pathogenic fungus each year, resulting in at least 1.7 million fatalities [1,2]. Fungal pathogens now outnumber drug-resistant *Mycobacterium tuberculosis* in terms of mortality, and they even outnumber malaria [3]. Mucormycosis is one of the most dangerous fungal diseases for humans. Mucormycosis is a disease caused by a group of fungi called mucormycetes, which includes various genera such as *Rhizopus*, *Mucor*, *Synsephalastrum*, *Absidia*, and *Cunninghamella* [4,5]. These fungi invade people with a history of diabetes, stem cell transplants, cancer, injection drug use, skin injury due to surgery, burns, or wounds [6–8]. Mucormycosis is a serious infection and needs to be treated with prescription antifungal medicine, usually amphotericin B, posaconazole, or isavuconazole [9].The increasing usage of antifungal medications has resulted in the emergence of fungal strains such as *Candida albicans* [10], *Lichtheimia corymbifera, R. microsporus, R. arrhizus,* and *M. circinelloids* that are resistant to the majority of antifungal treatments [11,12]. Most harmful fungi, similar to bacteria, have recently developed antibiotic resistance. To combat drug-resistant fungus, novel antifungal medicines based on current biotechnology must be investigated.

Nanoparticles (NPs) are a diverse class of materials that appeal to several researchers due to their tiny size (1–100 nm), exceptional properties, large surface area, enhanced reactivity, capacity to access the body easily, and multiple uses in modern science, including the industrial and medicinal sciences [13–20]. Selenium, silver, copper, magnesium, zinc, and titanium are some of the nanoparticles that have been reported [21–30]. AgNPs are widely utilized nanoparticles in numerous sectors of study, such as optical devices, ophthalmology, pharmaceutical, and the health sciences, for the creation of drug carriers, chemotherapeutic, nano-sensors, gene therapy, and other applications [31–36]. AgNPs have been shown to exhibit high antibacterial, anti-inflammatory, antibiofilm, and anticancer characteristics, with the antimicrobial capabilities of AgNPs being used to prevent infection against harmful microorganisms, such as eukaryotic microorganisms, bacteria, and viruses [37–40]. AgNPs are widely used in the biomedical control of diseases such as candidiasis [35] and aspergillosis [41] [42,43]. The efficacy of AgNPs as antimicrobial against microorganisms may be due to their link with enzymes, reactive oxygen species (ROS) levels, and changing structure contents that alter the membrane integrity and morphology [44–46]. Because of their antibacterial, anti-inflammatory, antibiofilm, anticoagulant, and anticancer properties, AgNPs are becoming a popular choice in the medical and biological fields [22,47,48]. Physical and chemical approaches for the synthesis of NPs are extensively utilized, but they have several downsides, such as the usage of high energy or dangerous chemicals, their high cost, and the generation of vast volumes of toxic byproducts that pollute the environment [49]. To overcome the constraints of physical and chemical procedures, low-cost, ecofriendly, simple, and nontoxic approaches that eliminate the use of hazardous and pricey solvents are required for metallic NPs manufacturing [50]. Biological synthesis of metal nanoparticles should be stable, biologically safe and ecofriendly [51–55]. Bacteria are one of the most significant groups of microorganisms, since they are employed in bioprocessing, enzyme manufacturing, acid generation, and nanotechnology, among other uses [56–58]. Microbial variety may be found in abundance in soil, and these microorganisms can be harnessed to benefit humans. Soil bacteria also require a low cost and low-nutrient medium for growth, making them an ideal option for use [47]. In this work, the bacterial strain *P. indica* S. Azhar was isolated from a soil sample and utilized to synthesize AgNPs in a simple, quick, and environmentally friendly manner. AgNPs were investigated and characterized by UV-vis, Ft-IR, TEM, and XRD. An attempt was made to investigate the antifungal and antioxidant activities of AgNPs against three mold strains (*Rhizopus microsporus, Mucor racemosus,* and *Syncephalastrum racemosum*) which cause mucormycosis in vitro.

2. Materials and Methods

2.1. Isolation and Identification of Strain S. Azhar

The bacterial strain *P. indica* S. Azhar was isolated from a soil sample collected from the garden of the Faculty of Science, Al-Azhar University, Cairo, Egypt. For the isolation of bacterial strains, nutrient agar medium was used. The 16S rRNA gene sequencing for bacteria was used to identify it genetically. The improved approach was used to extract bacterial genomic DNA. Single colonies were taken from agar plates and suspended in sterile deionized water in 50 μL. The cell suspension was then incubated in a water bath for 10 min at 97 °C before being centrifuged for 10 min at 15,000 rpm to separate the DNA-containing cell lysate. Using a genomic DNA template and two universal primers, PCR was utilized to amplify the 16S rRNA gene. The primers 27-f (5-AGAGTTTGATCCTGGCTCAG-3) and 1492-r were utilized (5-GGTTACCTTGTTACGACTT-3). Amounts of 1× PCR buffer, 0.5 mM $MgCl_2$, 2.5 U Taq-polymerase (QIAGEN, Hilden, Germany), 0.25 mM dNTP, 0.5 M of each primer, and 1 µL of isolated genomic DNA were added into the PCR mixture (50 µL). Sigma Scientific Services Company (Cairo, Egypt) used a Thermal Cycler to execute the PCR, which included a 3 min hot start at 94 °C, 30 cycles at 94 °C for 30 s, 55 °C for 30 s, and 72 °C for 1 min, and 10 min of gene extension at 72 °C. GATC Company used an ABI 3730x1 DNA sequencer to evaluate the sequencing (Ebersberg, Germany). Using the NCBI BLAST software, the acquired sequences were compared to those in the GenBank database. The bootstrap technique was used to create the phylogenetic tree.

2.2. *Extracellular Biosynthesis of AgNPs*

2.2.1. Bacterial Filtrate Preparation

P. indica cells were suspended in nutrient and injected in broth media for fermentation at 36 ± 2 °C for 48 h in an orbital shaker (120 rpm). The biomass was collected using filter paper No. 1 and rinsed in sterilized distilled water to eliminate any medium components before being suspended in 100 mL distilled water. At 34 ± 2 °C, the mixture was stirred for 24 h. Finally, the biomass filtrate was produced by passing it through Whatman filter paper No. 1 and centrifuging it for 5 min at 2000 rpm to sediment any remaining cell debris. The supernatant was used for AgNPs biosynthesis.

2.2.2. Biosynthesis of AgNPs by Biomass Filtrate

For the production of AgNPs, the previously prepared bacterial biomass filtrate of *P. indica* was employed as follows: in a 250 mL flask, 1 mM silver nitrate was combined with 100 mL biomass filtrate and incubated at 36 ± 2 °C for 24 h, and agitated at 150 rpm, as shown in Figure 1. Negative controls (cell filtrate) were also run along with the experiment.

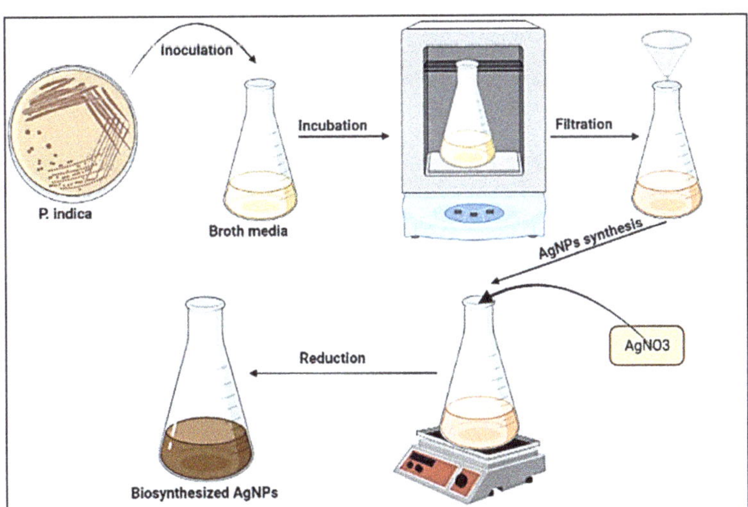

Figure 1. Process for the biosynthesis of AgNPs.

2.3. *Characterization of AgNPs*

The absorption characteristics of biogenically produced AgNPs were evaluated using UV-vis spectrophotometry (JENWAY 6305 Spectrophotometer, UK). Quartz cuvettes were used to investigate solutions containing AgNPs in the wavelength range 200–800 nm. Allocating the peak in the region of 400 to 500 nm indicated the existence of AgNPs. In a mortar, a known weight of AgNPs, 1 mg, was ground with dry 2.5 mg of KBr. The powder was then placed in a 2 mm micro-cup with a 2 mm internal diameter and loaded onto a FT-IR set at 26 °C ± 1 °C. The sample was scanned using infra-red in the range of 400–4000 cm^{-1}, utilizing Fourier transform-infrared spectrometty. To determine the size and form of nanoparticles, TEM (JEOL 1010, Tokyo, Japan) was used to characterize AgNPs. Drop-coating the AgNPs solution onto the carbon-coated copper grid and loading it onto a specimen holder were used to produce the sample. The sizes and shapes of AgNPs were validated using TEM micrographs. Stress analysis, residual–austenite quantitation, crystallite size/lattice, crystallite calculation, and materials analysis by overlaid X-ray diffraction patterns with a Shimadzu apparatus, using a nickel filter and Cu–Ka target (Shimadzu-Scientific Instruments (SSI), Kyoto, Japan), were obtained with the XRD 6000 series. The average crystallite size of AgNPs can also be measured utilizing the Debye-scherrer equation: $D = k\lambda/\beta \cos \theta$.

2.4. Antifungal Activity

Some genera of the Mucorales group were used for evaluation of the biosynthesized AgNPs as antifungal agent. *Rhizopus microsporus* (Accession no. MK623262.1), *Mucor racemosus* (Accession no. MG547571.1), and *Syncephalastrum racemosum* (Accession no. MK621186.1) were obtained from the Mycology Lab., Faculty of Science, Al-Azhar University. These three fungal strains were inoculated on malt extract agar (MEA) (Oxoid) plates, incubated for 3–5 days at 28 ± 2 °C, and then stored at 4 °C for future use [59–62]. Antifungal activity of AgNPs at 400 µg/mL were tested according to methods used by Dacrory et al. [5] with minor modifications. Briefly, 100 µL of AgNPs were added to plates, prepared previously on MEA media streaked with purified fungal strains (10^7 spores/mL). The plates were incubated for 2–3 days at 30 ± 2 °C. Minimum inhibitory concentration of AgNPs toward all tested fungal strains was assessed using the broth microdilution technique, according to the standard EUCAST methodology [63].

2.5. Antioxidant Activity

1,1-Diphenyl-2-picrylhydrazyl (DPPH) free radical-scavenging activity of the biosynthesized AgNPs was determined according to the method used by Khalil et al. [64], with minor modifications. Different concentrations of biosynthesized AgNPs (400–0.78 µg/mL) were used to determine the antioxidant activity. The DPPH solution (800 µL) was combined with 200 µL of the sample concentration and maintained at 25 °C in the dark for 30 min. After that, the absorbance was measured at 517 nm against a blank after centrifugation for 5 min at 13,000 rpm. The standard for this experiment was ascorbic acid. The following formula was used to determine antioxidant activity:

$$\text{Antioxidant activity } (\%) = \frac{\text{Control absorbance} - \text{Sample absorbance}}{\text{Control absorbance}} \times 100$$

2.6. In-Vitro Cytotoxicity

The cytotoxicity of AgNPs, at concentrations ranging from 31.25 to 1000 µg/mL, was tested against normal Vero cell lines obtained from ATCC using the MTT procedure [65] with minor modifications. As mentioned in Equation (1), the viability percentages were computed as follows:

$$\text{Viability } \% = \frac{\text{Test OD}}{\text{Control OD}} \times 100 \quad (1)$$

3. Results

3.1. Bacterial Strain Identification

Isolate S. Azhar's 16S rRNA gene sequence was 1140 bp, and the sequence was uploaded to NCBI (accession number MW820210). Strain S. Azhar has the greatest sequence similarity to *P. indica*, based on the 16S rRNA gene sequencing analysis. A neighbor-joining tree phylogenetic study further revealed that strain S. Azhar belonged to the Pseudomonas genus (Figure 2).

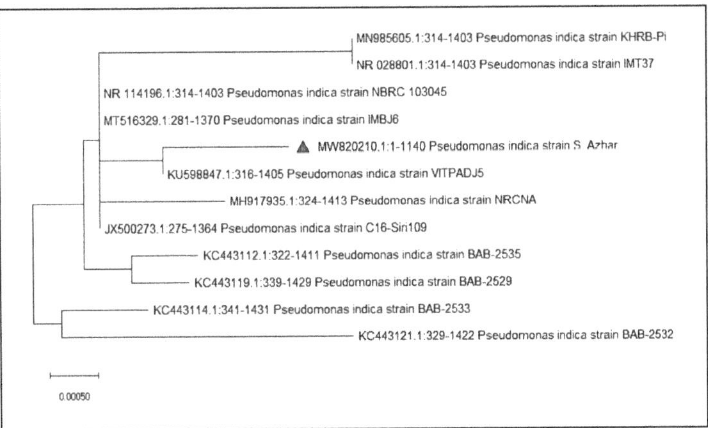

Figure 2. A neighbor-joining (NJ) tree based on 16S rRNA gene sequence analysis was used to show the phylogenetic relationships of the isolated strain S. Azhar with comparable type strains.

3.2. Synthesis and Characterization of AgNPs

In this study, biomass filtrate of *P. indica* was incubated with 1 mM $AgNO_3$ for 24 h in dark conditions. Appearance of brown color after contacting of the filtrate of *P. indica* S. Azhar with precursor ($AgNO_3$) at the reaction completion indicated AgNPs' formation; maximum absorbance peaks at 420 nm were seen in the UV-vis spectral analysis, which might correlate to spherical AgNPs, as seen in Figure 3A. Surface plasmon resonance (SPR) excitation might be responsible for the color alteration. After calcination, AgNPs were obtained as a black powder. The functional groups of those found in AgNPs were characterized using FT-IR analysis. As seen in Figure 3B. The occurrence of functional assemblies of biomolecules was discovered using FTIR wavelengths of 400 to 4000 cm^{-1}. Ten prominent peaks in the FTIR spectra of biosynthesized AgNPs were found at 474.4, 617.1, 1110.8, 1382.7, 1617.9, 1637.2, 2032.6, 2921.6, 3235.9' and 3415.3 cm^{-1} (Figure 3B). The peaks at 3235.9 and 3415.3 cm^{-1} correspond to the alcohol O–H stretching group or the secondary amine N–H stretching group. NH stretching of the protein's amide I band is shown by the bands at 1637.2and 1617.9 cm^{-1}. Alkyne stretch bands are represented by the peaks at 2032.6 and 2921.6 cm^{-1}. Furthermore, the bands seen at 1382.7 and 1110.8 cm^{-1} might be attributed to aromatic and aliphatic amine C-N stretching vibrations. Finally, the peaks at 617.1 and 474.4 cm^{-1} correspond to the bending of alkene (C=H) groups. As a result, the current work reveals that proteins or bacterial extracts attach quickly to AgNPs via the proteins' free amino or carboxyl groups; moreover, the acquired form of NPs changes as the protein binding with AgNPs varies.

The most effective approach for identifying morphological features, such as the size and form, of biosynthesized AgNPs is TEM examination. Figure 4 reveals the effective manufacture of spherical AgNPs using metabolites found in *P. indica* S. Azhar's filtrate, with typical sizes ranging from 2.4 to 53.5 nm. Furthermore, the biologically produced nanoparticles were evenly diffused, with no aggregation or morphological discrepancy.

XRD based AgNPs characterization exhibited five peaks at 2θ values: 38.2°, 44.46°, 64.22°, 77.52°, and 81.22°, which were assigned to planes 111, 200, 220, 311, and 222, respectively for AgNPs Figure 5A. The average sizes of crystallite Ag- particles were calculated using Scherrer's equation. In this context, the size of Ag particles ranged between 8 to 80 nm, which were the outputs from the analysis of the equation. In line with our clarification of the results, we reported the successful fabrication of crystallite, monoclinic-phase AgNPs at the same XRD diffraction planes utilizing metabolites of *P. indica* S. Azhar.

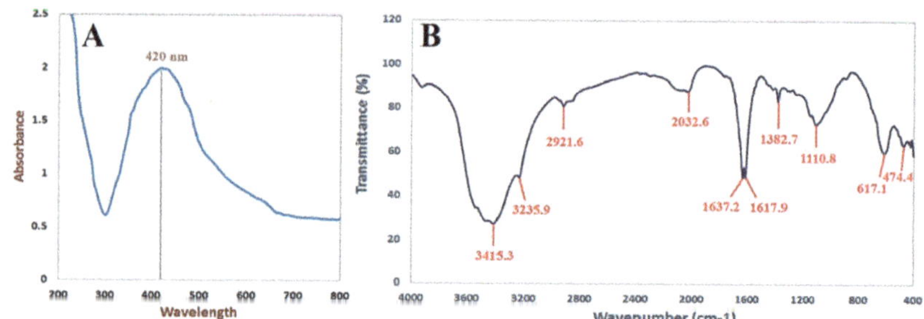

Figure 3. UV–Vis spectrophotometer (**A**) and FT-IR spectra (**B**) of AgNPs synthesized by *P. indica* S. Azhar.

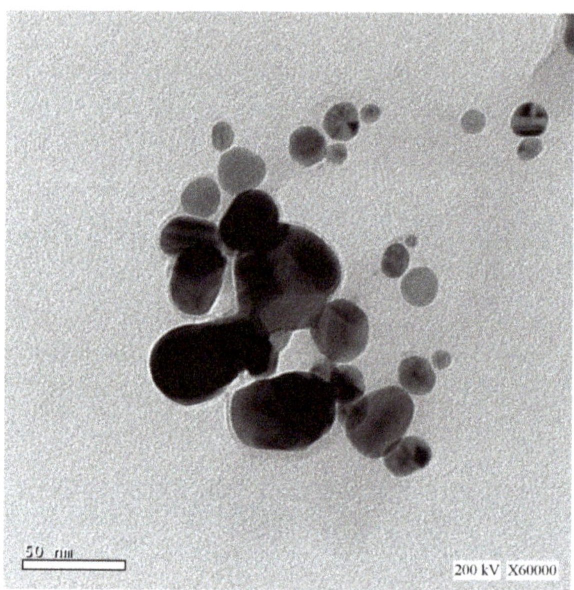

Figure 4. TEM image of AgNPs synthesized by of *P. indica* S. Azhar.

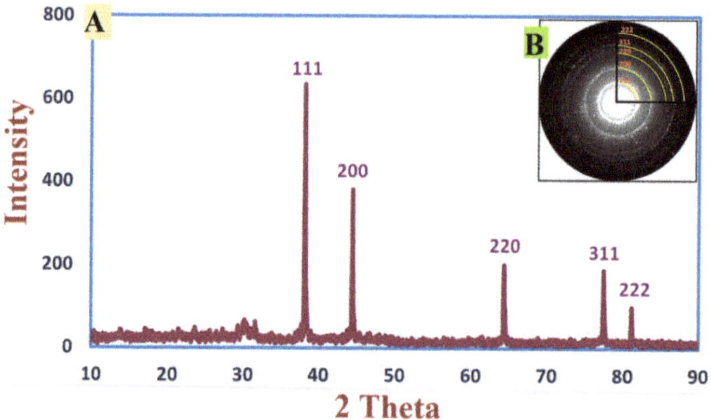

Figure 5. XRD pattern (**A**) and SAED pattern (**B**) of biosynthesized AgNPs.

3.3. Antifungal Activity

In this study, the antifungal activity of biosynthesized AgNPs against *R. microsporus*, *M. racemosus* and *S. racemosum* was carried out as shown in Figure 6. Results illustrated that the biosynthesized AgNPs exhibited outstanding antifungal activity against all tested fungal strains. Moreover, the antifungal activity was the highest toward *R. microsporus*, where the inhibition zone was 38 mm at a concentration of 400 µg/mL of AgNPs. Furthermore, AgNPs at concentration 400 µg/mL exhibited potential antifungal activity against *S. racemosum*, but this was lower than for *R. microsporus*, where the inhibition zone was 24 mm. On the other hand, the antifungal activity was the lowest toward *M. racemosus* where the inhibition zone was 19 mm, as shown in Figure 6A. Moreover, the MIC of AgNPs was assessed using the broth microdilution method against all tested fungal strains. Results revealed that the MIC of AgNPs was in the range of 50–100 µg/mL according to the fungal strain, where MIC was 50 µg/mL toward *R. microspores* and *S. racemosum*, while it was 100 µg/mL against *M. racemosus*, as shown in Figure 6B.

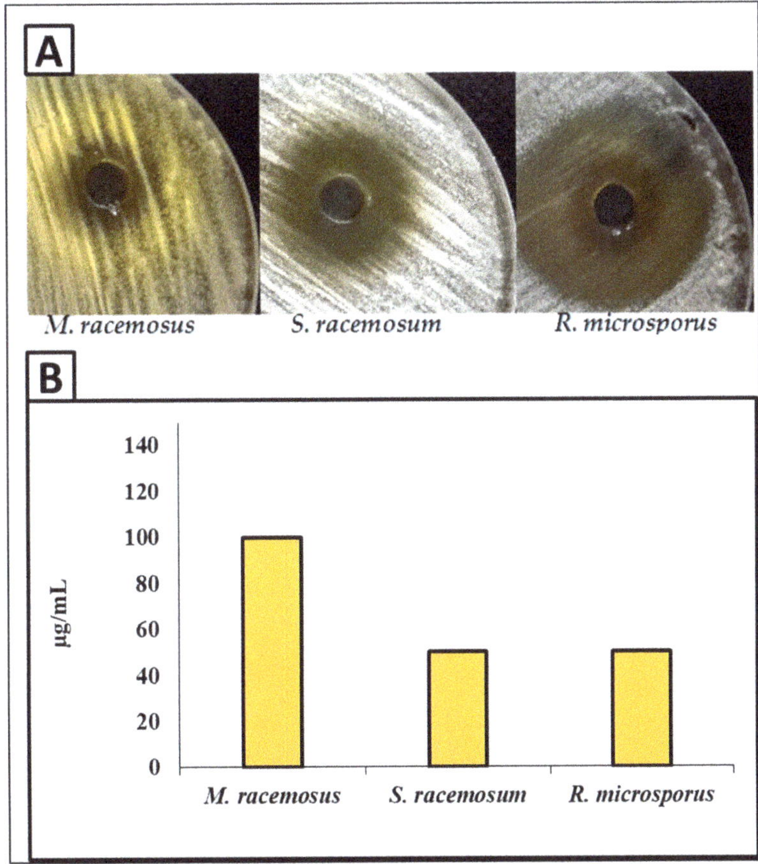

Figure 6. Antifungal activity of AgNPs at 400 µg/Ml using the agar well diffusion method (**A**) and the MIC of AgNPs using the broth microdilution method (**B**) toward *R. microsporus*, *S. racemosum*, and *M. racemosus*.

3.4. Antioxidant Activity

In this study, the antioxidant activity of AgNPs at different concentration (400–0.78 μg/mL) was evaluated using the DPPH method as shown in Figure 7. Results showed that AgNPs exhibited antioxidant activity, where the IC_{50} was 31 μg/mL when compared to ascorbic acid (0.79 μg/mL). In addition, antioxidant activity at 400, 200, 100, and 50 μg/mL were 96%, 86.6%, 72%, and 59.6%, respectively, and antioxidant activity decreased with decreasing concentrations, where the activity was 1% at 0.78 μg/mL. This suggests the possible application of AgNPs as an alternative antioxidant in the treatment of diseases that are caused due to free radicals.

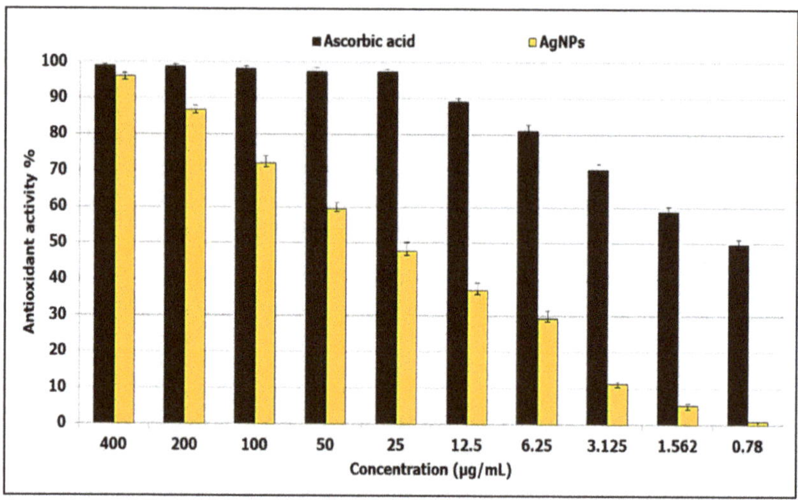

Figure 7. Antioxidant activity of AgNPs at different concentrations using DPPH method.

3.5. Cytotoxicity

Evaluation of the cytotoxicity of compounds in human normal cell lines is considered the first step to detect their safety [64]. Cytotoxicity of AgNPs at different concentrations was tested against the Vero cell line CCL-81, as illustrated in Figure 8. Results revealed that the IC_{50} of the biosynthesized AgNPs was 132.2 μg/mL, and the cell viability of Vero cells at two concentrations, 31.25 and 62.6 μg/mL, was 99% and 98%, respectively, which confirmed no cytotoxic effect of AgNPs at these two concentrations. In general, if the IC_{50} is $\geq$90 μg/mL, the material is classified as non-cytotoxic [66]. Therefore, this confirmed that the biosynthesized AgNPs are safe to use.

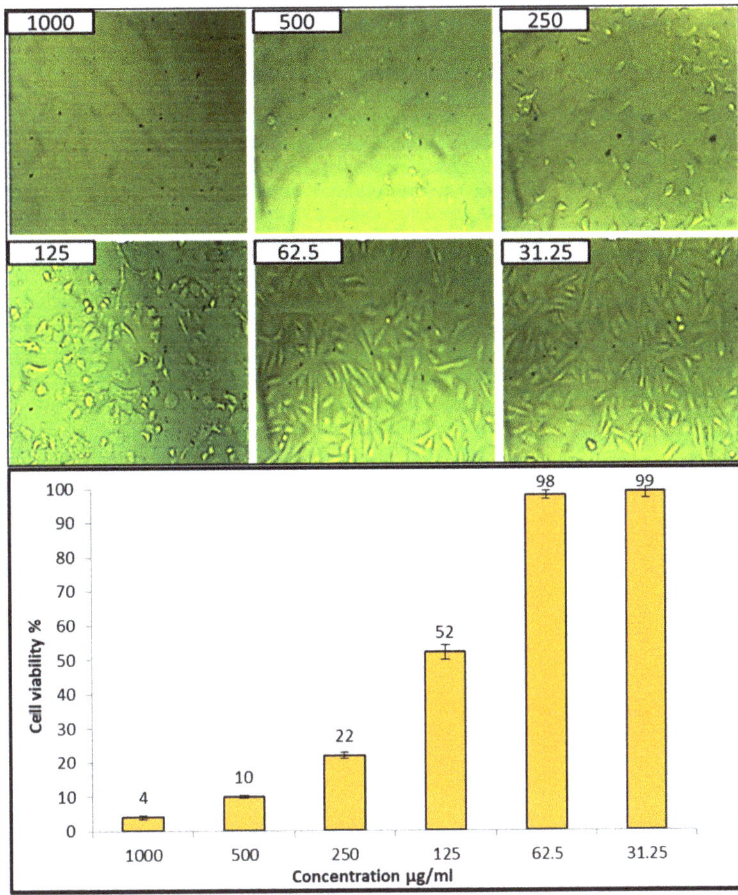

Figure 8. Cytotoxicity of AgNPs on Vero cell line.

4. Discussion

Green biosynthesis of AgNPs has risen in prominence as a potential alternative to chemical and physical techniques. Metabolites secreted by *P. indica* S. Azhar are effective in the formation of AgNPs, besides stabilizing the formed NPs. *P. indica* S. Azhar filtrate was utilized as a bio-reactor for the formation of AgNPs through harnessing bioactive macromolecules that are secreted from it. *P. indica* nanoparticles are currently in the early phases of development. *P. indica*-derived biogenic AgNPs are more appealing and less hazardous to the environment than other approaches. The proteins or enzymes involved in cell-free filtrates of *P. indica* that change nitrate to nitrite, and then reduced silver ions to silver in the metallic form, may be responsible for the synthesis of AgNPs. The absorbance peak at 420 nm in the UV-visible spectrum corresponded to the typical band of AgNPs generated by the *P. indica* cell free filtrate. The absorption peak at around 400–450 nm is attributed to AgNPs surface plasmon resonance, confirming the production of AgNPs [67]. Similarly, Alsharif et al. [68] observed that AgNPs generated by Bacillus cereus presented a single symmetric maximum at wavelength 420 nm, which is linked to spherical structure. The capping of biosynthesized AgNPs was discovered using Fourier transform-infrared spectroscopy. The FT-IR spectral analysis of AgNPs revealed N–H aliphatic amino group peaks at 3415.3 and 3235.9 cm^{-1}, whereas the $-CH_2$ group is indicated by the peak at

2921.6 cm^{-1} [69]. The (OH) stretch of the carboxyl group is shown by the spectrum peak at 2032.6 cm^{-1}. The binding of the amide I band of protein with the (N–H) stretch is associated with the peaks at 1637.2 and 1617.9 cm^{-1} [22]. The C–H and C–O stretching bands may be recognized at 1382.9 and 1110.8 cm^{-1}, respectively [47]. The relationship between Ag and physiologically active chemicals that are responsible for the production and stability of AgNPs as capping agents was investigated using FTIR. The infrared area of the spectrum is designated by the carboxylic acid group (CHOO–), obtained with an amine (NH_2) in the amino acid of proteins [70]. The most efficient approach for identifying morphological features, such as the size and form of produced AgNPs, is TEM examination. According to TEM data, the biosynthesized AgNPs had a spherical form with a size range of 2.4 to 53.5 nm. According to a previous study, researchers successfully formed spherical AgNPs with a size range of 6–50 nm using TEM, as seen in Alsharif et al. [68]. Other reports showed a difference in the shape and size of AgNPs biologically formed by bacteria [71,72]. The size range of AgNPs generated by *B. licheniformis* filtrate during incubation with 1 and 3 mM $AgNO_3$ was (3–130 nm) and (45–170 nm), respectively, according to Sarangadharan and Nallusamy [73]. When the biomass of *B. licheniformis* was incubated with 1 mM $AgNO_3$, the mean size of AgNPs reached around 50 nm in another investigation [74]. Specific peaks in the XRD spectra were used to illustrate the XRD pattern of the AgNPs biosynthesized by *P. indica*. The face center cubic (fcc) nano-structures of AgNPs (111), (200), (220), (311), and (222), were characterized by five diffraction peaks at 2θ values of 38.2°, 44.46°, 64.22°, 77.52°, and 81.22°, respectively, with an average size 8–80 nm. According to another study, the diffraction peaks at 2θ = 38.2°, 46.2°, 64.6°, and 77.6° exemplify the 111, 200, 220, and 311 Bragg's reflection of Ag nanoparticles' face-centered-cubic structure [75]. Previous research that demonstrated the creation of AgNPs utilizing microbes had comparable XRD results [69,76,77].

Mucormycosis disease is very dangerous to humans and is caused by members of the Mucorales genera, such as *Rhizopus, Mucor,* and *Sycephalastrum*. Therefore, the control of these fungi by biomaterials is required. The biosynthesized AgNPs in this study revealed promising antifungal activity toward Mucorales genera species such as *R. microsporus, M. racemosus, and S. racemosum*. Medda et al. [78] synthesized AgNPs using Aloe vera leaf extract, and found antifungal activity toward *Rhizopus* sp. and *Aspergillus* sp. Alananbeh et al. [79] revealed that promising antifungal activity of AgNPs toward *M. hemalis* and R. arrhizus. AgNPs might be critical in breaking down such resistance. AgNPs' efficacy can be attributed to a variety of mechanisms, including cell wall disintegration, surface protein degradation, nucleic acid damage caused by the formation and buildup of reactive oxygen and nitrogen species (ROS and free radicals), and proton pump blocking. AgNPs are thought to cause a buildup of silver ions, which obstructs respiration by causing intracellular ion efflux, causing harm to the electron transport system [80].

Oxidative stress is a condition in which the balance between a cell's antioxidative defense and oxidants is disturbed as a result of oxidant excess [78]. Furthermore, the presence of oxidants causes oxidative changes of biological systems at the molecular level (unsaturated bonds of lipids, proteins, DNA, and so on), resulting in damage and, ultimately, hastened cellular death [79]. Antioxidants are natural or manmade compounds that can help to prevent or postpone oxidative cell damage (ROS, RNS, free radicals, other unstable molecules) [79]. Therefore, antioxidant compounds are continuously required to resist oxidative stress. The biosynthesized AgNPs in the current study revealed antioxidant activity where the IC50 was 31 μg/mL. Previous studies have confirmed the antioxidant activity of AgNPs from different sources [81–83]. González-Ballesteros et al. [82] reported that AgNPs, silver nanoparticles, can be made from Lactobacillus brevis exopolysaccharides, and these were tested for antioxidant capabilities. At 100 g/mL, AgNPs had a moderate DPPH radical scavenging potential of 34.09% [83]. Evaluation of the cytotoxicity of compounds on human normal cell lines is considered the first step to detect their safety [64]. Therefore, the cytotoxicity of biosynthesized AgNPs was assessed, where results confirmed that AgNPs are safe in use.

5. Conclusions

In the current study, a fast, green, eco-friendly method was used to synthesize AgNPs by *P. indica* S. Azhar for controlling fungi that cause mucormycosis, as well as antioxidant activity. Results revealed that AgNPs were fabricated using metabolites of *P. indica*, then, the biosynthesized AgNPs were characterized using different modern techniques. Moreover, the biosynthesized AgNPs exhibited outstanding antifungal activity against *R. microsporus*, *M. racemosus*, and *S. racemosum*. Furthermore, results revealed antioxidant activity without any cytotoxicity on the Vero normal cell line. The results of this study warrant further in vivo experiments.

Author Contributions: Conceptualization, S.S.S. and A.H.H.; methodology, S.S.S. and A.H.H.; validation, S.S.S. and A.H.H.; formal analysis, S.S.S. and A.H.H.; investigation, S.S.S., O.M.A., A.M.R., K.A.A.-E. and A.H.H.; software S.S.S. and A.H.H.; resources, S.S.S. and A.H.H.; data curation, S.S.S. and A.H.H.; Funding O.M.A. and K.A.A.-E.; writing—original draft preparation, S.S.S. and A.H.H.; writing—review and editing, S.S.S., O.M.A., A.M.R., K.A.A.-E. and A.H.H.; visualization, S.S.S., O.M.A., A.M.R., K.A.A.-E. and A.H.H. All authors have read and agreed to the published version of the manuscript.

Funding: This study was supported by the Taif University Researchers Supporting Project (TURSP-2020/81) at Taif University in Taif, Saudi Arabia.

Institutional Review Board Statement: Not applicable.

Informed Consent Statement: Not applicable.

Data Availability Statement: The data used to support the conclusions of this study are accessible upon request from the corresponding author.

Acknowledgments: The authors express their sincere thanks to Faculty of Science (Boys), Al-Azhar University, Cairo, Egypt for providing the necessary research facilities.

Conflicts of Interest: The authors declare that they have no conflict of interest.

References

1. Chang, Y.-L.; Yu, S.-J.; Heitman, J.; Wellington, M.; Chen, Y.-L. New facets of antifungal therapy. *Virulence* **2017**, *8*, 222–236. [CrossRef] [PubMed]
2. Campoy, S.; Adrio, J.L. Antifungals. *Biochem. Pharmacol.* **2017**, *133*, 86–96. [CrossRef] [PubMed]
3. Brown, G.D.; Denning, D.W.; Gow, N.A.; Levitz, S.M.; Netea, M.G. White TC (2012) Hidden killers: Human fungal infections. *Sci. Transl. Med.* **2012**, *4*, 165rv113. [CrossRef] [PubMed]
4. Kontoyiannis, D.P.; Lewis, R.E. Agents of mucormycosis and entomophthoramycosis. In *Mandell, Douglas, and Bennett's Principles and Practice of Infectious Diseases*; Elsevier: Amsterdam, The Netherlands, 2014; pp. 2909–2919.
5. Dacrory, S.; Hashem, A.H.; Hasanin, M. Synthesis of cellulose based amino acid functionalized nano-biocomplex: Characterization, antifungal activity, molecular docking and hemocompatibility. *Environ. Nanotechnol. Monit. Manag.* **2021**, *15*, 100453. [CrossRef]
6. Roden, M.M.; Zaoutis, T.E.; Buchanan, W.L.; Knudsen, T.A.; Sarkisova, T.A.; Schaufele, R.L.; Sein, M.; Sein, T.; Chiou, C.C.; Chu, J.H.; et al. Epidemiology and outcome of zygomycosis: A review of 929 reported cases. Clinical infectious diseases: An official publication of the Infectious Diseases Society of America. *Clin. Infect. Dis.* **2005**, *41*, 634–653. [CrossRef]
7. Petrikkos, G.; Skiada, A.; Lortholary, O.; Roilides, E.; Walsh, T.J.; Kontoyiannis, D.P. Epidemiology and clinical manifestations of mucormycosis. Clinical infectious diseases: An official publication of the Infectious Diseases Society of America. *Clin. Infect. Dis.* **2012**, *54* (Suppl. 1), S23–S34. [CrossRef]
8. Walsh, T.J.; Gamaletsou, M.N.; McGinnis, M.R.; Hayden, R.T.; Kontoyiannis, D.P. Early clinical and laboratory diagnosis of invasive pulmonary, extrapulmonary, and disseminated mucormycosis (zygomycosis). Clinical infectious diseases: An official publication of the Infectious Diseases Society of America. *Clin. Infect. Dis.* **2012**, *54* (Suppl. 1), S55–S60. [CrossRef]
9. Skiada, A.; Lass-Floerl, C.; Klimko, N.; Ibrahim, A.; Roilides, E.; Petrikkos, G. Challenges in the diagnosis and treatment of mucormycosis. *Med. Mycol.* **2018**, *56* (Suppl. 1), S93–S101. [CrossRef]
10. Sanglard, D.; Odds, F.C. Resistance of Candida species to antifungal agents: Molecular mechanisms and clinical consequences. *Lancet Infect. Dis.* **2002**, *2*, 73–85. [CrossRef]

11. Dannaoui, E. Antifungal resistance in mucorales. *Int. J. Antimicrob. Agents* **2017**, *50*, 617–621. [CrossRef]
12. Pfaller, M.A. Antifungal drug resistance: Mechanisms, epidemiology, and consequences for treatment. *Am. J. Med.* **2012**, *125*, S3–S13. [CrossRef] [PubMed]
13. Salem, S.S.; Fouda, A. Green Synthesis of Metallic Nanoparticles and Their Prospective Biotechnological Applications: An Overview. *Biol. Trace Elem. Res.* **2021**, *199*, 344–370. [CrossRef] [PubMed]
14. Shaheen, T.I.; Fouda, A.; Salem, S.S. Integration of Cotton Fabrics with Biosynthesized CuO Nanoparticles for Bactericidal Activity in the Terms of Their Cytotoxicity Assessment. *Ind. Eng. Chem. Res.* **2021**, *60*, 1553–1563. [CrossRef]
15. Fouda, A.; Salem, S.S.; Wassel, A.R.; Hamza, M.F.; Shaheen, T.I. Optimization of green biosynthesized visible light active CuO/ZnO nano-photocatalysts for the degradation of organic methylene blue dye. *Heliyon* **2020**, *6*, e04896. [CrossRef] [PubMed]
16. Salem, S.S.; Fouda, M.M.G.; Fouda, A.; Awad, M.A.; Al-Olayan, E.M.; Allam, A.A.; Shaheen, T.I. Antibacterial, Cytotoxicity and Larvicidal Activity of Green Synthesized Selenium Nanoparticles Using *Penicillium corylophilum*. *J. Clust. Sci.* **2021**, *32*, 351–361. [CrossRef]
17. Elfeky, A.S.; Salem, S.S.; Elzaref, A.S.; Owda, M.E.; Eladawy, H.A.; Saeed, A.M.; Awad, M.A.; Abou-Zeid, R.E.; Fouda, A. Multifunctional cellulose nanocrystal /metal oxide hybrid, photo-degradation, antibacterial and larvicidal activities. *Carbohydr. Polym.* **2020**, *230*, 115711. [CrossRef] [PubMed]
18. Sharaf, O.M.; Al-Gamal, M.S.; Ibrahim, G.A.; Dabiza, N.M.; Salem, S.S.; El-ssayad, M.F.; Youssef, A.M. Evaluation and characterization of some protective culture metabolites in free and nano-chitosan-loaded forms against common contaminants of Egyptian cheese. *Carbohydr. Polym.* **2019**, *223*, 115094. [CrossRef]
19. Hasanin, M.; Al Abboud, M.A.; Alawlaqi, M.M.; Abdelghany, T.M.; Hashem, A.H. Ecofriendly synthesis of biosynthesized copper nanoparticles with starch-based nanocomposite: Antimicrobial, antioxidant, and anticancer activities. *Biol. Trace Elem. Res.* **2021**, 1–14. [CrossRef]
20. Hasanin, M.; Hashem, A.H.; Lashin, I.; Hassan, S.A.M. In vitro improvement and rooting of banana plantlets using antifungal nanocomposite based on myco-synthesized copper oxide nanoparticles and starch. *Biomass Convers. Biorefin.* **2021**, 1–11. [CrossRef]
21. Hashem, A.H.; Khalil, A.M.A.; Reyad, A.M.; Salem, S.S. Biomedical Applications of Mycosynthesized Selenium Nanoparticles Using *Penicillium expansum* ATTC 36200. *Biol. Trace Elem. Res.* **2021**, *199*, 3998–4008. [CrossRef]
22. Eid, A.M.; Fouda, A.; Niedbała, G.; Hassan, S.E.D.; Salem, S.S.; Abdo, A.M.; Hetta, H.F.; Shaheen, T.I. Endophytic streptomyces laurentii mediated green synthesis of Ag-NPs with antibacterial and anticancer properties for developing functional textile fabric properties. *Antibiotics* **2020**, *9*, 641. [CrossRef] [PubMed]
23. Hashem, A.H.; Salem, S.S. Green and ecofriendly biosynthesis of selenium nanoparticles using *Urtica dioica* (stinging nettle) leaf extract: Antimicrobial and anticancer activity. *Biotechnol. J.* **2021**, 2100432. [CrossRef] [PubMed]
24. Badawy, A.A.; Abdelfattah, N.A.H.; Salem, S.S.; Awad, M.F.; Fouda, A. Efficacy assessment of biosynthesized copper oxide nanoparticles (Cuo-nps) on stored grain insects and their impacts on morphological and physiological traits of wheat (*Triticum aestivum* L.) plant. *Biology* **2021**, *10*, 233. [CrossRef]
25. Mohamed, A.A.; Abu-Elghait, M.; Ahmed, N.E.; Salem, S.S. Eco-friendly Mycogenic Synthesis of ZnO and CuO Nanoparticles for In Vitro Antibacterial, Antibiofilm, and Antifungal Applications. *Biol. Trace Elem. Res.* **2021**, *199*, 2788–2799. [CrossRef]
26. Saied, E.; Eid, A.M.; Hassan, S.E.D.; Salem, S.S.; Radwan, A.A.; Halawa, M.; Saleh, F.M.; Saad, H.A.; Saied, E.M.; Fouda, A. The catalytic activity of biosynthesized magnesium oxide nanoparticles (Mgo-nps) for inhibiting the growth of pathogenic microbes, tanning effluent treatment, and chromium ion removal. *Catalysts* **2021**, *11*, 821. [CrossRef]
27. Abdelmoneim, H.E.M.; Wassel, M.A.; Elfeky, A.S.; Bendary, S.H.; Awad, M.A.; Salem, S.S.; Mahmoud, S.A. Multiple Applications of CdS/TiO_2 Nanocomposites Synthesized via Microwave-Assisted Sol–Gel. *J. Clust. Sci.* **2021**, 1–10. [CrossRef]
28. Hashem, A.H.; Abdelaziz, A.M.; Askar, A.A.; Fouda, H.M.; Khalil, A.M.A.; Abd-Elsalam, K.A.; Khaleil, M.M. *Bacillus megaterium*-Mediated Synthesis of Selenium Nanoparticles and Their Antifungal Activity against *Rhizoctonia solani* in Faba Bean Plants. *J. Fungi* **2021**, *7*, 195. [CrossRef]
29. Hashem, A.H.; Al Abboud, M.A.; Alawlaqi, M.M.; Abdelghany, T.M.; Hasanin, M. Synthesis of Nanocapsules Based on Biosynthesized Nickel Nanoparticles and Potato Starch: Antimicrobial, Antioxidant and Anticancer Activity. *Starch-Stärke* **2022**, *74*, 2100165. [CrossRef]
30. Mohamed, A.A.; Fouda, A.; Abdel-Rahman, M.A.; Hassan, S.E.-D.; El-Gamal, M.S.; Salem, S.S.; Shaheen, T.I. Fungal strain impacts the shape, bioactivity and multifunctional properties of green synthesized zinc oxide nanoparticles. *Biocatal. Agric. Biotechnol.* **2019**, *19*, 101103. [CrossRef]
31. Jain, N.; Jain, P.; Rajput, D.; Patil, U.K. Green synthesized plant-based silver nanoparticles: Therapeutic prospective for anticancer and antiviral activity. *Micro Nano Syst. Lett.* **2021**, *9*, 5. [CrossRef]
32. Abdel-Khalek, A.A.; Al-Quraishy, S.; Abdel-Gaber, R. Silver Nanoparticles Induce Time- and Tissue-Specific Genotoxicity in *Oreochromis niloticus*: Utilizing the Adsorptive Capacities of Fruit Peels to Minimize Genotoxicity. *Bull. Environ. Contam. Toxicol.* **2021**, 1–9. [CrossRef] [PubMed]
33. Nayak, S.; Manjunatha, K.B.; Goveas, L.C.; Rao, C.V.; Sajankila, S.P. Investigation of Nonlinear Optical Properties of AgNPs Synthesized Using Cyclea peltata Leaf Extract Post OVAT Optimization. *BioNanoScience* **2021**, *11*, 884–892. [CrossRef]

34. Beyene, H.D.; Werkneh, A.A.; Bezabh, H.K.; Ambaye, T.G. Synthesis paradigm and applications of silver nanoparticles (AgNPs), a review. *Sustain. Mater. Technol.* **2017**, *13*, 18–23. [CrossRef]
35. Elbahnasawy, M.A.; Shehabeldine, A.M.; Khattab, A.M.; Amin, B.H.; Hashem, A.H. Green biosynthesis of silver nanoparticles using novel endophytic *Rothia endophytica*: Characterization and anticandidal activity. *J. Drug Deliv. Sci. Technol.* **2021**, *62*, 102401. [CrossRef]
36. Hasanin, M.; Elbahnasawy, M.A.; Shehabeldine, A.M.; Hashem, A.H. Ecofriendly preparation of silver nanoparticles-based nanocomposite stabilized by polysaccharides with antibacterial, antifungal and antiviral activities. *BioMetals* **2021**, *34*, 1313–1328. [CrossRef]
37. Anees Ahmad, S.; Sachi Das, S.; Khatoon, A.; Tahir Ansari, M.; Afzal, M.; Saquib Hasnain, M.; Kumar Nayak, A. Bactericidal activity of silver nanoparticles: A mechanistic review. *Mater. Sci. Energy Technol.* **2020**, *3*, 756–769. [CrossRef]
38. Korkmaz, N.; Ceylan, Y.; Hamid, A.; Karadağ, A.; Bülbül, A.S.; Aftab, M.N.; Çevik, Ö.; Şen, F. Biogenic silver nanoparticles synthesized via *Mimusops elengi* fruit extract, a study on antibiofilm, antibacterial, and anticancer activities. *J. Drug Deliv. Sci. Technol.* **2020**, *59*, 101864. [CrossRef]
39. Pilaquinga, F.; Morey, J.; Torres, M.; Seqqat, R.; Piña, M.D.L.N. Silver nanoparticles as a potential treatment against SARS-CoV-2: A review. *WIREs Nanomed. Nanobiotechnol.* **2021**, *13*, e1707. [CrossRef]
40. Ciriminna, R.; Albo, Y.; Pagliaro, M. New Antivirals and Antibacterials Based on Silver Nanoparticles. *ChemMedChem* **2020**, *15*, 1619–1623. [CrossRef]
41. Bocate, K.P.; Reis, G.F.; de Souza, P.C.; Oliveira Junior, A.G.; Durán, N.; Nakazato, G.; Furlaneto, M.C.; de Almeida, R.S.; Panagio, L.A. Antifungal activity of silver nanoparticles and simvastatin against toxigenic species of *Aspergillus*. *Int. J. Food Microbiol.* **2019**, *291*, 79–86. [CrossRef]
42. Alam, A.; Tanveer, F.; Khalil, A.T.; Zohra, T.; Khamlich, S.; Alam, M.M.; Salman, M.; Ali, M.; Ikram, A.; Shinwari, Z.K.; et al. Silver nanoparticles biosynthesized from secondary metabolite producing marine actinobacteria and evaluation of their biomedical potential. *Antonie van Leeuwenhoek* **2021**, *114*, 1497–1516. [CrossRef] [PubMed]
43. Alomar, T.S.; AlMasoud, N.; Awad, M.A.; El-Tohamy, M.F.; Soliman, D.A. An eco-friendly plant-mediated synthesis of silver nanoparticles: Characterization, pharmaceutical and biomedical applications. *Mater. Chem. Phys.* **2020**, *249*, 123007. [CrossRef]
44. Shaheen, T.I.; Salem, S.S.; Fouda, A. Current Advances in Fungal Nanobiotechnology: Mycofabrication and Applications. In *Microbial Nanobiotechnology: Principles and Applications*; Lateef, A., Gueguim-Kana, E.B., Dasgupta, N., Ranjan, S., Eds.; Springer: Singapore, 2021; pp. 113–143. [CrossRef]
45. Aref, M.S.; Salem, S.S. Bio-callus synthesis of silver nanoparticles, characterization, and antibacterial activities via *Cinnamomum camphora* callus culture. *Biocatal. Agric. Biotechnol.* **2020**, *27*, 101689. [CrossRef]
46. Fouda, A.; Abdel-Maksoud, G.; Abdel-Rahman, M.A.; Salem, S.S.; Hassan, S.E.-D.; El-Sadany, M.A.-H. Eco-friendly approach utilizing green synthesized nanoparticles for paper conservation against microbes involved in biodeterioration of archaeological manuscript. *Int. Biodeterior. Biodegrad.* **2019**, *142*, 160–169. [CrossRef]
47. Huq, M.A. Green Synthesis of Silver Nanoparticles Using *Pseudoduganella eburnea* MAHUQ-39 and Their Antimicrobial Mechanisms Investigation against Drug Resistant Human Pathogens. *Int. J. Mol. Sci.* **2020**, *21*, 1510. [CrossRef] [PubMed]
48. Mohamed, A.A.; Fouda, A.; Elgamal, M.S.; EL-Din Hassan, S.; Shaheen, T.I.; Salem, S.S. Enhancing of cotton fabric antibacterial properties by silver nanoparticles synthesized by new Egyptian strain *Fusarium keratoplasticum* A1-3. In Proceedings of the 8th International Conference of The Textile Research Division (ICTRD 2017), National Research Centre, Cairo, Egypt, 25–27 September 2017; pp. 63–71.
49. Mohmed, A.A.; Saad, E.; Fouda, A.; Elgamal, M.S.; Salem, S.S. Extracellular biosynthesis of silver nanoparticles using *Aspergillus* sp. and evaluation of their antibacterial and cytotoxicity. *J. Appl. Life Sci. Int.* **2017**, *11*, 1–12. [CrossRef]
50. Tehri, N.; Vashishth, A.; Gahlaut, A.; Hooda, V. Biosynthesis, antimicrobial spectra and applications of silver nanoparticles: Current progress and future prospects. *Inorg. Nano-Met. Chem.* **2020**, *52*, 1–19. [CrossRef]
51. Zhang, D.; Ma, X.-L.; Gu, Y.; Huang, H.; Zhang, G.-W. Green Synthesis of Metallic Nanoparticles and Their Potential Applications to Treat Cancer. *Front. Chem.* **2020**, *8*, 799. [CrossRef] [PubMed]
52. Shehabeldine, A.M.; Hashem, A.H.; Wassel, A.R.; Hasanin, M. Antimicrobial and Antiviral Activities of Durable Cotton Fabrics Treated with Nanocomposite Based on Zinc Oxide Nanoparticles, Acyclovir, Nanochitosan, and Clove Oil. *Appl. Biochem. Biotechnol.* **2021**, 1–18. [CrossRef] [PubMed]
53. lashin, I.; Hasanin, M.; Hassan, S.A.M.; Hashem, A.H. Green biosynthesis of zinc and selenium oxide nanoparticles using callus extract of *Ziziphus spina-christi*: Characterization, antimicrobial, and antioxidant activity. *Biomass Convers. Biorefin.* **2021**, 1–14. [CrossRef]
54. El-Naggar, M.E.; Hasanin, M.; Hashem, A.H. Eco-Friendly Synthesis of Superhydrophobic Antimicrobial Film Based on Cellulose Acetate/Polycaprolactone Loaded with the Green Biosynthesized Copper Nanoparticles for Food Packaging Application. *J. Polym. Environment.* **2021**, 1–13. [CrossRef]
55. Abd Elkodous, M.; El-Husseiny, H.M.; El-Sayyad, G.S.; Hashem, A.H.; Doghish, A.S.; Elfadil, D.; Radwan, Y.; El-Zeiny, H.M.; Bedair, H.; Ikhdair, O.A. Recent advances in waste-recycled nanomaterials for biomedical applications: Waste-to-wealth. *Nanotechnol. Rev.* **2021**, *10*, 1662–1739. [CrossRef]
56. Angelin, J.; Kavitha, M. Exopolysaccharides from probiotic bacteria and their health potential. *Int. J. Biol. Macromol.* **2020**, *162*, 853–865. [CrossRef] [PubMed]

57. Adetunji, A.I.; Olaniran, A.O. Production and potential biotechnological applications of microbial surfactants: An overview. *Saudi J. Biol. Sci.* **2021**, *28*, 669–679. [CrossRef] [PubMed]
58. Moreno, V.M.; Álvarez, E.; Izquierdo-Barba, I.; Baeza, A.; Serrano-López, J.; Vallet-Regí, M. Bacteria as Nanoparticles Carrier for Enhancing Penetration in a Tumoral Matrix Model. *Adv. Mater. Interfaces* **2020**, *7*, 1901942. [CrossRef] [PubMed]
59. Fouda, A.; Khalil, A.; El-Sheikh, H.; Abdel-Rhaman, E.; Hashem, A. Biodegradation and detoxification of bisphenol-A by filamentous fungi screened from nature. *J. Adv. Biol. Biotechnol.* **2015**, *2*, 123–132. [CrossRef]
60. Hashem, A.H.; Hasanin, M.S.; Khalil, A.M.A.; Suleiman, W.B. Eco-green conversion of watermelon peels to single cell oils using a unique oleaginous fungus: *Lichtheimia corymbifera* AH13. *Waste Biomass Valoris.* **2019**, *11*, 5721–5732. [CrossRef]
61. Suleiman, W.; El-Sheikh, H.; Abu-Elreesh, G.; Hashem, A. Recruitment of Cunninghamella echinulata as an Egyptian isolate to produce unsaturated fatty acids. *Res. J. Pharm. Biol. Chem. Sci.* **2018**, *9*, 764–774.
62. Suleiman, W.; El-Skeikh, H.; Abu-Elreesh, G.; Hashem, A. Isolation and screening of promising oleaginous *Rhizopus* sp. and designing of Taguchi method for increasing lipid production. *J. Innov. Pharm. Biol. Sci.* **2018**, *5*, 8–15.
63. Rodriguez-Tudela, J.; Arendrup, M.; Arikan, S.; Barchiesi, F.; Bille, J.; Chryssanthou, E.; Cuenca-Estrella, M.; Dannaoui, E.; Denning, D.; Donnelly, J. Eucast Definitive Document E. Def 9.1: Method for the determination of broth dilution minimum inhibitory concentrations of antifungal agents for conidia forming moulds. *Def* **2008**, *9*, 1–13.
64. Khalil, A.; Abdelaziz, A.; Khaleil, M.; Hashem, A. Fungal endophytes from leaves of *Avicennia marina* growing in semi-arid environment as a promising source for bioactive compounds. *Lett. Appl. Microbiol.* **2021**, *72*, 263–274. [CrossRef] [PubMed]
65. Van de Loosdrecht, A.; Beelen, R.; Ossenkoppele, G.; Broekhoven, M.; Langenhuijsen, M. A tetrazolium-based colorimetric MTT assay to quantitate human monocyte mediated cytotoxicity against leukemic cells from cell lines and patients with acute myeloid leukemia. *J. Immunol. Methods* **1994**, *174*, 311–320. [CrossRef]
66. Ioset, J.-R.; Brun, R.; Wenzler, T.; Kaiser, M.; Yardley, V. *Drug Screening for Kinetoplastids Diseases: A Training Manual for Screening in Neglected Diseases*; DNDi and Pan-Asian Screening Network: Geneva, Switzerland, 2009.
67. Wypij, M.; Czarnecka, J.; Świecimska, M.; Dahm, H.; Rai, M.; Golinska, P. Synthesis, characterization and evaluation of antimicrobial and cytotoxic activities of biogenic silver nanoparticles synthesized from *Streptomyces xinghaiensis* OF1 strain. *World J. Microbiol. Biotechnol.* **2018**, *34*, 23. [CrossRef] [PubMed]
68. Alsharif, S.M.; Salem, S.S.; Abdel-Rahman, M.A.; Fouda, A.; Eid, A.M.; Hassan, S.E.D.; Awad, M.A.; Mohamed, A.A. Multi-functional properties of spherical silver nanoparticles fabricated by different microbial taxa. *Heliyon* **2020**, *6*, e03943. [CrossRef] [PubMed]
69. Salem, S.S.; El-Belely, E.F.; Niedbała, G.; Alnoman, M.M.; Hassan, S.E.D.; Eid, A.M.; Shaheen, T.I.; Elkelish, A.; Fouda, A. Bactericidal and in-vitro cytotoxic efficacy of silver nanoparticles (Ag-NPs) fabricated by endophytic actinomycetes and their use as coating for the textile fabrics. *Nanomaterials* **2020**, *10*, 2082. [CrossRef]
70. Alsamhary, K.I. Eco-friendly synthesis of silver nanoparticles by *Bacillus subtilis* and their antibacterial activity. *Saudi J. Biol. Sci.* **2020**, *27*, 2185–2191. [CrossRef]
71. Saeed, S.; Iqbal, A.; Ashraf, M.A. Bacterial-mediated synthesis of silver nanoparticles and their significant effect against pathogens. *Environ. Sci. Pollut. Res.* **2020**, *27*, 37347–37356. [CrossRef]
72. Ibrahim, E.; Zhang, M.; Zhang, Y.; Hossain, A.; Qiu, W.; Chen, Y.; Wang, Y.; Wu, W.; Sun, G.; Li, B. Green-Synthesization of Silver Nanoparticles Using Endophytic Bacteria Isolated from Garlic and Its Antifungal Activity against Wheat Fusarium Head Blight Pathogen *Fusarium graminearum*. *Nanomaterials* **2020**, *10*, 219. [CrossRef]
73. Sarangadharan, S.; Nallusamy, S. Biosynthesis and characterization of silver nanoparticles produced by *Bacillus licheniformis*. *Int. J. Pharma Med. Biol. Sci.* **2015**, *4*, 236.
74. Kalimuthu, K.; Babu, R.S.; Venkataraman, D.; Bilal, M.; Gurunathan, S. Biosynthesis of silver nanocrystals by *Bacillus licheniformis*. *Colloids Surf. B Biointerfaces* **2008**, *65*, 150–153. [CrossRef]
75. Fouad, H.; Hongjie, L.; Yanmei, D.; Baoting, Y.; El-Shakh, A.; Abbas, G.; Jianchu, M. Synthesis and characterization of silver nanoparticles using *Bacillus amyloliquefaciens* and *Bacillus subtilis* to control filarial vector Culex pipiens pallens and its antimicrobial activity. *Artif. Cells Nanomed. Biotechnol.* **2017**, *45*, 1369–1378. [CrossRef] [PubMed]
76. Du, J.; Singh, H.; Yi, T.-H. Biosynthesis of silver nanoparticles by *Novosphingobium* sp. THG-C3 and their antimicrobial potential. *Artif. Cells Nanomed. Biotechnol.* **2017**, *45*, 211–217. [CrossRef] [PubMed]
77. Zonooz, N.F.; Salouti, M. Extracellular biosynthesis of silver nanoparticles using cell filtrate of *Streptomyces* sp. ERI-3. *Scientia Iranica* **2011**, *18*, 1631–1635. [CrossRef]
78. Medda, S.; Hajra, A.; Dey, U.; Bose, P.; Mondal, N.K. Biosynthesis of silver nanoparticles from Aloe vera leaf extract and antifungal activity against *Rhizopus* sp. and *Aspergillus* sp. *Appl. Nanosci.* **2015**, *5*, 875–880. [CrossRef]
79. Alananbeh, K.; Al-Refaee, W.; Al-Qodah, Z. Antifungal effect of silver nanoparticles on selected fungi isolated from raw and waste water. *Indian J. Pharm. Sci.* **2017**, *79*, 559–567. [CrossRef]
80. Du, H.; Lo, T.-M.; Sitompul, J.; Chang, M.W. Systems-level analysis of *Escherichia coli* response to silver nanoparticles: The roles of anaerobic respiration in microbial resistance. *Biochem. Biophys. Res. Commun.* **2012**, *424*, 657–662. [CrossRef] [PubMed]
81. Azeez, L.; Lateef, A.; Adebisi, S.A. Silver nanoparticles (AgNPs) biosynthesized using pod extract of *Cola nitida* enhances antioxidant activity and phytochemical composition of *Amaranthus caudatus* Linn. *Appl. Nanosci.* **2017**, *7*, 59–66. [CrossRef]

82. González-Ballesteros, N.; Rodríguez-Argüelles, M.C.; Prado-López, S.; Lastra, M.; Grimaldi, M.; Cavazza, A.; Nasi, L.; Salviati, G.; Bigi, F. Macroalgae to nanoparticles: Study of *Ulva lactuca* L. role in biosynthesis of gold and silver nanoparticles and of their cytotoxicity on colon cancer cell lines. *Mater. Sci. Eng. C* **2019**, *97*, 498–509. [CrossRef]
83. Patra, J.K.; Baek, K.-H. Antibacterial Activity and Synergistic Antibacterial Potential of Biosynthesized Silver Nanoparticles against Foodborne Pathogenic Bacteria along with its Anticandidal and Antioxidant Effects. *Front. Microbiol.* **2017**, *8*, 167. [CrossRef]

Journal of *Fungi*

MDPI

Review

Metal Nanoparticles as Novel Antifungal Agents for Sustainable Agriculture: Current Advances and Future Directions

Aida R. Cruz-Luna [1], Heriberto Cruz-Martínez [2], Alfonso Vásquez-López [1,*] and Dora I. Medina [3,*]

1 Instituto Politécnico Nacional, CIIDIR-OAXACA, Hornos Núm 1003, Col. Noche Buena, Santa Cruz Xoxocotlán 71230, Mexico; luna_060877@hotmail.com

2 Tecnológico Nacional de México, Instituto Tecnológico del Valle de Etla, Abasolo S/N, Barrio del Agua Buena, Santiago Suchilquitongo 68230, Mexico; heri1234@hotmail.com

3 Tecnologico de Monterrey, School of Engineering and Sciences, Atizapan de Zaragoza 52926, Mexico

* Correspondence: avasquez@ipn.mx (A.V.-L.); dora.medina@tec.mx (D.I.M.)

Abstract: The use of metal nanoparticles is considered a good alternative to control phytopathogenic fungi in agriculture. To date, numerous metal nanoparticles (e.g., Ag, Cu, Se, Ni, Mg, and Fe) have been synthesized and used as potential antifungal agents. Therefore, this proposal presents a critical and detailed review of the use of these nanoparticles to control phytopathogenic fungi. Ag nanoparticles have been the most investigated nanoparticles due to their good antifungal activities, followed by Cu nanoparticles. It was also found that other metal nanoparticles have been investigated as antifungal agents, such as Se, Ni, Mg, Pd, and Fe, showing prominent results. Different synthesis methods have been used to produce these nanoparticles with different shapes and sizes, which have shown outstanding antifungal activities. This review shows the success of the use of metal nanoparticles to control phytopathogenic fungi in agriculture.

Keywords: metallic nanoparticles; agriculture; crop protection; antifungal activities; fungi

Citation: Cruz-Luna, A.R.; Cruz-Martínez, H.; Vásquez-López, A.; Medina, D.I. Metal Nanoparticles as Novel Antifungal Agents for Sustainable Agriculture: Current Advances and Future Directions. *J. Fungi* **2021**, *7*, 1033. https://doi.org/10.3390/jof7121033

Academic Editor: Kamel A. Abd-Elsalam

Received: 25 October 2021
Accepted: 25 November 2021
Published: 1 December 2021

1. Introduction

Since the beginning of agriculture, the biggest challenge has been pests and diseases produced by insects, bacteria, fungi, and other pathogens present in the environment [1–3]. This leads to large losses of crops, which are reflected in production with low profits, that is to say, earnings are directly affected [4,5]. Among the different pathogens, phytopathogenic fungi cause various diseases in agriculture [6]. Fungi have the versatility of adaptation to any medium and are capable of colonizing different substrates or media in extreme or precarious environmental conditions. They can affect different stages of the crop, from sowing to growth and production to postharvest [7,8].

Today, phytopathogenic fungi have mostly been controlled with chemical products, which are cheap and easy to obtain on the market [9,10]. However, due to their indiscriminate use, they have created several problems such as environmental pollution, diseases in humans and animals, and ecological imbalances [11,12]. In addition, the usage of chemical agents has resulted in fungi developing more resistance, becoming stronger against chemical products [13,14].

Currently, friendly and efficient alternatives for the environment are being used to control phytopathogen fungi, such as biological control [15,16], plant extracts [17], and essential oils [18–20]. Such alternatives have been beneficial and are therefore considered as a good choice. However, these alternatives have some challenges, such as the effect of delays, high acquisition costs, and constant applications that make them vulnerable [21,22].

Otherwise, another recently explored and applied route in agriculture is the use of nanomaterials, which have been successfully applied in other fields such as energy, medicine, and electronics [23–27]. Nanomaterials have become very important because their physicochemical properties are very different compared to bulk materials [28–30]. Furthermore, the shape, size, and composition of nanomaterials determine their physicochemical properties [28–30]. These peculiarities have made nanomaterials applicable in different areas. Specifically, in the field of agriculture, there are several nanomaterial applications, such as in the production, processing, storage, packaging, and transportation of agricultural products [31,32]. In production, nanomaterials offer ecological, efficient, and modern alternatives that can be very useful for the management of phytopathogenic diseases that can be used as bio-manufacturing agents, due to their easy handling and production [33,34].

Different nanomaterials have shown excellent antifungal activities; therefore, they are considered a good alternative to control phytopathogenic fungi [35–38]. Specifically, metal nanoparticles have been widely studied; consequently, they have been tested and led to significant results due to their excellent antifungal properties [39]. So far, numerous metal nanoparticles have been synthesized and used to control phytopathogenic fungi [40–47]. However, there is a current lack of critical and detailed reviews of current progress in the use of metal nanoparticles to control phytopathogenic fungi, as the currently available reviews only partially analyse the use of metal-based nanoparticles for controlling these pathogens [48,49].

Therefore, this review presents a comprehensive and detailed analysis of the current progress on the application of metal nanoparticles for controlling phytopathogenic fungi in agriculture. In the first instance, the possible mechanisms of action of nanoparticles on phytopathogenic fungi are reviewed. Afterwards, the progress on the use of metal nanoparticles as potential antifungal agents is reviewed in detail. Finally, conclusions and future directions are presented.

2. Mechanisms Involved in Antifungal Activity of Nanoparticles

The use of nanoparticles is a novel route to control phytopathogenic fungi in agriculture because they have shown high antifungal activity across a wide diversity of phytopathogenic fungi [50,51]. Several factors have an influence over their antifungal activity, such as the size distribution, shape, composition, crystallinity, agglomeration, and surface chemistry of the nanoparticles [52,53]. For example, small nanoparticles favor the surface area-to-volume ratio, which could promote their antifungal activity [54]. It is well-known that these mentioned factors can be modified and controlled through synthesis routes [55,56]. It has also been documented that the synthesis route can play an important role in the antifungal activity, as sometimes metal precursors or surfactants are not easy to remove from the nanoparticles. Therefore, the residues from the synthesis can modify the surface chemistry of the nanoparticles and consequently influence their antifungal activity [57]. Finally, another important factor is the species of phytopathogenic fungi, since each specie has a different morphological structure.

As mentioned before, several factors influence the antifungal activity of the nanoparticles. Therefore, it is necessary to know the interaction and action mechanism between the metal nanoparticles and the phytopathogenic fungi. At present, various possible antifungal action mechanisms of these nanoparticles have been proposed (see Figure 1).

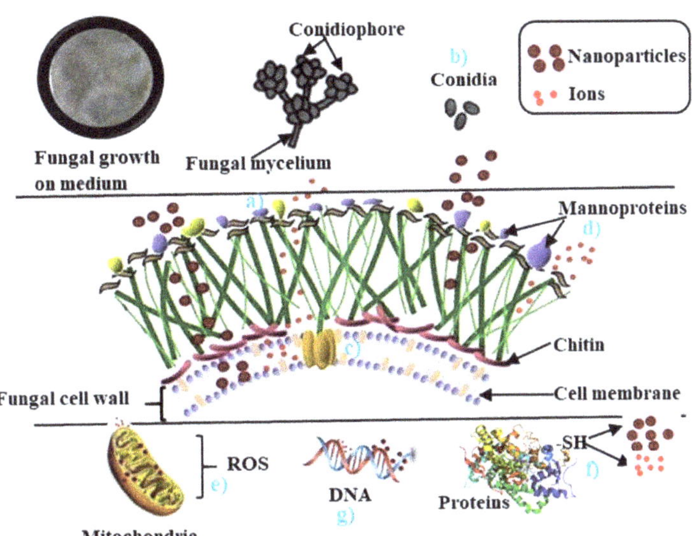

Figure 1. This is an illustration of the possible mechanisms of action of metal nanoparticles on phytopathogenic fungi. These are as follows: (**a**) ions are released by nanoparticles and bind to certain protein groups, which affect the function of essential membrane proteins and interfere with cell permeability. (**b**) The nanoparticles inhibit the germination of the conidia and suppress their development. (**c**) Nanoparticles and released ions disrupt electron transport, protein oxidation, and alter membrane potential. (**d**) They also interfere with protein oxidative electron transport. (**e**) They affect the potential of the mitochondrial membrane by increasing the levels of transcription of genes in response to oxidative stress (ROS). (**f**) ROS induces the generation of reactive oxygen species, triggering oxidation reactions catalyzed by the different metallic nanoparticles, causing severe damage to proteins, membranes, and deoxyribonucleic acid (DNA), and interfering with nutrient absorption. (**g**) The ions of the nanoparticles have a genotoxic effect that destroys DNA, therefore causing cell death [58–62].

3. Antifungal Properties of Metal Nanoparticles

Metal nanoparticles have been successfully applied to control different pathogens [63–65]. In this same direction, there are numerous studies on the use of metal nanoparticles to control phytopathogenic fungi in agriculture. Up to now, different nanoparticles have been used to control phytopathogenic fungi. For instance, Ag, Cu, Fe, Zn, Se, Ni, and Pd have shown outstanding antifungal properties. Therefore, a critical and detailed analysis of current advances in the use of metal nanoparticles on phytopathogenic fungi is presented.

3.1. Ag Nanoparticles

Ag nanoparticles have been extensively investigated in different scientific fields due to their antioxidant, antimicrobial, and anticancer properties as well as their characteristics of biocompatibility, easy production, relatively low cost, and non-toxicity, among others [66–72]. Due to these properties and their effective antifungal activities, Ag nanoparticles have also been the most investigated nanoparticles to control phytopathogenic fungi [73,74]. The main synthesis methods used to produce Ag nanoparticles to inhibit the growth of phytopathogenic fungi are the chemical and biological routes because they are easy to acquire and handle. In Figure 2, a generalized representation of the green or biological synthesis of metallic nanoparticles is illustrated. It can be observed that several factors can influence the synthesis of nanoparticles.

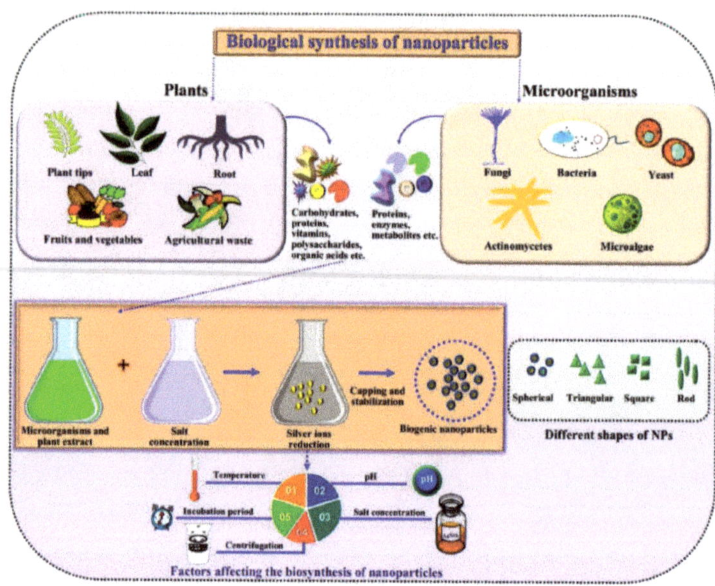

Figure 2. A generalized representation of the green synthesis of metallic nanoparticles [75].

For biological systems, many extracts of plants and fungi have been used in the synthesis of Ag nanoparticles [33,76–121]. In Table 1, the different extracts of plants and fungi that have been used to produce Ag nanoparticles are reported. In the case of the chemical route, several methods have been used to synthesize Ag nanoparticles, such as chemical reduction, sol-gel, and microemulsion [122–130]. To a lesser extent, physical methods have been used, such as high-voltage arc discharge and the irradiation method [131–133]. These different methods have made it possible to synthesize Ag nanoparticles with outstanding antifungal properties. Moreover, the biological syntheses present an additional benefit because they are environmentally friendly. Finally, it is interesting to note that several commercial Ag nanoparticles have been evaluated to inhibit the growth of phytopathogenic fungi, and have shown outstanding antifungal properties.

Table 1. Characteristics and antifungal evaluations of Ag nanoparticles.

Nanoparticle Properties			Antifungal Properties		Ref.
Synthesis Method	**Size (nm)**	**Shape**	**Specie of Fungi**	**Evaluation Method**	
Biological synthesis (*M. charantia* and *P. guajava*)	17 and 25.7	Spherical	*A. niger*, *A. flavus*, and *F. oxysporum*	In vitro	[76]
Biological synthesis (*M. azedarach*)	23	Spherical	*V. dahliae*	In vitro and in vivo	[77]
Biological synthesis (*A. indica*)	10–50	Spherical	*A. alternata*, *S. sclerotiorum*, *M. phaseolina*, *R. solani*, *B. cinerea*, and *C. lunata*	In vitro	[78]
Biological synthesis (*A. officinalis*, *T. vulgaris*, *M. pulegium*)	50	Spherical	*A. flavus* and *P. chrysogenum*	In vitro	[79]

Table 1. *Cont.*

Nanoparticle Properties			Antifungal Properties		Ref.
Synthesis Method	**Size (nm)**	**Shape**	**Specie of Fungi**	**Evaluation Method**	
Biological synthesis (*S. hortensis*)	-	-	*F. oxysporum*	In vitro	[80]
Biological synthesis (*O. fragrans*)	20	Spherical	*B. maydis*	In vitro	[81]
Biological synthesis (*P. glabra*)	29	Spherical	*R. nigricans*	In vitro	[82]
Biological synthesis (*W. somnifera*)	10–21	Spherical	*F. solani*	In vitro and in vivo	[83]
Biological synthesis (*P. vulgaris*)	12–16	Spherical	*Colletotrichum* sp., *F. oxysporum*, *F. acuminatum*, *F. tricinctum*, *F. graminearum*, *F. incarnatum*, *R. solani*, *S. sclerotiorum*, and *A. alternata*.	In vitro	[84]
Biological synthesis (*V. amygdalina*)	-	-	*F. oxysporum*, *F. solani*, and *C. canescent*	In vitro	[85]
Biological synthesis (*Z. officinale*)	75.3	Spherical	*A. alternata* and *C. lunata*	In vitro	[86]
Biological synthesis (*C. sinensis*)	-	-	*Irenopsis* spp., *Diaporthe* spp., and *Sphaerosporium* spp.	In vitro	[87]
Biological synthesis (*A. absinthium*)	-	-	*P. parasitica*, *P. infestans*, *P. palmivora*, *P. cinnamomi*, *P. tropicalis*, *P. capsici*, and *P. katsurae*	In vitro and in vivo	[88]
Biological synthesis (*M. parviflora*)	50.6	Spherical	*H. rostratum*, *F. solani*, *F. oxysporum*, and *A. alternata*	In vitro	[89]
Biological synthesis (Green and black teas)	10–20	Spherical	*A. flavus* and *A. parasiticus*	In vitro	[90]
Biological synthesis (*P. shell*)	10–50	Spherical and oval	*P. infestans* and *P. capsici*	In vitro	[91]
Biological synthesis (Ajwain and neem)	68	-	*C. musae*	In vitro and in vivo	[92]
Biological synthesis (*T. patula*)	15–30	Spherical	*C. chlorophyti*	In vitro and in vivo	[93]
Biological synthesis (*A. retroflexus*)	10–32	Spherical	*M. phaseolina*, *A. alternata*, and *F. oxysporum*	In vitro	[94]

Table 1. *Cont.*

Nanoparticle Properties			Antifungal Properties		Ref.
Synthesis Method	**Size (nm)**	**Shape**	**Specie of Fungi**	**Evaluation Method**	
Biological synthesis (*T. majus*)	35–55	Spherical	*A. niger*, *P. notatum*, *T. viridiae*, and *Mucor* sp.	In vitro	[95]
Biological synthesis (*T. foenum-graecum*)	20–25	Spherical	*A. alternata*	In vitro	[96]
Biological synthesis (Rice leaf)	3.7–29.3	Spherical	*R. solani*	In vitro	[97]
Biological synthesis (*P. urinaria*, *P. zeylanica*, and *S. dulcis*)	4–53	Various morphologies	*A. niger*, *A. flavus*, and *F. oxysporum*	In vitro	[98]
Biological synthesis (*C. globosum*)	11 and 14	Spherical	*F. oxysporum*	In vivo and in vitro	[99]
Biological synthesis (*T. longibrachiatum*)	10	Spherical	*F. verticillioides*, *F. moniliforme*, *P. brevicompactum*, *H. oryzae*, and *P. grisea*	In vitro	[100]
Biological synthesis (*A. terreus*)	5–30	Spherical	*A. flavus*	In vitro	[101]
Biological synthesis (*F. oxysporum*)	10–30	Spherical	*P. aphanidermatum*	In vitro and in vivo	[102]
Biological synthesis (*T. viride*)	12.7	Spherical	*A. solani*	In vitro	[103]
Biological synthesis (*F. solani*)	5–30	Spherical	*F. oxysporum*, *F. moniliform*, *F. solani*, *F. verticillioides*, *F. semitectum*, *A. flavus*, *A. terreus*, *A. niger*, *A. ficuum*, *P. citrinum*, *P. islandicum*, *P. chrysogenum*, *R. stolonifer*, *Phoma*, *A. alternata*, and *A. chlamydospora*	In vitro	[104]
Biological synthesis (*B. subtilis*)	16–20	Spherical	*A. alternate*, *A. niger*, *A. nidulans*, *C. herbarum*, *F. moniliforme*, *Fusarium* spp., *F. oxysporum*, and *T. harzianum*.	In vitro	[105]
Biological synthesis (*B. pseudomycoides*)	25–43	Spherical	*A. flavus*, *A. niger*, *A. tereus*, *P. notatum*, *R. olina*, *F. solani*, *F. oxysporum*, *T. viride*, *V. dahlia*, and *P. spinosum*	In vitro	[106]
Biological synthesis (*T. harzianum*)			*F. moniliforme*	In vitro	[107]
Biological synthesis (*Alternaria* sp.)	3–10	Spherical	*Alternaria* sp., *F. oxysporum*, *F. moniliforme*, and *F. tricinctum*.	In vitro	[108]

Table 1. *Cont.*

Nanoparticle Properties			Antifungal Properties		Ref.
Synthesis Method	Size (nm)	Shape	Specie of Fungi	Evaluation Method	
Biological synthesis (*Bacillus* sp.)	22.33–41.95	Spherical	*C. falcatum*	In vitro	[109]
Biological synthesis (*C. laurentii* and *R. glutinis*)	15–400	Spherical	*B. cinerea*, *P. expansum*, *A. niger*, *Alternaria* sp., and *Rhizopus* sp.	In vitro	[110]
Biological synthesis (*A. foetidus*)	20–40	Spherical	*A. niger*, *A. foetidus*, *A. flavus*, *F. oxysporum*, *A. oryzae*, and *A. parasiticus*	In vitro	[111]
Biological synthesis (*P. verrucosum*)	10–12	Spherical	*F. chlamydosporum* and *A. flavus*	In vitro	[112]
Biological synthesis (*N. oryzae*)	3–13	Spherical	*F. sambucinum*, *F. semitectum*, *F. sporotrichioides*, *F. anthophilium*, *F. oxysporum*, *F. moniliforme*, *F. fusarioids*, and *F. solani*	In vitro	[113]
Biological synthesis (*T. longibrachiatum*)	1–20	Spherical	*F. oxysporium*	In vitro	[114]
Biological synthesis (*A. versicolor*)	5–30	Spherical	*S. sclerotiorum* and *B. cinerea*	In vitro	[115]
Biological synthesis (*P. poae*)	19.8–44.9	Spherical	*F. graminearum*	In vitro	[116]
Biological synthesis (*Alternaria* spp.)	5–10	Spherical	*F. oxysporum*, *F. maniliforme*, *F. tricinctum*, and *Alternaria* sp.	In vitro	[117]
Biological synthesis (*I. hispidus*)	69.24	-	*Pythium* sp., *A. niger*, and *A. flavus*	In vitro	[118]
Biological synthesis (*S. griseoplanus*)	19.5–20.9	Spherical	*M. phaseolina*	In vitro	[119]
Biological synthesis (Sodium alginate)	6 and 40	Spherical	*C. gloeosporioides*	In vitro	[120]
Biological synthesis (*F. oxysporum*)	93 ± 11	Spherical	*A. flavus*, *A. nomius*, *A. parasiticus*, *A. ochraceus*, and *A. melleus*	In vitro	[121]
Biological synthesis (Glucose)	5–24	Spherical	*C. gloesporioides*	In vitro	[33]
Chemical synthesis	40–60	Spherical	*R. solani*	In vitro	[122]
Chemical synthesis	21 ± 2	Spherical	*C. gloeosporioides*	In vitro	[123]
Chemical synthesis	52	Spherical	*Phomopsis* sp.	In vitro	[124]

Table 1. *Cont.*

Nanoparticle Properties			Antifungal Properties		Ref.
Synthesis Method	**Size (nm)**	**Shape**	**Specie of Fungi**	**Evaluation Method**	
Chemical synthesis	30	Spherical	*F. graminearum, F. culmorum, F. sporotrichioides, F. langsethiae, F. poae, F. oxysporum, F. proliferatum,* and *F. verticillioides*	In vitro	[125]
Chemical synthesis	19–24	Spherical	*C. gloeosporioides*	In vitro	[126]
Chemical synthesis	25–32	-	*B. sorokiniana* and *A. brassicicola*	In vitro	[127]
Chemical synthesis	20	Spherical	*A. parasiticus*	In vitro	[128]
Chemical synthesis	100	Spherical	*M. phaseolina, S. sclerotiorum,* and *D. longicolla.*	In vitro	[129]
Chemical synthesis	-	-	*A. citri*	In vitro	[130]
Chemical synthesis	47	Spherical	*C. gloeosporioides*	In vitro	[134]
Commercial	7–25	-	*A. alternata, A. brassicicola, A. solani, B. cinerea, C. cucumerinum, C. cassiicola, C. destructans, D. bryoniae, F. oxysporum* f. sp. *cucumerinum, F. oxysporum* f. sp. *lycopersici, F. oxysporum, F. solani, Fusarium* sp., *G. cingulata, M. cannonballus, P. aphanidermatum P. spinosum,* and *S. lycopersici*	In vitro	[135]
Commercial	20–30	-	*B. sorokiniana* and *M. grisea*	In vitro and in vivo	[136]
Commercial	-	-	*R. solani, M. phaseolina, S. sclerotiorum, T. harzianum,* and *P. aphanidermatum*	In vitro and in vivo	[137]
Commercial	20	-	*S. homoeocarpa*	In vitro	[138]
Commercial	<100	-	*B. cinerea, A. alternata, M. fructicola, C. gloeosporioides, F. solani, F. oxysporum* f. sp. *Radicis Lycopersici,* and *V. dahliae*	In vitro and in vivo	[139]
Commercial	-	-	*R. solani, F. oxysporum, F. redolens, P. cactorum, F. hepática, G. frondosa, M. giganteus* and *S. crispa*	In vitro	[140]
Commercial	40–50	Spherical	*A. flavus*	In vitro	[141]
Commercial	20–30	-	*S. carvi*	In vitro and in vivo	[142]
Commercial	<100	-	*M. fructicola*	In vitro and in vivo	[143]
Commercial	4–8	-	*Colletotrichum*	In vitro and in vivo	[144]
Commercial	38	Spherical	*A. alternata* and *B. cinerea*	In vitro	[145]
Commercial	7–25	-	*S. cepivorum*	In vitro	[146]
Commercial	-	-	*B. cinerea*	In vitro and in vivo	[147]
Commercial	5–10	-	*R. solani*	In vitro and in vivo	[148]
Physical synthesis	5–65	Spherical	*F. culmorum*	In vitro	[131]
Physical synthesis	15–100	Spherical	*F. culmorum*	In vitro	[132]
Physical synthesis	5–15	Spherical	*P. capcisi*	In vitro and in vivo	[133]

As aforementioned, the characteristics of Ag nanoparticles such as shape, structure, and size play an important role in antifungal activity. According to Table 1, so far, most Ag nanoparticles synthesized by the different methods have been spherical, which may be because this kind of shape is easier to synthesize. In terms of size, they are polydisperse, which does not allow analysis in detail of the effect of the size of the nanoparticles on their antifungal activity. However, it is revealed that the smaller nanoparticles, between 10 and 30 nm, have greater antifungal effectiveness [76,77,90,94,99,104,108,126,133,148]. This is because the smaller nanoparticles penetrate or destroy the pathogen's cell membrane more quickly and thus unite the fungal hyphae and mycelium and deactivate these pathogens [99,108]. Ag nanoparticles ranging between 40 and 70 nm also show an inhibitory effect, destroying mycelium and spores and provoking the rupture of the membrane significantly [78,92,95,118,122,131]. Nevertheless, while the larger size has a good antifungal capacity, their penetration into the pathogen's membrane is slower, causing damage to mycelium and spores or the inhibition of fungal growth [110,121,129,132]. In Figure 3, severely damaged cell walls and hyphae with abnormal structures are shown in the presence of biosynthesized Ag NPs.

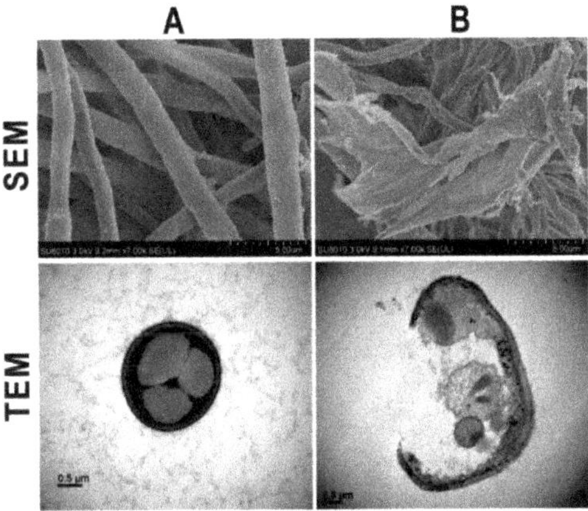

Figure 3. Microscopic images of SEM and TEM of *F. graminearum* in the absence (**A**) and presence (**B**) of the synthesized silver nanoparticle [116].

On the other hand, it has been reported that the concentration of nanoparticles can play an important role in antifungal activity [130]. Therefore, different concentrations of Ag nanoparticles have been evaluated. Several studies have shown that the concentration of Ag nanoparticles has an important role in antifungal activity [83,113,129,130,137,146]. Interestingly, low concentrations showed effectiveness in the suppression of fungi. For example, Ag nanoparticles synthesized with *M. charantia* and *P. guajava* extracts showed good antifungal capacity in concentrations of 20 ppm, inhibiting the growth of mycelium in fungi such as *A. niger*, *A. flavus*, and *F. oxysporum* [76]. A similar case occurred with Ag nanoparticles synthesized with *T. viride* extracts, which completely inhibited the growth of *A. solani* at low concentrations of 25 ppm [103]. In addition, excellent results were found in medium concentrations. For example, Ag nanoparticles synthesized with green and black tea were evaluated in four concentrations (i.e., 10, 25, 50, and 100 ppm) against *A. flavus* and *A. parasiticus*. The best results were obtained with doses of 100 ppm. Ag nanoparticles entered into the cell membrane, seriously affecting the respiratory chain, resulting in cell death [90]. A peculiarity was observed at very high concentrations of Ag nanoparticles (e.g., 500, 1000, 5000, and 10,000 ppm): with the increasing dose, the antifungal capacity presented a saturation of the Ag nanoparticles. According to the literature, this caused damage to the mycelium, such as oxidation, but not the complete inhibition of fungal pathogens [80,99,107]. Interestingly, some studies compare the antifungal activities of Ag nanoparticles with respect to chemical fungicides [109,144]. Ag nanoparticles showed similar results to chemical fungicides [109,144]. Therefore, the utilization of nanoparticles is a viable alternative to the use of chemical fungicides.

3.2. Cu Nanoparticles

The first study of Cu nanoparticles against fungi was reported by Giannousi et al. [149]. Since then, Cu nanoparticles have been considered a viable option for the treatment of fungal diseases [150,151]. Furthermore, Cu has several advantages: for instance, it is cheap, it is highly available, and its production in terms of nanoparticles is economical. Therefore, there are several studies on the use of Cu nanoparticles on phytopathogenic fungi [42,79,90,92,152–165]. The main synthesis methods to obtain Cu nanoparticles for the control of this pathogen are mentioned in Table 2. The chemical synthesis methods include chemical reduction and hydrothermal [158–164], whereas biological synthesis with different extracts of plants is widely used for its naturalness and its zero toxicity concerning the environment [42,90,92,154–156]. Finally, commercial nanoparticles, which are effective and easily acquired, have also been evaluated for the inhibition of phytopathogenic fungi [139,140,142,145,165].

Table 2. Characteristics and antifungal evaluations of Cu nanoparticles.

Nanoparticle Properties			**Antifungal Properties**		**Ref.**
Synthesis Method	**Size (nm)**	**Shape**	**Specie of Fungi**	**Evaluation Method**	
Biological synthesis (*Persea americana*)	42–90	Spherical	*A. flavus*, *A. fumigates*, and *F. oxysporum*.	In vitro	[42]
Biological synthesis (Ascorbic acid)	-	Spherical	*A. flavus* and *P. chrysogenum*	In vitro	[79]
Biological synthesis (Green and black teas)	26–40	Spherical	*A.flavus* and *A. parasiticus*.	In vitro	[90]
Biological synthesis (Ajwain and neem)	68	-	*C. musae*	In vitro	[92]
Biological synthesis (Ascorbic acid)	200–500	Faceted	*F. solani*, *Neofusicoccum* sp., and *F. oxysporum*.	In vitro	[152]
Biological synthesis (Ascorbic acid)	200–500	Faceted	*F. oxysporum* f. sp. *Lycopersici*	In vitro and in vivo	[153]
Biological synthesis (*C. paniculatus*)	5	Spherical	*F. oxysporum*	In vitro	[154]
Biological synthesis (*T. pinophilus*)	10	Spherical	*A. niger*, *A terreus*, and *A. fumigatus*	In vitro	[155]
Biological synthesis (*S. capillispiralis*)	3.6–59	Spherical	*Alternaria spp.*, *A. niger*, *Pythium* spp., and *Fusarium* spp.	In vitro	[156]
Biological synthesis (Ascorbic acid)	53–174	Spherical	*F. oxysporum* and *P. capsici*	In vitro	[157]
Chemical synthesis (Chemistry reduction)	20–50	Spherical	*Fusarium* sp.	In vitro	[158]
Chemical synthesis (Chemistry reduction)	-	-	*A. niger*	In vitro	[159]
Chemical synthesis (Hydrothermal)	14 ± 2	Spherical	*A. niger* and *A. oryzae*	In vitro	[160]
Chemical synthesis (Hydrothermal)	30–300	Spherical	*A. alternata*, *A solani*, *F. expansum*, and *Penicilliun* sp.	In vitro	[161]

Table 2. *Cont.*

Nanoparticle Properties			Antifungal Properties		Ref.
Synthesis Method	**Size (nm)**	**Shape**	**Specie of Fungi**	**Evaluation Method**	
Chemical synthesis (Chemistry reduction)	3–30	Spherical	*F. equiseti*, *F. oxysporum*, and *F. culmorum*	In vitro	[162]
Chemical synthesis (Chemistry reduction)	25–35	Spherical	*B. cinerea*	In vitro and in vivo	[163]
Chemical synthesis (Chemistry reduction)	14–37	Truncated octahedrons	*F.oxysporum*	In vitro	[164]
Commercial	25	-	*B. cinerea*, *A. alternata*, *M. fructicola*, *C. gloeosporioides*, *F. solani*, *F. oxysporum* f. sp. *Radicis Lycopersici*, and *V. dahliae*	In vitro and in vivo	[139]
Commercial	-	-	*R. solani*, *F. oxysporum*, *F. redolens*, *P. cactorum*, *F. hepática*, *G. frondosa*, *M. giganteus*, and *S. crispa*	In vitro	[140]
Commercial	20–30	-	*S. carvi*	In vitro and in vivo	[142]
Commercial	20	Spherical	*A. alternata* and *B. cinerea.*	In vitro	[145]
Commercial	25	-	*B. cinerea*	In vitro and in vivo	[165]

The studies carried out on Cu nanoparticles produced by the different synthesis methods have shown excellent antifungal activity in different species of phytopathogenic fungi. However, as in the case of Ag nanoparticles, there is a great diversity of sizes, which makes it difficult to analyze the size effect of Cu nanoparticles on antifungal activity (see Figure 4). In general, small nanoparticles range from 10 to 30 nm and penetrate the cell membrane more easily, causing a rupture and the leakage of cell contents [139,142,145,154,165]. Something similar occurs in medium-sized Cu nanoparticles (40 to 70 nm); however, by increasing their size, their fluidity in the membrane makes the growth and development of colonies of the pathogen impossible [90,92,158]. Finally, the large Cu nanoparticles (80 to >100 nm) inhibit the growth of mycelium and spores, thus demonstrating their antifungal capacity [152,153,161].

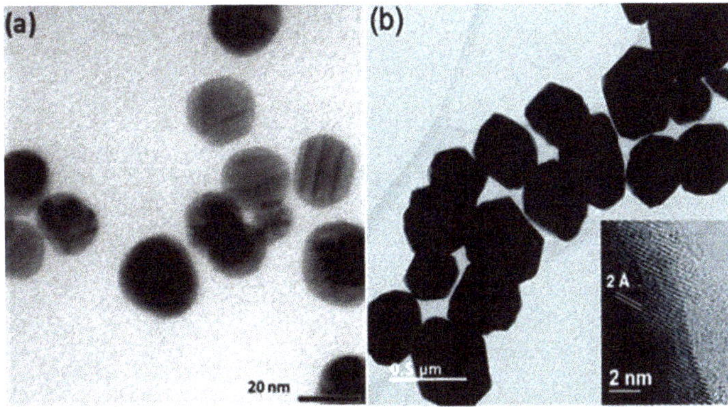

Figure 4. Cu nanoparticles synthesized with different shapes and sizes: (**a**) spherical shapes [158] and (**b**) faceted shapes [152].

Regarding the shape, the synthesized Cu nanoparticles are mainly spherical (see Figure 4a). That kind of shape has shown outstanding antifungal activities. According to several authors, spherical nanoparticles have the highest possibility of penetrating the membrane (and thus accessing the enzymes to initiate the cellular inhibition) faster [145,162]. Other shapes were also found, such as faceted ones with sizes in the range of 200–500 nm, which showed high effectiveness against *F. solani*, *Neofusicoccum* sp., and *F. oxysporum* (see Figure 4b) [152]. Another shape is the truncated octahedron structure (14 to 37 nm), which has been effective against *F. oxysporum* and caused its inhibition [164].

Another determining factor in inhibiting the growth of phytopathogenic fungi is the concentration of the Cu nanoparticles. To date, different concentrations (e.g., low, medium, and high) have been evaluated on phytopathogenic fungi. For example, low concentrations of Cu nanoparticles were evaluated against *F. oxysporum* at 0.1, 0.25, and 0.5 ppm. While the lowest concentration (0.1 ppm) promoted hard oxidative stress in the mycelium, the highest concentration (0.5 ppm) showed an antifungal capacity against *F. oxysporum* [164].

In addition, they have antifungal activities at medium concentrations (e.g., 5, 10, and 20 ppm). Cu nanoparticles demonstrated significant antifungal activity against *F. oxysporum* and *P. capsici*, which were inhibited by increasing the incubation time of the different concentrations. On the third day after their application, the inhibition increased slightly from 49% for 5 ppm to 63% for 20 ppm [157].

To cite another example, doses of 5, 15, 25, and 35 ppm were used against *R. solani*, *F. oxysporum*, *F. redolens*, *P. cactorum*, *F. hepática*, *G. frondosa*, *M. giganteus*, and *S. crispa*, demonstrating the antifungal capacity of Cu nanoparticles at a concentration of 35 ppm. In such a case, there was neither the growth of mycelium, nor the development of the pathogens studied [140]. Finally, for the highest concentrations of Cu nanoparticles, three different doses (300, 380, and 450 ppm) were evaluated. They were applied against *Fusarium* sp., demonstrating excellent antifungal capacity at the highest dose of 450 ppm [158]. Another study was carried out at four different high doses (i.e., 50, 100, 500, and 1000 ppm) against *B. cinerea*, *A. alternata*, *M. fructicola*, *C. gloeosporioides*, *F. solani*, *F. oxysporum*, and *V. dahlia*. In this study, Cu nanoparticles showed toxic activity at all concentrations and at the highest concentration of 1000 ppm they inhibited all phytopathogens [139]. In general, the Cu nanoparticles show antifungal capacity, affecting the phytopathogen intracellularly and extracellularly. Therefore, Cu nanoparticles are an excellent option for the control and management of different diseases of agronomic importance.

3.3. Other Metal Nanoparticles

As previously discussed, Ag and Cu nanoparticles are the most studied for the control of the growth of phytopathogenic fungi. However, other metal nanoparticles have been investigated as antifungal agents, such as Se [103,129,166], Ni [47,92], Mg [92], Pd [167], and Fe [90], which have shown promising results. Recently, Se nanoparticles were evaluated in vivo against *S. graminicola* in doses of 0 to 1000 ppm. To synthesize these nanoparticles, six strains of Trichoderma spp. (*T. asperellum*, *T. harzianum*, *T. atroviride*, *T. virens*, *T. longibrachiatum*, and *T. brevicompactum*) in the form of culture filtrate, cell lysate, and crude cell wall were used. The best result was found with *T. asperellum* in culture filtrate, demonstrating the antifungal capacity of Se nanoparticles [166]. In another report, Se nanoparticles were synthesized by the biological method using *T. viride* and they were evaluated at different concentrations (50, 100, 200, 300, 400, 500, 600, 700, and 800 ppm) against *A. solani* using the in vitro method. It was demonstrated that Se nanoparticles suppressed the growth of the fungus at 800 ppm [103]. Lastly, chemically synthesized Se nanoparticles were evaluated against *M. phaseolina*, *S. sclerotiorum*, and *D. longicolla* at different concentrations of 0.1, 0.5, 1, 5, 10, 50, and 100 ppm. The nanoparticles of Se inhibited *D. longicolla* from 10 ppm and up, and from 50 and 100 ppm for *M. phaseolina*. However, for *S. sclerotiorum*, the different concentrations of Se nanoparticles did not show any inhibition, allowing the growth and development of the pathogen [129].

Another metal that has been investigated for the control of phytopathogenic fungi is Ni. However, as in the case of Se nanoparticles, there are few studies available on the use of Ni nanoparticles aganist phytopathogenic fungi. In the first instance, commercial Ni nanoparticles were evaluated using in vitro and in vivo methods against two species of *F. oxysporum* at concentrations of 50 and 100 ppm. At a concentration of 100 ppm, the Ni nanoparticles significantly inhibited mycelial reproduction and the sporulation activities of the fungal pathogens under in vitro conditions. Meanwhile, under in vivo conditions, Ni nanoparticles at a concentration of 50 ppm reduced the severity of the disease by 58.4% and 57.0% in the cases of lettuce and tomato crops [47].

Finally, other nanoparticles investigated for the control of phytopathogenic fungi are Fe nanoparticles, highlighting the application of Fe nanoparticles synthesized by an ecological method using extracts of green and black tea leaves. Various concentrations (10, 25, 50, and 100 ppm) were evaluated against the fungi *A. flavus* and *A. parasiticus* in vitro. The results demonstrated a 43.5% inhibition with green tea extract and a 51.6% inhibition with black tea with doses of 100 ppm [90].

4. Conclusions and Future Directions

In this review, a critical and detailed analysis of the current progress on the application of metal-based nanoparticles for controlling phytopathogenic fungi in agriculture was presented. Based on this review, the following conclusions and future directions are proposed.

The progress achieved in the use of metal nanoparticles for the control of phytopathogenic fungi is outstanding since the studies developed so far clearly show that these nanoparticles can be an excellent alternative to chemical fungicides for the control of phytopathogenic fungi in agriculture.

Among the metallic nanoparticles, Ag nanoparticles have been the most studied as antifungal agents, followed by Cu nanoparticles. These nanoparticles have shown promising activity aganist different species of phytopathogenic fungi. Different synthesis methods have made it possible to produce nanoparticles with different shapes and sizes. However, the nanoparticles have been mainly spherical and polydisperse in size. Therefore, we consider it necessary to synthesize and evaluate nanoparticles of different shapes and size (e.g., octahedrons, icosahedrons, and faceted ones) and homogeneous in, since it is well known that these factors influence on antifungal activity.

For the rest of the metallic nanoparticles, such as Ni, Se, Mg, Pd, and Fe, there is little research. Therefore, it can be inferred that their antifungal properties are not well known, although the synthesis methods that have been tested for them have given good results. Hence, it is important to continue researching these metallic nanoparticles since there is a vast number of opportunities for researchers in this field.

Nowadays, the nanoparticles evaluated as antifungal agents have been mainly monometallic. Therefore, we consider it important to synthesize and evaluate bimetallic or trimetallic nanoparticles for the control of phytopathogenic fungi, since it has been documented that these nanoparticles have very different properties than monometallic nanoparticles.

According to this review, most of the studies were evaluated in vitro. However, it is important to apply the in vivo method to know the behavior of phytopathogens in the field. Applying the nanoparticles directly to the pathogens is preferable since the environments within the laboratory are different from those in the field. The lack of in vivo studies create a significant opportunity for the application of metal nanoparticles in the field of agriculture.

Author Contributions: Conceptualization, A.R.C.-L., H.C.-M., A.V.-L. and D.I.M.; formal analysis, A.R.C.-L., H.C.-M., A.V.-L. and D.I.M.; investigation, A.R.C.-L. and H.C.-M.; resources, H.C.-M., A.V.-L. and D.I.M.; data curation, A.R.C.-L. and H.C.-M.; writing—original draft preparation, A.R.C.-L.; writing—review and editing, H.C.-M., A.V.-L. and D.I.M.; supervision, A.V.-L. and D.I.M.; funding acquisition, H.C.-M., A.V.-L. and D.I.M. All authors have read and agreed to the published version of the manuscript.

Funding: The authors are grateful, for the funding sources provided by the Tecnológico Nacional de México and Instituto Politécnico Nacional through grant numbers 10800.21-P and SIP 20201078, respectively. The APC was funded by Tecnológico de Monterrey.

Institutional Review Board Statement: Not applicable.

Informed Consent Statement: Not applicable.

Data Availability Statement: Not applicable.

Conflicts of Interest: The authors declare no conflict of interest.

References

1. Karkhane, M.; Lashgarian, H.E.; Mirzaei, S.Z.; Ghaffarizadeh, A.; Cherghipour, K.; Sepahvand, A.; Marzban, A. Antifungal, antioxidant and photocatalytic activities of zinc nanoparticles synthesized by *Sargassum vulgare* extract. *Biocatal. Agric. Biotechnol.* **2020**, *29*, 101791. [CrossRef]
2. Malcolm, G.M.; Kuldau, G.A.; Gugino, B.K.; Jiménez-Gasco, M.D.M. Hidden host plant associations of soilborne fungal pathogens: An ecological perspective. *Phytopathology* **2013**, *103*, 538–544. [CrossRef]
3. Brauer, V.S.; Rezende, C.P.; Pessoni, A.M.; De Paula, R.G.; Rangappa, K.S.; Nayaka, S.C.; Gupta, V.K.; Almeida, F.; Brauer, V.S.; Rezende, C.P.; et al. Antifungal Agents in Agriculture: Friends and Foes of Public Health. *Biomolecules* **2019**, *9*, 521. [CrossRef]
4. Terra, A.L.M.; Kosinski, R.D.C.; Moreira, J.B.; Costa, J.A.V.; De Morais, M.G. Microalgae biosynthesis of silver nanoparticles for application in the control of agricultural pathogens. *J. Environ. Sci. Health Part B Pestic. Food Contam. Agric. Wastes* **2019**, *54*, 709–716. [CrossRef]
5. Almeida, F.; Rodrigues, M.L.; Coelho, C. The still underestimated problem of fungal diseases worldwide. *Front. Microbiol.* **2019**, *10*, 214. [CrossRef]
6. Blackwell, M. The Fungi: 1, 2, 3 ... 5.1 million species? *Am. J. Bot.* **2011**, *98*, 426–438. [CrossRef]
7. Sharma, R.; Singh, D.; Singh, R. Biological control of postharvest diseases of fruits and vegetables by microbial antagonists: A review. *Biol. Control* **2009**, *50*, 205–221. [CrossRef]
8. Spadaro, D.; Garibaldi, A.; Gullino, M.L. Control of *Penicillium expansum* and *Botrytis cinerea* on apple combining a biocontrol agent with hot water dipping and acibenzolar-S-methyl, baking soda, or ethanol application. *Postharvest Biol. Technol.* **2004**, *33*, 141–151. [CrossRef]
9. Fernández-Ortuño, D.; Torés, J.A.; De Vicente, A.; Pérez-García, A. Mechanisms of resistance to QoI fungicides in phytopathogenic fungi. *Int. Microbiol.* **2008**, *11*, 1. [CrossRef]
10. Pandey, S.; Giri, K.; Kumar, R.; Mishra, G.; Rishi, R.R. Nanopesticides: Opportunities in Crop Protection and Associated Environmental Risks. *Proc. Natl. Acad. Sci. India Sect. B Boil. Sci.* **2016**, *88*, 1287–1308. [CrossRef]
11. Adisa, I.O.; Pullagurala, V.L.R.; Peralta-Videa, J.R.; Dimkpa, C.O.; Elmer, W.H.; Gardea-Torresdey, J.L.; White, J.C. Recent advances in nano-enabled fertilizers and pesticides: A critical review of mechanisms of action. *Environ. Sci. Nano* **2019**, *76*, 2002–2030. [CrossRef]
12. Hahn, M. The rising threat of fungicide resistance in plant pathogenic fungi: Botrytis as a case study. *J. Chem. Biol.* **2014**, *7*, 133–141. [CrossRef]
13. Bolivar-Anillo, H.J.; Garrido, C.; Collado, I.G. Endophytic microorganisms for biocontrol of the phytopathogenic fungus *Botrytis cinerea*. *Phytochem. Rev.* **2020**, *19*, 721–740. [CrossRef]
14. Chattopadhyay, P.; Banerjee, G.; Mukherjee, S. Recent trends of modern bacterial insecticides for pest control practice in integrated crop management system. *3 Biotech* **2017**, *7*, 60. [CrossRef]
15. Bazioli, J.M.; Belinato, J.R.; Costa, J.H.; Akiyama, D.Y.; Pontes, J.G.D.M.; Kupper, K.C.; Augusto, F.; De Carvalho, J.E.; Fill, T.P. Biological control of citrus postharvest phytopathogens. *Toxins* **2019**, *11*, 460. [CrossRef]
16. Fira, D.; Dimkić, I.; Berić, T.; Lozo, J.; Stanković, S. Biological control of plant pathogens by Bacillus species. *J. Biotechnol.* **2018**, *285*, 44–55. [CrossRef] [PubMed]
17. Sales, M.D.C.; Costa, H.B.; Fernandes, P.M.B.; Ventura, J.A.; Meira, D.D. Antifungal activity of plant extracts with potential to control plant pathogens in pineapple. *Asian Pac. J. Trop. Biomed.* **2016**, *6*, 26–31. [CrossRef]
18. Santamarina, M.P.; Ibáñez, M.D.; Marqués, M.; Roselló, J.; Giménez, S.; Blázquez, M.A. Bioactivity of essential oils in phytopathogenic and post-harvest fungi control. *Nat. Prod. Res.* **2017**, *31*, 2675–2679. [CrossRef]
19. González, A.E.A.; Palou, E.; López-Malo, A. Antifungal activity of essential oils of clove (*Syzygium aromaticum*) and/or mustard (*Brassica nigra*) in vapor phase against gray mold (*Botrytis cinerea*) in strawberries. *Innov. Food Sci. Emerg. Technol.* **2015**, *32*, 181–185. [CrossRef]
20. Rosas-Díaz, J.; Vásquez-López, A.; Lagunez-Rivera, L.; Granados-Echegoyen, C.A.; Rodríguez-Ortiz, G. Etiología de la muerte prematura del cilantro (*Coriandrum sativum* L.) en Ocotlán de Morelos, Oaxaca, México y efecto de aceites esenciales in vitro sobre el patógeno. *Interciencia* **2017**, *42*, 591–596.
21. Chandler, D.; Bailey, A.; Tatchell, G.M.; Davidson, G.; Greaves, J.; Grant, W.P. The development, regulation and use of biopesticides for integrated pest management. *Philos. Trans. R. Soc. B Biol. Sci.* **2011**, *366*, 1987–1998. [CrossRef]

22. Raveau, R.; Fontaine, J.; Sahraoui, A.L.-H. Essential oils as potential alternative biocontrol products against plant pathogens and weeds: A review. *Foods* **2020**, *9*, 365. [CrossRef]
23. Utreja, P.; Verma, S.; Rahman, M.; Kumar, L. Use of Nanoparticles in medicine. *Curr. Biochem. Eng.* **2020**, *6*, 7–24. [CrossRef]
24. Nair, R.; Varghese, S.H.; Nair, B.G.; Maekawa, T.; Yoshida, Y.; Kumar, D.S. Nanoparticulate material delivery to plants. *Plant Sci.* **2010**, *179*, 154–163. [CrossRef]
25. Cruz-Martínez, H.; Rojas-Chávez, H.; Montejo-Alvaro, F.; Peña-Castañeda, Y.; Matadamas-Ortiz, P.; Medina, D. recent developments in graphene-based toxic gas sensors: A theoretical overview. *Sensors* **2021**, *21*, 1992. [CrossRef] [PubMed]
26. Cruz-Martínez, H.; Rojas-Chávez, H.; Matadamas-Ortiz, P.; Ortiz-Herrera, J.; López-Chávez, E.; Solorza-Feria, O.; Medina, D. Current progress of Pt-based ORR electrocatalysts for PEMFCs: An integrated view combining theory and experiment. *Mater. Today Phys.* **2021**, *19*, 100406. [CrossRef]
27. Payal, P. Role of nanotechnology in electronics: A review of recent developments and patents. *Recent Pat. Nanotechnol.* **2021**, *15*. [CrossRef] [PubMed]
28. Baletto, F.; Ferrando, R. Structural properties of nanoclusters: Energetic, thermodynamic, and kinetic effects. *Rev. Mod. Phys.* **2005**, *77*, 371–423. [CrossRef]
29. Ferrando, R.; Jellinek, J.; Johnston, R. Nanoalloys: From theory to applications of alloy clusters and nanoparticles. *Chem. Rev.* **2008**, *108*, 845–910. [CrossRef]
30. Cruz-Martínez, H.; Solorza-Feria, O.; Calaminici, P.; Medina, D. On the structural, energetic, and magnetic properties of M@Pd (M=Co, Ni, and Cu) core–shell nanoclusters and their comparison with pure Pd nanoclusters. *J. Magn. Magn. Mater.* **2020**, *508*, 166844. [CrossRef]
31. Mousavi, S.R.; Rezaei, M. Nanotechnology in agriculture and food production. *J. Appl. Environ. Biol. Sci.* **2011**, *1*, 414–419.
32. Baker, S.; Volova, T.; Prudnikova, S.; Satish, S.; Prasad, N. Nanoagroparticles emerging trends and future prospect in modern agriculture system. *Environ. Toxicol. Pharmacol.* **2017**, *53*, 10–17. [CrossRef]
33. Aguilar-Méndez, M.A.; Martín-Martínez, E.S.; Ortega-Arroyo, L.; Cobián-Portillo, G.; Sánchez-Espíndola, E. Synthesis and characterization of silver nanoparticles: Effect on phytopathogen Colletotrichum gloesporioides. *J. Nanopart. Res.* **2010**, *13*, 2525–2532. [CrossRef]
34. Abd-Elsalam, K. Special Issue: Fungal Nanotechnology. *J. Fungi* **2021**, *7*, 583. [CrossRef]
35. Shaikh, S.; Nazam, N.; Rizvi, S.M.D.; Ahmad, K.; Baig, M.H.; Lee, E.J.; Choi, I. Mechanistic insights into the antimicrobial actions of metallic nanoparticles and their implications for multidrug resistance. *Int. J. Mol. Sci.* **2019**, *20*, 2468. [CrossRef]
36. Kalia, A.; Abd-Elsalam, K.A.; Kuca, K. Zinc-based nanomaterials for diagnosis and management of plant diseases: Ecological safety and future prospects. *J. Fungi* **2020**, *6*, 222. [CrossRef]
37. Alghuthaymi, M.A.; Kalia, A.; Bhardwaj, K.; Bhardwaj, P.; Abd-Elsalam, K.A.; Valis, M.; Kuca, K. Nanohybrid antifungals for control of plant diseases: Current status and future perspectives. *J. Fungi* **2021**, *7*, 48. [CrossRef]
38. Kutawa, A.B.; Ahmad, K.; Ali, A.; Hussein, M.Z.; Wahab, M.A.A.; Adamu, A.; Ismaila, A.A.; Gunasena, M.T.; Rahman, M.Z.; Hossain, I. Trends in nanotechnology and its potentialities to control plant pathogenic fungi: A review. *Biology* **2021**, *10*, 881. [CrossRef]
39. Pereira, L.; Dias, N.; Carvalho, J.; Fernandes, S.; Santos, C.; Lima, N. Synthesis, characterization and antifungal activity of chemically and fungal-produced silver nanoparticles against Trichophyton rubrum. *J. Appl. Microbiol.* **2014**, *117*, 1601–1613. [CrossRef]
40. Abd-Elsalam, K.; Al-Dhabaan, F.A.; Alghuthaymi, M.; Njobeh, P.B.; Almoammar, H. Nanobiofungicides: Present concept and future perspectives in fungal control. In *Nano-Biopesticides Today and Future Perspectives*; Academic Press: Cambridge, MA, USA, 2019; pp. 315–351. [CrossRef]
41. Ali, S.M.; Yousef, N.M.H.; Nafady, N. Application of biosynthesized silver nanoparticles for the control of land snail eobania vermiculata and some plant pathogenic fungi. *J. Nanomater.* **2015**, *2015*, 218904. [CrossRef]
42. Rajeshkumar, S.; Rinitha, G. Nanostructural characterization of antimicrobial and antioxidant copper nanoparticles synthesized using novel Persea americana seeds. *OpenNano* **2018**, *3*, 18–27. [CrossRef]
43. Santhoshkumar, J.; Agarwal, H.; Menon, S.; Rajeshkumar, S.; Kumar, S.V. Green synthesis, characterization and applications of nanoparticles. Chaper 9. A biological synthesis of copper nanoparticles and its potential applications. *Micro Nano Technol.* **2019**, 199–221. [CrossRef]
44. Navale, G.R.; Late, D.J.; Shinde, S.S. Antimicrobial activity of ZnO nanoparticles/against pathogenic bacteria and fungi. *JSM Nanotechnol. Nanomed.* **2015**, *3*, 1033.
45. Ponnusamy, P.; Kolandasamy, M.; Viswanathan, E.; Balasubramanian, M.G. Antifungal activity of biosynthesised copper nanoparticles evaluated against red root-rot disease in tea plants. *J. Expo Nanosci.* **2016**, *11*, 1019–1031.
46. Borgatta, J.; Ma, C.; Hudson-Smith, N.; Elmer, W.; Pérez, C.D.P.; De La Torre-Roche, R.; Zuverza-Mena, N.; Haynes, C.L.; White, J.C.; Hamers, R.J. Copper Based Nanomaterials Suppress Root Fungal Disease in Watermelon (*Citrullus lanatus*): Role of particle morphology, composition and dissolution behavior. *ACS Sustain. Chem. Eng.* **2018**, *6*, 14847–14856. [CrossRef]
47. Ahmed, A.I.; Yadav, D.R.; Lee, Y.S. Applications of nickel nanoparticles for control of fusarium wilt on lettuce and tomato. *Int. J. Innov. Res. Sci. Eng. Technol.* **2016**, *5*, 7378–7385. [CrossRef]
48. Khezerlou, A.; Alizadeh-Sani, M.; Azizi-Lalabadi, M.; Ehsani, A. Nanoparticles and their antimicrobial properties against pathogens including bacteria, fungi, parasites and viruses. *Microb. Pathog.* **2018**, *123*, 505–526. [CrossRef] [PubMed]

49. Nayantara; Kaur, P. Biosynthesis of nanoparticles using eco-friendly factories and their role in plant pathogenicity: A review. *Biotechnol. Res. Innov.* **2018**, *2*, 63–73. [CrossRef]
50. Arciniegas-Grijalba, P.A.; Patiño-Portela, M.C.; Mosquera-Sánchez, L.P.; Guerrero-Vargas, J.A.; Rodríguez-Páez, J.E. ZnO nanoparticles (ZnO-NPs) and their antifungal activity against coffee fungus *Erythricium salmonicolor*. *Appl. Nanosci.* **2017**, *7*, 225–241. [CrossRef]
51. Medda, S.; Hajra, A.; Dey, U.; Bose, P.; Mondal, N.K. Biosynthesis of silver nanoparticles from Aloe vera leaf extract and antifungal activity against *Rhizopus* sp. and *Aspergillus* sp. *Appl. Nanosci.* **2014**, *5*, 875–880. [CrossRef]
52. Koduru, J.R.; Kailasa, S.K.; Bhamore, J.R.; Kim, K.-H.; Dutta, T.; Vellingiri, K. Phytochemical-assisted synthetic approaches for silver nanoparticles antimicrobial applications: A review. *Adv. Colloid Interface Sci.* **2018**, *256*, 326–339. [CrossRef]
53. Kasana, R.; Panwar, N.R.; Kaul, R.K.; Kumar, P. Biosynthesis and effects of copper nanoparticles on plants. *Environ. Chem. Lett.* **2017**, *15*, 233–240. [CrossRef]
54. Rai, M.; Ingle, A.P.; Paralikar, P.; Anasane, N.; Gade, R.; Ingle, P. Effective management of soft rot of ginger caused by *Pythium* spp. and *Fusarium* spp.: Emerging role of nanotechnology. *Appl. Microbiol. Biotechnol.* **2018**, *102*, 6827–6839. [CrossRef]
55. Srikar, S.K.; Giri, D.D.; Pal, D.B.; Mishra, P.K.; Upadhyay, S.N. Green synthesis of silver nanoparticles: A review. *Green Sustain. Chem.* **2016**, *06*, 34–56. [CrossRef]
56. Geoprincy, G.; Srri, B.V.; Poonguzhali, U.; Gandhi, N.N.; Renganathan, S.A. Review on green synthesis of silver nanoparticles. *Asian J. Pharm. Clin. Res.* **2013**, *6*, 8–12.
57. Alghuthaymi, M.A.; Almoammar, H.; Rai, M.; Said-Galiev, E.; Abd-Elsalam, K.A. Myconanoparticles: Synthesis and their role in phytopathogens management. *Biotechnol. Biotechnol. Equip.* **2015**, *29*, 221–236. [CrossRef]
58. Huerta-García, E.; Pérez-Arizti, J.A.; Márquez-Ramírez, S.G.; Delgado-Buenrostro, N.L.; Chirino, Y.I.; Iglesias, G.G.; López-Marure, R. Titanium dioxide nanoparticles induce strong oxidative stress and mitochondrial damage in glial cells. *Free Radic. Biol. Med.* **2014**, *73*, 84–94. [CrossRef]
59. Mikhailova, E.O. Silver nanoparticles: Mechanism of action and probable bio-application. *J. Funct. Biomater.* **2020**, *11*, 84. [CrossRef] [PubMed]
60. Zhao, X.; Zhou, L.; Rajoka, M.S.R.; Dongyan, S.; Jiang, C.; Shao, D.; Zhu, J.; Shi, J.; Huang, Q.; Yang, H.; et al. Fungal silver nanoparticles: Synthesis, application and challenges. *Crit. Rev. Biotechnol.* **2017**, *38*, 817–835. [CrossRef]
61. Kumari, M.; Shukla, S.; Pandey, S.; Giri, V.P.; Bhatia, A.; Tripathi, T.; Kakkar, P.; Nautiyal, C.S.; Mishra, A. Enhanced Cellular internalization: A bactericidal mechanism more relative to biogenic nanoparticles than chemical counterparts. *ACS Appl. Mater. Interfaces* **2017**, *9*, 4519–4533. [CrossRef]
62. Rana, A.; Yadav, K.; Jagadevan, S. A comprehensive review on green synthesis of nature-inspired metal nanoparticles: Mechanism, application and toxicity. *J. Clean. Prod.* **2020**, *272*, 122880. [CrossRef]
63. Król, A.; Pomastowski, P.; Rafińska, K.; Railean-Plugaru, V.; Buszewski, B. Zinc oxide nanoparticles: Synthesis, antiseptic activity and toxicity mechanism. *Adv. Colloid Interface Sci.* **2017**, *249*, 37–52. [CrossRef]
64. Consolo, V.F.; Torres-Nicolini, A.; Alvarez, V.A. Mycosinthetized Ag, CuO and ZnO nanoparticles from a promising Trichoderma harzianum strain and their antifungal potential against important phytopathogens. *Sci. Rep.* **2020**, *10*, 1–9. [CrossRef]
65. Akpinar, I.; Unal, M.; Sar, T. Potential antifungal effects of silver nanoparticles (AgNPs) of different sizes against phytopathogenic *Fusarium oxysporum* f. sp. radicis-lycopersici (FORL) strains. *SN Appl. Sci.* **2021**, *3*, 506. [CrossRef]
66. Jadhav, K.; Deore, S.; Dhamecha, D.; Hr, R.; Jagwani, S.; Jalalpure, S.; Bohara, R. Phytosynthesis of silver nanoparticles: Characterization, biocompatibility studies, and anticancer activity. *ACS Biomater. Sci. Eng.* **2018**, *4*, 892–899. [CrossRef]
67. Mosselhy, D.; El-Aziz, M.A.; Hanna, M.; Ahmed, M.A.; Husien, M.M.; Feng, Q. Comparative synthesis and antimicrobial action of silver nanoparticles and silver nitrate. *J. Nanopart. Res.* **2015**, *17*, 473. [CrossRef]
68. Huy, T.Q.; Huyen, P.T.; Le, A.-T.; Tonezzer, M. Recent advances of silver nanoparticles in cancer diagnosis and treatment. *Anti Cancer Agents Med. Chem.* **2020**, *20*, 1276–1287. [CrossRef]
69. Kharat, S.N.; Mendhulkar, V.D. Synthesis, characterization and studies on antioxidant activity of silver nanoparticles using Elephantopus scaber leaf extract. *Mater. Sci. Eng. C* **2016**, *62*, 719–724. [CrossRef]
70. Saravanakumar, A.; Peng, M.M.; Ganesh, M.; Jayaprakash, J.; Mohankumar, M.; Jang, H.T. Low-cost and eco-friendly green synthesis of silver nanoparticles using Prunus japonica (Rosaceae) leaf extract and their antibacterial, antioxidant properties. *Artif. Cells Nanomed. Biotechnol.* **2017**, *45*, 1165–1171. [CrossRef] [PubMed]
71. Khan, A.U.; Malik, N.; Khan, M.; Cho, M.H.; Khan, M.M. Fungi-assisted silver nanoparticle synthesis and their applications. *Bioprocess Biosyst. Eng.* **2018**, *41*, 1–20. [CrossRef] [PubMed]
72. Rajan, R.; Chandran, K.; Harper, S.L.; Yun, S.-I.; Kalaichelvan, P.T. Plant extract synthesized silver nanoparticles: An ongoing source of novel biocompatible materials. *Ind. Crop. Prod.* **2015**, *70*, 356–373. [CrossRef]
73. Gautam, N.; Salaria, N.; Thakur, K.; Kukreja, S.; Yadav, N.; Yadav, R.; Goutam, U. Green Silver Nanoparticles for Phytopathogen Control. *Proc. Natl. Acad. Sci. India Sect. B Boil. Sci.* **2020**, *90*, 439–446. [CrossRef]
74. Casagrande, M.G.; Germano-Costa, T.; Pasquoto-Stigliani, T.; Fraceto, L.F.; De Lima, R. Biosynthesis of silver nanoparticles employing *Trichoderma harzianum* with enzymatic stimulation for the control of *Sclerotinia sclerotiorum*. *Sci. Rep.* **2019**, *9*, 14351. [CrossRef]

75. Ali, A.; Ahmed, T.; Wu, W.; Hossain, A.; Hafeez, R.; Masum, M.I.; Wang, Y.; An, Q.; Sun, G.; Li, B. Advancements in plant and microbe-based synthesis of metallic nanoparticles and their antimicrobial activity against plant pathogens. *Nanomaterials* **2020**, *10*, 1146. [CrossRef]
76. Nguyen, D.H.; Vo, T.N.N.; Nguyen, N.T.; Ching, Y.C.; Thi, T.T.H. Comparison of biogenic silver nanoparticles formed by Momordica charantia and Psidium guajava leaf extract and antifungal evaluation. *PLoS ONE* **2020**, *15*, e0239360. [CrossRef] [PubMed]
77. Jebril, S.; Ben Jenana, R.K.; Dridi, C. Green synthesis of silver nanoparticles using Melia azedarach leaf extract and their antifungal activities: In vitro and in vivo. *Mater. Chem. Phys.* **2020**, *248*, 122898. [CrossRef]
78. Krishnaraj, C.; Ramachandran, R.; Mohan, K.; Kalaichelvan, P. Optimization for rapid synthesis of silver nanoparticles and its effect on phytopathogenic fungi. *Spectrochim. Acta Part A Mol. Biomol. Spectrosc.* **2012**, *93*, 95–99. [CrossRef]
79. Jafari, A.; Pourakbar, L.; Farhadi, K.; Mohamadgolizad, L.; Goosta, Y. Biological synthesis of silver nanoparticles and evaluation of antibacterialand antifungal properties of silver and copper nanoparticles. *Turk. J. Biol.* **2015**, *39*, 556–561. [CrossRef]
80. Abkhoo, J.; Panjehkeh, N. Evaluation of Antifungal Activity of Silver Nanoparticles on *Fusarium oxysporum*. *Int. J. Infect.* **2017**, *4*, e41126. [CrossRef]
81. Huang, W. Optimized Biosynthesis and Antifungal Activity of Osmanthus fragrans Leaf Extract-mediated Silver Nanoparticles. *Int. J. Agric. Biol.* **2017**, *19*, 668–672. [CrossRef]
82. Sahayaraj, K.; Balasubramanyam, G.; Chavali, M. Green synthesis of silver nanoparticles using dry leaf aqueous extract of *Pongamia glabra* Vent (Fab.), Characterization and phytofungicidal activity. *Environ. Nanotechnol. Monit. Manag.* **2020**, *14*, 100349. [CrossRef]
83. Prittesh, K.; Heena, B.; Rutvi, B.; Sangeeta, J.; Krunal, M. Synthesis and characterisation of silver nanoparticles using withania somnifera and antifungal effect against fusarium solani. *Int. J. Plant Soil Sci.* **2018**, *25*, 1–6. [CrossRef]
84. Ediz, E.; Kurtay, G.; Karaca, B.; Büyük, I.; Gökdemir, F.; Aras, S. Green synthesis of silver nanoparticles from *Phaseolus vulgaris* L extracts and investigation of their antifungal activities. *Hacet. J. Biol. Chem.* **2020**, *49*, 11–23. [CrossRef]
85. Oloyede, A.R.; Ayedun, E.O.; Sonde, O.I.; Akinduti, P.A. Investigation of antifungal activity of green-synthesized silver nanoparticles on phytopathogenic fungi. *Niger. J. Microbiol.* **2016**, *30*, 3323–3328.
86. Soundararajan, M.; Deora, N.; Lincoln, L.; Roopmani, P.; Gupta, S.; Shambu, R. Biogenic silver nanoparticles synthesised from Zingiber officinale and its antifungal properties. *Int. J. Biomed. Nanosci. Nanotechnol.* **2014**, *3*, 251. [CrossRef]
87. Obiazikwor, O.H.; Shittu, H.O. Antifungal activity of silver nanoparticles synthesized using Citrus sinensis peel extract against fungal phytopathogens isolated from diseased tomato (*Solanum lycopersicum* L.). *Issues Biol. Sci. Pharm. Res.* **2018**, *6*, 30–38. [CrossRef]
88. Ali, M.; Kim, B.; Belfield, K.D.; Norman, D.; Brennan, M.; Ali, G.S. Inhibition of Phytophthora parasitica and P. capsici by Silver Nanoparticles Synthesized Using Aqueous Extract of Artemisia absinthium. *Phytopathology* **2015**, *105*, 1183–1190. [CrossRef] [PubMed]
89. Al-Otibi, F.; Perveen, K.; Al-Saif, N.A.; Alharbi, R.I.; Bokhari, N.A.; Albasher, G.; Al-Otaibi, R.M.; Al-Mosa, M.A. Biosynthesis of silver nanoparticles using Malva parviflora and their antifungal activity. *Saudi J. Biol. Sci.* **2021**, *28*, 2229–2235. [CrossRef] [PubMed]
90. Asghar, M.A.; Zahir, E.; Shahid, S.M.; Khan, M.N.; Iqbal, J.; Walker, G. Iron, copper and silver nanoparticles: Green synthesis using green and black tea leaves extracts and evaluation of antibacterial, antifungal and aflatoxin B1 adsorption activity. *LWT* **2018**, *90*, 98–107. [CrossRef]
91. Velmurugan, P.; Sivakumar, S.; Young-Chae, S.; Seong-Ho, J.; Pyoung-In, Y.; Jeong-Min, S.; Sung-Chul, H. Synthesis and characterization comparison of peanut shell extract silver nanoparticles with commercial silver nanoparticles and their antifungal activity. *J. Ind. Eng. Chem.* **2015**, *31*, 51–54. [CrossRef]
92. Jagana, D.; Hegde, Y.R.; Lella, R. Green nanoparticles—A novel approach for the management of banana anthracnose caused by colletotrichum musae. *Int. J. Curr. Microbiol. Appl. Sci.* **2017**, *6*, 1749–1756. [CrossRef]
93. Sukhwal, A.; Jain, D.; Joshi, A.; Rawal, P.; Kushwaha, H.S. Biosynthesised silver nanoparticles using aqueous leaf extract of *Tagetes patula* L. and evaluation of their antifungal activity against phytopathogenic fungi. *IET Nanobiotechnol.* **2017**, *11*, 531–537. [CrossRef]
94. Bahrami-Teimoori, B.; Nikparast, Y.; Hojatianfar, M.; Akhlaghi, M.; Ghorbani, R.; Pourianfar, H.R. Characterisation and antifungal activity of silver nanoparticles biologically synthesised by *Amaranthus retroflexus* leaf extract. *J. Exp. Nanosci.* **2017**, *12*, 129–139. [CrossRef]
95. Valsalam, S.; Agastian, P.; Arasu, M.V.; Al-Dhabi, N.A.; Ghilan, A.-K.M.; Kaviyarasu, K.; Ravindran, B.; Chang, S.W.; Arokiyaraj, S. Rapid biosynthesis and characterization of silver nanoparticles from the leaf extract of *Tropaeolum majus* L. and its enhanced in-vitro antibacterial, antifungal, antioxidant and anticancer properties. *J. Photochem. Photobiol. B Biol.* **2018**, *191*, 65–74. [CrossRef] [PubMed]
96. Khan, A.U.; Khan, M.; Khan, M.M. Antifungal and antibacterial assay by silver nanoparticles synthesized from aqueous leaf extract of *Trigonella foenum-graecum*. *BioNanoScience* **2019**, *9*, 597–602. [CrossRef]
97. Kora, A.J.; Mounika, J.; Jagadeeshwar, R. Rice leaf extract synthesized silver nanoparticles: An in vitro fungicidal evaluation against *Rhizoctonia solani*, the causative agent of sheath blight disease in rice. *Fungal Biol.* **2020**, *124*, 671–681. [CrossRef]

98. Nguyen, D.H.; Lee, J.S.; Park, K.D.; Ching, Y.C.; Nguyen, X.T.; Phan, V.G.; Thi, T.T.H. Green Silver Nanoparticles Formed by Phyllanthus urinaria, Pouzolzia zeylanica, and Scoparia dulcis Leaf Extracts and the Antifungal Activity. *Nanomaterials* **2020**, *10*, 542. [CrossRef]
99. Madbouly, A.K.; Abdel-Aziz, M.S.; Abdel-Wahhab, M.A. Biosynthesis of nanosilver using Chaetomium globosum and its application to control Fusarium wilt of tomato in the greenhouse. *IET Nanobiotechnol.* **2017**, *11*, 702–708. [CrossRef]
100. Elamawi, R.M.; Al-Harbi, R.E.; Hendi, A.A. Biosynthesis and characterization of silver nanoparticles using *Trichoderma longibrachiatum* and their effect on phytopathogenic fungi. *Egypt. J. Biol. Pest Control.* **2018**, *28*, 28. [CrossRef]
101. Al-Othman, M.R.; Abd El-Aziz, A.R.M.; Mahmoud, M.A.; Eifan, S.A.; El-Shikh, M.S.; Majrashi, M. Application of silver nanoparticles as antifungal and antiaflatoxin B1 produced by *Aspergillus flavus*. *Dig. J. Nanomater. Bios.* **2014**, *9*, 151–157.
102. Elshahawy, I.; Abouelnasr, H.M.; Lashin, S.M.; Darwesh, O.M. First report of Pythium aphanidermatum infecting tomato in Egypt and its control using biogenic silver nanoparticles. *J. Plant. Prot. Res.* **2018**, *58*, 137–151. [CrossRef]
103. Ismail, A.-W.A.; Sidkey, N.M.; Arafa, R.A.; Fathy, R.M.; El-Batal, A.I. Evaluation of in vitro antifungal activity of silver and selenium nanoparticles against *Alternaria solani* caused early blight disease on potato. *Br. Biotechnol. J.* **2016**, *12*, 1–11. [CrossRef]
104. Abd El-Aziz, A.R.M.; AL-Othman, M.R.; Mahmoud, M.A.; Metwaly, H.A. Biosynthesis of silver nanoparticles using fusarium solani and its impact on grain borne fungi. *Dig. J. Nanomater. Biostruct.* **2015**, *10*, 655–662.
105. Bholay, A.D.; Nalawade, P.M.; Borkhataria, B.V. Fungicidal potential of biosynthesized silver nanoparticles against phytopathogens and potentiation of fluconazole. *World J. Pharm. Res.* **2013**, *1*, 12–15.
106. El-Saadony, M.T.; El-Wafai, N.A.; El-Fattah, H.I.A.; Mahgoub, S. Biosynthesis, Optimization and Characterization of Silver Nanoparticles Using a Soil Isolate of Bacillus pseudomycoides MT32 and their Antifungal Activity Against some Pathogenic Fungi. *Adv. Anim. Vet. Sci.* **2019**, *7*, 238–249. [CrossRef]
107. Amrinder, K.; Jaspal, K.; Anu, K.; Narinder, S. Effect of media composition on extent of antimycotic activity of silver nanoparticles against plant pathogenic fungus Fusarium moniliforme. *Plant Dis. Res.* **2016**, *31*, 1–5.
108. Win, T.T.; Khan, S.; Fu, P. Fungus- (*Alternaria* sp.) Mediated silver nanoparticles synthesis, characterization, and screening of antifungal activity against some phytopathogens. *J. Nanotechnol.* **2020**, *2020*, 8828878. [CrossRef]
109. Ajaz, S.; Ahmed, T.; Shahid, M.; Noman, M.; Shah, A.A.; Mehmood, M.A.; Abbas, A.; Cheema, A.I.; Iqbal, M.Z.; Li, B. Bioinspired green synthesis of silver nanoparticles by using a native *Bacillus* sp. strain AW1-2: Characterization and antifungal activity against Colletotrichum falcatum Went. *Enzym. Microb. Technol.* **2021**, *144*, 109745. [CrossRef]
110. Fernández, J.G.; Fernández-Baldo, M.A.; Berni, E.; Camí, G.; Durán, N.; Raba, J.; Sanz, M.I. Production of silver nanoparticles using yeasts and evaluation of their antifungal activity against phytopathogenic fungi. *Process. Biochem.* **2016**, *51*, 1306–1313. [CrossRef]
111. Roy, S.; Mukherjee, T.; Chakraborty, S.; Das, T.K. Biosynthesis, characterization & antifungal activity of silver nanoparticles synthesized by the fungus aspergillus FOETIDUS MTCC8876. *Dig. J. Nanomater. Biostruct.* **2013**, *8*, 197–205.
112. Yassin, M.A.; Elgorban, A.M.; El-Samawaty, A.E.-R.M.; Almunqedhi, B.M. Biosynthesis of silver nanoparticles using Penicillium verrucosum and analysis of their antifungal activity. *Saudi J. Biol. Sci.* **2021**, *28*, 2123–2127. [CrossRef]
113. Dawoud, T.M.; Yassin, M.A.; El-Samawaty, A.R.M.; Elgorban, A.M. Silver nanoparticles synthesized by Nigrospora oryzae showed antifungal activity. *Saudi J. Biol. Sci.* **2021**, *28*, 1847–1852. [CrossRef] [PubMed]
114. Elamawi, R.M.; Al-Harbi, R.E. Effect of biosynthesized silver nanoparticles on fusarium oxysporum fungus the cause of seed rot disease of faba bean, tomato and barley. *J. Plant Prot. Pathol.* **2014**, *5*, 225–237. [CrossRef]
115. Elgorban, A.M.; Aref, S.M.; Seham, S.M.; Elhindi, K.M.; Bahkali, A.H.; Sayed, S.R.; Manal, M.A. Extracellular synthesis of silver nanoparticles using Aspergillus versicolor and evaluation of their activity on plant pathogenic fungi. *Mycosphere* **2016**, *7*, 844–852. [CrossRef]
116. Ibrahim, E.; Zhang, M.; Zhang, Y.; Hossain, A.; Qiu, W.; Chen, Y.; Wang, Y.; Wu, W.; Sun, G.; Li, B. Green-synthesization of silver nanoparticles using endophytic bacteria isolated from garlic and its antifungal activity against wheat fusarium head blight pathogen *Fusarium graminearum*. *Nanomaterials* **2020**, *10*, 219. [CrossRef] [PubMed]
117. Win, T.T.; Khan, S.; Fu, P. Fungus (*Alternaria* sp.) mediated silver nanoparticles synthesis, characterization and application as phyto-pathogens growth inhibitor. *Res. Sq.* **2020**. [CrossRef]
118. Jaloot, A.S.; Owaid, M.N.; Naeem, G.A.; Muslim, R.F. Mycosynthesizing and characterizing silver nanoparticles from the mushroom Inonotus hispidus (Hymenochaetaceae), and their antibacterial and antifungal activities. *Environ. Nanotechnol. Monit. Manag.* **2020**, *14*, 100313. [CrossRef]
119. Vijayabharathi, R.; Sathya, A.; Gopalakrishnan, S. Extracellular biosynthesis of silver nanoparticles using Streptomyces griseoplanus SAI-25 and its antifungal activity against Macrophomina phaseolina, the charcoal rot pathogen of sorghum. *Biocatal. Agric. Biotechnol.* **2018**, *14*, 166–171. [CrossRef]
120. Xiang, S.; Ma, X.; Shi, H.; Ma, T.; Tian, C.; Chen, Y.; Chen, H.; Chen, X.; Luo, K.; Cai, L.; et al. Green synthesis of an alginate-coated silver nanoparticle shows high antifungal activity by enhancing its cell membrane penetrating ability. *ACS Appl. Bio Mater.* **2019**, *2*, 4087–4096. [CrossRef]
121. Bocate, K.P.; Reis, G.F.; de Souza, P.C.; Junior, A.G.O.; Durán, N.; Nakazato, G.; Furlaneto, M.C.; de Almeida, R.S.; Panagio, L.A. Antifungal activity of silver nanoparticles and simvastatin against toxigenic species of *Aspergillus*. *Int. J. Food Microbiol.* **2019**, *291*, 79–86. [CrossRef]

122. Elgorban, A.M.; El-Samawaty, A.E.-R.M.; Yassin, M.A.; Sayed, S.R.; Adil, S.F.; Elhindi, K.M.; Bakri, M.; Khan, M. Antifungal silver nanoparticles: Synthesis, characterization and biological evaluation. *Biotechnol. Biotechnol. Equip.* **2015**, *30*, 56–62. [CrossRef]
123. Muthuramalingam, T.R.; Shanmugam, C.; Gunasekaran, D.; Duraisamy, N.; Nagappan, R.; Krishnan, K. Bioactive bile salt-capped silver nanoparticles activity against destructive plant pathogenic fungi through in vitro system. *RSC Adv.* **2015**, *5*, 71174–71182. [CrossRef]
124. Mendes, J.E.; Abrunhosa, L.; Teixeira, J.A.; De Camargo, E.R.; De Souza, C.P.; Pessoa, J.D.C. Antifungal activity of silver colloidal nanoparticles against phytopathogenic fungus (*Phomopsis* sp.) in soybean seeds. *Int. J. Biol. Vet. Agric. Food Eng.* **2014**, *8*, 711–719. [CrossRef]
125. Tarazona, A.; Gómez, J.V.; Mateo, E.M.; Jiménez, M.; Mateo, F. Antifungal effect of engineered silver nanoparticles on phytopathogenic and toxigenic *Fusarium* spp. and their impact on mycotoxin accumulation. *Int. J. Food Microbiol.* **2019**, *306*, 108259. [CrossRef]
126. Nagaraju, R.S.; Sriram, R.H.; Achur, R. Antifungal activity of Carbendazim-conjugated silver nanoparticles against anthracnose disease caused by Colletotrichum gloeosporioides in mango. *J. Plant Pathol.* **2019**, *102*, 39–46. [CrossRef]
127. Kriti, A.; Ghatak, A.; Mandal, N. Inhibitory potential assessment of silver nanoparticle on phytopathogenic spores and mycelial growth of *Bipolaris sorokiniana* and *Alternaria brassicicola*. *Int. J. Curr. Microbiol. App. Sci.* **2020**, *9*, 692–699. [CrossRef]
128. Sedaghati, E.; Molaei, S.; Molaei, M.; Doraki, N. An evaluation of antifungal and antitoxigenicity effects of Ag/Zn and Ag nanoparticles on Aspergillus parasiticus growth and aflatoxin production. *PHJ* **2018**, *1*, 34–43. [CrossRef]
129. Vrandečić, K.; Ćosić, J.; Ilić, J.; Ravnjak, B.; Selmani, A.; Galić, E.; Pem, B.; Barbir, R.; Vrček, I.V.; Vinković, T. Antifungal activities of silver and selenium nanoparticles stabilized with different surface coating agents. *Pest Manag. Sci.* **2020**, *76*, 2021–2029. [CrossRef]
130. Farooq, M.; Ilyas, N.; Khan, I.; Saboor, A.; Khan, S.; Bakhtiar, M. Antifungal activity of plant extracts and silver nano particles against citrus brown spot pathogen (*Alternaria citri*). *IJOEAR* **2018**, *4*, 118–125.
131. Kasprowicz, M.J.; Kozioł, M.; Gorczyca, A. The effect of silver nanoparticles on phytopathogenic spores of Fusarium culmorum. *Can. J. Microbiol.* **2010**, *56*, 247–253. [CrossRef] [PubMed]
132. Gorczyca, A.; Pociecha, E.; Kasprowicz, M.J.; Niemiec, M. Effect of nanosilver in wheat seedlings and Fusarium culmorum culture systems. *Eur. J. Plant Pathol.* **2015**, *142*, 251–261. [CrossRef]
133. Luan, L.Q.; Xô, D.H. In vitro and in vivo fungicidal effects of γ-irradiation synthesized silver nanoparticles against Phytophthora capsici causing the foot rot disease on pepper plant. *J. Plant Pathol.* **2018**, *100*, 241–248. [CrossRef]
134. Nokkrut, B.-O.; Pisuttipiched, S.; Khantayanuwong, S.; Puangsin, B. Silver nanoparticle-based paper packaging to combat black anther disease in orchid flowers. *Coatings* **2019**, *9*, 40. [CrossRef]
135. Kim, S.W.; Jung, J.H.; Lamsal, K.; Kim, Y.S.; Min, J.S.; Lee, Y.S. Antifungal effects of silver nanoparticles (AgNPs) against various plant pathogenic fungi. *Mycobiology* **2012**, *40*, 53–58. [CrossRef]
136. Jo, Y.-K.; Kim, B.H.; Jung, G. Antifungal activity of silver ions and nanoparticles on phytopathogenic fungi. *Plant Dis.* **2009**, *93*, 1037–1043. [CrossRef]
137. Mahdizadeh, V.; Safaie, N.; Khelghatibana, F. Evaluation of antifungal activity of silver nanoparticles against some phytopathogenic fungi and *Trichoderma harzianum*. *JCP* **2015**, *4*, 291–300.
138. Li, J.; Sang, H.; Guo, H.; Popko, J.T.; He, L.; White, J.C.; Dhankher, O.P.; Jung, G.; Xing, B. Antifungal mechanisms of ZnO and Ag nanoparticles to *Sclerotinia homoeocarpa*. *Nanotechnology* **2017**, *28*, 155101. [CrossRef]
139. Malandrakis, A.A.; Kavroulakis, N.; Chrysikopoulos, C. Use of copper, silver and zinc nanoparticles against foliar and soil-borne plant pathogens. *Sci. Total. Environ.* **2019**, *670*, 292–299. [CrossRef]
140. Aleksandrowicz-Trzcińska, M.; Szaniawski, A.; Olchowik, J.; Drozdowski, S. Effects of copper and silver nanoparticles on growth of selected species of pathogenic and wood-decay fungi in vitro. *For. Chron.* **2018**, *94*, 109–116. [CrossRef]
141. Abdulrahaman, M.; Hussein, H.Z. Antifungal Effect of Silver Nanoparticles (AgNPs) Against Aspergillus flavus. *EC Microbiol.* **2017**, *6*, 63–66.
142. Zalewska, E.; Machowicz-Stefaniak, Z.; Król, E. Antifungal activity of nanoparticles against chosen fungal pathogens of caraway. *Acta Sci. Pol. Hortorum Cultus* **2016**, *15*, 121–137.
143. Malandrakis, A.A.; Kavroulakis, N.; Chrysikopoulos, C.V. Use of silver nanoparticles to counter fungicide-resistance in Monilinia fructicola. *Sci. Total. Environ.* **2020**, *747*, 141287. [CrossRef] [PubMed]
144. Lamsal, K.; Kim, S.W.; Jung, J.H.; Kim, Y.S.; Kim, K.S.; Lee, Y.S. Application of silver nanoparticles for the control of colletotrichum-species in vitro and pepper anthracnose disease in field. *Mycobiology* **2011**, *39*, 194–199. [CrossRef] [PubMed]
145. Ouda, S.M. Antifungal activity of silver and copper nanoparticles on two plant pathogens, *Alternaria alternata* and *Botrytis cinerea*. *Res. J. Microbiol.* **2014**, *9*, 34–42. [CrossRef]
146. Jung, J.-H.; Kim, S.-W.; Min, J.-S.; Kim, Y.-J.; Lamsal, K.; Kim, K.S.; Lee, Y.S. The Effect of nano-silver liquid against the white rot of the green onion caused by sclerotium cepivorum. *Mycobiology* **2010**, *38*, 39–45. [CrossRef] [PubMed]
147. Salem, E.A.; Nawito, M.A.S.; El-Raouf Ahmed, A.E.A. Effect of silver nano-particles on gray mold of tomato fruits. *J. Nanotechnol. Res.* **2019**, *1*, 108–118. [CrossRef]
148. Nejad, M.S.; Bonjar, G.H.S.; Khatami, M.; Amini, A.; Aghighi, S. In vitro and in vivo antifungal properties of silver nanoparticles against *Rhizoctonia solani*, a common agent of rice sheath blight disease. *IET Nanobiotechnol.* **2016**, *11*, 236–240. [CrossRef] [PubMed]

149. Giannousi, K.; Avramidis, I.; Dendrinou-Samara, C. Synthesis, characterization and evaluation of copper based nanoparticles as agrochemicals against Phytophthora infestans. *RSC Adv.* **2013**, *443*, 21743–21752. [CrossRef]
150. Elmer, W.; De La Torre-Roche, R.; Pagano, L.; Majumdar, S.; Zuverza-Mena, N.; Dimkpa, C.; Gardea-Torresdey, J.; White, J.C. Effect of Metalloid and Metal Oxide Nanoparticles on Fusarium Wilt of Watermelon. *Plant Dis.* **2018**, *102*, 1394–1401. [CrossRef]
151. Elmer, W.; White, J.C. The Future of Nanotechnology in Plant Pathology. *Annu. Rev. Phytopathol.* **2018**, *56*, 111–133. [CrossRef]
152. Pariona, N.; Mtz-Enriquez, A.I.; Sánchez-Rangel, D.; Carrión, G.; Paraguay-Delgado, F.; Rosas-Saito, G. Green-synthesized copper nanoparticles as a potential antifungal against plant pathogens. *RSC Adv.* **2019**, *9*, 18835–18843. [CrossRef]
153. Lopez-Lima, D.; Mtz-Enriquez, A.I.; Carrión, G.; Basurto-Cereceda, S.; Pariona, N. The bifunctional role of copper nanoparticles in tomato: Effective treatment for Fusarium wilt and plant growth promoter. *Sci. Hortic.* **2010**, *277*, 109810. [CrossRef]
154. Mali, S.C.; Dhaka, A.; Githala, C.K.; Trivedi, R. Green synthesis of copper nanoparticles using Celastrus paniculatus Willd. leaf extract and their photocatalytic and antifungal properties. *Biotechnol. Rep.* **2020**, *27*, e00518. [CrossRef] [PubMed]
155. Hasanin, M.; Al Abboud, M.A.; Alawlaqi, M.M.; Abdelghany, T.M.; Hashem, A.H. Ecofriendly Synthesis of Biosynthesized Copper Nanoparticles with Starch-Based Nanocomposite: Antimicrobial, Antioxidant, and Anticancer Activities. *Biol. Trace Element Res.* **2021**, 1–14. [CrossRef] [PubMed]
156. Hassan, S.E.-D.; Salem, S.S.; Fouda, A.; Awad, M.A.; El-Gamal, M.S.; Abdo, A. New approach for antimicrobial activity and bio-control of various pathogens by biosynthesized copper nanoparticles using endophytic actinomycetes. *J. Radiat. Res. Appl. Sci.* **2018**, *11*, 262–270. [CrossRef]
157. Pham, N.-D.; Duong, M.-M.; Le, M.-V.; Hoang, H.A.; Pham, L.-K. Preparation and characterization of antifungal colloidal copper nanoparticles and their antifungal activity against Fusarium oxysporum and Phytophthora capsici. *C. R. Chim.* **2019**, *22*, 786–793. [CrossRef]
158. Van Viet, P.; Nguyen, H.T.; Cao, T.M.; Van Hieu, L.; Pham, V. Fusarium antifungal activities of copper nanoparticles synthesized by a chemical reduction method. *J. Nanomater.* **2016**, *2016*, 1957612. [CrossRef]
159. Maqsood, S.; Qadir, S.; Hussain, A.; Asghar, A.; Saleem, R.; Zaheer, S.; Nayyar, N. Antifungal Properties of Copper Nanoparticles against Aspergillus niger. *Sch. Int. J. Biochem.* **2020**, *3*, 87–91. [CrossRef]
160. Seku , K.; Reddy, G.B.; Pejjai, B.; Mangatayaru Kotu, G.; Narasimha, G. Hydrothermal synthesis of copper nanoparticles, characterization and their biological applications. *Int. J. Nano Dimens.* **2018**, *9*, 7–14.
161. Nemati, A.; Shadpour, S.; Khalafbeygi, H.; Ashraf, S.; Barkhi, M.; Soudi, M.R. Efficiency of Hydrothermal Synthesis of Nano/Microsized Copper and Study on In Vitro Antifungal Activity. *Mater. Manuf. Process.* **2014**, *30*, 63–69. [CrossRef]
162. Bramhanwade, K.; Shende, S.; Bonde, S.; Gade, A.; Rai, M. Fungicidal activity of Cu nanoparticles against Fusarium causing crop diseases. *Environ. Chem. Lett.* **2016**, *14*, 229–235. [CrossRef]
163. Hashim, A.F.; Youssef, K.; Elsalam, K.A. Ecofriendly nanomaterials for controlling gray mold of table grapes and maintaining postharvest quality. *Eur. J. Plant Pathol.* **2019**, *154*, 377–388. [CrossRef]
164. Hermida-Montero, L.; Pariona, N.; Mtz-Enriquez, A.I.; Carrión, G.; Paraguay-Delgado, F.; Rosas-Saito, G. Aqueous-phase synthesis of nanoparticles of copper/copper oxides and their antifungal effect against Fusarium oxysporum. *J. Hazard. Mater.* **2019**, *380*, 120850. [CrossRef]
165. Malandrakis, A.A.; Kavroulakis, N.; Chrysikopoulos, C. Synergy between Cu-NPs and fungicides against *Botrytis cinerea*. *Sci. Total. Environ.* **2019**, *703*, 135557. [CrossRef]
166. Nandini, B.; Hariprasad, P.; Prakash, H.S.; Shetty, H.S.; Geetha, N. Trichogenic-selenium nanoparticles enhance disease suppressive ability of Trichoderma against downy mildew disease caused by Sclerospora graminicola in pearl millet. *Sci. Rep.* **2017**, *7*, 2612. [CrossRef] [PubMed]
167. Osonga, F.J.; Kalra, S.; Miller, R.M.; Isika, D.; Sadik, O.A. Synthesis, characterization and antifungal activities of eco-friendly palladium nanoparticles. *RSC Adv.* **2020**, *10*, 5894–5904. [CrossRef]

Journal of *Fungi*

Article

Trichoderma harzianum-Mediated ZnO Nanoparticles: A Green Tool for Controlling Soil-Borne Pathogens in Cotton

Shimaa A. Zaki [1,2], Salama A. Ouf [1], Fawziah M. Albarakaty [3,*], Marian M. Habeb [2], Aly A. Aly [2] and Kamel A. Abd-Elsalam [2,*]

1 Botany and Microbiology Department, Faculty of Science, Cairo University, Giza 12613, Egypt; shim.shimshim@yahoo.com (S.A.Z.); salama@sci.cu.edu.eg (S.A.O.)
2 Plant Pathology Research Institute, Agricultural Research Centre, Giza 12619, Egypt; marianmonir12@gmail.com (M.M.H.); aly.a.a@post.com (A.A.A.)
3 Department of Biology, Faculty of Applied Science, Umm Al-Qura University, Makkah Al Mukarramah P.O. Box 715, Saudi Arabia
* Correspondence: fmbarakati@uqu.edu.sa (F.M.A.); kamelabdelsalam@gmail.com (K.A.A.-E.)

Abstract: ZnO-based nanomaterials have high antifungal effects, such as inhibition of growth and reproduction of some pathogenic fungi, such as *Fusarium* sp., *Rhizoctonia solani* and *Macrophomina phaseolina*. Therefore, we report the extracellular synthesis of ZnONPs using a potential fungal antagonist (*Trichoderma harzianum*). ZnONPs were then characterized for their size, shape, charge and composition by visual analysis, UV–visible spectrometry, X-ray diffraction (XRD), Zeta potential, transmission electron microscopy (TEM), scanning electron microscopy (SEM) and energy-dispersive X-ray analysis (EDX). The TEM test confirmed that the size of the produced ZnONPs was 8–23 nm. The green synthesized ZnONPs were characterized by Fourier transform infrared spectroscopy (FTIR) studies to reveal the functional group attributed to the formation of ZnONPs. For the first time, trichogenic ZnONPs were shown to have fungicidal action against three soil–cotton pathogenic fungi in the laboratory and greenhouse. An antifungal examination was used to evaluate the bioactivity of the mycogenic ZnONPs in addition to two chemical fungicides (Moncut and Maxim XL) against three soil-borne pathogens, including *Fusarium* sp., *Rhizoctonia solani* and *Macrophomina phaseolina*. The findings of this study show a novel fungicidal activity in in vitro assay for complete inhibition of fungal growth of tested plant pathogenic fungi, as well as a considerable reduction in cotton seedling disease symptoms under greenhouse conditions. The formulation of a trichogenic ZnONPs form was found to increase its antifungal effect significantly. Finally, the utilization of biocontrol agents, such as *T. harzianum*, could be a safe strategy for the synthesis of a medium-scale of ZnONPs and employ it for fungal disease control in cotton.

Keywords: zinc oxide nanoparticles; *Gossypium barbadense*; *Fusarium* sp.; *Rhizoctonia solani*; *Macrophomina phaseolina*

Citation: Zaki, S.A.; Ouf, S.A.; Albarakaty, F.M.; Habeb, M.M.; Aly, A.A.; Abd-Elsalam, K.A. *Trichoderma harzianum*-Mediated ZnO Nanoparticles: A Green Tool for Controlling Soil-Borne Pathogens in Cotton. *J. Fungi* **2021**, *7*, 952. https://doi.org/10.3390/jof7110952

Academic Editor: Nuria Ferrol

Received: 27 September 2021
Accepted: 25 October 2021
Published: 10 November 2021

1. Introduction

Cotton (*Gossypium barbadense* L.) is a globally important crop that is extensively produced and traded, as well as one of Egypt's most valuable export crops [1]. Diseases of cotton seedlings are a worldwide problem caused by pathogenic soil-borne fungi. *Fusarium* spp. and *Rhizoctonia solani* are among the most pathogenic fungi present in cotton-producing regions in Egypt [2]. *R. solani* Kuhn, an anamorph of *Thanatephorus cucumeris* (Frank.) Donk [3], can cause pre-or post-emergence damping-off, seedling blight and root rot in cotton seedlings. *Fusarium* spp. are frequently obtained from infected cotton roots and classified as cotton seedling root pathogens [4]. *M. phaseolina* (*Tassi*) Goid infects over 100 families and 500 plant species all over the world [5,6]. *M. phaseolina* can cause charcoal rot in an abroad range of crops, such as sorghum, soybean, cotton, bean and corn, when conditions are favorable [7].

Bio-based NP synthesis has received a lot of interest in the last five years. It has eliminated difficult procedures necessary for NPs production utilizing microorganisms, such as fungi, bacteria and yeast, such as microbial cell culture upkeep, prolonged incubation time, several purification steps and so on [8]. Mycogenic nanoparticles offer advantages, including the formation of a capping from fungal biomolecules, which provides stability and can contribute to various biological activities [9]. ZnONPs were synthesized from fungal secondary metabolites of three monocultures of *Trichoderma* species including, *T. harzianum* and *T. reesei*. ZnONPs were biogenically produced using a cell filtrate of a strain of *T. harzianum* as a reducer and stabilizer agent [10]. Nevertheless, the microbial synthesis of ZnNPs remains unexplored [11]. *Monoascus purpureus*-mediated zinc oxide nanoparticles showed potent antifungal activity against six species of the most common food spoilage fungi [12].

Sustainable nanomaterials have become a promising option to control plant pathogenic fungi that are responsible for diseases in different crops. Crops treated with safe nanofungicides will acquire additional value because they are free of chemicals and effective at low doses [9]. They reduce food and feed spoilage and fungal pathogens and help protect human health and sustain the universal demand for high product quality [13,14]. Because of the broad range of uses of zinc oxide nanoparticles (ZnONPs), such as smart UV sensors, they have piqued researchers' attention [15], targeted drug delivery [16], antioxidant activity [17] biosensors [18], environmental remediation [19] and as a drought-tolerant agent as well as nutrient supply of crops [20]. Moreover, ZnONPs are characterized to be efficient against pathogenic fungi, mostly by their antimicrobial properties according to their photo-oxidizing and photocatalytic effects and considering infection control for the plant host [21]. Recently detailed reviews introduced the preparation methods and antifungal properties of ZnONPs and their possible antifungal mechanisms for plant diseases management and to improve food quality [22,23]. Under in vitro conditions, the biosynthesis of ZnONPs produced from *Trichodermas* spp. was used to suppress the development of *Xanthomonas oryzae* pv. *oryzae* [10]. An interactive protective impact of ZnONPs on seedling spray/seed soak followed by seedling and biocontrol treatments, *T. harzianum*, enhanced plant resistance to *R. solani*, the causal organisms of sunflower seedlings damping-off [24].

Antibacterial, antifungal, antiviral and anti-toxigenic activities against a range of phytopathogens may be achieved using zinc-based nanomaterials, which have targeted antimicrobial capabilities and low to negligible phytotoxic activities (Khan et al., 2021) [25]. Several applications methods may be used to employ these formulations in open fields or under greenhouse environments [22–24]. The use of fungi in the biogenic synthesis of ZnONPs has several benefits, including the production of a capping from fungal biomolecules, which provides stability and can contribute to different biological activities, such as the development of safe nanofungicides. Limited reports have used a *T. harzianum* cell filtrate for the production of ZnONPs as only a reducer and stabilizer agent [26]. Therefore, the main aims of the present study were to (1) synthesis a novel trichogenic-ZnONPs using an easy, eco-friendly, environmentally safe and costless approach, employing fungal metabolites from *T. harzianum* strains as a reducing agent and stabilizer to synthesize ZnONPs; (2) characterize the synthesized ZnONPs to confirm synthesis, structure, size and NPs morphology; (3) investigate the in vitro and in vivo antifungal activity of mycogenic ZnONPs against soil-borne pathogenic fungi including *R. solani*, *Fusarium* sp. and *M. phaseolina* isolated from infected soil in cotton-growing areas.

2. Materials and Methods

2.1. Preparation of Trichoderma Isolates Culture

Trichoderma sp. isolates were isolated from healthy cotton root rhizosphere soils. Various soil dilutions were cultivated on Rose Bengal Medium (Sigma–Aldrich, St. Louis, MO, USA) [27]. After 48 h, *Trichoderma* colonies were collected and cultivated on potato dextrose agar (PDA) (Sigma–Aldrich, St., Louis, MO, USA) media. Single spore isolation was used to isolate putative *Trichoderma* colonies [28]. To obtain a pure isolate, a single spore was transferred to a PDA medium. *Trichoderma* species were identified based on morphology technique in Assiut University Mycology Center (AUMC), Assuit, Egypt.

2.2. Preparation of Cultural Extract

The fungi were allowed to grow aerobically in a liquid medium containing KH_2PO_4 (7 g/L); K_2HPO_4 (2 g/L); $MgSO_4 \cdot 7H_2O$ (0.1 g/L); $(NH_2)SO_4$ (1 g/L);yeast extract (0.6 g/L); glucose (10 g/L), to prepare the biomass for biosynthesis experiments. The growing cultures were incubated in an orbital shaker and agitated at 150 rpm at 27 °C. The produced biomass was collected by sieving through a plastic sieve after 72 h of growth. After that, the biomass was thoroughly washed with sterile distilled water to remove any remaining broth medium components. At 28 °C for 48 h, *Trichoderma* biomass (20 g) was transferred to an Erlenmeyer flask containing 100 mL sterile deionized water. The biomass was agitated after incubation and the filtrate was obtained by filtering it using Whatman (Sigma–Aldrich, St., Louis, MO, USA) filter paper #1 [29].

2.3. Synthesis of Zinc Oxide Nanoparticles

Fifty millilitres $Zn(CH_3COO)_2.2H_2O$ salt (Molecular Biology Grade, Merck, Kenilworth, NJ, USA) was applied to 50mL fungal filtrate and incubated for 72 h at 150 rpm in an orbital shaker. A fungal biomass filtrate without the zinc acetate dihydrate solution served as a positive control, while the zinc acetate dihydrate with cell-free filtrate served as a negative control. After centrifugation at 10,000 rpm for 10 min, the pellet aggregated at the bottom of the flask was removed from the filtrate and lyophilized [30]. ZnONPs were dried overnight at 60 °C in an oven and used for further research to evaluate fungicidal activity. For NPs characterization, the developed ZnONPs were subjected to a variety of instrumental analytical techniques to describe physico-chemical proporties.

2.4. Characterization of Nanoparticles

2.4.1. Ultraviolet-Visible Spectrophotometer Analysis

A UV-vis spectrophotometer (T80 UV/vis spectrophotometer, PG Instruments Limited, Lutterworth, UK) was used to test the formation of reduced nanoparticles in colloidal solution. The supernatants' absorption spectra were measured between 200 and 800 nm. The formation of ZnONPs was detected by periodic sampling of aliquots (1 mL) of the aqueous portion in the range of 0 to 1100 nm using an ultraviolet-visible spectrophotometer after 2, 4 and 7 days. Distilled water was used as a blank.

2.4.2. Transmission Electron Microscopy (TEM)

Powder samples were put into a mortar before being mounted on a 200-mesh copper specimen grid with a film coating. At an accelerating voltage of 80 kV, TEM micrographs were taken on a Carl Zeiss Leo 912 AB OMEGA electron microscope (Carl Zeiss AG, Jena, Germany). After a drop of liquid ZnONPs was dried on the carbon-coated copper grids, a sample for analysis was prepared. Until loading onto a specimen holder, desiccators were used to dry TEM grid samples and keep them under vacuum. Image J 1.45 s software was used to assess the particle size distribution of nanoparticles.

2.4.3. Zeta Potential

A sample was prepared by dissolving ZnONPs powder in deionized water, then the sample was sonicated for 10 min using the Q500 sonicator. The zeta potential of an aqueous solution of ZnONPs was measured in Folded Capillary cell (DTS1070) with pH ranging from 2 to 11 by applying ±65 V across the electrodes by Zetatrac equipped with Mic rotrac FLEX Operating Software on Mansettingnano (Malvern Instruments, South borough, MA, USA).

2.4.4. X-ray Diffraction (XRD)

X-ray Diffraction (XRD) analysis was used to examine the structure of powder nanoparticles. Cu Kα radiation (λ = 1.54 Å) was used in the scattering range(2θ) of 0 to 80° at a scan rate of 0.03S1 on a D8-A25-Advance diffractometer (Bruker, Karlsruhe, Germany). As an internal standard for calibration, a standard silicon sample was used.

2.4.5. Scanning Electron Microscope (SEM)

Scanning electron microscopical analysis was made using a Tescan SEM (Tescanvega 3 SBU, Czech Republic) at an accelerating voltage of 20 kV. Samples were mounted on aluminium microscopy stubs using carbon tape, then coated with gold (Au) for 120 s using a Quorum Techniques Ltd. sputter coater (Q150t, Lewes, UK).

2.4.6. Energy Dispersive X-ray (EDX) Spectroscopy

On a JEOL(JEM-1230) electron microscope (Jeol, Tokyo, Japan), a drop of ZnONPs was put on carbon-coated copper grids and allowed to sit for 2 min and the excess solution was extracted with blotting paper and allowed to dry at room temperature.

2.4.7. Fourier TransformInfrared Spectroscopy (FTIR)

FTIR spectra were performed on a JASCO-4700 FTIR Spectrometer (Laser Spectroscopy Labs, UCI, Irvine, CA, USA) to detect the possible functional groups in biomolecules present in the fungal extract.

2.5. In Vitro Antifungal Activity of Synthesized Nanoparticles

To evaluate the antifungal effect of nanoparticles in vitro, *R. solani* (Rs9), *Fusarium* sp. (F10) and *M. phaseolina* isolate (M4) were grown in PDA medium at 35 °C for 7 days. Then, freshly prepared PDA containing different concentrations of synthesized ZnONPs (20, 40 and 100 µg/mL) were added. ZnONPs solutions were put in an ultrasonic bath for 15 min at a sonicating frequency of 37 kHz (Elmasonic S60, Elma, Singen, Germany) to disrupt nanoparticle aggregations. After the fungal media cooled to about 45 °C, the sonicated NPs were inoculated into the media, then the fungal culture medium was poured into Petri dishes. Five-millimetre disks of fungal inoculum were cut with a cork borer and inoculated at the center of the 9-cm-diameter Petri dish, incubated at 35 °C for 5–7 days. PDA plates free from ZnONPs cultured under the same conditions were used as controls. The linear growth of the fungi was measured [31].

2.6. Antifungal Activity under Greenhouse Conditions

In a greenhouse experiment, the effects of synthesized ZnONPs were evaluated against *R. solani* (RS9), *Fusarium* sp. (F10) and *M. phaseolina* (M4) on cotton cultivars Giza90 and Giza94. The pots, containing the autoclaved soil, were infested with two-week-old pathogen-sorghum cultures of *R. solani* (RS9), *Fusarium* sp. (F10) and *M. phaseolina* (M4) at a rate of 1, 5 and 50 g/kg soil, respectively. Cotton cultivars Giza90 and Giza94 seeds were surfaces sterilized with 10% sodium hypochlorite for 2 min before being washed in four changes of sterilized water. The sterilized seeds were subsequently immersed in a suspension of ZnONPs at 100 and 200 µg/mL concentrations for 12 h under static circumstances. The tested fungicides (Moncut (2 g/kg seeds) and Maxim XL (2 mL/L)) were added to slightly moist seeds of cotton cultivars Giza90 and Giza94. Table 1. Infested soil was poured into 15-cm pots, with 10 seeds sown in each pot. Sterilized sorghum grains were mixed fully with soil cotton seeds only in the control treatments. In infested control, infested soil at the rate of 1 g/kg of soil *R. solani* (RS9) and 50 g/kg of soil for *Fusarium* sp. (F10) and *M. phaseolina* (M4) with cotton seeds without any treatment was applied. At 28 °C, pots were dispersed on greenhouse benches randomly in a complete block design. There were three replicates (pots) for each treatment. Forty-five days after planting, plant height (cm/plant), dry weight (g/plant) and survival percentages were all measured [32].

Table 1. Fungicides and Trichogenic-ZnONPs applied in controlling damping-off of cotton seedlings under greenhouse conditions.

Treatment	Application Methods	Active Ingredients	Rate of Application
1-autoclaved soil	Seed dressing	Untreated	1 g for *R. solani* and 50 g for *Fusaruim* and *M. phaseolina* sterilized sorghum/kg soil
2-infested soil	Seed dressing	Untreated	1 g for *R. solani* and 50 g for *Fusaruim* and *M. phaseolina* infested sorghum/kg soil
3-Moncut	Seed dressing	Flutolanil	2 g/kg seeds
4-Maxim XL	Seed dressing	Fludioxonil, Mefanoxam	2 mL/L
5-ZnONPs	Seed dressing	Zinc Oxide	100 µg/mL
6-ZnONPs	Seed dressing	Zinc Oxide	200 µg/mL

2.7. Statistical Analysis

Data were subjected to statistical analysis of variance (ANOVA) via MSTAT-C software. The mean differences were compared by the least significant difference (LSD) test at $p \leq 0.05$.

3. Results

3.1. Trichoderma Isolates

A total of 50 *Trichoderma* strains (TC1–T50) were obtained from 22 soil samples. On the PDA medium, the colony's growth speed, conidiospore color, wheel pattern and pigment secretion were all studied. Then, they were identified by morphological methods, which identified 6 *Trichoderma* species.

3.2. Trichogenic Nanoparticles Synthesis

The synthesis of ZnONPs was detected by UV-vis and from all the strains screened, only four had the aptitude to synthesize ZnONPs. Three *Trichoderma* species (Tvivi, TC34 and TC28) were used for the biosynthesis of stable ZnONPs. Filtrates from each fungal strain were incubated with zinc acetate dihydrate for 72 h under dark conditions at 28 °C with agitation.

3.3. Physiochemical Characterization

3.3.1. UV-Vis Spectrophotometer

Figure 1 shows the UV-vis absorption spectrum of zinc oxide nanoparticles. The absorption spectrum was recorded for the synthesized ZnONPs sample by Tvivi, TC 34 and TC 28 after 2, 4 and 7 days after synthesis in the range of 0 to 1100 nm. The absorbance peak at 300 nm, which corresponded to the distinctive band of ZnONPs, was visible in the spectrum, for all tested *Trichoderma* isolates at all tested days (Figure 1). ZnONPs produced by Tvivi strain was chosen for further characterization.

3.3.2. Zeta Potential Analysis

The surface charges gained by ZnONPs were detected using zeta potential analysis, which may be used to learn more about the stability of the colloidal ZnONPs.In the present assay, we used a concentration of 40 µg/mL to measure zeta potential. As a result of this propensity, certain nanoparticles tended to agglomerate, reducing their surface area. ZnONPs must undergo prolonged ultrasonication in a water bath for at least 15 min to fix this problem. The result also signifies the presence of repulsive electrostatic forces among the synthesized ZnONPs, which leads to the monodispersity of the particles. In the present study, the zeta potential of ZnONPs was measured and was recorded as −24.0 mV (Figure 2A).

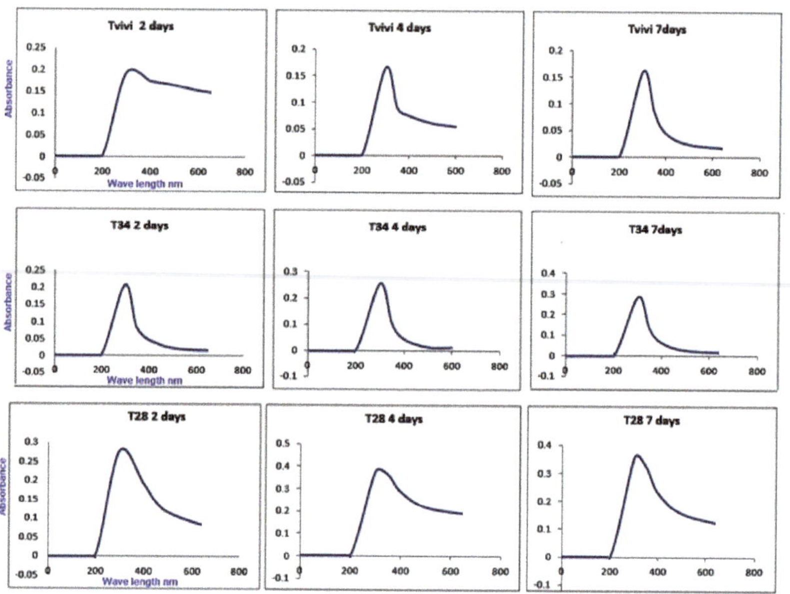

Figure 1. UV_vis spectrum of Trichogenic ZnONPs produced by Tvivi, T34 and T28 after 2, 4 and 7 days after synthesis.

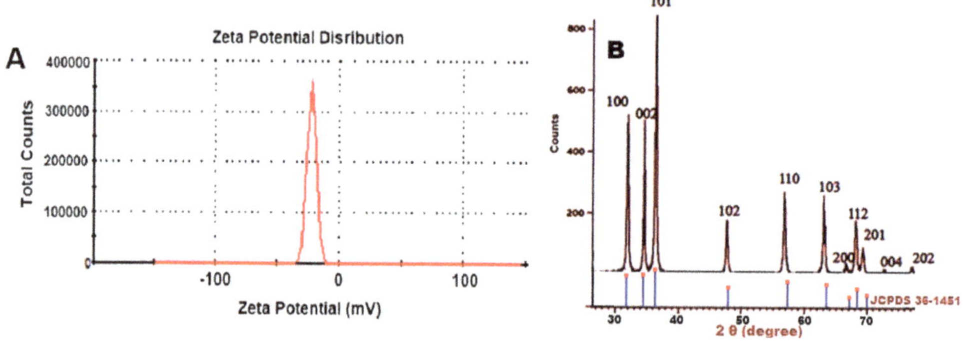

Figure 2. (**A**) Zeta potential analysis of synthesized ZnONPs. Trichoderma-mediated ZnONPs were spherical and rod-shaped and the potential value was found to be −24.0 mV. (**B**) X-ray diffraction pattern of Trichogenic ZnONPs. All peaks reveal the purity and crystalline nature. No traces of other impurity phases were detected.

3.3.3. X-ray Powder Diffractometer (XRD)

The XRD pattern of synthesized ZnONPs gave the diffraction peaks at (100), (002), (101), (102), (110), (103), (200), (112), (201), (004) and (202) planes, respectively, with the highest peak being the (101) plane (Figure 2B). The observed XRD peaks in the X-ray diffraction patterns of the ZnO samples were categorized by the hexagonal wurtzite structure of ZnO (JCPDS card 36-1451 data).

3.3.4. Transmission Electron Microscopy (TEM)

TEM was applied to know the actual size and shape of ZnONPs. The TEM image in the present study showed a mixture of hexagonal, spherical and rod-shaped a very small particles with a crystalline structure for the ZnONPs with an average size of 8–25 nm

(Figure 3A). The ZnONPs were individuals and agglomerated in clusters. Diffraction rings could be allocated as (100), (002) and (101) planes from the selected area diffraction (SAED) pattern of ZnONPs (Figure 3B), representing hexagonal structure coupled with the wurtzite-like structure of ZnONPs as shown in the XRD pattern.

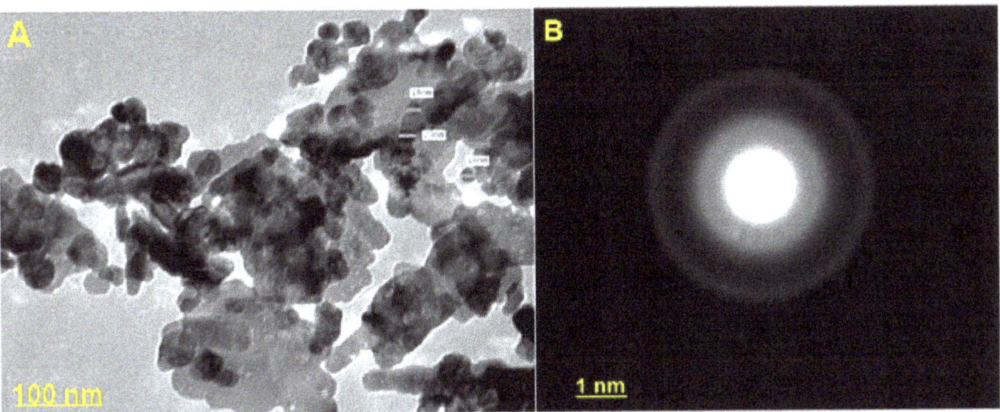

Figure 3. (**A**) Transmission electron microscopy (TEM) image of synthesized ZnO-NPs; the inset shows the corresponding particle size distribution and shape. (**B**): Selected Area Electron Diffraction (SAED) of Trichogenic-ZnONPs.

3.3.5. Scanning Electron Microscopy (SEM)

SEM is a high-resolution surface imaging approach that uses an electron beam to obtain information on nanostructures and other materials at the microscopic level. SEM analysis of synthesized ZnONPs exhibited clear spherical, rod and hexagonal shapes with and well-distributed ZnONPs with aggregation. Figure 4 presents a microscopic image of the obtained ZnONPs shown at different magnifications. The studies of the nanomaterial showed different sizes of the particles in a range comprising 24 to 50 nm. The microstructure of nanocrystalline ZnO had a skeletal form resulting from the process of coagulation (Figure 4A,B). The ZnONPs (Figure 4C,D).

3.3.6. Energy Dispersive X-ray Analysis (EDX)

An EDX spectrum was used on the ZnONPs to determine the amount of metal and oxides in the sample. The EDX spectrum of the produced NPs was recorded in the spot-profile mode from one of the densely populated ZnONPs areas, as shown in Figure 5. The synthesis of ZnONPs is represented by distinct peaks observed for zinc and oxygen and carbon atoms. The atomic percentages of the elements inset of Figure 5 indicated zinc as the dominant element, representing more than 72.49% of the entire composition, with oxygen representing 27.51%, indicating that the ZnONPs were extremely pure.

3.3.7. Fourier Transforms Infrared Spectroscopy (FTIR) Analysis

The interfaces between zinc oxide and bioactive components of fungal extract were discovered using FTIR on green synthesized ZnONPs. It was carried out to discover the organic functional groups or potential biomolecules involved in the production of ZnONPs. In the present results, FTIR spectrum showed 3398, 3233, 2912, 1640, 1629, 1561, 1461, 1018, 576 and 533 cm^{-1} (Table 2). In FTIR spectrum, the peak observed at 3398 cm^{-1} corresponded to OH stretching vibrations and 3233, peak observed at 3323 responding to C-H stretch of alkenyl and 1640 corresponded to C=O stretching 1629 responding to –C=C– aromatic stretching of fungal biomass and 1561 responding to C=C stretch in the aromatic ring and C=O stretch in polyphenols and 1461 corresponded to C-N stretch of amide-I in protein and 1018 responding to C-O stretching in amino acid, while 576 and 533 corresponded Zn-O stretching and hexagonal phase ZnO respectively.

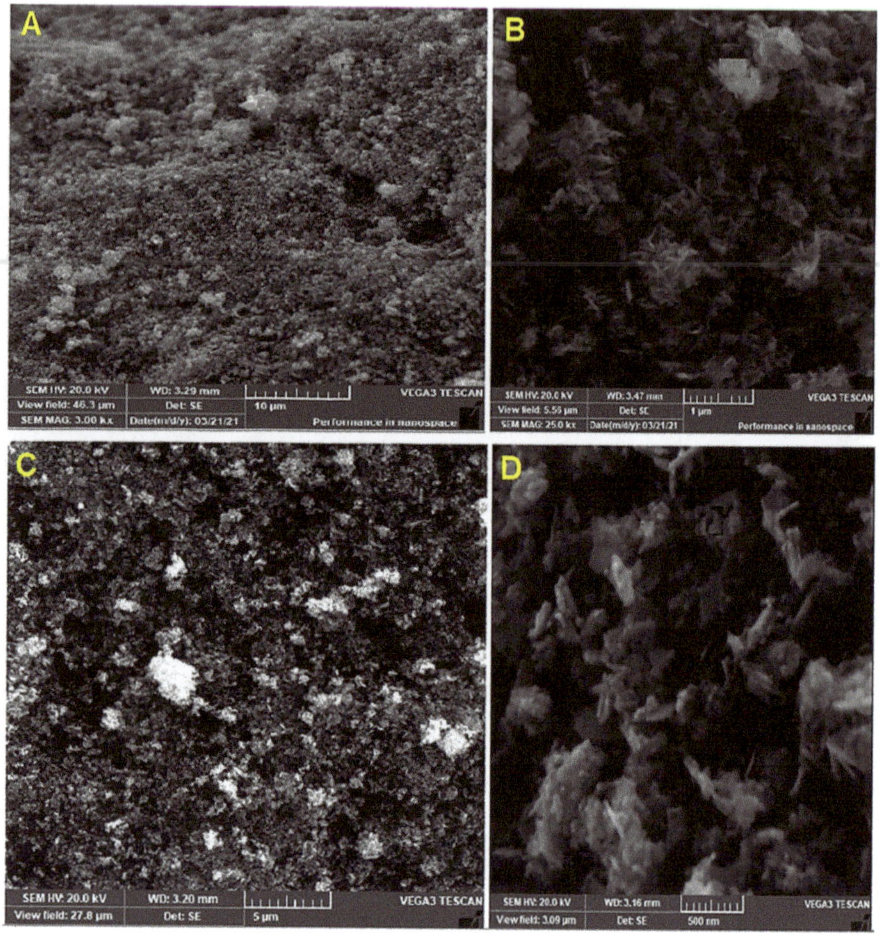

Figure 4. Scanning electron microscope micrographs at different magnifications, (**A**) 10, (**B**) 5, (**C**) 1μm and (**D**) 500 nm of Trichogenic-ZnONPs.

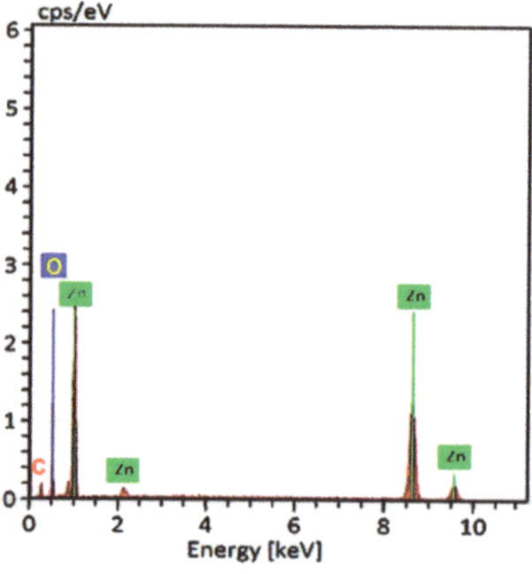

Figure 5. Elemental and energy dispersive X-ray spectroscopic analysis of Trichogenic-ZnONPs.

Table 2. Functional Group present in the Trichogenic ZnONPs analyzed by FTIR.

	Frequency (cm) $^{-1}$	Abs	Possible Assignment
1	3398.9229	96.32	OH stretching vibrations
2	3233.0747	97.1775	C-H stretch of alkanyle
3	2912.9492	97.6787	The C-H stretch in alkanes
4	2854.1311	97.3885	O H stretch in a carboxylic acid
5	1640.1611	99.027	C=O stretching
6	1629.5546	99.0942	–C=C– aromatic stretching of fungal biomass
7	1613.1626	98.9696	H-O-H binding vibration
8	1561.094	98.5039	C=C stretch in the aromatic ring and C=O stretch in polyphenols
9	1556.2728	98.4978	C=C/amine—NH stretching
10	1461.778	98.8046	C-N stretch of amide-I in protein
11	1382.7108	99.117	Acetate group stretching
12	1034.6226	96.4734	O-H Asymmetric stretching
13	1018.2305	96.3747	C-O stretching in amino acid
14	929.5211	97.4376	C-N stretching amine
15	773.3152	98.6183	C-N stretching amine
16	576.6116	95.2955	Zn-O stretching
17	533.2211	88.8474	hexagonal phase ZnO

3.4. In Vitro Antifungal Activity of Synthesized ZnONPs

The potentiality of ZnONPs for controlling *R. solani* (RS9), *Fusarium* sp. (F10) and *M. phaseolina* (M4) was tested by plating the fungal culture media supplemented with (control), 20, 40 and 100 µg/mL of ZnONPs and the diameter of the mycelium growth was measured after 7 days. ZnONPs caused a significant reduction in the mycelia growth of *R. solani* (RS9), *Fusarium* sp. (F10) and *M. phaseolina* (M4) by all concentrations. As shown in (Figure 6), under the effect of ZnONPs treatments, the mycelial diameter of *R. solani* (RS9), *Fusarium* sp. (F10) and *M. phaseolina* (M4) was reduced by 100% at all the tested concentrations.

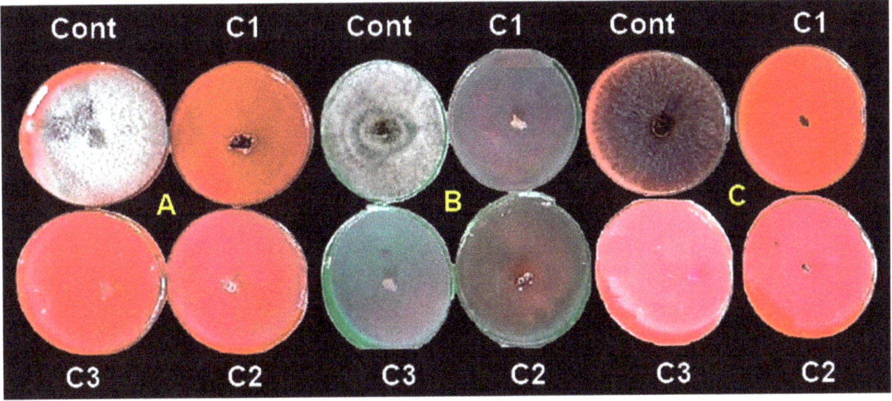

Figure 6. The inhibitory effect of mycelia growth on F10 (**A**), Rs9 (**B**), M4 (**C**) on potato dextrose agar medium containing ZnONPs at concentrations: Control, (**C1**) 20, (**C2**) 40 and (**C3**) 100 µg/mL after 7 days.

3.5. Effect of ZnONPs against Cotton Damping-Off Disease under Greenhouse Conditions

The effects of ZnONPs on the cotton seedling disease were studied in a greenhouse experiment. In a greenhouse, four treatments were tested for their ability to reduce cotton seedling disease produced by three pathogenic isolates (F10, Rs9 and M4). The number of surviving seedlings increased. Analysis of variance (ANOVA) (Table 3) showed that the treatment was a very highly significant source of variation ($p = 0.00$) of all the tested variables. Fungus × treatment interaction was a very highly significant source of variation

(p = 0.00) only in the case of plant height. The effect of fungus was a non-significant source of variation in all the tested variables.

Table 3. Analysis of variance of the effect of some fungi, treatments and their interaction on some growth variables of cotton seedlings of Giza90 grown in soil infested under greenhouse conditions.

Growth Variables and Sources of Variation	D.F	Mean Square	F Value	$p \geq$ F
Survival				
Replicates	2	118.51	0.51	0.61
Fungi (F)	2	78.65	0.34	0.72
Treatments (T)	5	3944.66	16.91	0.00
F × T	10	127.82	0.55	0.84
Error	34	23.22		
Plant height				
Replicates	2	3.72	0.31	0.74
Fungi (F)	2	17.95	1.49	0.24
Treatments (T)	5	244.83	20.25	0.00
F × T	10	50.89	4.21	0.00
Error	34	12.09		
Dry weight	2			
Replicates	2	0.22	0.90	0.42
Fungi (F)	5	0.32	1.30	0.29
Treatments (T)	10	4.13	16.7	0.00
F × T	34	0.23	0.94	0.51
Error	2	0.25		

Since there was no fungus × treatment interaction on survival (Table 4), the general mean was used to compare treatment means. Seeds of Giza90 treated with ZnONPs (200 µg/mL) showed the maximum efficiency in controlling disease regardless of fungus (91.111% survival). The difference between the general means of the tested fungi was non-significant.

Table 4. Effect of some fungi, treatments and their interaction on survival percentage of cotton seedlings of Giza90 grown in infested soil under greenhouse conditions.

Treatment	ZnO/Survival/Giza90 (%)							
	F10		Rs9		M4		Mean	
	%	Transformed [a]	%	Transformed [a]	%	Transformed [a]	%	Transformed [a]
ZnONPs (µg/mL)	76.667	61.910	80.000	63.440	53.333	46.910	70.000	57.420
ZnONPs (µg/mL)	96.667	83.853	93.333	81.147	83.333	70.077	91.111	78.359
Maxim XL (2 mL)	86.667	72.293	86.667	72.783	90.000	75.000	87.778	73.359
Moncut (2 g)	86.667	72.783	83.333	70.763	83.333	66.147	84.444	69.898
Infested soil	30.000	28.077	16.667	15.000	26.667	30.293	24.445	24.457
Autoclaved soil	30.000	28.077	16.667	15.000	26.667	30.293	24.445	24.457
Mean	77.778	65.653	76.111	64.498	71.667	61.596	75.185	63.916

LSD ($p \leq 0.05$) (transformed data) for treatments = 14.25. LSD ($p \leq 0.05$) for fungus is non-significant. [a] Percentage of data were transformed into arcsine angles before carrying out the analysis of variance to produce an approximately constant variance.

Because cultivar × treatment interaction was significant on plant height (Table 5), an interaction LSD was calculated to compare treatment means within each tested fungus. All treatments were effective in controlling disease compared to the infested control. The high concentrations of ZnONPs showed the maximum efficiency in controlling the disease for all tested fungi (25.193, 26.433 and 24.767 cm) for F10, Rs9 and M4, respectively. Since there was no fungus × treatment interaction on dry weight (Table 6), the general mean was used to compare treatment means. Seeds of Giza90 treated with ZnONPs (200 µg/mL) showed the maximum efficiency in controlling disease regardless of fungus (2.200 g). The difference between the general means of the tested fungi was non-significant.

Table 5. Effect of some fungi, treatments and their interaction on plant height of cotton seedlings of Giza90 grown in infested soil under greenhouse conditions.

	ZnO/Plant Height (cm)/Giza90			
Treatment	**F10**	**Rs9**	**M4**	**Mean**
ZnONPs (100 µg/mL)	17.267	21.797	18.387	19.150
ZnONPs (200 µg/mL)	25.193	26.433	24.767	25.464
Maxim XL (2 mL)	18.573	21.843	15.917	18.778
Moncut (2 g)	20.147	16.627	21.137	19.304
Infested soil	7.150	4.800	20.303	10.751
Autoclaved soil	24.307	24.833	23.847	24.329
Mean	18.773	19.389	20.726	19.629

LSD ($p \leq 0.05$) for fungus × treatment = 5.62.

Table 6. Effect of some fungi, treatments and their interaction on the dry weight of cotton seedlings of Giza90 grown in soil infested under greenhouse conditions.

	ZnO/Dry Weight (g)/Giza90			
Treatment	**F10**	**Rs9**	**M4**	**Mean**
ZnONPs (100 µg/mL)	1.700	1.493	1.370	1.521
ZnONPs (200 µg/mL)	2.250	2.377	1.973	2.200
Maxim XL (2 mL)	2.073	1.527	1.137	1.579
Moncut (2 g)	1.660	1.833	1.137	1.543
Infested soil	0.187	0.133	0.377	0.232
Autoclaved soil	1.660	2.050	2.090	1.933
Mean	1.588	1.569	1.347	1.501

LSD ($p \leq 0.05$) for treatments = 0.47. LSD ($p \leq 0.05$) for fungus is non-significant.

Analysis of variance (ANOVA) of Table 7 showed that treatment was a very highly significant source of variation (p = 0.00) of all the tested variables. Fungus and fungus × treatment interaction was non-significant sources of variation of all the tested variables.

Table 7. Analysis of variance of the effect of some fungi, treatments and their interaction on some growth variables of cotton seedlings of Giza94 grown in soil infested under greenhouse conditions.

Growth Variables and Sources of Variation	D.F.	Mean Square	F. Value	$p \geq$ F
Survival				
Replicates	2	556.02	2.79	0.08
Fungi (F)	2	19.16	0.10	0.91
Treatments (T)	5	4674.53	23.45	0.00
F × T	10	215.36	1.08	0.40
Error	34	199.32		
Plant height				
Replicates	2	38.05	1.37	0.27
Fungi (F)	2	23.36	0.84	0.44
Treatments (T)	5	320.18	11.52	0.00
F × T	10	36.91	1.33	0.26
Error	34	27.80		
Dry weight	2			
Replicates	2	0.42	1.32	0.28
Fungi (F)	5	0.14	0.46	0.64
Treatments (T)	10	4.99	15.77	0.00
F × T	34	0.33	1.04	0.43
Error	2	0.32		

Since there were no effects of fungus × treatment interaction on survival (Table 8), the general mean was used to compare treatment means. Seeds of Giza94 treated with Moncut

(2 g) showed the maximum efficiency in controlling disease regardless of fungus (88.889% survival) followed by Maxim XL (2 mL) and ZnONPs (200 μg/mL). The difference between the general means of the tested fungi was non-significant (Figure 7).

Table 8. Effect of some fungi, treatments and their interaction on survival percentage of cotton seedlings of Giza94 grown in soil infested under greenhouse conditions.

ZnO/Survival/Giza94 (%)								
Treatment	**F10**		**Rs9**		**M4**		**Mean**	
	%	**Transformed [a]**	%	**Transformed [a]**	%	**Transformed [a]**	%	**Transformed [a]**
ZnONPs (100 μg/mL)	36.667	36.930	10.000	11.070	43.333	41.057	30.000	29.686
ZnONPs (200 μg/mL)	73.333	59.693	73.333	59.693	76.667	61.223	74.444	60.203
Maxim XL (2 mL)	80.000	67.860	76.667	65.840	76.667	61.910	77.778	65.203
Moncut (2 g)	83.333	70.077	93.333	77.707	90.000	75.000	88.889	74.261
Infested soil	16.667	19.223	30.000	28.077	6.667	12.293	17.778	19.864
Autoclaved soil	86.667	72.783	86.667	72.293	83.333	66.147	85.556	70.408
Mean	62.778	54.428	61.667	52.447	62.778	52.938	62.408	53.271

LSD ($p \leq 0.05$) (transformed data) for treatments = 13.18. LSD ($p \leq 0.05$) for fungus is non-significant. [a] Percentage data were transformed into arcsine angles before carrying out the analysis of variance to produce an approximately constant variance.

Since there was no fungus × treatment interaction on plant height and dry weight (Tables 9 and 10), the general mean was used to compare treatment means. Seeds of Giza94 treated with ZnONPs (200 μg/mL) showed the maximum efficiency in controlling disease regardless of fungus (24.300 cm and 2.094 g). The difference between the general means of the tested fungi was non-significant.

Table 9. Effect of some fungi, treatments and their interaction on plant height of cotton seedlings of Giza94 grown in infested soil under greenhouse conditions.

ZnO/Plant Height (cm)/Giza94				
Treatment	**F10**	**Rs9**	**M4**	**Mean**
ZnONPs (100 μg/mL)	19.527	4.887	14.830	13.081
ZnONPs (200 μg/mL)	25.377	22.557	24.967	24.300
Maxim XL (2 mL)	19.113	22.713	18.997	20.274
Moncut (2 g)	21.333	22.653	17.973	20.653
Infested soil	11.917	11.533	11.000	11.483
Autoclaved soil	26.730	26.950	25.507	26.396
Mean	20.666	18.549	18.879	19.365

LSD ($p \leq 0.05$) for treatments = 4.92., LSD ($p \leq 0.05$) for fungus is non-significant.

Table 10. Effect of some fungi, treatments and their interaction on the dry weight of cotton seedlings of Giza94 grown in infested soil under greenhouse conditions.

ZnO/Dry Weight (g)/Giza94				
Treatment	**F10**	**Rs9**	**M4**	**Mean**
ZnONPs (100 μg/mL)	1.427	0.507	1.563	1.166
ZnONPs (200 μg/mL)	2.253	2.170	1.860	2.094
Maxim XL(2 mL)	1.537	1.753	1.490	1.593
Moncut (2 g)	1.697	1.497	1.360	1.518
Infested soil	0.403	0.707	0.537	0.549
Autoclaved soil	2.837	3.020	2.270	2.709
Mean	1.692	1.609	1.513	1.605

LSD ($p \leq 0.05$) for treatments = 0.53. LSD ($p \leq 0.05$) for fungus is non-significant.

Figure 7. Cotton Seedlings cultivars Giza90 and Giza 93 obtained by sowing uncoated cotton seeds in sterilized soil infested with three fungal pathogens including, *Fusarium*, *R. solani* and *M. phaseolina* as a negative control, uncoated seeds sown in sterilized soil as a positive control, treated seeds with two fungicides (Maxim XL, Moncut) sown in infested soil and coated seeds with ZnONPs (100, 200 μg/mL) in infested soil. Photos were taken after 45 days under standard growth conditions in greenhouse conditions.

4. Discussion

Nanoparticles derived from Trichoderma are still in the early stages of research. Mycogenic ZnONPs utilizing *Trichoderma* sp. are more compelling and less harmful to the environment than other methods. Therefore, ZnONPs produced utilizing a cell-free aqueous filtrate of *T. harzianum* were shown to have strong antifungal efficacy against the soil-borne pathogen complexes in cotton in this investigation. To confirm whether the synthesized nanoparticles were still stable for one week or changed in the UV results, the absorption spectrum was recorded for the synthesized ZnO sample by Tvivi, T34 and T28 after 2, 4 and 7 days after synthesis. The UV-visible spectrum showed the absorbance peak at 300 nm corresponding to the characteristic band of zinc oxide nanoparticles for all screened Trichoderma isolates at all tested periods. The obtained UV-vis spectrophotometer results were in agreement with Dobrucka et al. [33], who reported that the maximum absorption of about 310 nm, which is a characteristic band of pure ZnO, verified the presence of ZnONPs biologically with the use of the extract of *Chelidonium majus*. Furthermore, there was no additional peak in the spectrum, confirming that the produced products were pure ZnO [34,35]. In addition, Perveen et al. [36] reported that UV-visible spectroscopy investigation showed a peak at 300 nm, which corresponded to the wavelength of ZnO quantum dots' surface plasmon resonance. The UV-vis spectra of ZnONPs synthesized by *A. niger* show that at 390 nm, ZnONPs have a high absorption spectra [37]. Jamdagni et al. [38] found that the UV spectrum range of ZnONPs is 320–390 nm, which is a similar result.

The XRD diffraction peaks were 31.84°, 34.52°, 36.33°, 47.63°, 56.71°, 62.96°, 68.13°, 69.18°, 70.16°, 73.21° and 78.56°, which agreed with Sadatzadeh et al., Yedurkar et al. and Malaikozhundanet al. [39–41]. The peaks showed the characteristic hexagonal wurtzite structure of ZnO (JCPDS card no. 36–1451) [42]. The Wurtzite structure was prevalent because it is stable in ambient conditions. It also revealed that the synthesized nanopowder was impurity-free because it lacked any XRD peaks other than zinc oxide peaks. The XRD diffraction peaks matched well with Wurtzite ZnO of the Joint Committee on Powder Diffraction Standards (JCPDS) Card number 36–1451 and were in good accord with the reported literature [43].

The magnitude of the Zeta potential (−30 mV to +30 mV) indicates the potential stability of the colloidal system [44–46]. The Zeta potential is related to the nanoparticles' stability in the solution. The larger zeta potential values represent a lower degree of aggregation that leads to a higher degree of stability of nanoparticles and a smaller z-averaged hydrodynamic diameter. At lower zeta values, the nanoparticles flocculate early and the stability of the nano-suspension reduces [44]. The Zeta potential of ZnONPsin the present study was –24.0 mV, which provided evidence that the fabricated nanoparticles were moderately stable, which led to the monodispersity of the particles. The result was in agreement with Divya et al. [45], who showed a zeta potential of −5.36 mV. Furthermore, Zakharova [46] reported a zeta potential of 9 mV of ZnONPs had high antimicrobial efficacy and increased ZnONPs toxicity. It has been proven that some nanoparticles have a tendency to aggregate and that this process of aggregation reduces the surface area of nanoparticles. To solve this problem, ZnONPs require extensive ultrasonication in the water bath for a minimum of 15 min.

The results were in agreement with González et al. [47], who reported that TEM analysis of the synthesized ZnONPs showed spherical, hexagonal and rod shapes. Pillai et al. [48] reported that the synthesized ZnONPs from an aqueous extract of *Beta vulgaris* were spherical with a size of nearly 20 ± 2 nm. Morphology of bio nanoparticles produced from *Cinnamomum tamala* was rod-shaped, the particles size within the range 30 ± 3 nm.

TEM results of biosynthesized ZnONPs by *Anacardium occidentale* leaf extract confirmed the hexagonal structure with an average particle size of 33 nm [49]. Our results are in harmony with Ruddaraju et al. [50]. The results were in agreement with Ruddaraju et al. and Javed et al. [50,51]. The SEM images described surface topological details of different nano-objects based on the electron density of the surface due to their higher resolution and bigger field depth [52]. The agglomeration of ZnONPs might be attributed either due to its polarity and electrostatic attraction between ZnONPs or due to the high surface energy of ZnONPs. The high surface energy of ZnONPs could be originated from an aqueous synthetic medium [53–55]. TEM is used to magnify an image by using electromagnetic lenses to magnify an electron beam that travels through thin specimens in a nearly parallel manner. The objective lens is the principal electromagnetic lens. For example, an SEM picture generally shows bigger agglomerated particles, but TEM images have a greater resolution. This means that TEM is superior to SEM in terms of its ability to measure the nanoparticles' size and has a greater resolution than SEM [54]. EDX analysis is a chemical microanalysis technique that is used in conjunction with SEM to evaluate elemental composition by detecting X-rays released from the sample during electron beam bombardment [55]. EDX analysis was in good agreement with XRD results. The EDX results of the present study were in agreement with several reports [36,39,40]. The FTIR spectrum revealed 3398, 3233, 2912, 1640, 1629, 1561, 1461, 1018, 576 and 533 cm^{-1} in the current study. The peak at 1640 corresponded to C=O stretching of the functional group. The peak in the range 1556 corresponded to C=C/amine—NH stretching of the aromatic compound [55].

The wide peak at 3233 cm^{-1} may be attributed to an alkenyl group's C-H stretch, whereas 2104 cm^{-1} was moved to –C≡C– stretching vibrations [40]. Secondary metabolites found in *C. roseus* have been linked to the conversion of zinc acetate dihydrate to zinc oxide nanoparticles. The FTIR spectrum showed peaks at 3233, 2104, 1640, 1556, 1399, 1086, 926, 773, 849, 715, 1035, 482, 410 cm^{-1} [56]. Due to stretching alkenyl groups formed by zinc acetate salts and their reduction in ZnONPs, the FTIR spectra peak showed high-intensity broadband of 3233 cm^{-1} [37,56]. According to the results of our FTIR analysis, *Trichoderma*-mediated ZnONPs were synthesized using two distinct processes: reduction and capping. On the surfaces of both the biosynthesized ZnONPs that function as reducing and stabilizing agents, FTIR examination indicated the presence of proteins, amino acids, polyphenols, carboxyl and hydroxyl groups. ZnONPs are characterized by their strong aromatic ring and carboxylic acid appearance in the FTIR bands. According to the results of the FTIR analysis described various mycochemicals such as phenolic, proteins, amino acids, aldehydes, ketone and other functional groups were involved in the reduction, capping and stabilization of zinc oxide NPs (Figure 8).

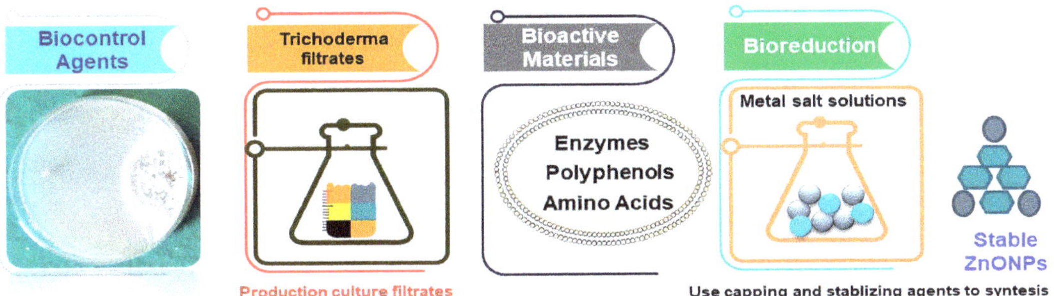

Figure 8. A schematic illustration depicting the methods used by *Trichoderma harzianum* strains to produce green zinc oxide nanoparticles.

In in vitro assay, in addition to inhibiting the vegetative mycelial growth of phytopathogenic fungi, zinc-mediated nanoparticles or composites can kill spores or inhibit spore germination (sporostatic/sporicidal activities) at low concentrations, such as a significant decrease in fungal growth of *B. cinerea* and *P. expansum* shown on ZnONPs (3 mM/L concentration) treatment [57]. Yehia and Ahmed [58] reported the antifungal efficiency of ZnONPs investigated against *F. oxysporum*. The maximum inhibition of mycelial growth was seen at (12 mg/L) when *F. oxysporum* growth was inhibited by 77 percent. HPLC quantification was used to study the influence of ZnONPs on the mycotoxin fusaric acid. The amount of fusaric acid was lowered from 39.0 to 0.20 mg/g. Scanning electron microscopy showed evident deformation in mycelia that had been treated with ZnONPs in *F. oxysporum*, which may cause growth inhibition.

In the present work, in vitro assay, zero fungal growth was investigated with concentrations starting from 20 μg/mL of ZnONPs. Fungicidal properties against three pathogenic fungi were explored in our study. Due to the current dearth of understanding of different aspects of fungal disease biology, these antifungal properties are currently restricted. Lahuf et al. [24] found that a concentration of 15 mg/mL led to complete inhibition (100%) of *R. solani*; however, lower doses of ZnONPs (10, 5 and 2.5 mg/mL) resulted in lower levels of inhibition of *R. solani*, by 83.21, 71.03 and 57 percent inhibition, respectively. Furthermore, it was discovered that ZnONPs have fungistatic rather than deadly fungicidal effects on *R. solani*. The ZnONPs fungicidal properties revealed that they were diffusible via the growing media [59]. Shen et al. and Raghupathi et al. [60,61] documented the antifungal effects of ZnONPs on microbial populations. It was suggested that zero-valent metal nanoparticles might successfully permeate pathogenic microorganism cell membranes through the lipid bilayer because of their reduced hydrophobicity due to the absence of surface charge [62]. ZnO showed obvious destruction of the cell walls and plasmolysis of the internal organs of the tested fungi [63]. In vitro studies against *F. oxysporum*, *R. solani* and *Sclerotium rolfsii* revealed that a mixture of *Trichoderma asperellum* and chitosan nanoparticles was better than *Trichoderma* alone and carbendazim 0.1% in suppressing pathogen mycelial growth [64].

In the current study, under greenhouse conditions, the results of the disease management studies of zinc oxide NPs, at two different concentrations (100 and 200 μg/mL), seed treatments for efficacy in the control of damping-off in cotton, compared to Maxim XL and Moncut chemical fungicides, indicated that ZnONPs (200 μg/mL), gave the maximum efficiency in disease control, compared to other treatments in Giza90 for all growth parameters (survival, plant height and dry weight). However, in the case of Giza94 cultivars, ZnONPs (200 μg/mL) NPs were not the best treatment in disease control in the case of survival only. However, it increased the survival significantly compared to infested control, but it was the best treatment in the case of plant height and dry weight. These results indicated that ZnONPs behavior was affected by the cultivar and it may need to be used at different optimum concentrations according to cotton cultivars to give the maximum survival during further future studies. The ZnONPs may form an antifungal layer around cotton seeds that protects cotton seedlings from the three pathogenic fungi. When ZnONPswas used as an antifungal agent against *R. solani* at concentrations of 30, 60 and 90 g mL^{-1}, the second and the third concentration raised the percentages of Giza90 seedlings that survived to 85 and 86%, respectively, compared to 43.5 percent persisted seedlings at the concentration of 30 g m^{-1} [65]. González-Merino et al. [66] evaluated the antifungal activity of ZnONPs against *F.oxysporum* on tomato plants under greenhouse conditions. ZnONPs from 1500 to 3000 g/mL achieved the best plant height with a range of 166.0 to 175.40 cm, a severity of 0.40–0.80 and a disease incidence of 20–40%. In a pot experiment, foliar spraying of ZnONPs was more successful than seed priming in enhancing plant dry weight and controlling the *Pectobacterium betavasculorum*, *Meloidogyne incognita* and *R. solani*, causal disease complex of beetroot (*Beta Vulgaris* L.) [67]. Nevertheless, most ZnONPs may have accumulated on the seed's exterior surface, with only a few particles moving into the stele and available for biodistribution and bioaccumulation, making seed priming less effective than foliar spray [68]. ZnONPs are believed to interact with pathogens through mechanical enfolding, which could be one of the main mechanisms of ZnONPs toxicity against *R. solani* [69].

5. Conclusions

This study used the biological control agent *T. harzianum* as a stabilizing agent for the green synthesis of biogenic ZnONPs with a relatively small size of 8–23 nm. UV-visible spectroscopy, XRD, zeta potential, TEM, SEM and EDX were used to validate the synthesis and structure, as well as to characterize size distribution, zeta potential, morphology and so on. Moreover, their antifungal activity against soil-borne pathogens like *R. solani* (RS9), *Fusarium* sp. (F10) and *M. phaseolina* (M4) were demonstrated in vitro and under greenhouse conditions. Trichogenic-mediated ZnONPs inhibited hyphal development in three cotton seedlings, indicating that they are effective against fungal infections. As a consequence of the aforementioned findings, it may be inferred that some *T. harzianum* strains produce a variety of proteins and enzymes, obviating the need for chemical reducers and stabilizers. As a result, the biological method for the production of ZnONPs utilizing *T. harzianum* has been presented in this work. The application of ZnONPs in the form of nanofungicides in agroecosystems has yet to be completely investigated and further study on risk assessment is still needed.

Author Contributions: Conceptualization, K.A.A.-E.; and S.A.Z.; methodology, S.A.Z.; investigation, S.A.Z. and M.M.H.; resources, F.M.A.; writing—original draft preparation, writing—review and editing, A.A.A., S.A.O. and K.A.A.-E.; visualization, S.A.Z. and M.M.H.; supervision, S.A.O. and K.A.A.-E.; funding acquisition, K.A.A.-E. and F.M.A. All authors have read and agreed to the published version of the manuscript.

Funding: This research received no external funding.

Conflicts of Interest: The authors declare no conflict of interest.

References

1. Kew Royal Botanic Gardens. Cultivation of Cotton in Egypt. (*Gossypium barbadense*, L.). *Bull. Misc. Inf. Kew* **1897**, *122–123*, 102–104.
2. Asran-Amal, A.; Abd-Elsalam, K.A.; Omar, M.R.; Aly, A.A. Antagonistic potential of *Trichoderma* spp. against *Rhizoctonia solani* and use of M13 microsatellite-primed PCR to evaluate the antagonist genetic variation. *J. Plant Dis. Prot.* **2005**, *112*, 550–561. [CrossRef]
3. Fulton, N.D.; Awaddle, B.; Thomas, J.A. Influence of Planting Date on Fungi Isolated from Diseased Cotton Seedlings. *Plant Dis. Rep.* **1956**, *40*, 556–558.
4. Colyer, P.D. Frequency and Pathogenicity of *Fusarium* spp. Associated with Seedling Diseases of Cotton in Louisiana. *Plant Dis.* **1988**, *72*, 400–402. [CrossRef]
5. Mihail, J.D.; Taylor, S.J. Interpreting Variability among Isolates of *Macrophominaphaseolina* in Pathogenicity, Pycnidium Production and Chlorate Utilization. *Can. J. Bot.* **1995**, *73*, 1596–1603. [CrossRef]
6. Srivastava, A.K.; Singh, T.; Jana, T.K.; Arora, D.K. Induced Resistance and Control of Charcoal Rot in *Cicer Arietinum* (Chickpea) by *Pseudomonas fluorescens*. *Can. J. Bot.* **2001**, *79*, 787–795. [CrossRef]
7. Mihail, J.D. Macrophomina. In *Methods for Research on Soil-Borne Phytopathogenic Fungi*; Singleton, L.L., Mihail, J.D., Rush, C.M., Eds.; American Phytopathological Society Press: St. Paul, MN, USA, 1992; p. 134.
8. Khan, S.A.; Shahid, S.; Mahmood, T.; Lee, C.-S. Contact Lenses Coated with Hybrid Multifunctional Ternary Nanocoatings (Phytomolecule-Coated ZnO nanoparticles: Gallic Acid: Tobramycin) for the Treatment of Bacterial and Fungal Keratitis. *Acta Biomater.* **2021**, *128*, 262–276. [CrossRef]
9. Abd-Elsalam, K.A.; Hashim, A.F.; Alghuthaymi, M.A.; Said-Galiev, E. Nanobiotechnological Strategies for Toxigenic Fungi and Mycotoxin Control. In *Food Preservation*; Academic Press: Cambridge, MA, USA, 2017; pp. 337–364. [CrossRef]
10. Shobha, B.; Lakshmeesha, T.R.; Ansari, M.A.; Almatroudi, A.; Alzohairy, M.A.; Basavaraju, S.; Alurappa, R.; Niranjana, S.R.; Chowdappa, S. Mycosynthesis of ZnO Nanoparticles Using *Trichoderma* spp. Isolated from Rhizosphere Soils and Its Synergistic Antibacterial Effect against *Xanthomonas oryzae* pv. *oryzae*. *J. Fungi* **2020**, *6*, 181. [CrossRef]
11. Yusof, H.M.; Mohamad, R.; Zaidan, U.H.; Rahman, N.A.A. Microbial synthesis of zinc oxide nanoparticles and their potential application as an antimicrobial agent and a feed supplement in animal industry: A review. *J. Anim. Sci. Biotechnol.* **2019**, *10*, 1–22. [CrossRef]
12. Ammar, H.A.; Alghazaly, M.S.; Assem, Y.; AbouZeid, A.A. Bioengineering and Optimization of PEGylated Zinc Nanoparticles by Simple Green Method Using *Monascus purpureus* and Their Powerful antifungal Activity Against the Most Famous Plant Pathogenic Fungi, Causing Food Spoilage. *Environ. Nanotechnol. Monit. Manag.* **2021**, *16*, 100543. [CrossRef]
13. Abd-Elsalam, K.A.; Khokhlov, A.R. Eugenol Oil Nanoemulsion: Antifungal Activity against *Fusarium oxysporum* f. sp. *vasinfectum* and Phytotoxicity on Cottonseeds. *Appl. Nanosci.* **2015**, *5*, 255–265. [CrossRef]
14. Alghuthaymi, M.A.; Almoammar, H.; Rai, M.; Said-Galiev, E.; Abd-Elsalam, K.A. Myconanoparticles: Synthesis and Their Role in Phytopathogens Management. *Biotechnol. Biotechnol. Equip.* **2015**, *29*, 221–236. [CrossRef] [PubMed]
15. Sosna-Głębska, A.; Sibiński, M.; Szczecińska, N.; Apostoluk, A. UV–Visible Silicon Detectors with Zinc Oxide Nanoparticles Acting as Wave length Shifters. *Mater. Today* **2020**, *20*, 1763. [CrossRef]

16. Fahimmunisha, B.A.; Ishwarya, R.; AlSalhi, M.S.; Devanesan, S.; Govindarajan, M.; Vaseeharan, B. Green Fabrication, Characterization and Antibacterial Potential of Zinc Oxide Nanoparticles Using *Aloe Socotrina* Leaf Extract: A Novel Drug Delivery Approach. *J. Drug Deliv. Sci. Technol.* **2020**, *55*, 101465. [CrossRef]
17. Lingaraju, K.; Naika, H.R.; Manjunath, K.; Basavaraj, R.B.; Nagabhushana, H.; Nagaraju, G.; Suresh, D. Biogenic Synthesis of Zinc Oxide Nanoparticles Using *Rutagraveolens* (L.) and Their Antibacterial and Antioxidant Activities. *Appl. Nanosci.* **2016**, *6*, 703–710. [CrossRef]
18. Hwa, K.-Y.; Subramani, B. Synthesis of Zinc Oxide Nanoparticles on Graphene-Carbon Nanotube Hybrid for Glucose Biosensor Applications. *Biosens. Bioelectron.* **2014**, *62*, 127–133. [CrossRef] [PubMed]
19. Singh, J.; Dutta, T.; Kim, K.H.; Rawat, M.; Samddar, P.; Kumar, P. Green'synthesis of Metals and Their Oxide Nanoparticles: Applications for Environmental Remediation. *J. Nanobiotechnol.* **2018**, *16*, 84. [CrossRef]
20. Pokhrel, L.R.; Dubey, B. Evaluation of Developmental Responses of Two Crop Plants Exposed to Silver and Zinc Oxide Nanoparticles. *Sci. Total Environ.* **2013**, *452–453*, 321–332. [CrossRef]
21. Sur, D.H.; Mukhopadhyay, M. Role of Zinc Oxide Nanoparticles for Effluent Treatment Using *Pseudomonas putida* and *Pseudomonas aureofaciens*. *Bioprocess Biosyst. Eng.* **2019**, *42*, 187–198. [CrossRef] [PubMed]
22. Sun, Q.; Li, J.; Le, T. Zinc Oxide Nanoparticle as a Novel Class of Antifungal Agents: Current Advances and Future Perspectives. *J. Agric. Food Chem.* **2018**, *66*, 11209–11220. [CrossRef]
23. Kalia, A.; Abd-Elsalam, K.A.; Kuca, K. Zinc-Based Nanomaterials for Diagnosis and Management of Plant Diseases: Ecological Safety and Future Prospects. *J. Fungi* **2020**, *6*, 222. [CrossRef]
24. Lahuf, A.A.; Kareem, A.A.; Al-Sweedi, T.M.; Alfarttoosi, H.A. Evaluation the Potential of Indigenous Biocontrol Agent Trichoderma Harzianum and Its Interactive Effect with Nanosized ZnO Particles against the Sunflower Damping-off Pathogen. *Rhizoctoniasolani.* In Proceedings of the IOP Conference Series: Earth and Environmental Science, Banda Aceh, Indonesia, 21–22 August 2019; IOP Publishing: Bristol, UK, 2019; Volume 365, p. 012033. [CrossRef]
25. Khan, S.A.; Shahid, S.; Lee, C.-S. Green Synthesis of Gold and Silver Nanoparticles Using Leaf Extract of *ClerodendrumInerme;* Characterization, Antimicrobial and Antioxidant Activities. *Biomolecules* **2020**, *10*, 835. [CrossRef]
26. Consolo, V.F.; Torres-Nicolini, A.; Alvarez, V.A. Mycosinthetized Ag, CuO and ZnO Nanoparticles from a Promising *Trichoderma harzianum* Strain and Their Antifungal Potential against Important Phytopathogens. *Sci. Rep.* **2020**, *10*, 20499. [CrossRef] [PubMed]
27. Zhou, C.; Guo, R.; Ji, S.; Fan, H.; Wang, J.; Wang, Y.; Liu, Z. Isolation of *Trichoderma* from Forestry Model Base and the Antifungal Properties of Isolate TpsT17 toward *Fusarium oxysporum*. *Microbiol. Res.* **2020**, *231*, 126371. [CrossRef]
28. Dou, K.; Gao, J.; Zhang, C.; Yang, H.; Jiang, X.; Li, J.; Li, Y.; Wang, W.; Xian, H.; Li, S.; et al. *Trichoderma* biodiversity in major ecological systems of China. *J. Microbiol.* **2019**, *57*, 668–675. [CrossRef]
29. Elamawi, R.M.; Al-Harbi, R.E.; Hendi, A.A. Biosynthesis and Characterization of Silver Nanoparticles Using *Trichoderma Longibrachiatum* and Their Effect on Phytopathogenic Fungi. *Egypt. J. Biol. Pest Control* **2018**, *28*, 28. [CrossRef]
30. Rajan, A.; Cherian, E.; Baskar, G. Biosynthesis of zinc oxide nanoparticles using *Aspergillus fumigatus* JCF and its antibacterial activity. *Int. J. Mod. Sci. Technol.* **2016**, *1*, 52–57.
31. Henam, S.D.; Ahmad, F.; Shah, M.A.; Parveen, S.; Wani, A.H. Microwave Synthesis of Nanoparticles and Their Antifungal Activities. *Spectrochim. Acta A Mol. Biomol. Spectrosc.* **2019**, *213*, 337–341. [CrossRef] [PubMed]
32. Abd-Elsalam, K.A.; Vasil'kov, A.Y.; Said-Galiev, E.E.; Rubina, M.S.; Khokhlov, A.R.; Naumkin, A.V.; Shtykova, E.V.; Alghuthaymi, M.A. Bimetallic Blends and Chitosan Nanocomposites: Novel Antifungal Agents against Cotton Seedling Damping-Off. *Eur. J. Plant Pathol.* **2018**, *151*, 57–72. [CrossRef]
33. Dobrucka, R.; Dlugaszewska, J.; Kaczmarek, M. Cytotoxic and Antimicrobial Effects of Biosynthesized ZnO Nanoparticles Using of *Chelidonium majus* Extract. *Biomed. Microdevices* **2018**, *20*, 5. [CrossRef]
34. Wahab, R.; Ansari, S.G.; Kim, Y.S.; Seo, H.K.; Kim, G.S.; Khang, G.; Shin, H.-S. Low-Temperature Solution Synthesis and Characterization of ZnO Nano-Flowers. *Mater. Res. Bull.* **2007**, *42*, 1640–1648. [CrossRef]
35. Wahab, R.; Ansari, S.G.; Kim, Y.S.; Song, M.; Shin, H.-S. The Role of PH Variation on the Growth of Zinc Oxide Nanostructures. *Appl. Surf. Sci.* **2009**, *255*, 4891–4896. [CrossRef]
36. Perveen, R.; Shujaat, S.; Qureshi, Z.; Nawaz, S.; Khan, M.I.; Iqbal, M. Green versus Sol-Gel Synthesis of ZnO Nanoparticles and Antimicrobial Activity Evaluation against Panel of Pathogens. *J. Mater. Res. Technol.* **2020**, *9*, 7817–7827. [CrossRef]
37. Gao, Y.; Arokia Vijaya Anand, M.; Ramachandran, V.; Karthikkumar, V.; Shalini, V.; Vijayalakshmi, S.; Ernest, D. Biofabrication of Zinc Oxide Nanoparticles from *Aspergillus niger*, Their Antioxidant, Antimicrobial and Anticancer Activity. *J. Clust. Sci.* **2019**, *30*, 937–946. [CrossRef]
38. Jamdagni, P.; Khatri, P.; Rana, J.S. Green Synthesis of Zinc Oxide Nanoparticles Using Flower Extract of *Nyctanthes Arbor-Tristis* and Their Antifungal Activity. *J. King Saud Univ. Sci.* **2018**, *30*, 168–175. [CrossRef]
39. Sadatzadeh, A.; Charati, F.R.; Akbari, R.; Moghaddam, H.H. Green Biosynthesis of Zinc Oxide Nanoparticles via Aqueous Extract of Cottonseed. *J. Mater. Environ. Sci.* **2018**, *9*, 2849–2853.
40. Yedurkar, S.; Maurya, C.; Mahanwar, P. Biosynthesis of Zinc Oxide Nanoparticles Using Ixora Coccinea Leaf Extract—A Green Approach. *Open J. Synth. Theory Appl.* **2016**, *5*, 1–14. [CrossRef]
41. Malaikozhundan, B.; Vinodhini, J. Nanopesticidal Effects of *Pongamiapinnata* Leaf Extract Coated Zinc Oxide Nanoparticle against the Pulse Beetle, *Callosobruchus maculatus*. *Mater. Today Commun.* **2018**, *14*, 106–115. [CrossRef]

42. Soares, V.A.; Xavier, M.J.S.; Rodrigues, E.S.; de Oliveira, C.A.; Farias, P.M.A.; Stingl, A.; Ferreira, N.S.; Silva, M.S. Green Synthesis of ZnO Nanoparticles Using Whey as an Effective Chelating Agent. *Mater. Lett.* **2020**, *259*, 126853. [CrossRef]
43. Pandao, M.R.; Sajid, M. Synthesis and Characterization of Nano Zinc Oxide for Linseed. *J. Pharmacogn. Phytochem.* **2021**, *10*, 846–848.
44. Baddar, Z.E.; Matocha, C.J.; Unrine, J.M. Surface coating effects on the sorption and dissolution of ZnO nanoparticles in soil. *Environ. Sci. Nano* **2019**, *6*, 2495–2507. [CrossRef]
45. Divya, M.; Govindarajan, M.; Karthikeyan, S.; Preetham, E.; Alharbi, N.S.; Kadaikunnan, S.; Khaled, J.M.; Almanaa, T.N.; Vaseeharan, B. Antibiofilm and Anticancer Potential of β-Glucan-Binding Protein-Encrusted Zinc Oxide Nanoparticles. *Microb. Pathog.* **2020**, *141*, 103992. [CrossRef] [PubMed]
46. Zakharova, O.; Kolesnikov, E.; Vishnyakova, E.; Strekalova, N.; Gusev, A. Antibacterial activity of ZnO nanoparticles: Dependence on particle size, dispersion media and storage time. In Proceedings of the IOP Conference Series: Earth Environmental Science, Voronezh, Russia, 4–5 October 2018; IOP Publishing: Bristol, UK, 2019; Volume 226, p. 012062.
47. González, S.C.E.; Bolaina-Lorenzo, E.; Pérez-Trujillo, J.J.; Puente-Urbina, B.A.; Rodríguez-Fernández, O.; Fonseca-García, A.; Betancourt-Galindo, R. Antibacterial and Anticancer Activity of ZnO with Different Morphologies: A Comparative Study. *3 Biotech* **2021**, *11*, 68. [CrossRef]
48. Pillai, A.M.; Sivasankarapillai, V.S.; Rahdar, A.; Joseph, J.; Sadeghfar, F.; Anuf, A.R.; Rajesh, K.; Kyzas, G.Z. Green Synthesis and Characterization of Zinc Oxide Nanoparticles with Antibacterial and Antifungal Activity. *J. Mol. Struct.* **2020**, *1211*, 128107. [CrossRef]
49. Zhao, C.; Zhang, X.; Zheng, Y. Biosynthesis of Polyphenols Functionalized ZnO Nanoparticles: Characterization and Their Effect on Human Pancreatic Cancer Cell Line. *J. Photochem. Photobiol. B* **2018**, *183*, 142–146. [CrossRef]
50. Ruddaraju, L.K.; Pammi, S.V.N.; Pallela, P.N.V.K.; Padavala, V.S.; Kolapalli, V.R.M. Antibiotic Potentiation and Anti-Cancer Competence through Bio-Mediated ZnO Nanoparticles. *Mater. Sci. Eng. C Mater. Biol. Appl.* **2019**, *103*, 109756. [CrossRef] [PubMed]
51. Javed, R.; Rais, F.; Fatima, H.; Haq, I.U.; Kaleem, M.; Naz, S.S.; Ao, Q. Chitosan Encapsulated ZnO Nanocomposites: Fabrication, Characterization and Functionalization of Bio-Dental Approaches. *Mater. Sci. Eng. C Mater. Biol. Appl.* **2020**, *116*, 111184. [CrossRef]
52. Elumalai, K.; Velmurugan, S.; Ravi, S.; Kathiravan, V.; Adaikala Raj, G. Bio-Approach: Plant Mediated Synthesis of ZnO Nanoparticles and Their Catalytic Reduction of Methylene Blue and Antimicrobial Activity. *Adv. Powder Technol.* **2015**, *26*, 1639–1651. [CrossRef]
53. Teulon, J.M., Godon, C.; Chantalat, L.; Moriscot, C.; Cambedouzou, J.; Odorico, M.; Ravaux, J.; Podor, R.; Gerdil, A.; Habert, A.; et al. On the Operational Aspects of Measuring Nanoparticle Sizes. *Nanomaterials* **2018**, *9*, 18. [CrossRef]
54. Nain, V.; Kaur, M.; Sandhu, K.S.; Thory, R.; Sinhmar, A. Development, Characterization and Biocompatibility of Zinc Oxide Coupled Starch Nanocomposites from Different Botanical Sources. *Int. J. Biol. Macromol.* **2020**, *162*, 24–30. [CrossRef] [PubMed]
55. Rastogi, L.; Arunachalam, J. Sunlight Based Irradiation Strategy for Rapid Green Synthesis of Highly Stable Silver Nanoparticles Using Aqueous Garlic (*Allium sativum*) Extract and Their Antibacterial Potential. *Mater. Chem. Phys.* **2011**, *129*, 558–563. [CrossRef]
56. Gupta, M.; Tomar, R.S.; Kaushik, S.; Mishra, R.K.; Sharma, D. Effective Antimicrobial Activity of GreenZnONano Particles of *Catharanthus roseus*. *Front. Microbiol.* **2018**, *9*, 2030. [CrossRef] [PubMed]
57. He, L.; Liu, Y.; Mustapha, A.; Lin, M. Antifungal Activity of Zinc Oxide Nanoparticles against *Botrytis Cinerea* and *Penicillium expansum*. *Microbiol. Res.* **2011**, *166*, 207–215. [CrossRef] [PubMed]
58. Yehia, R.S.; Ahmed, O.F. In Vitro Study of the Antifungal Efficacy of Zinc Oxide Nanoparticles against *Fusarium oxysporum* and *Peniciliumexpansum*. *Afr. J. Microbiol. Res.* **2013**, *7*, 1917–1923. [CrossRef]
59. Barsainya, M.; Pratap Singh, D. Green Synthesis of Zinc Oxide Nanoparticles by *Pseudomonas aeruginosa* and Their Broad-Spectrum Antimicrobial Effects. *J. Pure Appl. Microbiol.* **2018**, *12*, 2123–2134. [CrossRef]
60. Shen, Z.; Chen, Z.; Hou, Z.; Li, T.; Lu, X. Ecotoxicological Effect of Zinc Oxide Nanoparticles on Soil Microorganisms. *Front. Environ. Sci. Eng.* **2015**, *9*, 912–918. [CrossRef]
61. Raghupathi, K.R.; Koodali, R.T.; Manna, A.C. Size-Dependent Bacterial Growth Inhibition and Mechanism of Antibacterial Activity of Zinc Oxide Nanoparticles. *Langmuir* **2011**, *27*, 4020–4028. [CrossRef]
62. Hulkoti, N.I.; Taranath, T.C. Biosynthesis of Nanoparticles Using Microbes—A Review. *Colloids Surf. B Biointerfaces* **2014**, *121*, 474–483. [CrossRef] [PubMed]
63. Farouk, A.M. Improvement of Bean Resistance to Damping-Off and Root-Rot Diseases. Ph.D. Thesis, Cairo University, Cairo, Egypt, 2016.
64. Boruah, S.; Dutta, P. Fungus Mediated Biogenic Synthesis and Characterization of Chitosan Nanoparticles and Its Combine Effect with *Trichoderma asperellum* against *Fusarium oxysporum, Sclerotium rolfsii* and *Rhizoctonia solani*. *Indian Phytopathol.* **2021**, *74*, 81–93. [CrossRef]
65. Al-Dhabaan, F.A.; Shoala, T.; Ali, A.A.; Alaa, M.; Abd-Elsalam, K.; Abd-Elsalam, K. Chemically-Produced Copper, Zinc Nanoparticles and Chitosan-Bimetallic Nanocomposites and Their Antifungal Activity against Three Phytopathogenic Fungi. *Int. J. Agric. Technol.* **2017**, *13*, 753–769.

66. González-Merino, A.M.; Hernández-Juárez, A.; Betancourt-Galindo, R.; Ochoa-Fuentes, Y.M.; Valdez-Aguilar, L.A.; Limón-Corona, M.L. Antifungal Activity of Zinc Oxide Nanoparticles in *Fusarium oxysporum—Solanum lycopersicum* Pathosystem under Controlled Conditions. *J. Phytopathol.* **2021**, *169*, 533–544. [CrossRef]
67. Khan, M.R.; Siddiqui, Z.A. Role of Zinc Oxide Nanoparticles in the Management of Disease Complex of Beetroot (*Beta vulgaris* L.) Caused by *Pectobacterium betavasculorum, Meloidogyne incognita* and *Rhizoctonia solani. Hortic. Environ. Biotechnol.* **2021**, *62*, 225–241. [CrossRef]
68. Khan, M.; Siddiqui, Z.A. Zinc Oxide Nanoparticles for the Management of *Ralstoniasolanacearum, Phomopsis Vexans* and *Meloidogyne Incognita* Incited Disease Complex of Eggplant. *Indian Phytopathol.* **2018**, *71*, 355–364. [CrossRef]
69. Khan, M.; Khan, A.U.; Alam, J.; Parveen, A.; Moon, I.-S.; Alam, M.; Cabral-Pinto, M.M.; Ahamed, M.; Yadav, V.K.; Yadav, K. Molecular Docking of Biosynthesized Zinc Oxide Nanoparticles to Screen Their Impact on Fungal Pathogen of Carrot Plant. *Res. Sq.* **2021**. [CrossRef]

MDPI
St. Alban-Anlage 66
4052 Basel
Switzerland
Tel. +41 61 683 77 34
Fax +41 61 302 89 18
www.mdpi.com

Journal of Fungi Editorial Office
E-mail: jof@mdpi.com
www.mdpi.com/journal/jof

www.ingramcontent.com/pod-product-compliance
Lightning Source LLC
LaVergne TN
LVHW071523170726
843515LV00008B/2052

* 9 7 8 3 0 3 6 5 7 7 2 1 0 *